PROGRESS IN BRAIN RESEARCH

VOLUME 124

CEREBELLAR MODULES: MOLECULES, MORPHOLOGY AND FUNCTION

Other volumes in PROGRESS IN BRAIN RESEARCH

Volume 100: Neuroscience: From the Molecular to the Cognitive, by F.E. Bloom (Ed.) – 1994, ISBN 0-444-81678-X.
Volume 101: Biological Function of Gangliosides, by L. Svennerholm et al. (Eds.) – 1994, ISBN 0-444-81658-5.
Volume 102: The Self-Organizing Brain: From Growth Cones to Functional Networks, by J. van Pelt, M.A. Corner, H.B.M. Uylings and F.H. Lopes da Silva (Eds.) – 1994, ISBN 0-444-81819-7.
Volume 103: Neural Regeneration, by F.J. Seil (Ed.) – 1994, ISBN 0-444-81727-1.
Volume 104: Neuropeptides in the Spinal Cord, by F. Nyberg, H.S. Sharma and Z. Wiesenfeld-Hallin (Eds.) – 1995, ISBN 0-444-81719-0.
Volume 105: Gene Expression in the Central Nervous System, by A.C.H. Yu et al. (Eds.) – 1995, ISBN 0-444-81852-9.
Volume 106: Current Neurochemical and Pharmacological Aspects of Biogenic Amines, by P.M. Yu, K.F. Tipton and A.A. Boulton (Eds.) – 1995, ISBN 0-444-81938-X.
Volume 107: The Emotional Motor System, by G. Holstege, R. Bandler and C.B. Saper (Eds.) – 1996, ISBN 0-444-81962-2.
Volume 108: Neural Development and Plasticity, by R.R. Mize and R.S. Erzurumlu (Eds.) – 1996, ISBN 0-444-82433-2.
Volume 109: Cholinergic Mechanisms: From Molecular Biology to Clinical Significance, by J. Klein and K. Löffelholz (Eds.) – 1996, ISBN 0-444-82166-X.
Volume 110: Towards the Neurobiology of Chronic Pain, by G. Carli and M. Zimmermann (Eds.) – 1996, ISBN 0-444-82149-X.
Volume 111: Hypothalamic Integration of Circadian Rhythms, by R.M. Buijs, A. Kalsbeek, H.J. Romijn, C.M.A. Pennartz and M. Mirmiran (Eds.) – 1996, ISBN 0-444-82443-X.
Volume 112: Extrageniculostriate Mechanisms Underlying Visually-Guided Orientation Behavior, by M. Norita, T. Bando and B.E. Stein (Eds.) – 1996, ISBN 0-444-82347-6.
Volume 113: The Polymodal Receptor: A Gateway to Pathological Pain, by T. Kumazawa, L. Kruger and K. Mizumura (Eds.) – 1996, ISBN 0-444-82473-1.
Volume 114: The Cerebellum: From Structure to Control, by C.I. de Zeeuw, P. Strata and J. Voogd (Eds.) – 1997, ISBN 0-444-82313-1.
Volume 115: Brain Function in Hot Environment, by H.S. Sharma and J. Westman (Eds.) – 1998, ISBN 0-444-82377-8.
Volume 116: The Glutamate Synapse as a Therapeutical Target: Molecular Organization and Pathology of the Glutamate Synapse, by O.P. Ottersen, I.A. Langmoen and L. Gjerstad (Eds.) – 1998, ISBN 0-444-82754-4.
Volume 117: Neuronal Degeneration and Regeneration: From Basic Mechanisms to Prospects for Therapy, by F.W. van Leeuwen, A. Salehi, R.J. Giger, A.J.G.D. Holtmaat and J. Verhaagen (Eds.) – 1998, ISBN 0-444-82817-6.
Volume 118: Nitric Oxide in Brain Development, Plasticity, and Disease, by R.R. Mize, T.M. Dawson, V.L. Dawson and M.J. Friedlander (Eds.) – 1998, ISBN 0-444-82885-0.
Volume 119: Advances in Brain Vasopressin, by I.J.A. Urban, J.P.H. Burbach and D. De Wied (Eds.) – 1999, ISBN 0-444-50080-4.
Volume 120: Nucleotides and their Receptors in the Nervous System, by P. Illes and H. Zimmermann (Eds.) – 1999, ISBN 0-444-50082-0.
Volume 121: Disorders of Brain, Behavior and Cognition: The Neurocomputational Perspective, by J.A. Reggia, E. Ruppin and D. Glanzman (Eds.) – 1999, ISBN 0-444-50175-4.
Volume 122: The Biological Basis for Mind Body Interactions, by E.A. Mayer and C.B. Saper (Eds.) – 2000, ISBN 0-444-50049-9.
Volume 123: Peripheral and Spinal Mechanisms in the Neural Control of Movement, by M.D. Binder (Ed.) – 1999, ISBN 0-444-50288-2.

PROGRESS IN BRAIN RESEARCH

VOLUME 124

CEREBELLAR MODULES: MOLECULES, MORPHOLOGY AND FUNCTION

EDITED BY

N.M. GERRITS, T.J.H. RUIGROK and C.I. DE ZEEUW

Department of Anatomy, Erasmus University, 3000 DR Rotterdam, The Netherlands

ELSEVIER

AMSTERDAM – LAUSANNE – NEW YORK – OXFORD – SHANNON – SINGAPORE – TOKYO

2000

ELSEVIER SCIENCE B.V.
Sara Burgerhartstraat 25
P.O. Box 211, 1000 AE Amsterdam, The Netherlands

First edition 2000

Library of Congress Cataloging in Publication Data
A catalog record from the Library of Congress has been applied for.

ISBN: 0-444-50108-8 (volume)
ISBN: 0-444-80104-9 (series)

∞ The paper used in this publication meets the requirements of ANSI/NISO Z39.48-1992 (Permanence of Paper).
Printed in The Netherlands.

List of Contributors

C.D. Aizenman, Department of Neuroscience, Johns Hopkins University, School of Medicine, 725 N. Wolfe Street, Baltimore, MD 21205, USA

R. Anelli, Northwestern University, Institute for Neuroscience, 320 E. Superior Street, Chicago, IL 60611, USA

R. Apps, Department of Physiology, School of Medical Sciences, University of Bristol, Bristol BS8 1TD, UK

G. Cheron, Laboratory of Movement Biomechanics, Université Libre de Bruxelles, ISEPK, 28 Avenue Paul Héger, CP 168, 1000 Brussels, Belgium and Laboratory of Neurophysiology, Université de Mons, Mons, Belgium

D. Cohen, Department of Neurobiology, Life Sciences Institute and Center for Neural Computation, Hebrew University, Givat Ram, Jerusalem 91904, Israel

P. D'Alcantara, Brain Research Unit CP601 and IRIBHN, Université Libre de Bruxelles, 1000 Brussels, Belgium

E. De Schutter, Laboratory for Theoretical Neurobiology, Born-Bunge Foundation, University of Antwerp, Universiteitsplein 1, B2-610 Antwerp, Belgium

C.I. De Zeeuw, Department of Anatomy, Erasmus University Rotterdam, P.O. Box 1738, 3000 DR Rotterdam, The Netherlands

A. Devor, Department of Neurobiology, Life Sciences Institute, The Hebrew University, Jerusalem 91904, Israel

M.R. Diño, Northwestern University, Institute for Neuroscience, 320 E. Superior Street, Chicago, IL 60611, USA

J.-P. Draye, Laboratory of Movement Biomechanics, Université Libre de Bruxelles, 1000 Brussels, Belgium

L.M. Eisenman, Department of Pathology, Anatomy and Cell Biology, Thomas Jefferson University, Philadelphia, PA 19107, USA

M.A. Frens, Department of Physiology, Erasmus University Rotterdam, P.O. Box 1738, 3000 DR Rotterdam, The Netherlands

M. Garwicz, Department of Physiological Sciences, Lund University, Sölvegatan 19, S-223 62 Lund, Sweden

N.M. Gerrits, Department of Anatomy, Erasmus University, 3000 DR Rotterdam, The Netherlands

J.C. Glover, Department of Anatomy, University of Oslo, Institute of Basic Medical Sciences, N-0317, Oslo 3, Norway

K. Hashimoto, Department of Physiology, Kanazawa University School of Medicine, Takara-machi, Kanazawa 920-8640, Japan

G. Hesslow, Department of Physiological Sciences, Section for Neurophysiology, Lund University, Sölvegatan 19, S-223 62 Lund, Sweden

S.M. Highstein, Department of Otolaryngology, Box 8115, Washington University School of Medicine, 4566 Scott Avenue, St. Louis, MO 63110, USA

Y. Hirata, Department of Otolaryngology, Box 8115, Washington University School of Medicine, 4566 Scott Avenue, St. Louis, MO 63110, USA

E.J. Huang, Department of Neuroscience, Johns Hopkins University, School of Medicine, 725 N. Wolfe Street, Baltimore, MD 21205, USA

Y. Inoue, Division of Biology 147-75, California Institute of Technology, Pasadena, CA 91125, USA

M. Ivarsson, University College London, Department of Anatomy and Developmental Biology, Gower Street, Foster Court, London WC1E 6BT, UK

H. Jiang, Department of Pharmacology and Physiology, University of Rochester, Rochester, NY 14642-8711, USA

A. Jones, MRC Laboratory of Molecular Biology, MRC Centre, Hills Road, Cambridge, CB2 2QH, UK

K. Jun, National CRI Center for Calcium and Learning, and Department of Life Science, Pohang University of Science and Technology, Pohang 790-784, Korea

M. Kano, Department of Physiology, Kanazawa University School of Medicine, Takaramachi, Kanazawa 920-8640, Japan

W.M. Kistler, Center for Neuromimetic Systems of the Swiss Federal Institute of Technology, EPFL-DI, CH-1015 Lausanne, Switzerland

H. Kurihara, Department of Anatomy, Hokkaido University School of Medicine, Sapporo 060-0807, Japan

D.J. Linden, Department of Neuroscience, Johns Hopkins University, School of Medicine, 725 N. Wolfe Street, Baltimore, MD 21205, USA

S.G. Lisberger, Howard Hughes Medical Institute, Department of Physiology, W.M. Keck Foundation Center for Integrative Neuroscience, and Sloan Center for Theoretical Neurobiology, University of California, San Francisco, CA 94143, USA

A. Lohof, Laboratoire de Neurobiologie, CNRS URA 1857, Ecole Normale Supérieure, Paris, France

R. Maex, Laboratory for Theoretical Neurobiology, Born-Bunge Foundation, University of Antwerp, Universiteitsplein 1, B-2610 Antwerp, Belgium

P.B. Manis, Department of Neuroscience and Department of Otolaryngology - Head and Neck Surgery, Johns Hopkins University, School of Medicine, 725 N. Wolfe Street, Baltimore, MD 21205, USA

P. Mann-Metzer, Department of Neurobiology, Life Sciences Institute and Center for Neural Computation, Hebrew University, Givat Ram, Jerusalem 91904, Israel

A.L. Mathoera, Department of Physiology, Erasmus University Rotterdam, P.O. Box 1738, 3000 DR Rotterdam, The Netherlands

D. Merlo, MRC Laboratory of Molecular Biology, MRC Centre, Hills Road, Cambridge, CB2 2QH, UK

M. Meyer, Department of Neurochemistry, Max Planck Institute for Psychiatry, Am Klopferspitz 18a, D-82152 Planegg, Germany

E. Mugnaini, Northwestern University, Institute for Neuroscience, 320 E. Superior Street, Chicago, IL 60611, USA

M.G. Nunzi, Northwestern University, Institute for Neuroscience, 320 E. Superior Street, Chicago, IL 60611, USA

S. Offermanns, Division of Biology 147-75, California Institute of Technology, Pasadena, CA 91125, USA

M. Parmentier, IRIBHN, Université Libre de Bruxelles, 1000 Brussels, Belgium
M. Paterlini, MRC Laboratory of Molecular Biology, MRC Centre, Hills Road, Cambridge, CB2 2QH, UK
J.L. Raymond, Department of Neurobiology, Stanford University School of Medicine, Fairchild D-251, Stanford, CA 94305, USA
T.J.H. Ruigrok, Department of Anatomy, Erasmus University Rotterdam, P.O. Box 1738, 3000 DR Rotterdam, The Netherlands
S.N. Schiffmann, Brain Research Unit CP601, Université Libre de Bruxelles, 1000 Brussels, Belgium
K. Schilling, Anatomisches Institut, Rheinische Friedrich-Wilhelms-Universität, Nussallee 10, D-53115 Bonn, Germany
S. Schurmans, IRIBHN, Université Libre de Bruxelles, 1000 Brussels, Belgium
H.-S. Shin, National CRI Center for Calcium and Learning, and Department of Life Science, Pohang University of Science and Technology, Pohang 790-784, Korea
Y. Shinoda, Department of Systems Neurophysiology, School of Medicine, Tokyo Medical and Dental University, 1-5-45, Yushima, Bunkyo-ku, Tokyo 113-8519, Japan
M.I. Simon, Division of Biology 147-75, California Institute of Technology, Pasadena, CA 91125, USA
I. Sugihara, Department of Systems Neurophysiology, School of Medicine, Tokyo Medical and Dental University, 1-5-45, Yushima, Bunkyo-ku, Tokyo 113-8519, Japan
Y. Sugiuchi, Department of Systems Neurophysiology, School of Medicine, Tokyo Medical and Dental University, 1-5-45, Yushima, Bunkyo-ku, Tokyo 113-8519, Japan
P. Svensson, Department of Physiological Sciences, Section for Neurophysiology, Lund University, Sölvegatan 19, S-223 62 Lund, Sweden
T.M. Teune, Department of Anatomy, Erasmus University Rotterdam, P.O. Box 1738, 3000 DR Rotterdam, The Netherlands
J. Van der Burg, Department of Anatomy, Erasmus University Rotterdam, P.O. Box 1738, 3000 DR Rotterdam, The Netherlands
J. Van der Moer, Department of Anatomy, Erasmus University Rotterdam, P.O. Box 1738, 3000 DR Rotterdam, The Netherlands
J. Van der Steen, Department of Physiology, Erasmus University Rotterdam, P.O. Box 1738, 3000 DR Rotterdam, The Netherlands
J. L. Van Hemmen, Department of Physics, Technical University of Munich, James-Franck-Strasse, D-85748 Garching/Munich, Germany
A. Volny-Luraghi, Laboratory for Theoretical Neurobiology, Born-Bunge Foundation, University of Antwerp, Universiteitsplein 1, B2610 Antwerp, Belgium
J. Voogd, Department of Anatomy, Erasmus University Rotterdam, P.O. Box 1738, 3000 DR Rotterdam, The Netherlands
B.P. Vos, Laboratory for Theoretical Neurobiology, Born-Bunge Foundation, University of Antwerp, Universiteitsplein 1, B-2610 Antwerp, Belgium
M. Watanabe, Department of Anatomy, Hokkaido University School of Medicine, Sapporo 060-0807, Japan
W. Wisden, MRC Laboratory of Molecular Biology, MRC Centre, Hills Road, Cambridge, CB2 2QH, UK
D. Wu, Department of Pharmacology and Physiology, University of Rochester, Rochester, NY 14642-8711, USA
H.-S. Wu, Department of Systems Neurophysiology, School of Medicine, Tokyo Medical and Dental University, 1-5-45, Yushima, Bunkyo-ku, Tokyo 113-8519, Japan

Y. Wu, Department of Pharmacology and Physiology, University of Rochester, Rochester, NY 14642-8711, USA
Y. Yarom, Department of Neurobiology, Life Sciences Institute and Center for Neural Computation, Hebrew University, Givat Ram, Jerusalem 91904, Israel

List of Participants

Apps, R., The Medical School, Department of Physiology, University of Bristol, University Walk, Bristol BS8 1TD, United Kingdom. E-mail: r.apps@bristol.ac.uk

Atkins, M.J., Department of Physiology, University of Lund, Sölvegatan 19, S-223 62 Lund, Sweden

Attwell, P.J.E., Department of Anatomy and Development Biology, University College London, Gower Street, London WC1E 6BT, United Kingdom

Axelrad, H., Laboratory of Neurophysiology, Medical Faculty Pitié-Salpetriere, 91 Bd de l'Hôpital, 75013 Paris 13, France. E-mail: hax@biomath.jussieu.fr

Baker, M.R., Department of Anatomy, Downing Street, Cambridge, CB2 3DY, United Kingdom

Belton, T., Department of Pharmacology and Physiology, University of Chicago, Surgery and Brain Res Inst, J233, 5801 St., Ellis Ave, Chicago, IL 60637, USA. E-mail: tbelton@midway.uchicago.edu

Bower J., Division of Biology, California Institute of Technology, Pasadena, CA 91125, USA. E-mail: jbower@bbb.caltech.edu

Cheron G., Laboratory of Neurosciences, Université de Mons-Hinaut, Place du Parc 20, 7000 Mons, Belgium

Cohen, D., Department of Neurobiology, Institute of Life Sciences, Hebrew University, Givat Ram, Jerusalem 91904, Israel

De Callatay, A., XL Knowledge Laboratory, Avenue General de Gaulle 49/11, Brussels, B-1050, Belgium. E-mail: 106144.3376@compuserve.com

De Schutter, E., Bron-Bunge Foundation, University of Antwerp, Universiteitsplein 1, 2610 Antwerp, Belgium. E-mail: erik@bbf.uia.ac.be

De Zeeuw, C.I., Department of Anatomy, Erasmus University, P.O. Box 1738, 3000 DR Rotterdam, The Netherlands. E-mail: dezeeuw@anat.fgg.eur.nl

Eisenman, L.M., Department of Anatomy, Pathology and Cell Biology, Jefferson Medical College, Philadephia, PA 19107, USA. E-mail: eisenma1@jeflin.tju.edu

Frens, M.A., Department of Physiology, Erasmus University, P.O. Box 1738, 3000 DR Rotterdam, The Netherlands. E-mail: frens@fys1.fgg.eur.nl

Galjart, N., Department of Cell Biology and Genetics, Erasmus University, P.O. Box 1738, 3000 DR Rotterdam, The Netherlands. E-mail: galjart@ch1.fgg.eur.nl

Garwicz, M., Department of Physiology, University of Lund, Sölvegatan 19, S-223 62 Lund, Sweden. E-mail: martin.garwicz@mphy.lu.se

Gerrits, N.M., Department of Anatomy, Erasmus University, P.O. Box 1738, 3000 DR Rotterdam, The Netherlands. E-mail: gerrits@anat.fgg.eur.nl

Gibson, A.R., Division of Neurobiology, Barrow Neurological Institute, 305 St Thomas Road, Phoenix, AZ 85013, USA. E-mail: agibson@mha.chw.edu

Glover, J.C., Department of Anatomy, Institute of Basic Medical Sciences, University of Oslo, P.O. Box 1105, Blindern, N-0317, Oslo 3, Norway. E-mail: joel@pons.uio.no

Goossens, H.H.L.M., Department of Physiology, Erasmus University, P.O.Box 1738, 3000 DR Rotterdam, The Netherlands. E-mail: goossens@fys1.fgg.eur.nl

Hesslow, G., Department of Physiology, University of Lund, Sölvegatan 19, S-223 62 Lund, Sweden. E-mail: germund.hesslow@mphy.lu.se

Highstein, S.M., Department of Otolaryngology, Washington University School of Medicine, 4911 Barnes Hospital Plaza, St. Louis, MO 63110, USA. E-mail: highstein@medicine.wustl.edu

Hoogenraad, C.C., Department of Cell Biology and Genetics, Erasmus University, P.O. Box 1738, 3000 DR Rotterdam, The Netherlands

Ito, M., Riken Brain Science Institute, 2-1 Hirosawa, Wako-shi, Saitama 351-01, Japan. E-mail: masao@postman.riken.go.jp

Ivarsson M., Department of Physiology, University of Lund, Sölvegatan 19, S-223 62 Lund, Sweden. E-mail: magnus.ivarsson@mphy.lu.se

Kano, M., Department of Physiology, Kanazawa University School of Medicine, 13-1 Takara-machi, Kanazawa 920-8640, Japan. E-mail: mkano@med.kanazawa-u.ac.jp

King J.S., Department of Cell Biology, Neurobiol. and Anatomy, Ohio State University, 333 W 10th Ave., Columbus OH 43221, USA. E-MAIL: king11@osu.edu

Koekoek, S.K.E., Department of Anatomy, Erasmus University, P.O. Box 1738, 3000 DR Rotterdam, The Netherlands

Kogan, A., Department of Neurobiology, Institute of Life Sciences, Hebrew University, Givat Ram, Jerusalem 91904, Israel. E-mail: annakn@lobster.ls.huji.ac.il

Leergaard, T.B., Department of Anatomy, Institute of Basic Medical Sciences, P.O. Box 1105, Blindern, N-0317, Oslo 3, Norway. E-mail: trygve.leergaard@basalmed.uio.no

Linden, D.J., Department of Neuroscience, Johns Hopkins University, School of Medicine, 725 N Wolfestr., Baltimore, MD 21205, USA. E-mail: david.linden@qmail.bs.jhu.edu

Llinas, R., Department of Physiology and Neuroscience, New York University Medical Center, 555 1st Avenue, New York, NY 10016, USA. E-mail: clarkm01@mchip00.med.nyu.edu

Mann-Metzer, P., Department of Neurobiology, Institute of Life Sciences, Hebrew University, Givat Ram, Jerusalem 91904, Israel

Mathoera, A.L., Department of Physiology, Erasmus University, P.O. Box 1738, 3000 DR Rotterdam, The Netherlands

McCrea, R.A., Department of Pharmacology and Physiology, University of Chicago, Surgery and Brain Res Inst, J233 5801 St Ellis Avenue, Chicago, IL 60637, USA. E-mail: rammccrea@midway.uchicago.edu

Mugnaini, E., Institute of Neuroscience, Northwestern University, 5-305 Ward Building, 303 E Chicago Avenue, Chicago, IL 60611, USA. E-mail: e-mugnaini@nwu.edu

Nitschke, M.F., Department of Neurology, Medical University, Ratzeburger Allee 160, D 23538 Luebeck, Germany. E-mail: nitschke_m@neuro.mu-luebeck.de

Oberdick, J.D., Ohio State University, Biotechnology Center, 190 Rightmire Hall, 1060 Carmack Road, Columbus, OH 43210, USA. E-mail: oberdick.1@osu.edu

Raymond, J.L., Keck Ctr Integrative Neuroscience, UCSF, Box 0444, San Francisco, CA 94143, USA. E-mail: jenr@stanford.edu

Ruigrok, T.J.H., Department of Anatomy, Erasmus University, P.O. Box 1738, 3000 DR Rotterdam, The Netherlands. E-mail: ruigrok@anat.fgg.eur.nl

Santamaria, F., Division of Biology, California Institute of Technology, Pasadena, CA 91125, USA. E-mail: fidel@bbb.caltech.edu

Schilling, K., Anatomisches Institut, Rheinische Friedrich-Wilhelms-Universitaet, Nussallee 10, D-53115 Bonn, Germany. E-mail: karl.schilling@uni-bonn.de

Shinoda, Y., Department of Physiology, School of Medicine, Tokyo Medical and Dental University, 1-5-45 Yushima, Bunkyo-Ku, Tokyo 113, Japan. E-mail: yshinoda.phy-1@med.tmd.ac.jp

Simpson, J.I., Department of Physiology and Neuroscience, New York University Medical Center, 555 1st Avenue, New York, NY 10016, USA. E-mail: simpsj01@m-chip00.med.nyu.edu

Sotelo, C., INSERM U-106, Hôptital de la Salpetriere, 47 Boulevard de l'Hôpital, 75651 Paris, Cedex 13, France. E-mail: u106@inserm.fr

Stein, J., Department of Physiology, University of Oxford, Parks Road, Oxford OX1 3PT, UK. E-mail: john.stein@physiol.ox.ac.uk

Svensson, P., Department of Physiology and Neuroscience, University of Lund, Sölvegatan 19, S 223 62 Lund, Sweden. E-mail: par.svensson@mphy.lu.se

Van Alphen, A.M., Department of Anatomy, Erasmus University, P.O.Box 1738, 3000 DR Rotterdam, The Netherlands

Vos, B.P., Bron-Bunge Foundation, University of Antwerp, Universiteitsplein 1, 2610 Antwerp, Belgium. E-mail: bart@bbf.uia.ac.be

Wisden, B., Neurobiology Division, MRC-LMB Centre, Hills Road, Cambridge CB2 2QH, UK. E-mail: bw@mrc-lmb.cam.ac.uk

Yarom, Y., Department of Neurobiology, Institue of Life Sciences, Hebrew University, Givat Ram, Jerusalem 91904, Israel. E-mail: yarom@vms.huji.ac.il

Yeo, C., Department of Anatomy and Development Biology, University College London, Gower Street, London WC1E 6BT, United Kingdom. E-mail: c.yeo@ucl.ac.uk

Preface

J. Voogd

The present volume of Progress in Brain Research contains the proceedings of the second meeting on the Cerebellum entitled *Cerebellar Modules: Molecules, Morphology and Function,* which was held in August 1998, in Rotterdam, The Netherlands, to mark the retirement of Jan Voogd as chairman of the Department of Anatomy at the Erasmus University of Rotterdam. Unlike the previous meeting, this symposium reflects a specific point in time in a continuous line of research.

For over three decades Jan Voogd engulfed himself and his co-workers in what sometimes looked like idiosyncratic details of cerebellar anatomy. He was able to extract from his findings the basic principles of cerebellar development and structure, and convert them into his concept of functional modular organization. For those who worked with him it is no surprise that he did not drown in all those details. Apart from being an extremely acute observer and having a sharp analytical mind, he was also a gifted scientific artist who could visualize details and explain his concept in either simple drawings or elaborate three-dimensional reconstructions.

Voogd's work is a prime example of how detailed neuroanatomy can provide a framework in a systems-neuroscience approach. Our insight into the anatomy of the modular organization, illustrating the extreme hardwiring of the cerebellum, seems in several aspects far removed from results obtained through other disciplines. Particularly cell biology techniques have opened a Pandora's box of functional details, which have not yet been linked satisfactorily to the cerebellar modules. Simultaneously, progressively more refined techniques in anatomy and electrophysiology have expanded our insight into the function of neuronal assemblies. Nonetheless, a consensus on the issue of cerebellar function is still not in sight.

The present volume contains important contributions by leading cerebellar scientists, in part as a tribute to Voogd's achievements, but more as the result of continuous innovative efforts to strengthen the multidisciplinary approach in this field of research. We are confident that the ideas and concepts put forward in this volume will inspire both old and new students to participate in unraveling the cerebellar enigma.

N.M. Gerrits

Acknowledgements

This volume largely represents the material that was presented at the Symposium *Cerebellar Modules: Molecules, Morphology and Function* held August 25 - 28, 1998 in Rotterdam, The Netherlands. The Editors acknowledge the generous financial support of the Royal Dutch Academy of Sciences (KNAW), the Faculty of Medicine and Health Sciences of the Erasmus University Rotterdam (EUR), the Vereniging Trustfonds of the EUR, the Gerrit-Jan Mulder Stichting, Elsevier Science B.V. (Amsterdam), Blackwell Science (Oxford), Carl Zeiss B.V. (Weesp) and PAES Nederland BV (Zoeterwoude). The organization of the symposium was made possible through the excellent and energetic help of the technicians and staff of the Department of Anatomy (EUR) and of the Erasmus Medical Congress Organization.

N.M. Gerrits, T.J.H. Ruigrok and C.I. De Zeeuw

Contents

III. Modules and circuits

IV. Models and learning

SECTION I

Development

N.M. Gerrits, T.J.H. Ruigrok and C.I. De Zeeuw (Eds.)
Progress in Brain Research, Vol 124

CHAPTER 1

Neuroepithelial 'compartments' and the specification of vestibular projections

Joel C. Glover*

Department of Anatomy, University of Oslo, Institute of Basic Medical Sciences, N-0317, Oslo 3, Norway

Introduction

The vestibulo-ocular reflex, with its disynaptic, open loop organization and linear vectorial relationship between sensory organs and effector muscles, has long been a favorite of systems neurophysiologists interested in the study of sensorimotor integration. The same attributes provide an appealing subject for the neuroembryologist interested in the generation of synaptic specificity. How do vestibular interneurons receive synapses from vestibular afferents, and distribute synapses to extraocular motoneurons, with the specificity necessary to engender the reflex arc with the behaviorally appropriate translation of head movement to compensatory eye movement?

Our goal is to elucidate the mechanisms that pattern the neuron groups involved in vestibulo-ocular and other vestibular reflexes and that specify their connectivity. Here, I review our studies of the anatomical organization of these neuron groups at 11 days of embryonic development, and chronicle their appearance and patterning from earlier stages up to this stage. At 11 days of embryonic development, the brain is well-developed, the major nuclei of the hindbrain can be recognized (Tan and Le Douarin, 1991), and the organization of the vestibular projections is very similar to that in adult chickens (Wold, 1978a,b). At the same time, 11 days is proximate enough to the formative stages of the vestibular system to provide a convenient as well as appropriate endpoint for developmental studies.

To obtain our description of vestibular projections in the 11 day chicken embryo, we have made use of four main techniques: (1) retrograde axonal tracing, to determine the location of vestibular neurons that project to particular targets; (2) anterograde axonal tracing, to determine the patterns of synaptic termination exhibited by defined groups of vestibular neurons; (3) neurotransmitter immunohistochemistry, and (4) retrograde transport of radiolabeled neurotransmitters or neurotransmitter analogs, to determine the potential neurotransmitter phenotypes of the vestibular neurons.

Clustering and connectivity of vestibulo-ocular neurons: a hodological mosaic

The second order vestibular neurons in the hindbrain were originally classified by anatomists into nuclei based on cytoarchitectonic attributes. Four major nuclei are identified in mammals: the superior, lateral, medial, and descending (or inferior), each with a specific spatial domain. In attempts to elucidate function, the afferent and efferent connections of these nuclei have been studied with both anatomical and physiological techniques. The picture is complex. On the afferent side, each nucleus receives inputs from multiple vestibular end organs and from a variety of central

*Corresponding author. e-mail: joel@pons.uio.no

structures. Each nucleus also has diverse efferent projections that may encompass both ascending and descending components involved in vestibulo-ocular and vestibulospinal reflexes, respectively. Nevertheless, for both the afferent and efferent connections there are clear examples of topographic specificity within nuclei (reviewed in Highstein and McCrea, 1988).

Cytoarchitectonic criteria have also been applied to the vestibular system of avians. Here, the description is slightly more complex than in mammals, due to the presence of a fifth major nucleus, the tangential nucleus (Peusner and Morest, 1977c; Wold, 1976). Otherwise the cytoarchitectonic organization is similar to that of mammals (reviewed in Wold, 1976). Again, the pattern of afferent and efferent connections is heterogeneous, but not without examples of topographic specificity (Wold, 1978a,b; Evinger and Erichsen, 1986; Labandeira-Garcia et al., 1989; Cox and Peusner, 1990a,b; Arends et al., 1991).

In our studies of the vestibular system of the chicken embryo, we have taken a different tack. Instead of using cytoarchitecture to classify the vestibular neurons, we have used hodology; that is, the pattern by which the vestibular neurons project their axons.

By injecting retrograde axonal tracers such as horseradish peroxidase or fluorescent dextran-amines into the cervical spinal cord, we showed that vestibulospinal neurons project in two specific pathways, namely the lateral vestibulospinal tract (LVST) and the medial longitudinal fascicle (MLF). Any given vestibulospinal neuron projects along one of three specific axonal trajectories, either ipsilaterally in the LVST, ipsilaterally in the MLF, or contralaterally in the MLF (Glover and Petursdottir, 1988). The vestibulospinal neurons that project along a given trajectory are clustered in a unique spatial domain, segregated to a large extent from the other domains (Fig. 1a). The largest of these vestibulospinal clusters (in terms of the number of constituent neurons) projects ipsilaterally in the LVST, and has therefore been termed the LVST group. The smallest and next largest pools project ipsilaterally and contralaterally, respectively, in the MLF. Since together these constitute what is traditionally known as the medial vestibulospinal tract (MVST), they have been termed the ipsilateral and contralateral MVST (iMVST and cMVST) groups (Glover and Petursdottir, 1991).

The same approach was used to locate the vestibular neurons that project to the oculomotor and trochlear motor nuclei, and to define the pathways and trajectories on which their axons project (Petursdottir, 1990). This population of vestibulo-ocular neurons (which does not include those that project to the abducens nucleus) also projects in two specific pathways. One of these is the MLF, and the other, evidently the homologue of the mammalian brachium conjunctivum (BC), curves around the rostral aspect of the fourth ventricle, crosses the midline ventral to the MLF at a level just caudal to the trochlear nucleus, and then ascends on the contralateral side just beneath the MLF (Jansen, 1991). A third tract that carries vestibulo-ocular projections in mammals is the ascending tract of Deiters (ATD), which runs parallel and immediately lateral to the MLF, inserting between the MLF and BC at rostral levels (Highstein and Reisine, 1981; Markham et al., 1986). As far as we know, this tract has not been identified as a separate entity in birds, though this by no means rules out the existence of avian axon populations that are homologous to those carried by the ATD in mammals (see below).

The vestibulo-ocular neurons of the chicken embryo project in the MLF and the BC along one of three trajectories: ipsilaterally in the MLF, contralaterally in the MLF, and contralaterally in the BC. Those projecting in each trajectory are clustered in a unique spatial domain, largely segregated from the other domains (Fig. 1b, Fig. 2). Caudal to the VIIIth nerve there is a coherent cluster that projects exclusively ipsilaterally in the MLF. Lying largely dorsal to it is another cluster that projects exclusively contralaterally in the MLF. This latter cluster stretches from the midline to the dorsolateral aspect of the medulla. At its medial extreme it is coextensive with abducens nucleus interneurons that are known to innervate the contralateral medial rectus motoneurons in the oculomotor nuclear complex of mammals (Baker and Highstein, 1978). Using a simplified version of Petursdottir's (1990) nomenclature, I refer to these as the ipsilateral caudal (iC) and contralateral caudal (cC) vestibulo-ocular groups.

Rostral to the VIIIth nerve there are two discriminable clusters of vestibulo-ocular neurons. One of these, a slender but tightly-packed longitudinal column of neurons, projects ipsilaterally in the MLF. The other is more extensive mediolaterally, lies more rostral and dorsal, even within the cerebellar peduncle, and projects contralaterally in the BC. Again, using a simplified version of Petursdottir's (Petursdottir, 1990) nomenclature, I refer to these as the ipsilateral rostral (iR) and the contralateral rostral (cR) vestibulo-ocular groups, respectively.

The organization of vestibulo-ocular groups in the 11 day embryo is virtually identical to that reported by Labandeira-Garcia et al. (1989) in the 8 week hatchling chick, following retrograde tracing

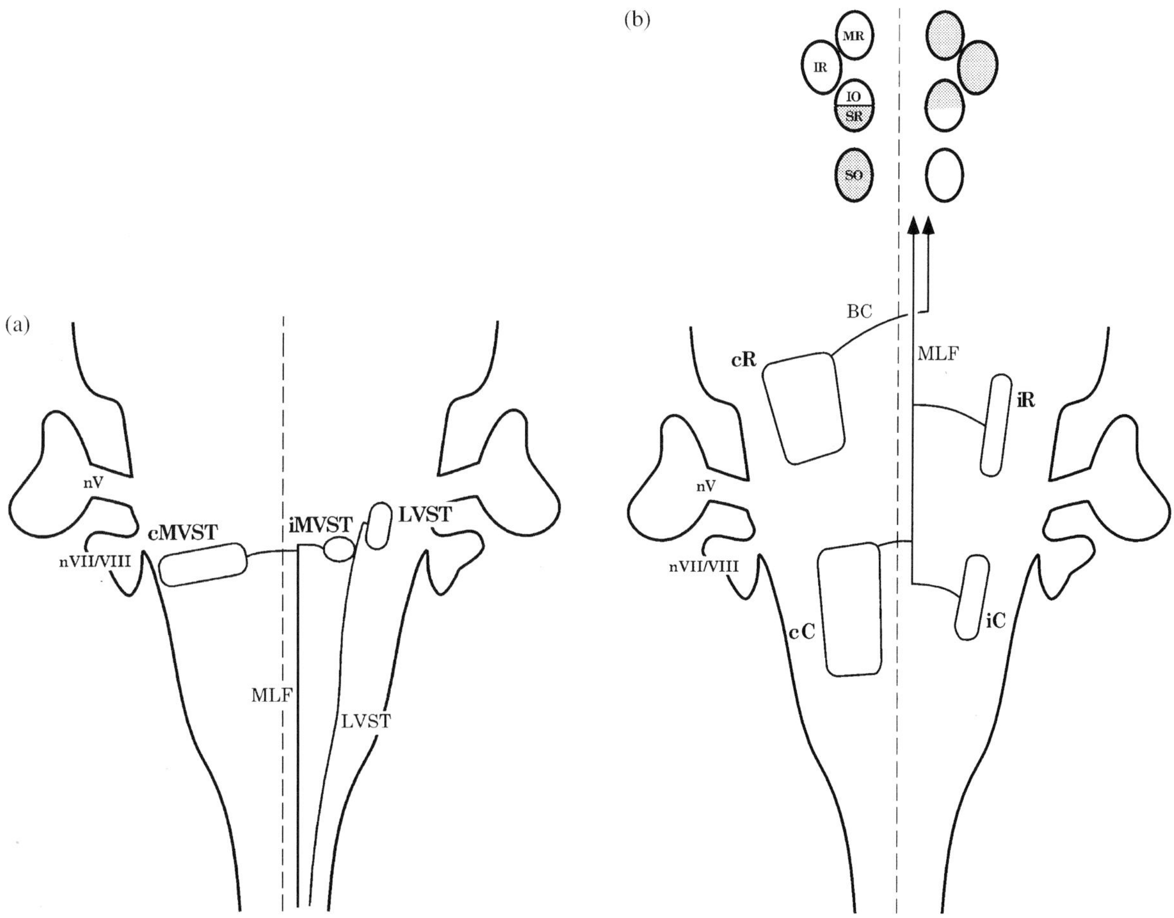

Fig. 1. Schematic representations of vestibulospinal (A) and vestibulo-ocular (B) neuron pools and their axonal trajectories in horizontal projections of the hindbrain. Spatial relationships are approximate and exaggerate the segregation seen in the horizontal plane in the 11-day chicken embryo (compare with Figure 2), but give a realistic representation of the segregation observed at earlier embryonic stages. The difference in apparent segregation between early and late stages is due largely to the morphometric changes in the hindbrain pursuant to elaboration of the pontine and cervical flexures. Abbreviations: nV (trigeminal nerve), nVII/VIII (facial/cochlear-vestibular nerves), MLF (medial longitudinal fascicle), cMVST (contralateral medial vestibulospinal tract neuron group), iMVST(ipsilateral medial vestibulospinal tract neuron group), LVST (lateral vestibulospinal tract and neuron group), MR (medial rectus motoneuron pool), IR (inferior rectus motoneuron pool), IO (inferior oblique motoneuron pool), SR (superior rectus motoneuron pool), SO (superior oblique motoneuron pool), BC (brachium conjunctivum), cR (contralateral rostral neuron group), iR (ipsilateral rostral neuron group), cC (contralateral caudal neuron group), iC (ipsilateral caudal neuron group).

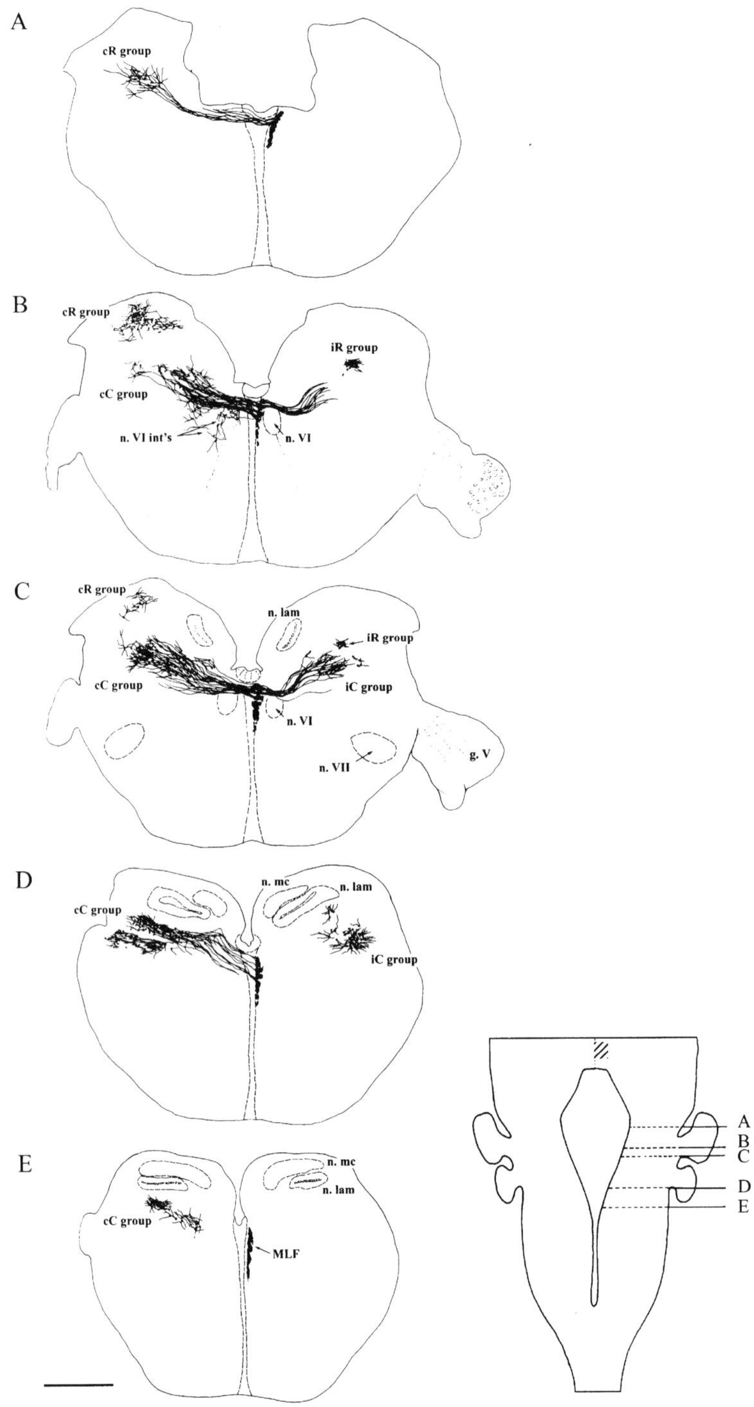

Fig. 2. The spatial organization of vestibulo-ocular neuron pools shown in camera lucida drawings of a series of transverse sections through the hindbrain of an 11-day chicken embryo in which HRP had been injected into the MLF unilaterally just caudal to the trochlear nucleus. The rostrocaudal level of each section is shown in the horizontal projection (inset) as is the site of HRP injection (hatching, inset).

selectively from the oculomotor nuclear complex. In the hatchling, the cytoarchitectonic divisions of the vestibular complex are apparent, allowing a direct comparison of the classical vestibular nuclei to the groups that we have defined by hodology. The iC group is localized within the descending nucleus, while the iR group lies as a tight cluster in the central region of the superior nucleus. The cC group is not localized to a single vestibular nucleus, but rather extends through the medial, descending, and tangential nuclei. The cR group is not contained within the vestibular nuclear complex at all, but rather lies well dorsal to the superior and Deiters' nuclei. Labandeira-Garcia et al. (1989) argue that this group most likely corresponds to the vestibular cell group 'y' of mammals (Carpenter and Cowie, 1985).

By applying differentiable retrograde tracers to the vestibulospinal and vestibulo-ocular projections in the same embryo, Petursdottir (1990) examined directly the spatial relationship between vestibulospinal and vestibulo-ocular neuron groups. From these experiments it is clear that the ipsilaterally-projecting vestibulospinal and vestibulo-ocular neurons are largely segregated along the longitudinal axis, with the LVST and iMVST groups sandwiched between the iC and iR groups (Fig. 3).

The organization of vestibulospinal and vestibulo-ocular neurons into coherent clusters with characteristic and specific axonal trajectories thus bears a close resemblance to the organization of motoneuron pools. Do the vestibular neuron 'pools' also have specific termination patterns? To address this issue, the projections from the vestibulo-ocular groups have been traced anterogradely, either individually or in defined combinations. Jansen (1991) labeled anterogradely the axonal projections from some of the vestibulo-ocular groups using the lipophilic axonal tracer DiI in formaldehyde-fixed preparations of the brain stem. Glover and Rinde (1995) labeled the vestibulo-ocular projections and their oculomotor neuron targets with different fluorescent dextran-amines in the same embryo in vitro. The results indicate principal terminations from the cR group in the contralateral inferior oblique and superior rectus pools, from the iR group in the ipsilateral superior oblique and inferior rectus pools, from the iC group in the ipsilateral

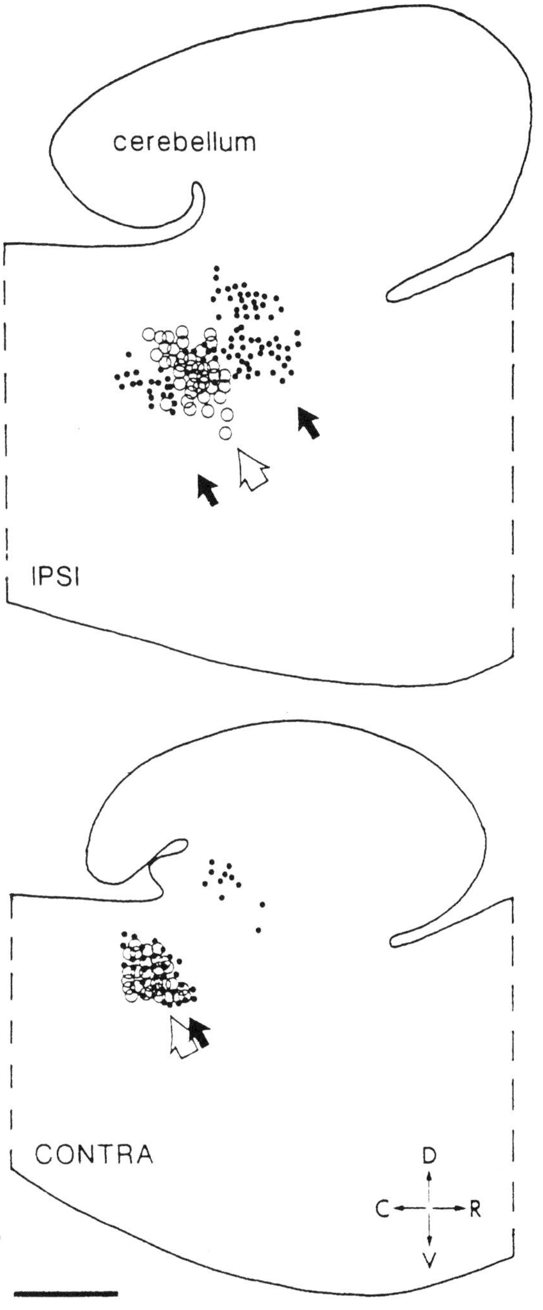

Fig. 3. Vestibulospinal neurons (circles, open arrows) and vestibulo-ocular neurons (dots, filled arrows) that project ipsilaterally are segregated along the rostrocaudal axis (top), whereas those that project contralaterally are intermingled (bottom). Plots from camera lucida drawings of parasagittal sections. Rostral is to right.

medial rectus, inferior oblique/superior rectus pair, and superior oblique pools, and from the cC group in the contralateral medial rectus, superior oblique, and inferior rectus pools (Fig. 4). Since the cC group includes the abducens interneurons, the source of excitatory input to contralateral medial rectus motoneurons in mammals, it seems likely that the non-abducens portion of the cC group is the source of the terminations in the inferior rectus and superior oblique pools. In addition to these principle termination patterns, Jansen (1991) noted the presence of minor terminations that have not been indicated in Fig. 4. These include a termination by the cR group in the contralateral medial rectus pool, and a termination by the iR group in the ipsilateral inferior oblique/superior rectus pool.

An important limitation to these anterograde tracing studies is that the motoneurons were labeled retrogradely en masse from the oculomotor nerve. This reveals the full extent of the motoneuron pools and their combined dendritic fields. It is also sufficient to discriminate the domains of the different pools, but not of the different dendritic fields, which can extend well beyond the confines of each pool. Moreover, given the minor terminations noted by Jansen (1991), the above description, though a good representation of the predominant termination patterns, should not be taken to exclude other potential connections of functional significance (see below).

Assessment of probable neurotransmitter phenotype

The selective termination patterns of the vestibulo-ocular groups within the trochlear and oculomotor motoneuron pools suggest that these groups might be activated in specific combinations to drive conjugate eye movements of specific directions. An important consideration, therefore, is the functional sign of each termination. Although there are some exceptions, most notably the ipsilateral excitatory ATD projection to the medial rectus pool (Highstein and Reisine, 1981; Markham et al., 1986), in general the ipsilateral and contralateral vestibulo-ocular projections in mammals are, respectively, inhibitory and excitatory (Precht, 1979). This would suggest that in the chicken the iR and iC groups are inhibitory and the cR and cC groups are excitatory. Our studies of potential neurotransmitter phenotype (Storm-Mathisen, et al., 1990; Glover et al., in preparation) support this notion.

By combining immunohistochemistry for the inhibitory transmitter GABA with retrograde labeling of the various neuron groups, we showed that the iR group is clearly GABA-immunoreactive (Fig. 5). Another method we have used to examine neurotransmitter phenotype is to apply radiolabeled transmitters or transmitter analogs to the target extraocular motoneuron pools. Through transmitter-specific uptake mechanisms in the presynaptic terminals, the radiolabeled material can be sequestered by the vestibulo-ocular axons and transported retrogradely to their cell bodies. The transmitter uptake mechanisms are reportedly specific enough that neurons using a particular transmitter are labeled selectively by the radiolabeled transmitter or transmitter analog. Although indirect, this method can provide corroborative evidence that a particular neuron group uses a particular neurotransmitter. In these experiments, we used tritiated GABA or tritiated nipicotenic acid for labeling putative GABAergic neurons, and tritiated aspartate for labeling putative glutamatergic or aspartatergic neurons. The results were somewhat variable, almost certainly reflecting variability in the efficacy and precision of the radiolabel application. Nevertheless, they indicate that the iR group is labeled by tritiated GABA, while the cR group and abducens interneurons (part of the cC group) are labeled by tritiated aspartate (Glover et al., in preparation).

A tentative functional classification

Taken together with the results of axonal tracing experiments, the data on the probable neurotransmitter phenotypes of the vestibulo-ocular groups accord well with the description that has been obtained in mammals. First of all, they are consistent with the general finding that ipsilateral vestibulo-ocular projections are principally inhibitory, while contralateral vestibulo-ocular projections are principally excitatory (Precht, 1979). Specifically, they provide evidence that the iR group is GABAergic while the cR group and at least the abducens interneurons and rostral portion

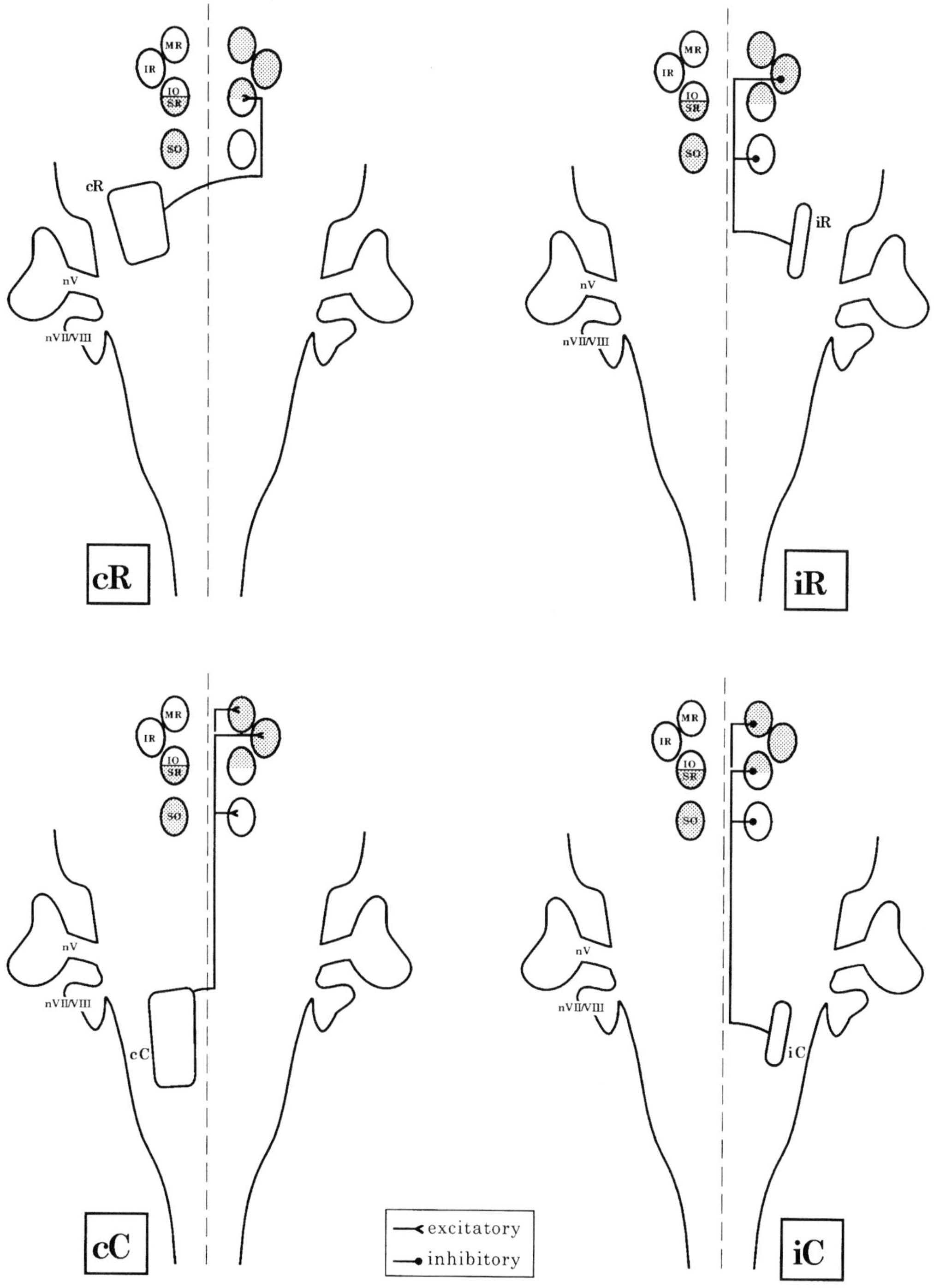

Fig. 4. Principle termination patterns of the individual vestibulo-ocular pools.

of the cC group are glutamatergic. If we assume this relationship to hold for the iC group and the remainder of the cC group, we can propose the simple connectivity scheme shown in Fig. 6. In this scheme, each of the motoneuron pools receives both an excitatory and an inhibitory input. Moreover, each of the hodologically defined vestibulo-ocular groups has a termination pattern that is consistent with a potentially specific action. For example, the non-abducens part of the cC group excites the superior oblique and inferior rectus motoneurons innervating their respective muscles in opposite eyes, whereas the iR group inhibits the same motoneuron pools. Likewise, the cR group excites the inferior oblique and superior rectus motoneurons innervating their respective muscles in opposite eyes, whereas the iC group inhibits the same motoneuron pools. Each of these actions can be considered partially synergistic in the generation of vertical or rotatory eye movements, and they can be coupled to provide the antagonism necessary to conjugate such movements in the two eyes. The cC and iC groups, moreover, respectively provide excitation and inhibition to the medial rectus pool, and therefore have an antagonistic relationship in the generation of horizontal eye movements.

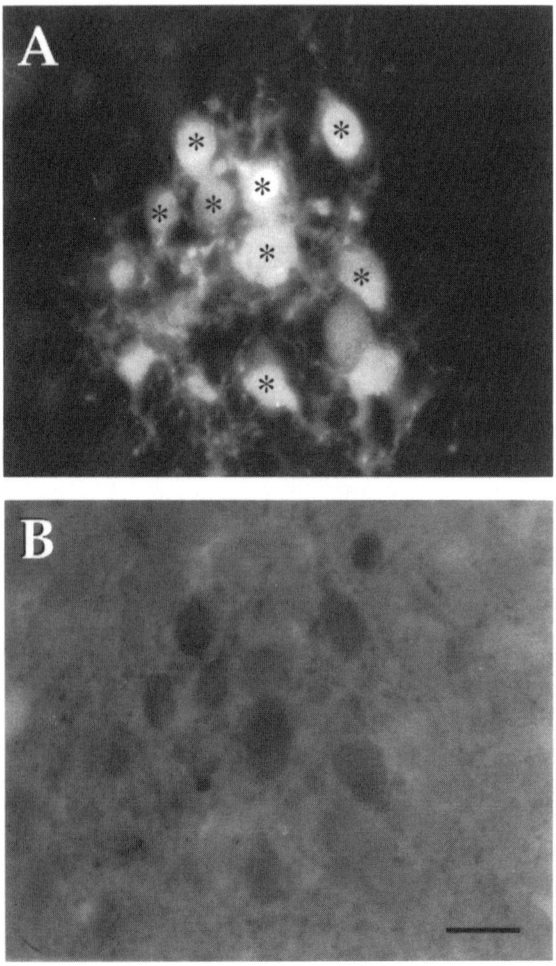

Fig. 5. GABA-immunoreactivity in retrogradely labeled iR vestibulo-ocular neurons. Upper panel shows a photographic positive of a transverse section through the iR group, with individual neurons retrogradely labeled with a fluorescent dextran-amine. Neurons indicated with an asterisk are GABA-immunoreactive, as seen in the lower panel, where dark cell profiles are stained with an antiserum to GABA.

Support for a functional pooling of vestibular neurons comes also from the organization of the vestibulospinal projections. It has long been known that the LVST and MVST have different spinal targets and functional roles (Wilson and Peterson, 1978). Recently, Shinoda et al. (1996) have shown that the functional sign of individual vestibulospinal neurons is strongly correlated with their descending trajectory: those projecting in the LVST are excitatory, those projecting in the contralateral MVST are also excitatory (although there are a few exceptions), whereas those projecting in the ipsilateral MVST are inhibitory. In the MVST we again see the dualism of ipsilateral inhibitory versus contralateral excitatory actions that is so characteristic of the vestibulo-ocular projections.

Although the scheme shown in Fig. 6 is appealing, it must be emphasized that there is still some lack of clarity in its organization, specifically regarding the cC and iC groups. The cC group as a whole excites the medial rectus, superior oblique, and inferior rectus pools. Since the rostral part of the cC group contains the abducens interneurons that in many species are known to selectively excite the medial rectus pool, it is tempting to divide the cC group into two components, dedicated respectively to horizontal and vertical/rotatory eye movements. This distinction in termination pattern has yet to be demonstrated directly in the chicken, however. The iC group as a whole terminates in the medial rectus, inferior oblique, superior rectus, and superior oblique pools. Can this group also be subdivided into components respectively dedicated to horizontal and vertical/rotatory eye movements? This is not yet clear. There is no obvious anatomical subdivision of the iC group that might be correlated with a further hodological subdivision, so additional tracing experiments will be required to resolve this question. The functional sign of the iC projection is also questionable. In mammals, the medial rectus pool does not appear to receive a direct inhibitory input, but rather two excitatory inputs, one contralateral from the abducens interneurons, and one ipsilateral via the ATD from neurons in the ventral region of the lateral vestibular nucleus (Highstein and Reisine, 1981). A parsimonious resolution would be if the iC projection to the medial rectus pool were in fact excitatory and thus homologous to the ATD projection of mammals, despite coursing in the MLF of the chicken embryo. Another quirk in the scheme is the projection from the iC group to the superior oblique pool, which would appear to be redundant, since inhibition to the superior oblique pool evidently is supplied by the iR group. One possibility is that the projection from the iC group to the superior oblique pool is one of the minor projections not shown in Figs 4 and 6. Given that semicircular canal and extraocular muscle orientations are not precisely aligned, and that the disparities vary as a function of eye laterality (Simpson and Graf, 1981; Ezure and Graf, 1984a), such terminations as these probably represent the auxiliary connections that are theoretically required to compensate for spatial incongruence in the vestibulo-ocular sensorimotor transformation (Ezure and Graf, 1984b; Graf et al., 1993).

Lastly, it is not known to what extent *individual* neurons in each group provide concerted or sepa-

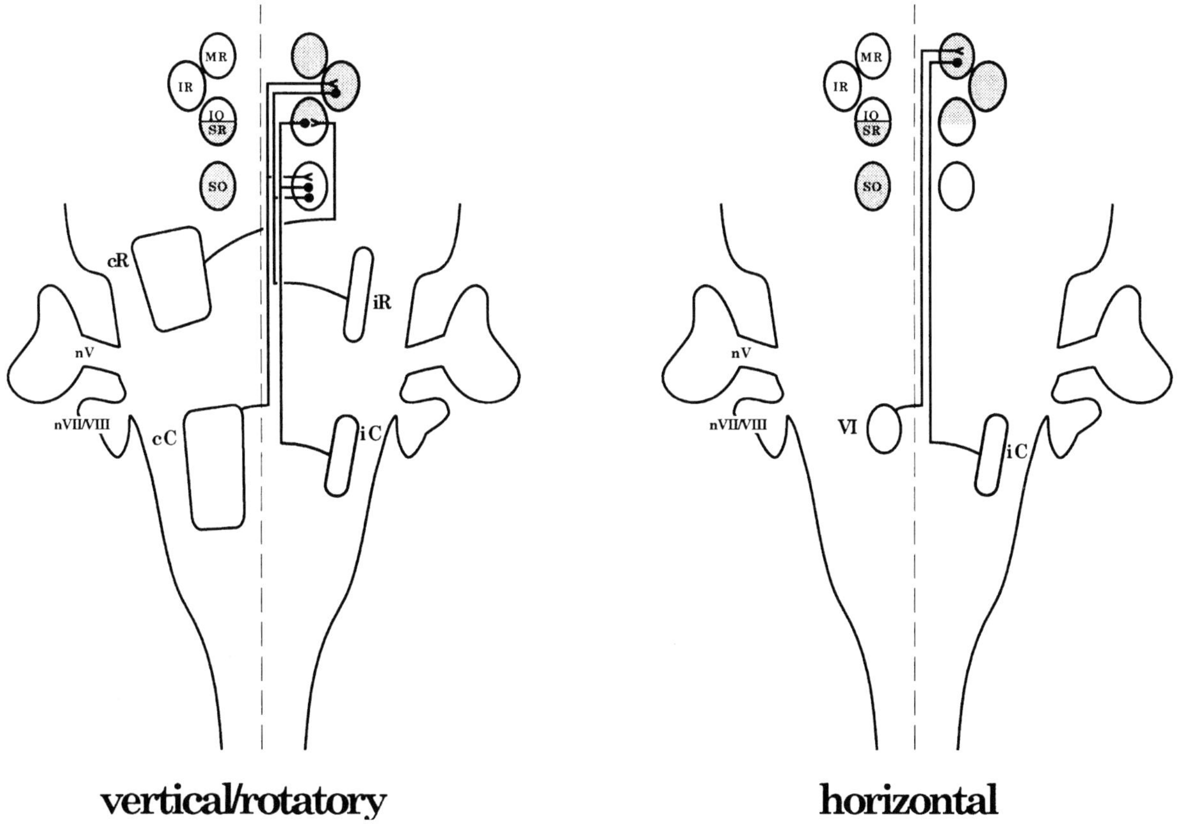

Fig. 6. Proposed connectivity patterns for vertical/rotatory and horizontal eye movements.

rate innervation to the motoneuron pools innervated by the group, nor is it known whether all neurons in each group use the same transmitter. The pattern by which the neurons in each group are recruited by the different vestibular end organs is also unknown. Until these issues are resolved, the functional specificity of each group cannot be defined unambiguously.

Despite these caveats, the information obtained to date shows that the vestibular projections in the 11-day chicken embryo are organized along the general vertebrate plan. More significantly, the description presented here includes a feature of vestibular organization that appears not to have been elucidated fully in other species: namely, that vestibular projection neurons are compartmentalized into a hodological mosaic in much the same way as their motoneuron targets in the spinal cord and mesencephalon. It is important to realize that the hodologically defined groups of vestibular neurons do not correspond directly to any of the classically defined vestibular nuclei. They are defined exclusively on the basis of location and axonal trajectory, making each of them topographically unique within the system of second-order vestibular projections. They represent a separate level of organization that is poorly related to cytoarchitectonics but that is evidently more closely tied to functional relationships.

Developmental significance of the hodological mosaic

Connectivity patterns such as those shown in Figs 4 and 6, wherein groups of neurons and their targets are synaptically coupled in a predictable topographic relationship, are not uncommon in the nervous system and have received the close scrutiny of neuroembryologists interested in the mechanisms that generate synaptic specificity. One of the most intensely studied connectivity patterns is the topographic mapping of motoneuron pools onto peripheral target muscles. There is now compelling evidence that spinal motoneurons are prespecified to recognize particular pathways in the developing limb, such that their axons are directed unerringly to a particular muscle (Eisen, 1994). This specification process involves position-specific signals within and impinging on the neural tube that determine the functional 'identity' of different motoneuron pools (Matise and Lance-Jones 1996; Tsuchida et al., 1994; Appel et al., 1995; Ensini et al., 1998; Lin et al., 1998). Operating either on the motoneurons themselves or on their progenitors, such signals presumably regulate the expression of genes that code for the membrane receptors used by axon growth cones to discriminate among pathway cues in the periphery. In this way, the position of a motoneuron or its progenitor at some critical timepoint would be a predictor of the motoneuron's future axon trajectory and termination pattern.

An obvious question is whether this scheme can be generalized to other neuron types, and specifically the vestibular system. Do vestibular neurons project unerringly along the appropriate pathways within the brain stem and spinal cord to their proper targets? If so, are vestibular neuron groups specified by virtue of their position to project along these pathways?

The vestibular neurons in the chicken embryo are generated starting on day 2 of embryonic development (McConnell and Sechrist, 1980), and their differentiation presumably begins shortly thereafter. By 3.5 days of development the axons from some vestibular groups have extended up to several hundred microns and can be labeled retrogradely with axonal tracers. At this time the hindbrain neural tube is partitioned into transient but morphologically overt neuromeres, termed rhombomeres (Källén, 1955; Vaage, 1969; Lumsden and Keynes, 1989), and the vestibulospinal and vestibulo-ocular groups present as distinct, segregated clusters that are delimited in part by rhombomere boundaries (Glover, 1989, 1993; Glover and Petursdottir, 1991) (Figs 7 and 8). For example, the LVST group is localized to a lateral domain lying within rhombomere 4 (r4) with slight spillover into r5, the iC group is localized to the same lateral domain in r5 and r6, and the cC group is localized to a more medial domain in r5 and r6 (Fig. 8A, B). Thus, the LVST and iC groups comprise a continuous longitudinal column through r4, r5, and r6, whose constituent neurons exhibit marked differences in pathway choice at about the r4/5 boundary (Fig. 8C). Similarly, the cC and iC groups lie within a

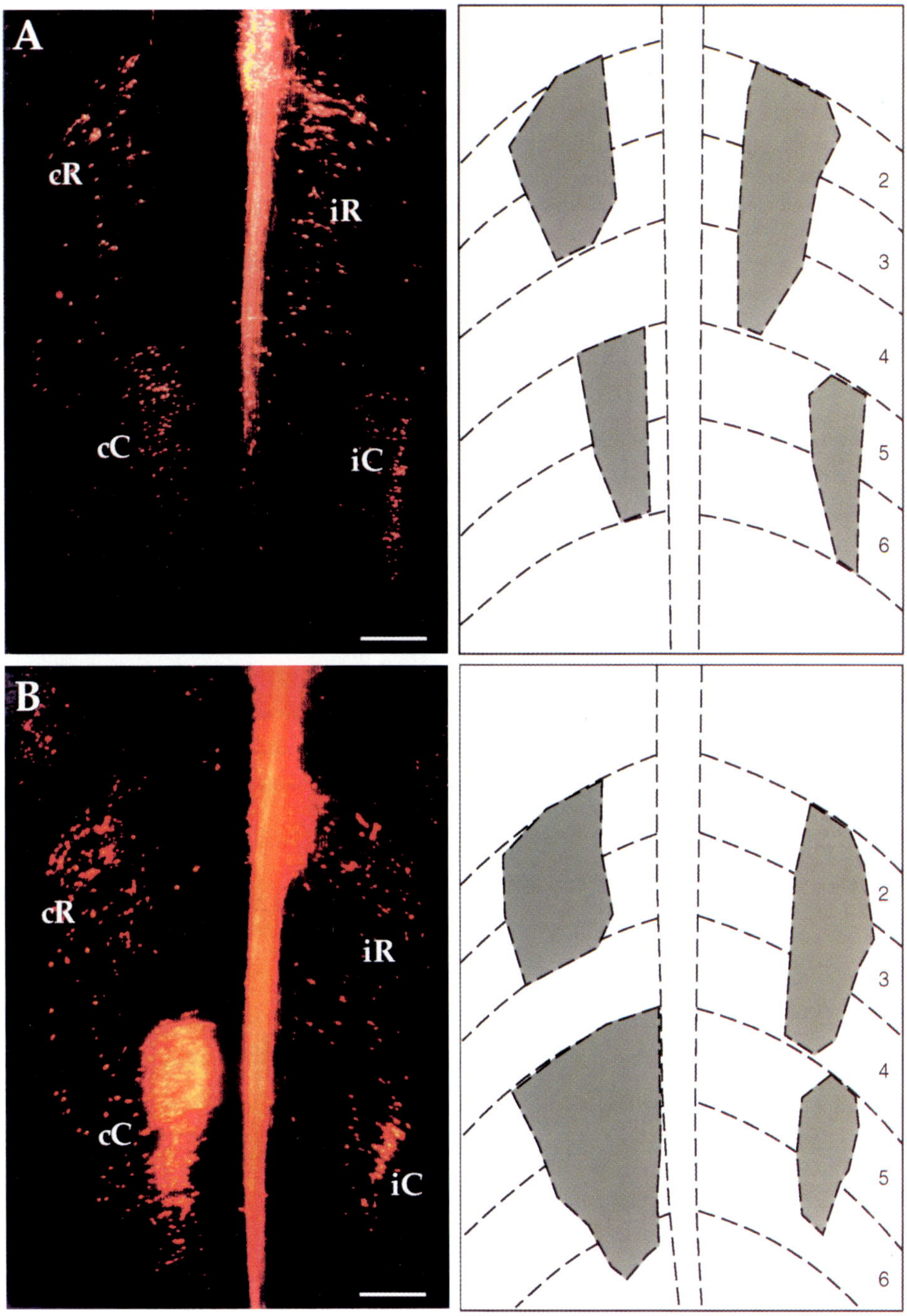

Fig. 7. The spatial domains of the vestibulo-ocular interneuron groups at early embryonic stages. Panels A and B show wholemounts of the hindbrain viewed from the ventral surface, at d4 and d5, respectively. The vestibulo-ocular neuron groups have been retrogradely labeled after unilateral injection of rhodamine dextran-amine into the medial longitudinal fascicle in the lower mesencephalon. The vertical dashed lines represent the lateral borders of the midline floor plate, on each side of which courses the medial longitudinal fascicle. The schematic drawing to the right of each panel shows qualitative estimates of the domains containing the neurons of each group with respect to the rhombomeres, which are numbered consecutively.

mediolateral band whose constituent neurons exhibit marked differences in pathway choice at a specific, though covert, longitudinal limit. Thus, there is a close correlation between the mediolateral and rhombomeric position of a neuron and the pathway choices exhibited by its axon. Position is related to differences in the axonal projections of the groups that in effect define them as ipsilaterally vs. contralaterally projecting and vestibulospinal vs. vestibulo-ocular.

The rostral vestibulo-ocular groups also occupy specific domains that are largely segregated from each other, the iR group lying medial and ventral to the cR group (Fig. 7). These two groups are less

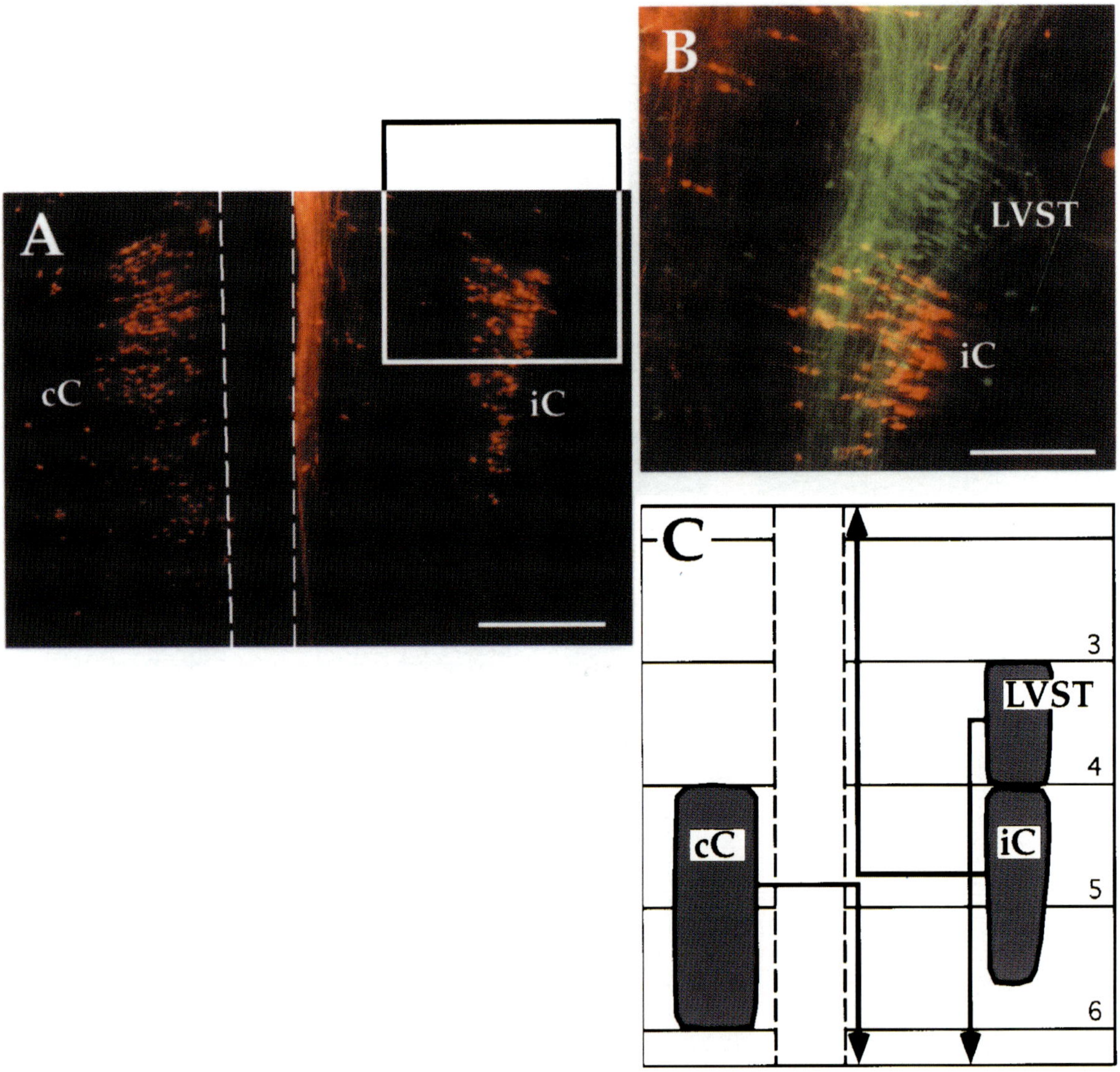

Fig. 8. (A) Comparison of the spatial domains of the iC and cC groups, both labeled retrogradely with rhodamine dextran-amine. The vertical dashed lines represent the lateral borders of the midline floor plate, on each side of which courses the medial longitudinal fascicle. (B) Inset from (A) showing a direct comparison of the spatial domains of the iC vestibulo-ocular group and the LVST vestibulospinal group. The two groups have been differentially labeled retrogradely with rhodamine dextran-amine (iC, red) and fluorescein dextran-amine (LVST, green). (C) A schematic illustration of the relationship between the rhombomeric domains and the axonal pathways of different neuron groups constituting a lateral longitudinal column.

tightly clustered than the iC and cC groups, so it is more difficult to define the limits of their domains and to assess how sharply they are segregated. By qualitative judgement the iR group lies predominantly within r2, but also spreads into r3 and r4, whereas the cR group lies predominantly within r2 but also spreads into r3 and slightly into r1 (Fig. 7).

Since some vestibular groups are difficult to identify at the stages when rhombomeres are visible, we substantiated the relationship between vestibular groups and the rhombomeres by combining fate mapping of the rhombomeres with retrograde labeling of the vestibular groups later. This substantiated the general scheme outlined above, but emphasized that not all of the vestibular groups are sharply registered with rhombomere boundaries (Díaz et al., 1998).

One of the difficulties in assessing the spatial pattern of the vestibular neuron groups is that the pattern is not static. Neurons migrate, and the domains of neuron groups can change. Such changes can include translocations, expansions, and contractions. For example, comparing Figs 7A and 7B shows that the iR group tends to translocate laterally, and the cC group tends to expand laterally and caudally, between 5 days and 6 days of development. Movement of vestibular neurons after they project their axons is probably an important component of the cytoarchitectonic changes that occur in the vestibular groups with further development. Thus, in contrast to the relatively sharp boundaries between the domains of vestibular groups at early stages, there is more overlap at later stages (Glover and Petursdottir, 1988, 1991; Petursdottir, 1990). Following the progress of these changes can reveal how the final distribution of vestibular neurons arises as the hindbrain matures, and can serve to validate the identification of groups at early stages. For example, the greater mediolateral and caudal extent of the cC group compared to the iC group on day 6 (Fig. 7B) neatly matches the situation in the 11 day chicken embryo and the hatchling chick (see Fig. 2).

The rhombomeres have been shown to be developmental pseudocompartments whose boundaries restrict the movement of cells within the plane of the neural tube and whose domains are characterized by specific patterns of regulatory gene expression (Keynes and Krumlauf, 1994). Many of the genes that are expressed in the rhombomeres at the stages when vestibular neurons are generated and begin to differentiate code for transcription factors, proteins that bind to specific DNA sequences and can thereby regulate transcription. This can result either in enhancement or repression of transcription. In addition to exhibiting rhombomere-specific patterns of expression, transcription factor genes and other developmental regulatory genes are expressed at specific locations within the transverse plane (see, for example, Lumsden and Krumlauf, 1996), such that the neural tube exhibits regionally specific gene expression along the longitudinal, mediolateral, and dorsoventral axes. Moreover, the patterns of gene expression can exhibit dynamic temporal patterns.

Thus, neurons lying within different rhombomeres and at different mediolateral positions within a given rhombomere must experience stereotypically different patterns of gene expression at some stage of their genesis and differentiation. This suggests that the spatiotemporal disposition of vestibular neurons regulates their differentiation, potentially specifying such properties as axon pathway choice, termination pattern, and neurotransmitter phenotype. Support for this hypothesis can be obtained by correlating the expression of specific genes with the domains and timing of vestibular group determination. Along these lines, we have shown that several genes, including the homeobox gene HoxB1, the paired box gene Pax 7, and the nuclear retinoid receptor RXRalpha, have expression patterns that can be related to the differentiation of specific vestibular groups (Fig. 9).

Besides the spatiotemporal pattern of transcription factors within the neural tube, other mechanisms exist that could contribute to the patterning of the vestibular groups. For example, (Lumsden et al., 1994) have observed that clones of neurons derived from single progenitor cells in the hindbrain neural tube are in some cases homogeneous with respect to axonal trajectory. An intriguing example is a clone of neurons observed in the lateral region of r4 whose axons all descend in a lateral longitudinal fascicle (the LLF). The

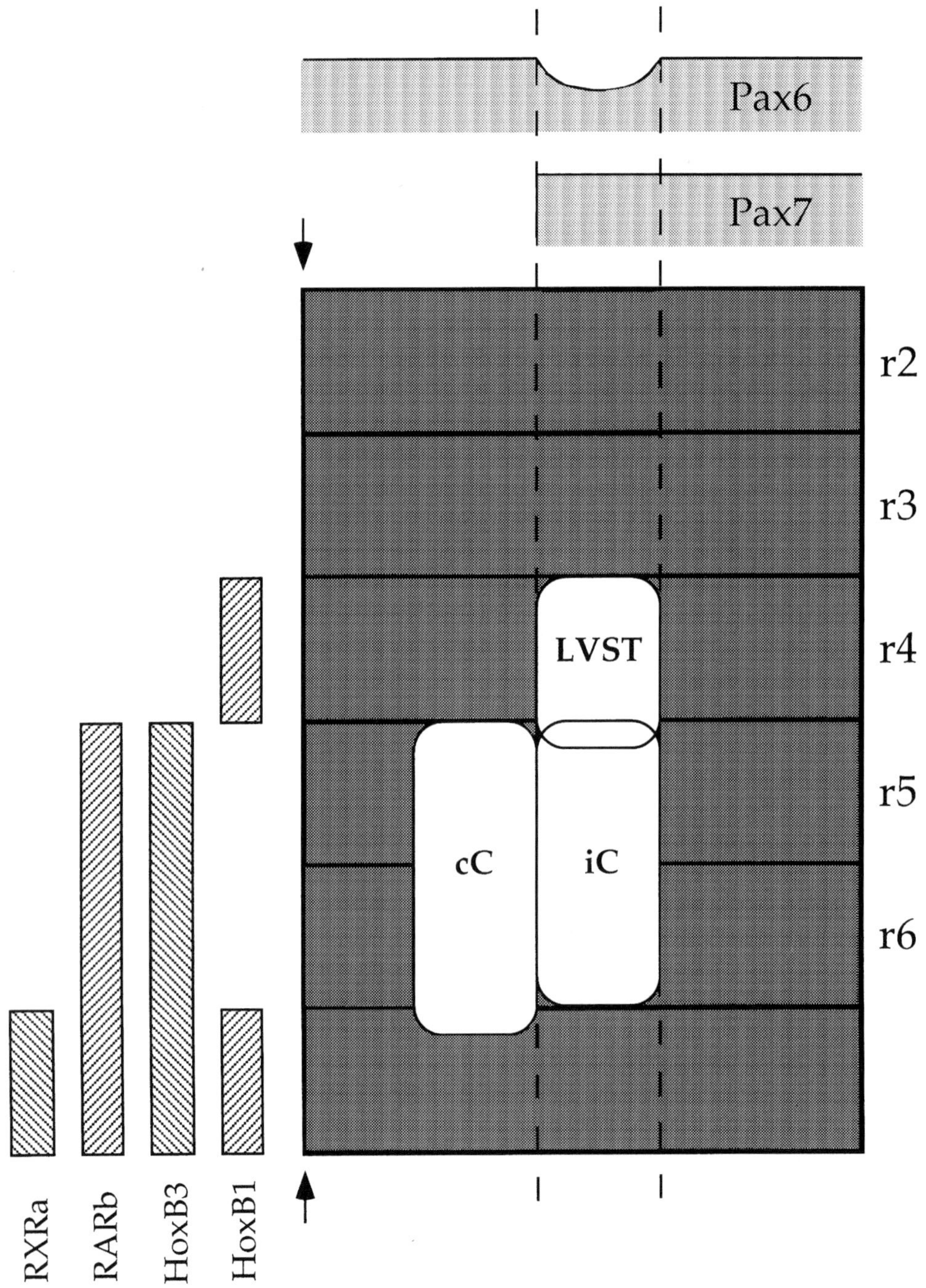

Fig. 9. Schematic representation of the expression domains of several transcription factors as they relate to the longitudinal and mediolateral domains of the LVST, iC, and cC groups at early stages. One side of the hindbrain has been drawn as a rectangular field and divided into rhombomeres (r); the midline is indicated by arrows. Abbreviations: RXR_a (retinoid X receptor alpha), RAR_b (retinoic acid receptor beta), Hox (homeobox), Pax (paired box).

resemblance to the LVST group is compelling. One implication is that the LVST group originates from one or a few progenitors, and that its axon pathway choice is determined by heredity, not by position. In other words, axon pathway choice would not be determined by the position of the neurons per se but by some factor inherited from the progenitor cell that instructs the neurons to project along the trajectory appropriate for the LVST group. The observed correlation between position and pathway choice would in this case simply be the consequence of the position of the progenitor cell. Of course, the expression of the instructive hereditary factor could be determined in the progenitor cell as a function of its position, which would still be an example of positional determination. This can only be assesed by finding the positions at which specific progenitors are situated. Alternatively, pathway choice may actually be determined in the neurons themselves as a function of their position, in which case the clonal homogeneity of pathway choice would simply reflect the similarity in the positions of its constituent neurons. Experimental manipulation of position will be required to distinguish between these alternatives.

Another possibility to explain the correlation between position and pathway choice is that the latter causes the former (Glover and Petursdottir, 1991). That is, as a consequence of extending its axon into a particular pathway, a neuron receives retrograde signals that instruct it to migrate to a particular position. Some spinal motoneurons (Tsuchida et al., 1994) and commissural interneurons (Eide and Glover, 1996) have indeed been shown to project their axons along specific pathways prior to migrating to their ultimate positions. However, if this process were involved in organizing the vestibular groups, then one might expect to see a greater degree of overlap of group domains at earlier stages. The opposite seems to be the case. This underscores the importance of extending our observations to the earliest stages possible in future studies.

Anterograde tracing of vestibulo-ocular axons during outgrowth and target contact

That vestibular groups can be identified by retrograde labeling at early stages of axon outgrowth demonstrates the selective nature of that outgrowth: the axons from a given group do not extend into inappropriate pathways, a behavior that would necessitate a later phase of retraction to establish the ultimate pattern of projections. For example, the initial trajectories of the iC axons and of the LVST axons are in both cases mediad, bringing them in close proximity to the LLF. But only the LVST axons turn to course in the LLF; the iC axons apparently ignore this potential pathway and continue towards the midline. Selective axon pathway choice evidently distinguishes the iC and cC groups as well. On reaching the midline, the iC vestibulo-ocular axons project ipsilaterally from the outset, whereas the cC axons project contralaterally from the outset, apparently ignoring the ipsilateral MLF.

This interpretation has a caveat, however. The selective *anterograde* labeling required to differentially assay the behavior of these axons as they initially contact the available pathways is technically difficult, and has not been performed. Although we cannot rule out a certain degree of trial and error at first contact, any indecisiveness on the part of the axons must be resolved well before they have projected more than a few rhombomere lengths longitudinally, since retrograde labeling at that time reveals no obvious errors.

To determine whether the vestibulo-ocular groups establish specific termination patterns as selectively as they evidently do their axon trajectories, we have performed anterograde labeling at somewhat *later* stages, prior to and during the innervation of the extraocular motoneuron targets. Anterograde labeling of the iR and cR groups has demonstrated a selective innervation of the appropriate motoneuron pools at least as early as day 9, shortly after the axons reach the target area (Jansen, 1991). The process has been followed more closely for the iC and cC groups by Glover and Rinde (1995), who labeled the oculomotor nerve retrogradely, and the combined projection from the iC and cC groups anterogadely, using different fluorescent tracers in the same embryo. This allowed the simultaneous visualization of the axon terminal collaterals and their potential motoneuron targets.

The iC and cC axons reach the level of the oculomotor complex by about day 5, at which time a substantial number of immature motoneurons can be labeled retrogradely from the oculomotor nerve. The ensuing process of innervation is highly dynamic, owing to several dramatic changes in the motoneurons. These are initially organized in a single cluster ipsilateral to the labeled nerve, and have not yet differentiated dendrites. Over the next several days, this single cluster gradually resolves into subgroups corresponding to the separate oculomotoneuron pools. Concurrently, some of the motoneurons migrate across the midline to establish the contralaterally projecting superior rectus pool (Puelles, 1978; Naujoks-Manteuffel et al., 1991; Glover and Rinde, 1995). In addition, all of the motoneurons begin to elaborate dendrites that extend beyond the confines of the pools of origin. On this dynamic backdrop, the iC and cC vestibulo-ocular axons make their entrance by the extension of collateral branches. The process is illustrated in Fig. 10. Terminal collaterals first appear late on day 6, at which time they extend obliquely toward the dorsomedial pole of the still coherent motoneuron cluster, with some veering across the midline in the vicinity of the migrating superior rectus motoneurons. By day 7, the number of collaterals has increased, the ipsilateral collateral projection to the dorsomedial pole of the motoneuron cluster especially is strengthened, and the motoneurons are beginning to separate into pools, showing that the dorsomedial pole in fact corresponds to the medial rectus pool. By day 9 the medial rectus pool is clearly resolved and is heavily invested with terminal collaterals. Additional collaterals extend into the inferior rectus and the overlapping inferior oblique and superior rectus pools. This establishes the predominant features of the mature pattern of termination for the combined iC and cC axons (see Fig. 4).

The early extension of the iC and cC terminal collaterals is clearly selective. They aim initially for a specific region of the developing motoneuron cluster, focusing on that region even as later collaterals to neighboring regions arise. The region corresponding to the Edinger–Westphal nucleus is essentially ignored at all stages. A reasonable interpretation of the sequence of events is that iC and cC axons predetermined to innervate medial rectus motoneurons establish the first collaterals, which are either selectively attracted by or selectively recognize the differentiating medial rectus motoneurons. Somewhat later, iC and cC axons predetermined to innervate the other motoneuron pools extend collaterals into these.

Although by appearances the initially selective innervation pattern exhibited by these vestibulo-ocular axons suggests a process of selective attraction or recognition, it is important to realize that alternative nonselective mechanisms may exist. One possibility involves the mechanism of chemoattractant secretion, which has been implicated in the steering of axon growth in several other parts of the nervous system (Goodman, 1996). For example, if the extension of any collateral is contingent on the release of a general chemoattractant common to all motoneurons, then a careful timing of

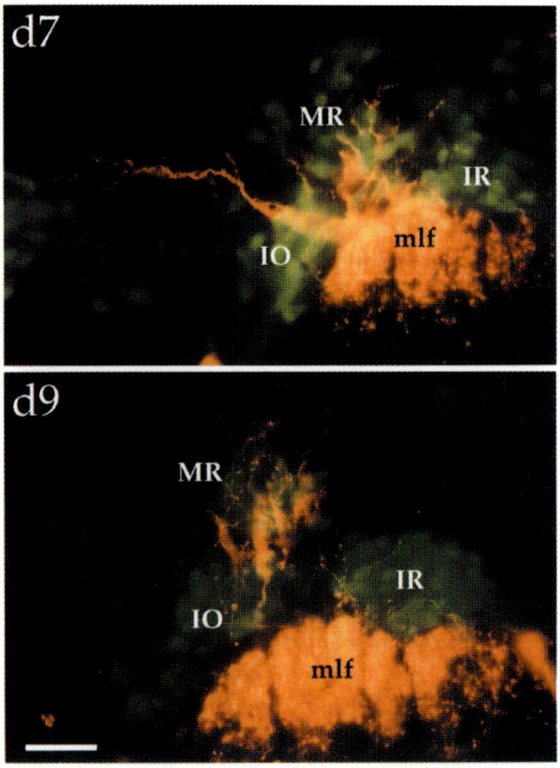

Fig. 10. The growth of vestibulo-ocular collaterals (labeled red) into specific regions of the oculomotor nuclear complex (labeled green) at the indicated developmental stages (embryonic day 7 and day 9) as seen in transverse sections. Abbreviations: mlf (medial longitudinal fascicle), MR (medial rectus pool), IO (inferior oblique pool), IR (inferior rectus pool).

the onset of the ability of different motoneurons to release the attractant, the arrival of different vestibulo-ocular axons at the target area, and the onset and termination of their capacity to respond to the attractant could produce the observed behavior. Experimental manipulations are required to resolve such issues.

Summary

The implication that there exist coherent vestibulo-ocular neuron pools with specific functions may provide new insight into how conjugate eye movements are synthesized within the vestibulo-ocular reflex. The systematic relationship between pool position and synergistic principle terminations, the 'hodological mosaic' suggests, moreover, a determinate groundplan established by developmental mechanisms operative at early stages in the hindbrain neuroepithelium. From such a groundplan, evolutionary and use-dependent modifications could mold connectivity patterns functionally appropriate for each species and individual. How the expression of developmentally regulatory genes contributes to establishing the mosaic organization of the vestibular system is the current focus of our research.

References

Appel, B., Korzh, V., Glasgow, E., Thor, S., Edlund, T., Dawid, I.B. and Eisen, J.S. (1995) Motoneuron fate specification revealed by patterned LIM homeobox gene expression in embryonic zebrafish. *Development*, 121: 4117–4125.

Arends, J.J., Allan, R.W. and Zeigler, H.P. (1991) Organization of the cerebellum in the pigeon (Columba livia): III. Corticovestibular connections with eye and neck premotor areas. *J. Comp. Neurol.*, 306: 273–289.

Baker, R. and Highstein, S.M. (1978) Vestibular projections to medial rectus subdivision of oculomotor nucleus. *J. Neurophysiol.*, 41: 1629–1646.

Boyle, R., Goldberg, J.M. and Highstein, S.M. (1990) Vestibular-nerve inputs to vestibulospinal and vestibulo–ocular neurons of the squirrel monkey. In: A. Berthoz, W. Graf and P.P. Vidal (Eds), *The Head-neck Sensory-motor System*, New York: Wiley. pp. 255–265.

Cabot, J.B., Reiner, A. and Bogan, N. (1982) Avian bulbospinal pathways: anterograde and retrograde studies of cells of origin, funicular trajectories and laminar terminations. *Prog. Brain Res.*, 57: 79–108.

Carpenter, M.B. and Cowie, R.J. (1985) Transneuronal transport in the vestibular and auditory systems of the squirrel monkey and the arctic ground squirrel. I. Vestibular system. *Brain Res.*, 358: 249–263.

Chalepakis, G., Tremblay, P. and Gruss, P. (1992) Pax genes, mutants and molecular function. *J. Cell Sci. – Suppl.*, 16: 61–67.

Correia, M.J., Eden, A.R., Westlund, K.N. and Coulter, J.D. (1983) A study of some of the ascending and descending vestibular pathways in the pigeon (*Columba livia*) using anterograde transneuronal autoradiography. *Brain Res.*, 278: 53–61.

Cox, R.G. and Peusner, K.D. (1990a) Horseradish peroxidase labeling of the central pathways in the medulla of the ampullary nerves in the chicken, *Gallus gallus*. *J. Comp. Neurol.*, 297: 564–581.

Cox, R.G. and Peusner, K.D. (1990b) Horseradish peroxidase labeling of the efferent and afferent pathways of the avian tangential vestibular nucleus. *J. Comp. Neurol.*, 296: 324–341.

Díaz, C., Puelles, L., Marín, F. and Glover, J.C. (1998) The relationship between rhombomeres and vestibular neuron populations as assessed in quail-chicken chimeras. *Dev. Biol.*, 202: 14–28.

Eide, A.L. and Glover, J.C. (1996) The development of an identified spinal commissural interneuron population in an amniote: The neurons of the avian Hofmann nuclei. *J. Neurosci.*, 16: 5749–5761.

Eisen, J.S. (1994) Development of motoneuronal phenotype. *Ann. Rev. Neurosci.*, 17: 1–30.

Ensini, M., Tsuchida, T.N., Belting, H.G. and Jessell, T.M. (1998) The control of rostrocaudal pattern in the developing spinal cord: specification of motor neuron subtype identity is initiated by signals from paraxial mesoderm. *Development*, 125: 969–982.

Evinger, C. and Erichsen, J.T. (1986) Transsynaptic retrograde transport of fragment C of tetanus toxin demonstrated by immunohistochemical localization. *Brain Res.*, 380: 383–388.

Ezure, K. and Graf, W. (1984a) A quantitative analysis of the spatial organization of the vestibulo–ocular reflexes in lateral- and frontal-eyed animals – I. Orientation of semicircular canals and extraocular muscles. *Neurosci.*, 12: 85–93.

Ezure, K. and Graf, W. (1984b) A quantitative analysis of the spatial organization of the vestibulo–ocular reflexes in lateral- and frontal-eyed animals – II. Neuronal networks underlying vestibulo-oculomotor coordination. *Neurosci.*, 12: 95–109.

Glover, J.C. (1989) Rhombomeres and longitudinal boundaries in the chicken embryo subdivide vestibular neurons according to axonal pathway. *Soc. Neurosci. Abstr.*, 15: 959.

Glover, J.C. (1993) The development of brain stem projections to the spinal cord in the chicken embryo. *Brain Res. Bull.*, 30: 265–271.

Glover, J.C. and Petursdottir, G. (1988) Pathway specificity of reticulospinal and vestibulospinal projections in the 11-day chicken embryo. *J. Comp. Neurol.*, 270: 25–38.

Glover, J.C. and Petursdottir, G. (1991) Regional specificity of developing reticulospinal, vestibulospinal and vestibulo–ocular projections in the chicken embryo. *J. Neurobiol.*, 22: 353–376.

Glover, J.C. and Rinde, A. (1995) Initially selective termination patterns exhibited by vestibulo–ocular projections in the chicken embryo. *Soc. Neurosci. Abstr.*, 21: 803.

Goodman, C.S. (1996) Mechanisms and molecules that control growth cone guidance. *Ann. Rev. Neurosci.*, 19: 341–377.

Goodman, C.S. and Shatz, C.J. (1993) Developmental mechanisms that generate precise patterns of neuronal connectivity. *Cell*, 72 Suppl. 77–98.

Graf, W., Baker, J. and Peterson, B.W. (1993) Sensorimotor transformation in the cat's vestibuloocular reflex system. I. Neuronal signals coding spatial coordination of compensatory eye movements. *J. Neurophysiol.*, 70: 2425–2441.

Graf, W. and Ezure, K. (1986) Morphology of vertical canal related second order vestibular neurons in the cat. *Exp. Brain Res.*, 63: 35–48.

Graf, W., McCrea, R.A. and Baker, R. (1983) Morphology of posterior canal related secondary vestibular neurons in rabbit and cat. *Exp. Brain Res.*, 52: 125–138.

Gross, G.H. and Oppenheim, R.W. (1985) Novel sources of descending input to the spinal cord of the hatching chick. *J. Comp. Neurol.*, 232: 162–179.

Highstein, S.M. and McCrea, R.A. (1988) The anatomy of the vestibular nuclei. *Rev. Oculomot. Res.*, 2: 177–202.

Highstein, S.M. and Reisine, H. (1981) The ascending tract of Deiters' and horizontal gaze. *Ann. N.Y. Acad. Sci.*, 374: 102–111.

Isu, N., Uchino, Y., Nakashima, H., Satoh, S., Ichikawa, T. and Watanabe, S. (1988) Axonal trajectories of posterior canal-activated secondary vestibular neurons and their coactivation of extraocular and neck flexor motoneurons in the cat. *Exp. Brain Res.*, 70: 181–191.

Isu, N. and Yokota, J. (1983) Morphophysiological study on the divergent projection of axon collaterals of medial vestibular nucleus neurons in the cat. *Exp. Brain Res.*, 53: 151–162.

Jansen, J.K. (1991) A note on the development of the vestibulo–ocular pathway in the chicken. *Anat. Embryol.*, 184: 305–311.

Källén, B. (1955) Regeneration in the hindbrain of neural tube stages of chick embryos. *Anat. Record*, 123: 169–182.

Keynes, R. and Krumlauf, R. (1994) Hox genes and regionalization of the nervous system. *Ann. Rev. Neurosci.*, 17: 109–132.

Labandeira-Garcia, J.L., Guerra-Seijas, M.J., Labandeira-Garcia, J.A. and Jorge-Barreiro, F.J. (1989) Afferent connections of the oculomotor nucleus in the chick. *J. Comp. Neurol.*, 282: 523–534.

Lin, J.H., Saito, T., Anderson, D.J., Lance-Jones, C., Jessell, T.M. and Arber, S. (1998) Functionally related motor neuron pool and muscle sensory afferent subtypes defined by coordinate ETS gene expression. *Cell*, 95: 393–407.

Lumsden, A., Clarke, J.D., Keynes, R. and Fraser, S. (1994) Early phenotypic choices by neuronal precursors, revealed by clonal analysis of the chick embryo hindbrain. *Development*, 120: 1581–1589.

Lumsden, A. and Keynes, R. (1989) Segmental patterns of neuronal development in the chick hindbrain. *Nature*, 337: 424–428.

Lumsden, A. and Krumlauf, R. (1996) Patterning the vertebrate neuraxis. *Science*, 274: 1109–1115.

Marín, F. and Puelles, L. (1995) Morphological fate of rhombomeres in quail/chick chimeras: a segmental analysis of hindbrain nuclei. *Eur.J. Neurosci.*, 7: 1714–1738.

Matise, M.P. and Lance-Jones, C. (1996) A critical period for the specification of motor pools in the chick lumbosacral spinal cord. *Development*, 122: 659–669.

Markham, C.H., Furuya, N., Bak, I.J. and Ornitz, E.M. (1986) Synaptic connections of horizontal canal mediated ascending Deiters tract axons on medial rectus motoneurons in cat. *Auris Nasus Larynx*, 13 Suppl. 2: S1–14.

McConnell, J.A. and Sechrist, J.W. (1980) Identification of early neurons in the brainstem and spinal cord: I. An autoradiographic study in the chick. *J. Comp. Neurol.*, 192: 769–783.

McCrea, R.A., Strassman, A. and Highstein, S.M. (1987a) Anatomical and physiological characteristics of vestibular neurons mediating the vertical vestibulo–ocular reflexes of the squirrel monkey. *J. Comp. Neurol.*, 264: 571–594.

McCrea, R.A., Strassman, A., May, E. and Highstein, S.M. (1987b) Anatomical and physiological characteristics of vestibular neurons mediating the horizontal vestibulo–ocular reflex of the squirrel monkey. *J. Comp. Neurol.*, 264: 547–570.

McCrea, R.A., Yoshida, K., Evinger, C. and Berthoz, A. (1981) The location, axonal arborization and termination sites of eye-movement related secondary vestibular neurons demonstrated by intra-axonal HRP injections in the alert cat. In: A.F. Fuchs and W. Becker (Eds), *Progress in Oculomotor Research*, New York: Elsevier, pp. 379–386.

Minor, L.B., McCrea, R.A. and Goldberg, J.M. (1990) Dual projections of secondary vestibular axons in the medial longitudinal fasciculus to extraocular motor nuclei and the spinal cord of the squirrel monkey. *Exp. Brain Res.*, 83: 9–21.

Naujoks-Manteuffel, C., Sonntag, R. and Fritzsch, B. (1991) Development of the amphibian oculomotor complex: evidences for migration of oculomotor motoneurons across the midline. *Anat. Embryol.*, 183: 545–552.

Petursdottir, G. (1990) Vestibulo-ocular projections in the 11-day chicken embryo: pathway specificity. *J. Comp. Neurol.*, 297: 283–297.

Peusner, K.D. (1980) Developmental changes of synaptic endings in the nucleus vestibularis tangentialis of the chicken: degeneration of spoon endings. *Dev. Neurosci.*, 3: 266–276.

Peusner, K.D. (1981) An ultrastructural study of the development of synaptic endings in the nucleus vestibularis tangentialis of the chick embryo. *Neurosci.*, 6: 2335–2350.

Peusner, K.D. (1984a) The development of synapses and 'spoon' synaptic terminal space in the tangential vestibular nucleus: a quantitative electron microscope study. *J. Comp. Neurol.*, 230: 372–385.

Peusner, K.D. (1984b) Development of synaptic organization in the tangential vestibular nucleus: a quantitative electron microscope study. *J. Comp. Neurol.*, 230: 386–392.

Peusner, K.D. and Morest, D.K. (1977a) A morphological study of neurogenesis in the nucleus vestibularis tangentialis of the chick embryo. *Neurosci.*, 2: 209–227.

Peusner, K.D. and Morest, D.K. (1977b) Neurogenesis in the nucleus vestibularis tangentialis of the chick embryo in the absence of the primary afferent fibers. *Neurosci.*, 2: 253–270.

Peusner, K.D. and Morest, D.K. (1977c) The neuronal architecture and topography of the nucleus vestibularis tangentialis in the late chick embryo. *Neurosci.*, 2: 189–207.

Precht, W. (1979) Vestibular mechanisms. *Ann. Rev. Neurosci.*, 2: 265–289.

Puelles, L. (1978) A Golgi-study of oculomotor neuroblasts migrating across the midline in chick embryos. *Anat. Embryol.*, 152: 205–215.

Ramón y Cajal, S. (1908) Sur un noyau special du nerf vestibulaire des poissons et des oiseaux. *Trab del Lab del invest biol de la Universidad de Madrid*, 6: 210–224.

Roelink, H., Porter, J.A., Chiang, C., Tanabe, Y., Chang, D.T., Beachy, P.A. and Jessell, T.M. (1995) Floor plate and motor neuron induction by different concentrations of the amino-terminal cleavage product of sonic hedgehog autoproteolysis. *Cell*, 81: 445–455.

Sanders, E.B. (1929) A consideration of certain bulbar, midbrain and cerebellar centers and fiber tracts in birds. *J. Comp. Neurol.*, 49: 155–222.

Shinoda, Y., Sugiuchi, Y., Futami, T., Kakei, S., Izawa, Y. and Na, J. (1996) Four convergent patterns of input from the six semicircular canals to motoneurons of different neck muscles in the upper cervical cord. *Ann. N.Y. Acad. Sci.*, 781: 264–275.

Simpson, J.I. and Graf, W. (1981) Eye-muscle geometry and compensatory eye movements in lateral-eyed and frontal-eyed animals. *Ann. N.Y. Acad. Sci.*, 374: 20–30.

Storm-Mathisen, J., Jansen, J.K.S., Glover, J.C. and Fischer, B.O. (1990) The vestibulo–ocular (V–O) pathway in the chicken embryo. *Soc. Neurosci. Abstr.*, 16: 903.

Tan, K. and Le Douarin, N.M. (1991) Development of the nuclei and cell migration in the medulla oblongata. Application of the quail-chick chimera system. *Anat. Embryol.*, 183: 321–343.

Tosney, K.W., Hotary, K.B. and Lance-Jones, C. (1995) Specifying the target identity of motoneurons. *Bioessays*, 17: 379–382.

Tsuchida, T., Ensini, M., Morton, S.B., Baldassare, M., Edlund, T., Jessell, T.M. and Pfaff, S.L. (1994) Topographic organization of embryonic motor neurons defined by expression of LIM homeobox genes. *Cell*, 79: 957–970.

Uchino, Y. and Hirai, N. (1984) Axon collaterals of anterior semicircular canal-activated vestibular neurons and their coactivation of extraocular and neck motoneurons in the cat. *Neurosci. Res.*, 1: 309–325.

Uchino, Y., Sasaki, M., Isu, N., Hirai, N., Imagawa, M., Endo, K. and Graf, W. (1994) Second-order vestibular neuron morphology of the extra-MLF anterior canal pathway in the cat. *Exp. Brain Res.*, 97: 387–396.

Vaage, S. (1969) The segmentation of the primitive neural tube in chick embryos (*Gallus domesticus*). *Adv. Anat. Embryol. Cell Biol.*, 41: 1–88.

Wilkinson, D.G. (1995) Genetic control of segmentation in the vertebrate hindbrain. *Perspect. Dev. Neurobiol.*, 3: 29–38.

Wilkinson, D.G. and Krumlauf, R. (1990) Molecular approaches to the segmentation of the hindbrain. *Trends Neurosci.*, 13: 335–339.

Wilson, V.J. and Peterson, B.W. (1978) Peripheral and central substrates of vestibulospinal reflexes. *Physiol. Rev.*, 58: 80–105.

Wold, J.E. (1975) The vestibular nuclei in the domestic hen (*Gallus domesticus*). II. Primary afferents. *Brain Res.*, 95: 531–543.

Wold, J.E. (1976) The vestibular nuclei in the domestic hen (*Gallus domesticus*). I. Normal anatomy. *Anat. Embryol.*, 149: 29–46.

Wold, J.E. (1978a) The vestibular nuclei in the domestic hen (*Gallus domesticus*) III. Ascending projections to the mesencephalic eye motor nuclei. *J. Comp. Neurol.*, 179: 393–405.

Wold, J.E. (1978b) The vestibular nuclei in the domestic hen (*Gallus domesticus*). IV. The projection to the spinal cord. *Brain Behav. Evol.*, 15: 41–62.

Wold, J.E. (1979a) The vestibular nuclei in the domestic hen (*Gallus domesticus*). *Exp. Brain Res.*, 34: 217–232.

Wold, J.E. (1979b) The vestibular nuclei in the domestic hen (*Gallus domesticus*). VII. Afferents from the spinal cord. *Arch. Ital. Biol.*, 117: 30–44.

Wold, J.E. (1981) The vestibular nuclei in the domestic hen (*Gallus domesticus*). VI. Afferents from the cerebellum. *J. Comp. Neurol.*, 201: 319–341.

Yamada, T., Placzek, M., Tanaka, H., Dodd, J. and Jessell, T.M. (1991) Control of cell pattern in the developing nervous system: polarizing activity of the floor plate and notochord. *Cell*, 64: 635–647.

N.M. Gerrits, T.J.H. Ruigrok and C.I. De Zeeuw (Eds.)
Progress in Brain Research, Vol 124

CHAPTER 2

Antero-posterior boundaries and compartments in the cerebellum: evidence from selected neurological mutants

Leonard M. Eisenman*

Department of Pathology, Anatomy and Cell Biology, Thomas Jefferson University, Philadelphia, PA 19107, USA

Tribute to Professor Voogd

It is altogether fitting that we gathered together at Erasmus University to celebrate the life work of Professor Jan Voogd. His contributions to the understanding of the morphological heterogeneity of the cerebellum are now classics in the field. What he has accomplished with basic staining methods complemented by modern neuroanatomical methods has demonstrated once again the incredible power of keen observation by a knowledgeable individual. His contributions are seminal to our current thinking about cerebellar organization. All the modern techniques including immunostaining, receptor binding, transgenic manipulations, knock-outs and -ins, etc., have attested to the validity of his observations and the interpretations he ascribed to those observations. The cerebellum will never be the same again. The deceptive histological homogeneity of the cerebellar cortical architecture has succumbed to the intricate and complicated heterogeneity of parasagittal compartments interlacing with antero-posterior compartments. In dedication to Professor Voogd, and with a true sense of admiration and warm friendship, I offer my contribution to this celebratory volume. Professor Voogd spent much of his effort on demonstrating the existence and organization of parasagittal compartments. This contribution is complementary in that it deals with the evidence for antero-posterior compartments in cerebellar architecture as demonstrated in a few spontaneous neurological murine mutations.

Classic antero-posterior boundaries

One of the first things observed about the morphological characteristics of the cerebellum is that it is organized into lobules. These lobules are, for the most part, oriented in the mediolateral plane and are separated by fissures. The now classic studies of Larsell (Larsell, 1952; Larsell and Jansen, 1972) were instrumental in demonstrating similar lobular organization in different animal species. They argued for this common plan in spite of the obvious differences in the overall size of the cerebellum, the size of particular cerebellar lobules, and the organization and arrangement of these lobules in different species. Their studies highlighted the constancy across species, of the anterior, posterior and flocculonodular lobes separated by primary and posterolateral fissures, respectively. These subdivisions are often referred to as the spino-, ponto, and vestibulo-cerebellum referring to the major source of mossy fiber afferents, respectively, to each of these cerebellar regions. More recently, a number of lines of investigation have shed some doubt on the registry between these prominent fissures and the boundaries between anatomical and functional compartments within the antero-posterior dimension of the cerebellum. This evidence comes from

*Corresponding author. Tel.: 215–503–1686; Fax: 215–923–3808; e-mail: leonard.eisenman@mail.tju.edu

the careful analysis of (a) cerebellar development by Korneliussen (1967, 1968) and others (b) the terminal distribution of different mossy fiber afferents (c) phenotypes of a number of spontaneous neurological mutants and (d) the expression of different molecular markers and homeodomain genes both in normal developing mice and in trangenic and knockout mice. The reader is referred to recent published reviews that deal with cerebellar development and compartmentation (Hatten and Heintz, 1995; Herrup and Kuemerle, 1997; Hawkes and Eisenman, 1997). In this chapter I would like to focus on the evidence for antero-posterior compartments derived from the analysis of the mutant mice *meander tail*, *Lurcher* and *weaver*.

Meander tail (mea/mea)

Originally described by Hollander and Waggie (1977) and more fully documented by Ross et al. (1990), the *meander tail* mutant mouse exhibits a phenotype consisting of a crooked tail and an abnormal cerebellum. This recessive mutation is located on chromosome 4 and the adult homozygous animal exhibits a behavioral ataxia that is most probably associated with the cerebellar malformation.

The cerebellar malformation is quite unique in that it primarily affects the anterior region, which is almost entirely agranular (Fig. 1B). The posterior region exhibits relatively normal cytoarchitecture that continues up to a sharp boundary with the agranular cortex in the anterior region. This anterior cortex extends mediolaterally and the sharp boundary demarcating abnormal from normal cortex is present at the interface along this cortical margin. This anterior cortex is also characterized by a lack of normal cortical lamination including ectopic multilayering of Purkinje cells (PC). Interestingly, the defect does not appear to affect in any obvious fashion the molecular layer interneurons, the stellate and basket cells. These interneurons can be identified by their parvalbumin immunoreactivity.

Analysis of the development of the *meander tail* cerebellum indicates that the abnormality is already apparent at embryonic day (E) 16 as a reduction in

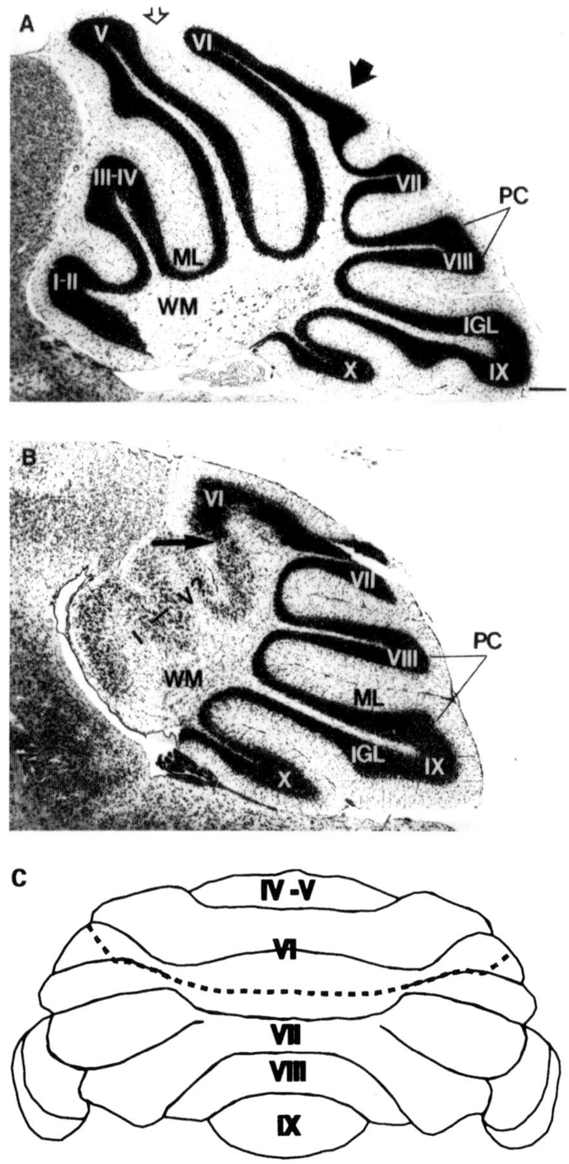

Fig. 1. (A) Photomicrograph of a parasagittal section through the vermal region of a wildtype mouse cerebellum stained with cresyl echt violet. (B) Photomicrograph of a parasagittal section through the vermal region of a *mea/mea* mouse cerebellum stained with cresyl echt violet. Note the sharp boundary (large arrow) separating the agranular region anterior (left and down) from the relatively normal cytoarchitecture seen in the posterior region (above and right). (C) Drawing of a dorsal view of the wildtype mouse cerebellum. The dotted line indicates the approximate posterior boundary to the region that would contain an equivalent number of Purkinje cells to that counted in the *meander tail* agranular anterior region.

the thickness of the external granular layer (egl), the germinal layer for the granule cells that normally populate the internal granular layer (Napieralski and Eisenman, 1993). Examination of the newborn *meander tail* cerebellum reveals that the egl in the anterior region is deficient and that this deficiency is present up to and a bit posterior to the initial indentation that marks the site of the primary fissure. Since this fissure is traditionally taken as the boundary between the anterior and posterior lobes of the cerebellum, it suggests that even at this early age the developmental defect in this mutant occupies a region that is more extensive than the traditionally described anterior lobe as defined by the primary fissure.

In an attempt to further delineate the relative position of this posterior boundary to the defective region of the *meander tail* cerebellum, we quantified the number of Purkinje cells in the cerebellum of the homozygous, heterozygous and wildtype mouse cerebellum (Napieralski and Eisenman, 1996). For this purpose cerebella from all three genotypes were sectioned in all three planes and immunoreacted with an antibody against calbindin, a calcium binding protein that, within the cerebellum, is specific to Purkinje cells. The results of this analysis indicated that there was approximately equal numbers of Purkinje cells in the cerebellum of all three genotypes. Thus, unlike the situation in other neurological mutants where there is a reduction in the number of Purkinje cells that accompanies agranularity, in the *meander tail* the number of Purkinje cells is not altered. The cell counts in the wildtype and heterozygous animals were then separated into counts reflective of the number of Purkinje cell in the anterior lobe or the posterior lobe as defined by the position of the primary fissure. A similar quantitative separation was made in the homozygous *meander tail* except here the sharp area of transition between the normal cortical cytoarchitecture posterior and the abnormal agranular cortex anterior was used as the point of separation. This analysis indicated there were many more Purkinje cells within the abnormal anterior region of the *meander tail* cerebellum relative to the anterior lobe of the wildtype or heterozygous mouse. This suggests that the gene-targeted area in the *meander tail* is more encompassing than just the anterior lobe as defined by the primary fissure in the wildtype mouse. To determine approximately how extensive this area is we calculated the area in the wildtype cerebellum that this number of Purkinje cells would populate starting from the ventral hilus and proceeding anteriorly and superiorly around the cerebellum. This was done by determining the average number of Purkinje cells per unit length of the Purkinje cell line in the wildtype animal. Using this approach it became clear that this number of Purkinje cells would populate the anterior lobe plus a good portion of the anterior part of lobule VI of the wildtype cerebellum (Fig. 1C). This finding strongly suggests that a boundary in this approximate location is recognized by the *meander tail* gene, resulting in the very clear and distinct compartmentalized phenotype. In concert with other data it strongly suggest that this is a unique anterior-posterior boundary of the cerebellum.

Lurcher (Lc/+)

This spontaneous semi-dominant mutation, first described by Phillips (1960) is lethal in the homozygous condition. However, in the heterozygote it leads to the premature death of all Purkinje cells in the cerebellum. Recently the mutation in *Lurcher* has been identified in a δ2 glutamate receptor gene (Zuo et al., 1997)

In the cerebellum of the Lc/+ development proceeds relatively normally until day postnatal 8–9 when Purkinje cells begin to die. These cells undergo apoptosis at a rapid rate over the next few weeks and by the second postnatal month all Purkinje cells are gone. Subsequent to the loss of Purkinje cells there is a secondary loss of cerebellar granule cells and inferior olive neurons. In spite of this severe loss of one of the principal targets of the olivocerebellar projection, the Purkinje cells, and the severe reduction in the number of olivary neurons, the normal topographic organization of the projection is retained (Heckroth and Eisenman, 1988). During development of this projection the olivocerebellar 'climbing' fibers contact the Purkinje cell somas in a normal fashion. However as development proceeds they never enter the molecular layer to entwine the developing Purkinje cell dendritic trees as occurs in normal mouse development (Heckroth et al., 1990). In the adult *Lurcher* the olivocerebellar fibers terminate in the area where the Purkinje cell somas had resided. Interestingly, the relationship between the granule cell 'parallel' fibers and the Purkinje cell dendrites develops in an apparently normal fashion until the latter cells die.

It had been shown in the normal adult rat that there is a close relationship between the topography of the olivocerebellar projection and the distribution of the Purkinje cell specific antigen, zebrin I (formerly, mabQ113, Gravel et al., 1987). Since the olivocerebellar topography is retained in the adult *Lurcher*, we performed an analysis of the development of zebrin II expression in the *Lurcher* to see whether its development progressed normally up until the time of Purkinje cell death (Tano et al., 1992). Zebrin II has an identical distribution pattern in the rat as does zebrin I (Brochu et al., 1990; Lannoo et al., 1991). In the adult mouse cerebellum zebrins are Purkinje cell-specific and are found in alternating bands of Purkinje cells (Eisenman and Hawkes, 1993). In some lobules, in particular vermal lobules X, parts of VI–VII and in the flocculus and paraflocculus zebrin II expression is evident in all Purkinje cells. It had been demonstrated previously, that in the normal rat the development of zebrin expression within Purkinje cells occurs in a number of stages (Leclerc et al., 1988). Initially, there is no zebrin expression in Purkinje cells. Then at postnatal day 5 (P5) zebrin appears and is restricted to the posterior lobules, X, IX, and part of VIII. Zebrin expression is upregulated in most Purkinje cell in the vermis during the period of P7 through P12. This upregulation of expression continues so that by P15 all Purkinje cells of the cerebellum are zebrin positive. Then interestingly, as development proceeds, zebrin expression is down-regulated in specific parasagittal bands so that by P25 the adult pattern of alternating zebrin-positive and -negative bands are apparent. Thus, in normal rat development there is multistage pattern of zebrin expression which first produces regional zebrin expression, which then extends throughout the cerebellum and then expression diminishes in restricted parasagittal bands in most lobules. Remarkably, for the most part these zebrin+ and − bands are concordant with the parasagittal compartments initially described by Professor Voogd using the Hagqvist stain for myelin (Voogd, 1967, 1969).

In wildtype and *Lurcher* mice, the developmental analysis of zebrin II expression indicated an initial similar sequence of expression (Tano et al., 1992). However, in *Lurcher*, the further development of zebrin expression is altered. After an initial period of no expression, it appears in the posterior lobules, including lobules X, IX and part of VIII at approximately P5. This pattern is very similar to that seen in normal mouse development. However, unlike the normal mouse, the expansion of zebrin expression does not occur. Purkinje cells are still present in other lobules of the vermis and hemisphere, as evidenced by calbindin immunostaining, but they fail to initiate expression of zebrin at the appropriate developmental time. It is interesting that this arrest occurs coincident with a stage seen in the normal mouse. This suggests that the genetic defect in *Lurcher* may recognize a developmental boundary within lobule VIII that separates anteroposterior compartments in the posterior lobe. This region of the cerebellum includes the nodulus, uvula and part of the pyramis. Thus, it includes the flocculo-nodular lobe and a portion of the posterior lobe that is anterior to the posterolateral fissue. However, these data do not preclude the possibility that the region posterior to this boundary consists of more than one compartment. While this boundary in lobule VIII does not correspond to the anterior boundary of either the primary or secondary vestibular-cerebellar projection, there is some correspondence between this boundary in lobule VIII and that seen with other molecular markers of the cerebellum.

In a recent elegant study, of which I am sorry to say I was not aware when I made my presentation, the gene mutation in Lc/+ has been identified as G to A transitions that change a highly conserved alanine to a threonine residue in transmembrane domain III of the mouse δ2 glutamate receptor gene (GluRδ2, Zuo et al., 1997). Electrophysiological analysis indicates that Lc/+ Purkinje cells exhibit a very high membrane conductance and a depolarized resting potential that suggests the presence of a large, constitutive inward current, the major portion of which is carried by Na^+ ions. During development, GluRδ2 expression is upregulated and redistributed to the Purkinje cell-parallel fiber synapses concurrently with the onset of Lc/+ cerebellar phenotypic abnormalities (Takayama et al, 1996). These findings suggest that excessive exposure to glutamate may lead to excitotoxicity and apoptotic death among PC (Zuo et al., 1997).

How does this exciting new information about the gene mutation relate to our results concerning the developmental stages and arrest in zebrin expression in Lc/+. It is clear that in both the wildype and Lc/+ mice there appears to be a boundary in lobule VIII. Is there some relationship between this boundary and the action of the *Lurcher* gene? It is clear that by 2–3 months postnatal, all PC in the *Lurcher* are gone (Caddy and Biscoe, 1979). However some preliminary data from our laboratory concerning the differential timing of PC loss in different cerebellar lobules suggest that PC in the posterior compartment, where zebrin expression occurs during development, are some of the last PC to die (unpublished studies). In addition to zebrin, other molecules characterize this subpopulation (Neustadt et al., 1988; Frantz et al., 1994; Rotter et al., 1995). These molecular attributes may provide some, albeit temporary, protection from the action of the *Lurcher* gene.

Weaver (wv/wv)

This mutant has received a great deal of attention. First due to the interesting and complex phenotype it presents and more recently, because of the identification of the gene defect as a single base pair substitution in a gene coding for a G-protein-coupled, inward rectifying potassium channel of the GIRK2 family (Patil et al., 1995). The complex phenotype includes cerebellar, substantia nigra and spermatogenic defects. In the cerebellum, there is a graded degree of hypogranularity, very severe in the medial cerebellum, but much less so in the lateral cerebellum (Herrup and Trenkner, 1987). The cerebellum is much reduced in size but retains a semblance of normal foliation medially with abnormal elongation of some lobules in the lateral cerebellum. Chimeric analysis has demonstrated that in the cerebellum, granule cells and PC are the primary targets of the *weaver* gene (Goldowitz, 1989; Smeyne and Goldowitz, 1989, 1990). Analysis of the expression of the GIRK2 gene suggests it is expressed in all granule and Purkinje cells (Slesenger et al., '96). Therefore, explaining the specificity and regional variation in cell vulnerability has become a challenge for developmental biologists.

As it has become apparent in the above discussion of *meander tail* and *Lurcher* mutants, regional variation in vulnerability to the actions of mutant genes is a common finding. Following up on this theme, we examined the phenotype of the *weaver* in more detail, using standard histological stains and zebrin immunohistochemistry (Eisenman et al., 1998). There had been a number of studies describing the abnormal cerebellar lobulation, cell deletions and cellular heterotopias (Blatt and Eisenman, 1985; Herrup and Trenkner, 1987). Our analysis added to these by identifying subtle histological changes in the antero-porterior dimension especially, but not exclusively, in the medial cerebellum of the *weaver*. Differences in both the degree of granularity and positioning of remaining granule and Purkinje cells suggested regional variations and boundaries to these variations, one of which was similar to that seen in the *meander tail*. Anteriorly, the *weaver* cerebellum contains some granule cells but most of these are ectopically positioned within Purkinje and molecular layers. This region is also characterized by a clear multilayering of Purkinje cells. This type of abnormal cytoarchitecture exists throughout the anterior region up to a point that appears to be in lobule VI. Posterior to this, the cytoarchitecture changes in what appears to be a much reduced lobule VII, VIII and perhaps IX. In this region there is an almost complete deletion of granule cells. The few small cells that are present, are most likely stellate and basket cells, as determined by parvalbumin immunostaining. Because there is a severe size reduction in the *weaver* cerebellum and somewhat atypical lobulation, it is difficult to determine with any degree of precision whether the lobules seen in the *weaver* are equivalent to those seen in the wildtype cerebellum. With this caveat in mind, it appears that an additional histological transition occurs in the region of lobule IX–X. Posterior to this boundary, there is a slight increase in granule cells, which are found primarily within a much reduced internal granule cell layer. The relative number of granule cells is perhaps similar to that seen in the anterior region, but the positioning of these cells is different, i.e. among and superficial to the PC anteriorly, and deep to the PC, in this most posterior region. It may be that this posterior transition boundary is an additional one to that seen in the wildtype and *Lurcher* mice, vis a vis zebrin developmental staging and zebrin expression, respectively (Fig. 2).

Analysis of zebrin immunostainig in the adult *weaver* cerebellum reveals some striking modifications to the normal pattern. In the anterior vermal region there is normally a midline band and two additional bands on either side. In the *weaver* there is only one such lateral band. Thus, the intermediate P2 (identification of this band is not a certainty) is missing. In addition to this striking modification in the zebrin pattern there is also a missing, or much reduced central cluster of zebrin positive PC in lobule VI. These cells normally populate the posterior part of lobule VI and anterior part of VII. This is part of the cerebellar region that is much reduced in size in the *weaver.*

Thus, the data in adult *weaver,* both in terms of cytoarchitectonic heterogeneity and zebrin patterning, point to the existence of boundaries which segregate antero-posterior compartments which are differentially affected by the *weaver* mutation. One boundary is similar to that described for the *meander tail* mutant discussed above.

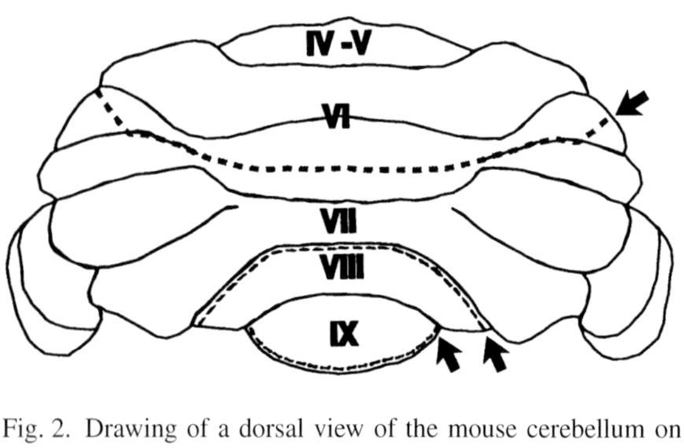

Fig. 2. Drawing of a dorsal view of the mouse cerebellum on which are indicated the boundaries that were identified in the spontaneous mutants *meander tail*, *Lurcher*, and *weaver*. These are indicated by arrows. The boundary in lobule IX is actually closer to the base of the posterolateral fissure (not seen) which divides lobules IX and X.

Other evidence

Consistent with these findings in a few spontaneous neurological mutants are studies in other mutants (Thomas et al., 1991; Hess and Wilson, 1991), developmental analyses of the expression of a number of mouse homologues of Drosophila homeodomain genes (Millen et al., 1995), and the expression of other molecules both during development and in the adult (Neustadt et al., 1988; Rotter et al., 1995; Tojo et al., 1995). There is also some correspondence between these molecular compartments and the areas of termination of spinocerebellar afferents to the anterior lobe and secondary vestibular cerebellar afferents to the posterior compartment.

Conclusion

Professor Voogd's remarkable contributions to our perception and understanding of the cerebellum have forever changed the neuroscientist's thinking about this 'head ganglion of the proprioceptive system', as Sherringtion (1947) called it. The homogeneity in cytoarchitecture has been complicated by heterogeneity in connections, molecular architecture, and gene function. It is clear now that this heterogeneity exists not only in the parasagittal plane but also in the orthogonal, antero-posterior dimension. The grid established by these intersecting compartments sets the stage for a myriad of smaller units that have unique anatomical and molecular signatures, and therefore, provide the substrate for the different functional domains of the cerebellum. It remains for future detailed analyses of the cerebellar heterogeneity to delineate the relationship between each compartment and the function for which it is in whole, or in part, responsible.

References

Blatt, G.J. and Eisenman, L.M. (1985) A qualitative and quantitative light microscopic study of the inferior olivary complex of normal, *reeler* and *weaver* mutant mice. *J. Comp. Neurol.*, 232: 117–128.

Brochu, G., Maler, L. and Hawkes, R. (1990) Zebrin II: a polypeptide antigen expressed selectively by Purkinje cells reveals compartments in rat and fish cerebellum. *J. Comp. Neurol.*, 291: 538–552.

Caddy, K.W. and Biscoe, T.J. (1979) Structural and quantitative studies on the normal C3H and *Lurcher* mutant mouse. *Phil. Trans. R. Soc. Lond.*, 287: 167–201.

Eisenman, L.M. and Hawkes, R. (1993) Antigenic compartmentation in the mouse cerebellar cortex: Zebrin and HNK–1 reveal a complex, overlapping molecular topography. *J. Comp. Neurol.*, 335: 586–605.

Eisenman, L.M., Gallagher, E. and Hawkes, R. (1998) Regionalization defects in the *Weaver* mouse cerebellum. *J. Comp. Neurol.*, 394: 431–444.

Frantz, G.D., Weimann, J.M., Levin, M.E. and McConnell, S.K. (1994) Otx1 and Otx2 define layers and regions in developing cerebral cortex and cerebellum. *J. Neurosci.*, 14: 5725–5740.

Goldowitz, D. (1989) The *weaver* granuloprival phenotype is due to intrinsic action of the mutant locus in granule cells: evidence from homozygous *weaver* chimeras. *Neuron*, 2: 1565–1575.

Gravel, C., Eisenman, L.M., Sassaville, R. and Hawkes, R. (1987) Parasagittal organization of the rat cerebellar cortex: a direct correlation between antigenic Purkinje cell bands revealed by mabQ113 and the organization of the olivocerebellar projection. *J. Comp. Neurol.*, 263: 294–310.

Hatten, M.E. and Heintz, N. (1995) Mechanisms of neural patterning and specification in the developing cerebellum. *Ann. Rev. Neurosci.*, 18: 385–408.

Hawkes, R. and Eisenman, L.M. (1997) Stripes and zones: the origins of regionalization in the adult cerebellum. *Perspect. Dev. Neurobiol.*, 5: 95–104.

Heckroth, J.A. and Eisenman, L.M. (1988) The olivocerebellar projection in 'lurcher' mutant mice. *Neurosci. Lett.*, 85: 199–204.

Heckroth, J.A., Goldowitz, D. and Eisenman, L.M. (1990) Olivocerebellar maturation in normal and *Lurcher* mutant mice: defective development in *Lurcher*. *J. Comp. Neurol.*, 291: 415–430.

Herrup, K. and Kuemerle, B. (1997) The compartmentation of the cerebellum. *Ann. Rev. Neurosci.*, 20: 61–90.

Herrup, K. and Trenkner, E. (1987) Regional differences in cytoarchitecture of the *weaver* cerebellum suggest a new model for *weaver* gene action. *Neuroscience*, 23: 871–885.

Hess, E.J. and Wilson, W.C. (1991) *Tottering* and *leaner* mutations perturb transient developmental expression of tyrosine hyproxylase in embryologically distinct Purkinje cells. *Neuron*, 6: 123–132.

Hollander, W.F. and Waggie, K.S. (1977) Meander tail: a recessive mutant located on chromosome 4 of the mouse. *J. Hered.*, 68: 403–406.

Korneliussen, H.K. (1967) Cerebellar corticogenesis in Cetacea, with special reference to regional variations. *J. Hirnforsch.*, 9: 151–186.

Korneliussen, H.K. (1968) On the ontogenetic development of the cerebellum (nuclei, fissures and cortex) of the rat, with special reference to regional variations in corticogenesis. *J. Hirnforsch.*, 10: 379–412.

Lannoo, M.J., Brochu, G., Maler, L. and Hawkes, R. (1991) Zebrin II immunoreactivity in the rat and in the weakly electric teleost Eigenmannia (Gymnotiformes) reveals three modes of Purkinje cell development. *J. Comp. Neurol.*, 310: 215–233.

Larsell, O. (1952) The morphogenesis and adult pattern of the lobules and tissues of the cerebellum of the white rat. *J. Comp. Neurol.*, 97: 281–356.

Larsell, O. and Jansen, J. (1972) The Comparative Anatomy and Histology of the Cerebellum. The Human Cerebellum, Cerebellar Connections and the Cerebellar Cortex. University of Minnesota Press, Minneapolis, IX, 268 pp.

Leclerc, N., Gravel, C. and Hawkes, R. (1988) Development of parasagittal zonation in the rat cerebellar cortex: mabQ113 antigenic bands are created postnatally by the suppression of antigen expression in a subset of Purkinje cells. *J. Comp. Neurol.*, 273: 399–420.

Millen, K.J., Hui, C-C. and Joyner, A.L. (1995) A role for En–2 and other murine homologues of Drosophila segment polarity genes in regulating positional information in the developing cerebellum. *Development*, 121: 3935–3945.

Napieralski, J.A. and Eisenman, L.M. (1993) Developmental analysis of the external granular layer in the *meander tail* mutant mouse: do cerebellar microneurons have independent progenitors. *Dev. Dynam.*, 197: 244–254.

Napieralski, J.A. and Eisenman, L.M. (1996) Further evidence for a unique developmental compartment in the cerebellum of the *meander tail* mutant mouse as revealed by the quantitative analysis of Purkinje cells. *J. Comp. Neurol.*, 364: 718–728

Neustadt, A., Frostholm, A. and Rotter, A. (1988) Topographical distribution of muscarinic cholinergic receptors in the cerebellar cortex of the mouse, rat, guinea pig and rabbit: a species comparison. *J. Comp. Neurol.*, 272: 317–330.

Patil, N., Cox, D.R., Bhat, D., Faham, M., Meyero, R.M. and Peterson, A.S. (1995) A potassium channel mutation in *weaver* mice implicates membrane excitability in granule cell differentiation. *Nature Genet.*, 2: 126–129.

Phillips, R.J.S. (1960) *Lurcher* a new gene in linkage group XI of the house mouse. *J. Genet.*, 57: 35–42.

Ross, M.E., Fletcher, C., Mason, C.A., Hatton, M.E. and Heintz, N. (1990) *Meander tail* reveals a discrete developmental unit in the mouse cerebellum. *Proc. Nat. Acad. Sci. USA*, 87: 4189–4192.

Rotter, A., McAndrew, P.A., Burghes, A.H.M., Chiu, I.-M., Evans, J.E., Sutter, M. and Frostholm, A. (1995) Expression of aFGF and a novel protein tyrosine phosphatase mRNA follows a rostrocaudal boundary in the murine cerebellar cortex. *Soc. Neurosci.*, 21:1191.

Sherrington, C. (1947) *The Integrative Action of the Nervous System*. 2nd edn. Cambridge University Press, Cambridge.

Slesenger, P.A., Patil, J., Liao, Y.N., Jan, L.Y. and Cox, D.R. (1996) Functional effects of the mouse weaver mutation on G protein-gated inwardly rectifying K^+ channels. *Neuron*, 16: 321–331.

Smeyne, R.T. and Goldowitz, D. (1989) Development and death of external granular layer cells in the *weaver* mouse cerebellum: a quantitative study. *J. Neurosci.*, 9: 1608–1620.

Smeyne, R.T. and Goldowitz, D. (1990) Purkinje cell loss is due to a direct action of the *weaver* gene in Purkinje cells: evidence from chimeric mice. *Dev. Brain Res.*, 52: 211–218.

Takayama, C., Nakagawa, S. Watanabe, M., Mishina, M. and Inoue, Y. (1996) Developmental changes in expression and distribution of the glutamate receptor channel δ2 subunit according to Purkinje cell maturation. *Dev. Brain Res.*, 92: 147–155.

Tano, D., Niapieralski, J.A., Eisenman, L.M., Messer, A., Plummer, J. and Hawkes, R. (1992) Novel developmental boundary in the cerebellum revealed by zebrin expression in the *Lurcher* (Lc/+) mutant mouse. *J. Comp. Neurol.*, 323: 128–136.

Thomas, K.R., Musci, T.S., Neumann, P.E. and Capecchi, M. (1991) *Swaying* is a mutant allele of the proto-oncogene *WNT-1*. *Cell*, 67: 969–976.

Tojo, H., Takami, K., Kaisho, Y., Nakata, M., Abe, T., Shiho, O. and Igarashi, K. (1995) Neurotrophin–3 is expressed in the posterior lobe of mouse cerebellum, but does not affect the cerebellar development. *Neurosci. Lett.*, 192: 162–172.

Voogd, J. (1967) Comparative aspects of the structure and fibre connections of the mammalian cerebellum. *Prog. Brain Res.*, 25: 94–134.

Voogd, J. (1969) The importance of fiber connections in the comparative anatomy of the mammalian cerebellum. In: R. Llinas (Ed.), *Neurobiology of Cerebellar Evolution and Development*. AMA-ERF Institute for Biomedical Research, Chicago, pp. 493–541.

Zuo, J., De Jager, P.L., Takahashi, K.A., Jiang, W., Linden, D.J. and Heintz, N. (1997) Neurodegeneration in *Lurcher* mice caused by mutation in the δ2 glutamate receptor gene. *Nature*, 388: 769–773.

N.M. Gerrits, T.J.H. Ruigrok and C.I. De Zeeuw (Eds.)
Progress in Brain Research, Vol 124

CHAPTER 3

Climbing fiber synapse elimination during postnatal cerebellar development requires signal transduction involving Gαq and phospholipase Cβ4

K. Hashimoto[1–3], M. Watanabe[4], H. Kurihara[4], S. Offermanns[5], H. Jiang[6], Y. Wu[6], K. Jun[7], H-S. Shin[7], Y. Inoue[5], D. Wu[6], M.I. Simon[5] and M. Kano[1,2,8]*

[1] *Department of Physiology, Kanazawa University School of Medicine, Takara-machi, Kanazawa 920–8640, Japan*
[2] *CREST, Japan Science and Technology Corporation*
[3] *Department of Physiology, Jichi Medical School, Minamikawachi-machi, Tochigi-ken 329–0498, Japan*
[4] *Department of Anatomy, Hokkaido University School of Medicine, Sapporo 060–0807, Japan*
[5] *Division of Biology 147–75, California Institute of Technology, Pasadena, CA 91125, USA*
[6] *Department of Pharmacology and Physiology, University of Rochester, Rochester, NY 14642–8711, USA*
[7] *National CRI Center for Calcium and Learning, and Department of Life Science, Pohang University of Science and Technology, Pohang 790–784, Korea.*
[8] *Laboratory for Cellular Neurophysiology, Brain Science Institute, RIKEN, Wako-shi, Saitama 351–0198, Japan*

Introduction

The climbing fiber (CF)-Purkinje cell (PC) synapse of the cerebellum provides a good model to study how mature and functional synaptic connections are formed during postnatal development (Crépel, 1982). In new-born rodents, cerebellar PCs are innervated by multiple CFs. Elimination of supernumerary CFs occurs during postnatal development, and the mature state is attained at around postnatal day 20, in which most PCs are innervated by single CFs. Previous studies have demonstrated that multiple CF innervation persists in the cerebella of adult animals that are devoid of parallel fibers (PFs), the other major excitatory inputs to PCs. These results suggest that the developmental change from multiple- to mono-CF innervation is dependent on the formation of PF to PC synapses (Crépel, 1982; Ito, 1984). We have previously reported that mutant mice deficient in metabotropic glutamate receptor subtype 1 (mGluR1) and those deficient in the γ subunit of protein kinase C (PKCγ) are impaired in CF synapse elimination, although their PF–PC synapses are functional and morphologically normal (Kano et al., 1995; Kano et al., 1997). mGluR1 is known to couple to two similar members of the heterotrimeric Gq family, Gαq and Gα11 (Nakanishi, 1994; Pin and Duvoisin, 1995). They share 88% amino acid sequence identity and are expressed in almost every cell type (Strathmann and Simon, 1990; Wilkie et al., 1991). The Gq family is known to activate phospholipase Cβ (PLCβ) that catalyzes the hydrolysis of phosphatidylinositol 4,5-bisphosphate (PIP2) to produce two second messengers, inositol 1,4,5-trisphosphate (IP_3) and diacylglycerol (DG) (Exton, 1996). IP_3 mobilizes Ca^{2+} from intracellular stores (Berridge, 1993), whereas DG activates PKC together with Ca^{2+} (Nishizuka, 1992). Molecular cloning has

*Corresponding author. Tel.: +81–76–265–2170;
Fax: +81–76–234–4224;
e-mail: mkano@med.kanazawa-u.ac.jp

identified four isoforms (PLCβ1–4) that have been shown to be activated by the Gq class of G proteins (Exton, 1996). Among these four PLCβ isoforms, PLCβ4 is reported to be enriched in cerebellar Purkinje cells (PCs) (Tanaka and Kondo, 1994; Roustan et al., 1995; Watanabe et al., 1998).

The receptors activating Gq family members including mGluR1 do not discriminate between Gαq and Gα11 (Exton, 1996). There appears to be little difference between the abilities of both G protein α subunits to regulate PLCβ isoforms (Exton, 1996). No pharmacological tools are available that specifically block either of the two Gq family members, or one of the four PLCβ isoforms. It is, therefore, necessary to take genetic approaches to identify the Gq family member and PLCβ isoform that are specifically involved in CF synapse elimination. We have generated two strains of mice deficient in Gαq and PLCβ4, respectively, by inactivating the genes encoding these molecules in mouse embryonic stem cells (Offermanns et al., 1997a; Kano et al., 1998). This chapter will review our recent electrophysiological, morphological and behavioral data on these mutant mice. The data strongly suggest that Gαq and PLCβ4 transduce signals within PCs from mGluR1 to PKCγ that are required for CF synapse elimination during the third postnatal week.

Impaired CF synapse elimination in Gαq mutant mice

Gαq mutant mice are ataxic

Homozygous Gαq mutant mice showed increased mortality during the first postnatal day, presumably due to internal bleeding during birth trauma that results from a defect in platelet aggregation (Offermanns et al., 1997b). Sixty to seventy percent of homozygous Gαq mutant newborns survived the postnatal period and developed clear signs of motor coordination deficit that became obvious about three weeks after birth. The animals displayed awkward and jerky movements, loss of balance during rearing and occasionally fell on their side (Offermanns et al., 1997a).

Morphology of Gαq mutant cerebellum is normal

No obvious morphological defects could be observed on examination of the peripheral and central nervous system of Gαq homozygous mutant mice. The cerebellum of the Gαq mutant mouse was indistinguishable in size and foliation from that of the wild-type mouse at two months of age (Figs 1A, F). In the mutant and wild-type mice, the cerebellar cortex showed a tri-laminar organization consisting of the molecular layer, PC layer, and granule cell (GC) layer (Figs 1B, G). No appreciable differences were found in thickness of these cortical layers, nor in cell density of the PCs and GCs. The morphology of PC dendrites was examined by immunohistochemistry with antibody against spot 35/calbindin, a cytosolic Ca^{2+}-binding protein specific to PCs in the cerebellum. In both mice, a thick primary dendrite extended from the apical pole of soma toward the pial surface, and branched elaborately into the secondary and tertiary dendrites (Figs 1C, H). The tertiary dendrites in the mutant mouse were studded with spines as numerous as in the wild-type mouse. Bergmann glial (BG) fibers were visualized by immunohistochemistry with anti-glial fibrillary acidic protein (GFAP) antibody. No significant differences were found between the mutant and wild-type mice, as to the linear alignment of GFAP-immunopositive fibers toward the pial surface and their density in the molecular layer (Figs 1D, I).

The formation of PF–PC synapses was examined by electron microscopy at two months of age. The molecular layer of the mutant cerebellum contained numerous profiles of the PF terminals and PC spines, as did that of the wild-type cerebellum (Figs 1E, J). In both mice, the PF–PC synapses constituted the majority of the synaptic structures in the molecular layer, and were surrounded by cell processes of the BGs. When counted on electron micrographs at two months of age (Weibel, 1979), profile numbers of the PF–PC synapses per 100 μm^2 were 24.8 ± 1.5 (mean ± SEM., $n = 10$) and 21.0 ± 1.1 ($n = 10$) for two mutant mice, and 21.4 ± 1.4 ($n = 10$) and 19.7 ± 1.2 ($n = 10$) for two wild-type mice, showing no significant difference between the mutant and wild-type mice. These morphological observations indicate that deletion

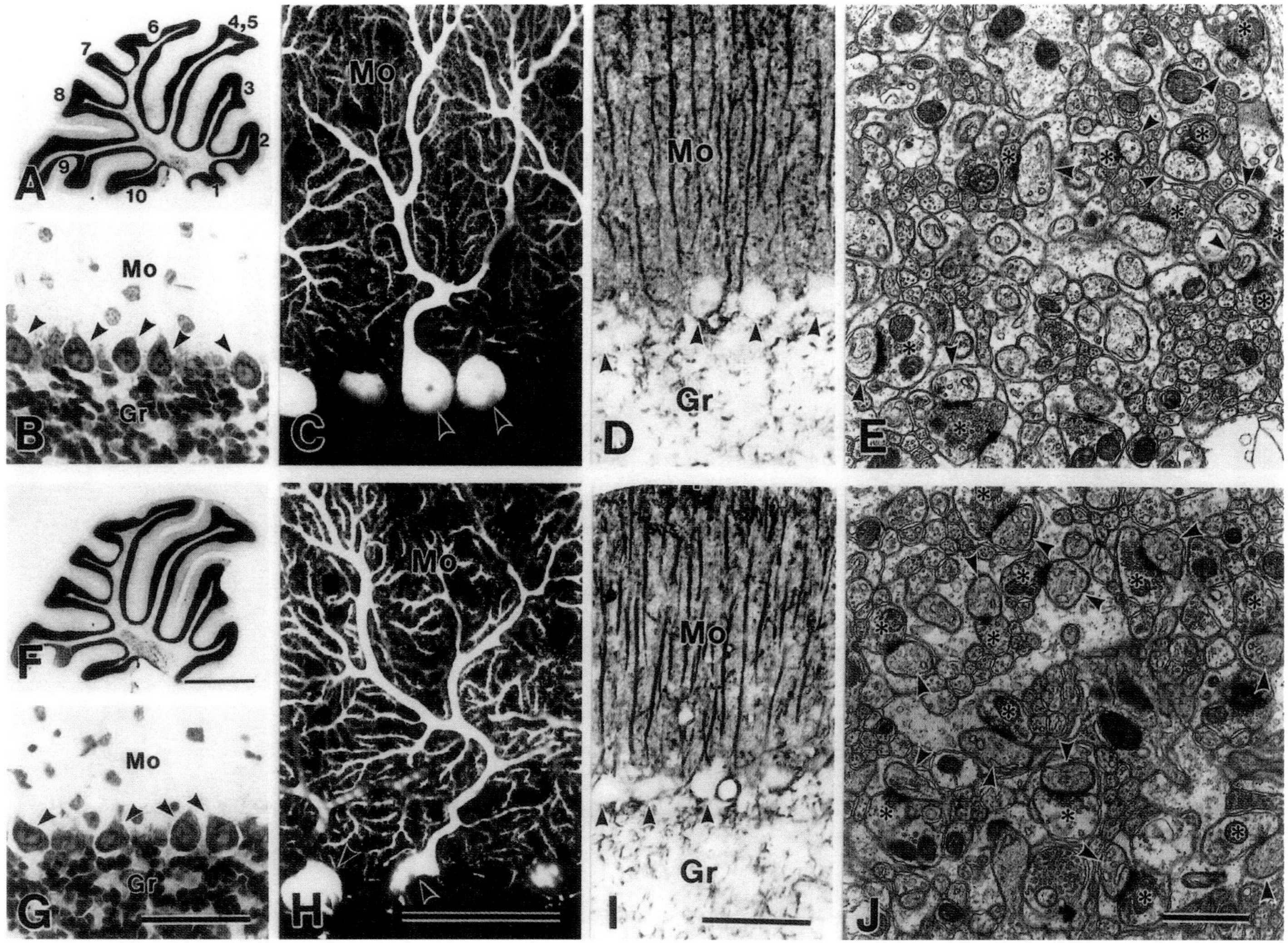

Fig. 1. Cerebellar anatomy of the Gαq mutant and wild-type mice. Gαq mutant (A–E) and wild-type (F–J) mice were studied at 2 months of age. (A, F) Toluidine blue-stained images of parasagittal microslicer sections through the cerebellar vermis. The lobule number is indicated as 1–10. Rostral is to the right, and dorsal is to the top. (B, G) High power views of the cerebellar cortex. Arrowheads indicate the cell bodies of Purkinje cells. (C, H) Confocal scanning microscopic images of PC dendrites immunostained with spot 35/calbindin antibody. (D, I) Bergmann glia fibers immunostained with GFAP antibodies. (E, J) Electron micrographs of the molecular layer. Arrowheads indicate PC spines, and asterisks indicate PF terminals. Gr, granular layer; Mo, molecular layer. Scale bars, F, 1 mm; G, H, I, 50 μm; J, 1 μm

of the Gαq gene did not affect gross development of the cerebellar anatomy, cell production, cytodifferentiation, and formation of PF–PC synapses.

Gαq mutant PCs are multiply-innervated by CFs

In sagittal cerebellar slices prepared from 22–109 day old (P22–P109) wild-type mice or age-matched Gαq mutant animals, CFs were stimulated in the GC layer, and evoked responses in single PCs were recorded using patch-clamp techniques in the whole-cell configuration (Edwards et al., 1989; Konnerth et al., 1990; Llano et al., 1991). To search all CFs innervating the recorded PC, the stimulation electrode was moved systematically in the GC layer and the stimulus intensity was gradually increased (pulse width 0.1 ms, strength 0–100V) at each stimulation site. When a CF was stimulated, a clearly discernible excitatory postsynaptic current (EPSC) was elicited in an all-or-none fashion (e.g. Fig. 2A, upper traces). In some PCs, more than two discrete CF–EPSCs were elicited at one stimulus site with different stimulus thresholds (e.g. Fig. 2A, lower traces). The number of CFs innervating the recorded PC was estimated by the number of discrete CF–EPSC steps elicited in that PC (Kano et al., 1995; Kano et al., 1997; Offermanns et al., 1997a; Kano et al., 1998). Thus as summarized in the frequency distribution histogram (Fig. 2B, left panel), 82.6%, 14.8% and 2.6% of wild-type PCs recorded in the present study (P22–P59) are innervated by one, two and three CFs, respectively. On the other hand, in Gαq mutant PCs, 59.6% had single CF–EPSC steps, whereas 28.2% and 12.1% had two and three discrete steps, respectively (Fig. 2B, left panel).

The difference of CF innervation between the wild-type and Gαq mutant mice during P22–P59 could be due to the retardation of general cerebellar maturation in the Gαq mutant mice. To assess this possibility, we examined CF–EPSCs in older mice (P70–P109) (Fig. 2B, right panel). The differences between the frequency distributions of PCs during P22–P59 and those during P70–P109 were not significant for either the wild-type or the mutant mice ($p > 0.05$, χ^2 test).

The cerebellum is not necessarily homogeneous in terms of distribution of various signaling molecules. Such heterogeneity may influence CF synapse elimination during postnatal cerebellar development. We examined possible interlobular difference in CF innervation by plotting the recording sites of PCs on the standard sagittal plane. However, we did not find any significant interlobular difference (Fig. 3). In Gαq mutant mice, multiply innervated PCs were distributed evenly throughout the ten cerebellar lobules (Fig. 3B) with significantly higher percentages when compared to the wild-type mice (Fig. 3A). This contrasts with the highly heterogeneous distribution of multiply innervated PCs in the PLCβ4 mutant cerebellum (see Fig. 7).

Next, we followed the developmental course of CF innervation. During the first postnatal week (P1–P7), the majority of PCs were multiply innervated by CFs in both wild-type and Gαq mutant mice (Fig. 4A). However, the frequency distribution of the number of CF–EPSC steps per PC in the mutant mice was not significantly different from that in the wild-type mice ($p > 0.05$, χ^2 test). During the second postnatal week (P8–P14), the percentage of PCs with multiple CF–EPSC steps decreased markedly in both wild-type and mutant mice (Fig. 4B). The distributions were not significantly different between the two strains ($p > 0.05$, χ^2 test). In contrast, the frequency distributions between the two strains during the third postnatal week (P15–P21) were markedly different (Fig. 4C) ($p < 0.01$, χ^2 test). This indicates that a higher percentage of the PCs was multiply innervated by CFs in the Gαq mutant than in the wild-type mice. These results suggest that regression of multiple CF innervation initially occurs normally but the process is specifically impaired during the third postnatal week in the Gαq mutant mice.

Other electrophysiological parameters of Gαq mutant EPSCs are normal

The kinetics of CF–EPSCs were normal in the Gαq mutant mice. There is no significant difference in either the 10%–90% rise time or the decay time constant between the Gαq mutant and wild-type mice. Short-term synaptic plasticity examined by applying paired stimuli was normal. CF–EPSCs

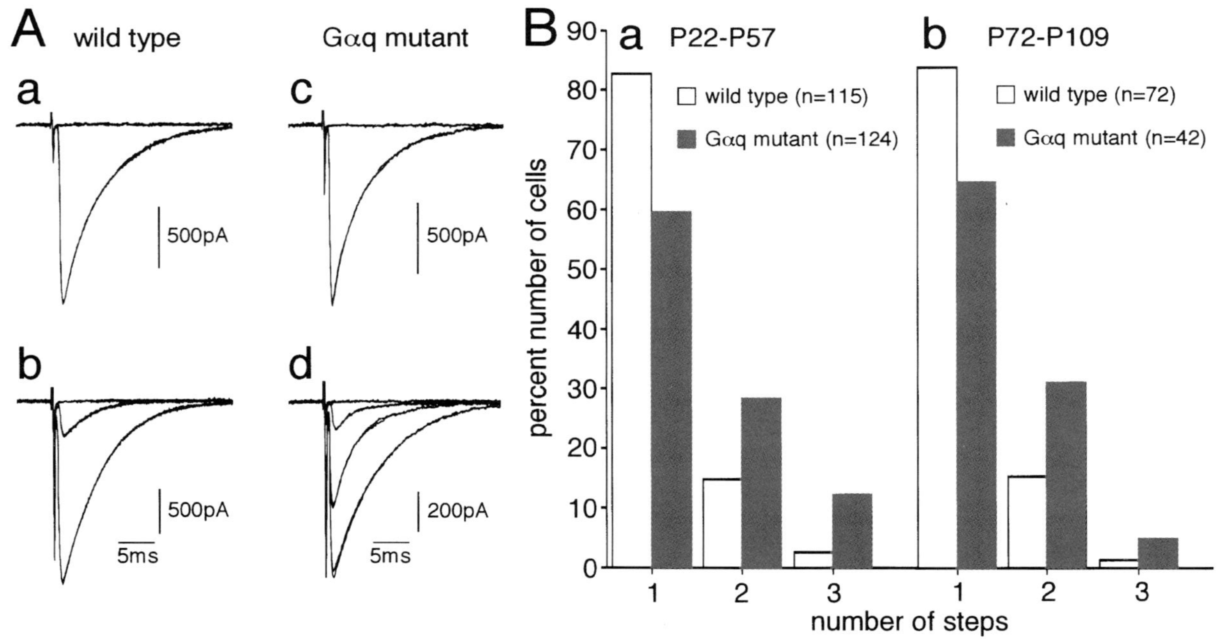

Fig. 2. Multiple CF innervation of PCs in mature Gαq mutant mice. (A) EPSCs elicited by CF stimulation in a wild-type (a, b) and a Gαq mutant (c, d) PCs. One or two traces are superimposed at each threshold intensity. Stimuli were applied at 0.2 Hz. Holding potentials was − 10 mV for a, b, c and 0 mV for d. (B) Summary histograms showing number of discrete steps of CF–EPSCs of the wild-type (open colums) and Gαq mutant (hatched colums) PCs. Data obtained from mice at P22–P57 (a) and at P72–P109 (b), respectively. All cells were studied blind to the mouse genotype (from Offermanns et al., 1997).

showed prominent paired-pulse depression in both monoinnervated and multiply innervated PCs derived from the Gαq mutant and wild-type mice. The magnitudes of paired-pulse depression were not significantly different between the wild-type and Gαq mutant mice at varying interpulse intervals. We also verified that the current-voltage relations of EPSCs were linear in the mono innervated and double innervated PCs from the wild-type and Gαq mutant. Furthermore, CF–EPSCs in both wild-type and Gαq mutant mice were not affected by an NMDA receptor blocker, DL–2-amino–5-phosphonopentanoate (AP5, 100 μM), but were totally suppressed by an AMPA receptor antagonist, 6-cyano–7-nitroquinoxaline–2,3-dion (CNQX, 10 μM). These results indicate that the CF–EPSCs of mature PCs from wild-type and Gαq mutant mice are exclusively mediated by the non-NMDA subtype of glutamate receptors.

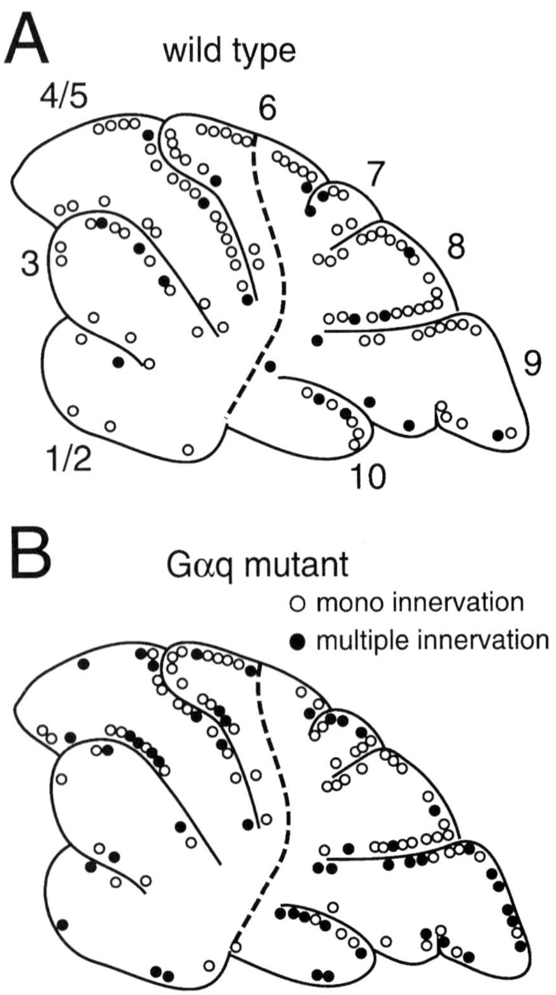

Fig. 3. Distribution of multiply-innervated PCs in Gαq mutant cerebellum. Spatial distribution of mono-innervated (open circles) and multiply-innervated (filled circles) PCs of the wild-type (A) and Gαq mutant (B) cerebella plotted on the standard midsagittal plane of the vermis. Data were obtained from 7 wild-type ($n = 115$ PCs) and 8 Gαq mutant ($n = 123$ PCs) cerebella. Note that multiply-innervated PCs were evenly distributed in the rostral and caudal regions of the cerebellum.

The nature of EPSCs elicited by stimulation of PFs was examined in mice at P50–P54. There was no significant difference between the wild-type and Gαq mutant mice in the kinetics of PF–EPSCs. Furthermore, in both wild-type and Gαq mutant PCs, PF–EPSCs were little affected by AP5 (100 μM), but were totally suppressed by CNQX (10 μM). Taken together, the results suggest that basic properties of PF–PC excitatory synaptic transmission are largely normal in the Gαq mutant mice.

Impaired CF synapse elimination in PLCβ4 Mutant mice

PLCβ4 mutant mice are ataxic

The PLCβ4 mutant mice were viable and fertile. However, they exhibited various signs of ataxia, which became obvious about 2–3 weeks after birth. These include general uncoordinated movements, body swaying while moving, ataxic gait, and intention tremors. Examination of the footprint pattern during walking showed that PLCβ4 mutant mice walk with shorter strides (Fig. 5A) and cannot walk for an extended time on a straight path. When placed on a thin rod, the PLCβ4 mutant mice managed to stay on it as long as the wild-type mice. However, the PLCβ4 mutant mice could not place their hind feet on the thin rod. They maintained themselves on their bellies (Fig. 5B). When they traversed the rod, they could not coordinate the placement of their hind feet. These abnormalities were not seen with the forelimbs of the PLCβ4 mutant or with the wild-type mice. The abnormalities of the PLCβ4 mutant mice are not due to muscle weakness, because they showed normal

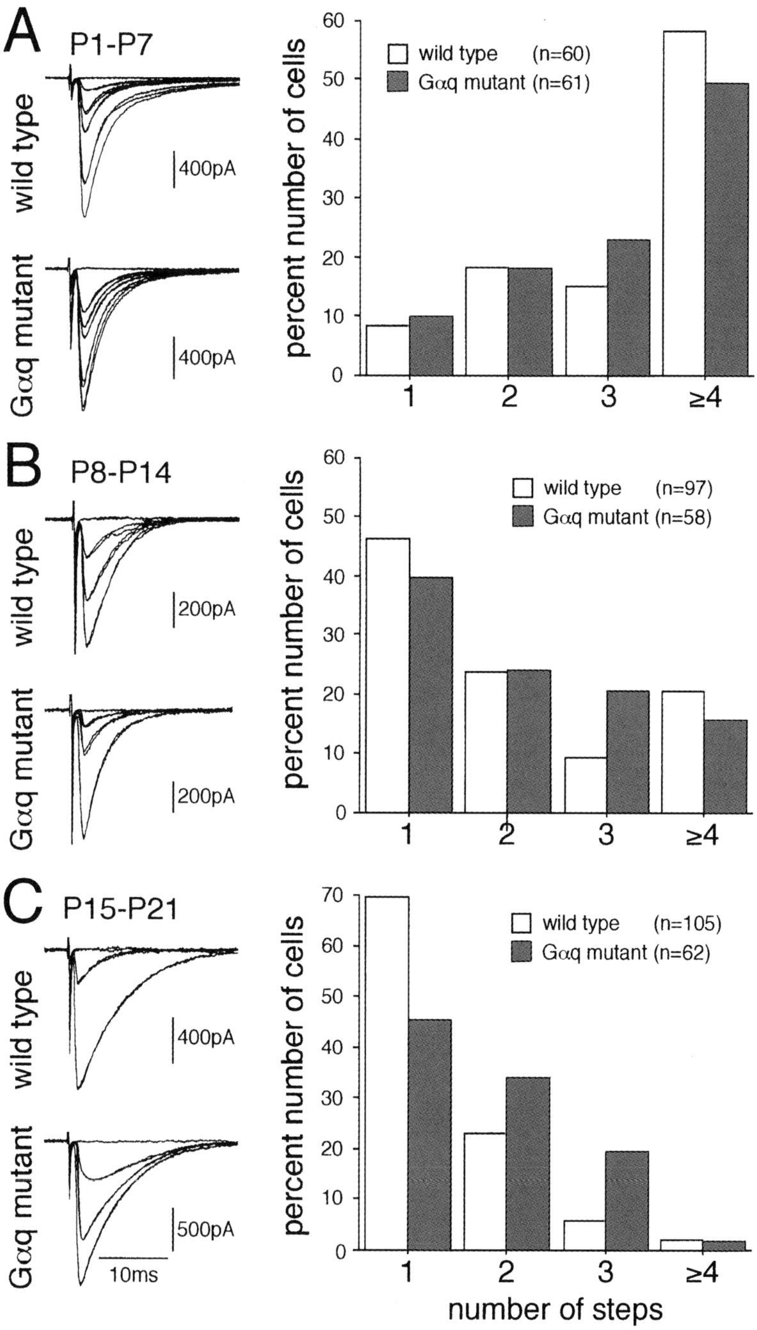

Fig. 4. Early postnatal development of CF innervation. (A) (left): CF–EPSCs of the wild-type (P7, holding potential − 30 mV) and Gαq mutant (P6, holding potential − 60 mV) PCs. (right): Summary graph showing numbers of discrete steps of CF–EPSCs of the wild-type (open colums) and mutant (hatched colums) PCs from mice at P1–P7. (B) similar to A, but examples of CF–EPSCs of the wild-type (P10, holding potential − 20 mV) and Gαq mutant (P10, holding potential − 20 mV) PCs and summary graph of the data from mice at P8–P14. (C) similar to A and B, but examples of CF–EPSCs of the wild-type (P21, holding potential − 10 mV) and Gαq mutant (P18, holding potential − 10 mV) PCs and summary graph of the data from mice at P15–P21 (from Offermanns et al., 1997).

grasping responses. To quantify the differences in motor performance between the wild-type and the PLCβ4 mutant mice, we used the rotorod test (Fig. 5C). The duration of balancing on the rod of the PLCβ4 mutant was significantly shorter than that of wild-type mice (Fig. 5C). These results indicate that PLCβ4 mutant mice exhibit clear motor discoordination.

Morphology of PLCβ4 mutant cerebellum is normal

The cerebellum of the PLCβ4 mutant mice showed similar size and foliation to that of wild-type mice at two months of age. No appreciable differences were found in the thickness of respective cortical layers and in GC density (Figs 6A, B). In both strains of mice, cell bodies of PCs were aligned in a monolayer at similar intervals, and their dendritic structures were indistinguishable (Figs 6C, D). The PF–PC synapses were examined by electron microscopy at two months of age. In both strains of mice, the molecular layer contained numerous profiles of PC spines, round or oval swellings having a few fragments of smooth endoplasmic reticulum and well-developed postsynaptic density (Figs 6E, F). Most of the PC spines formed contacts with PF terminals, which contained a number of clear round synaptic vesicles (Figs 6E, F). The PF–PC synapses were surrounded by cell processes of the BG (Figs 6E, F). When counting the PF–PC synapses on electron micrographs, the profile number per 100μm^2 was 21.3 ± 1.3 for the wild-type mice and 20.1 ± 1.7 for the mutant mice, showing no significant differences ($n = 3$; t-test, $p = 0.27$).

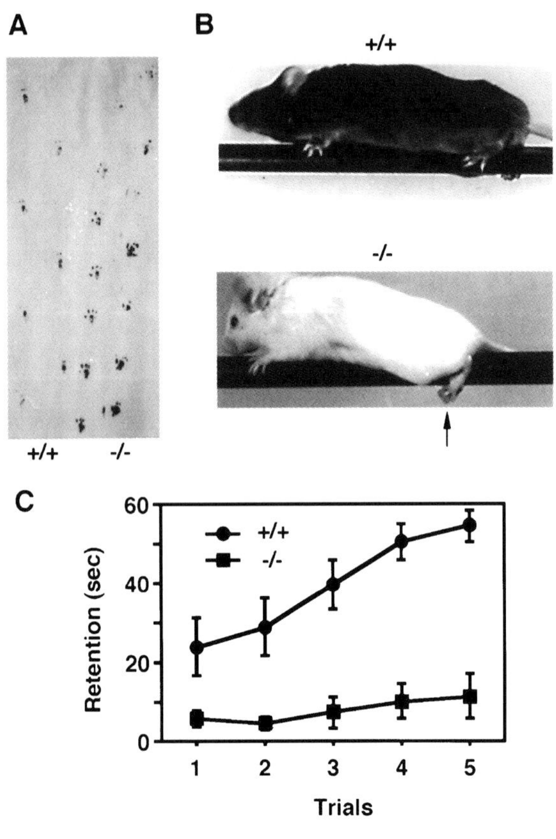

Fig. 5. Behavioral analysis of PLCβ4 mutant mice. (A) Hindlimb footprint pattern during walking. The PLC β4-null mutant mice (mu) walked with shorter and uneven paces in comparison with the wild-type mice. (B) Postures of the mice on a thin bar. The wild-type mice manage to stand on the thin bar, whereas the PLC β4-null mutant mice are unable to place their hindlimbs on the bar. They rather cling to the bar and maintain themselves on their bellies. (C) Running rotorod test (12 rpm). $F[1, 19] = 35.30$, $p < 0.001$.

Persistent multiple CF innervation of PLCβ4 mutant Purkinje cells in the rostral cerebellum

We recorded CF–EPSCs from 35–55 day old (P35–P55) wild-type mice or age-matched PLCβ4 mutant PCs and estimated the number of CFs innervating the recorded PCs. We found that 82.4% (89 out of 108) and 17.6% (19 out of 108) of PCs (P35–P55) were innervated by one (e.g. Fig. 7A upper trace) and two CFs (e.g. Fig. 7A lower trace), respectively, in wild-type PCs. On the other hand, in PLCβ4 mutant PCs, 54.7% (47 out of 86) had single CF–EPSC steps (e.g. Fig. 7B upper trace), whereas 27.9% (24 out of 86), 15.1% (13 out of 86) and 2.3% (2 out of 86) had two, three and four discrete steps, respectively (e.g. Fig. 7B lower trace). In total, significantly higher percentages of PCs remain multiply innervated in PLCβ4 mutant mice than in wild-type mice ($p < 0.01$, χ^2 test).

We looked for possible interlobular differences in CF innervation by plotting the recording sites of PCs on the standard sagittal plane. In wild-type mice, PCs with mono CF innervation and those with multiple CF innervation were evenly distributed (Fig. 7C). By marked contrast, in PLCβ4

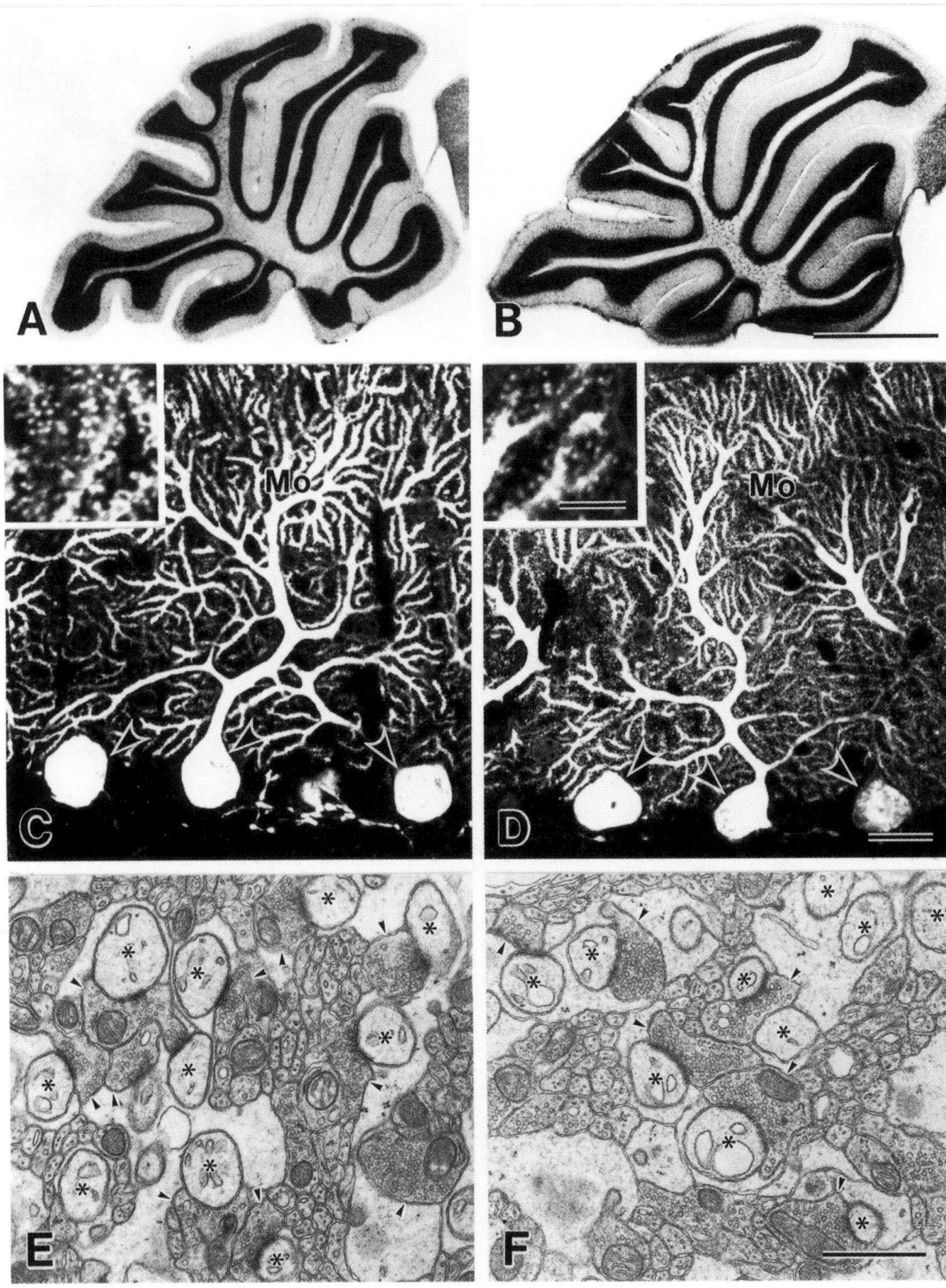

Fig. 6. Cerebellar anatomy of the PLCβ4 mutant and wild-type mice. PLCβ4 mutant (A, C, E) and wild-type (B, D, F) mice were studied at two months of age. (A, B) Toluidine blue-stained images of parasagittal sections through the cerebellar vermis. (C, D) Confocal scanning microscopic images of PC dendrites immunostained with spot 35/calbindin antibody. Arrowheads indicate PC somata. (E, F) Electron micrographs of the molecular layer. Asterisks indicate PC spines, and arrowheads indicate PF terminals. Mo, molecular layer. Scale bars, A, 1 mm; B, 50 μm; inset in B, 10 μm; C, 1 μm.

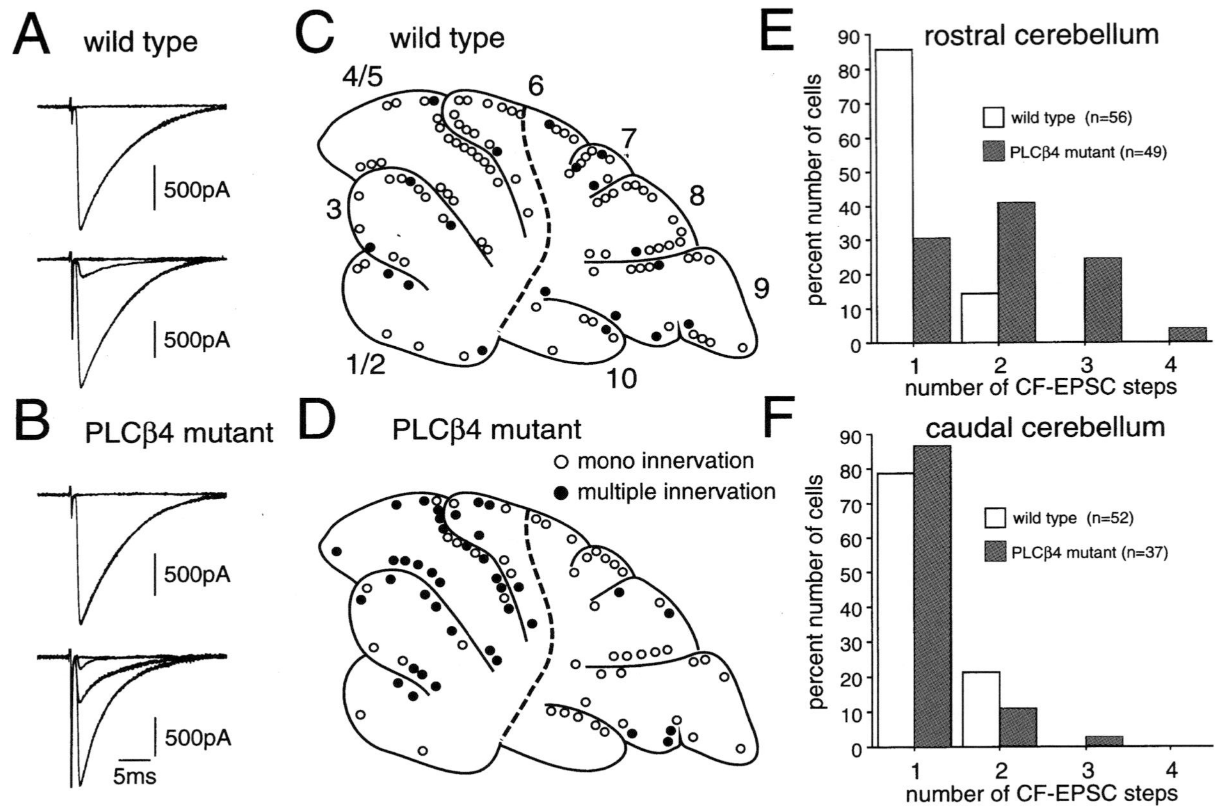

Fig. 7. Persistent multiple CF innervation of PCs confined to the rostral cerebellum in mature PLCβ4 mutant mice. (A, B) EPSCs elicited by stimulation of CFs in the GC layer in a wild-type (A) and a PLCβ4 mutant (B) PCs. Records were taken from mice at postnatal day 50 (P50 upper panel in A), P80 (lower panel in A), P46 (upper panel in B) and p46 (lower panel in B), respectively. Stimuli were applied at 0.2 Hz. One or two traces are superimposed at each threshold intensity. Holding potentials was − 10 mV. (C, D) Spatial distribution of mono-innervated (open circles) and multiply-innervated (filled circles) PCs of the wild-type (C) and PLCβ4 mutant (D) cerebella. Recording sites of PCs obtained from 6 wild-type ($n = 108$ PCs) and 6 PLCβ4 mutant ($n = 86$ PCs) cerebella were plotted on the standard midsagittal plane of the vermis. The broken line indicates the demarcation of the border of the rostral and caudal cerebellum that corresponds to the border of the two cerebellar regions in terms of PLCβ4 and PLCβ3 mRNA expression patterns (see Fig. 8). (E, F) Summary histograms showing the number of discrete steps of CF–EPSCs of the wild-type (open columns) and PLCβ4 mutant (hatched colums) PCs sampled in the rostral (E) and caudal (F) cerebellum. Data obtained from mice at P35–P55. The wild-type and mutant PCs were almost all analyzed before knowledge of the mouse genotype was revealed (from Kano et al., 1998).

mutant mice, multiply-innervated PCs were much more numerous in the rostral half (lobule 1 to the rostral half of lobule 6) than in the caudal half (the caudal half of lobule 6 to lobule 10) of the cerebellum (Fig. 7D). The border does not coincide with the classical anatomical boundary, the primary fissure. It rather corresponds to the newly found cerebellar boundary in some mutant mice (Napieralski and Eisenman, 1996) or that determined by the staining patterns of biochemical markers, such as zebrins (Hawkes and Gravel, 1991). The frequency distribution histogram constructed from data in the rostral cerebellum (Fig. 7E) clearly shows that the difference between the wild-type mice and PLCβ4 mutant mice was highly significant ($p < 0.0001$, χ^2-test for independent samples). By marked contrast, in the caudal cerebellum, the frequency distribution of mono innervated PLCβ4 mutant PCs was similar to that of wild-type PCs (Fig. 7F). These results indicate that PLCβ4 mutant mice have a defect in CF synapse elimination that is primarily manifested in the rostral half of the cerebellar vermis.

mRNA expressions of PLCβ3 and β4 are reciprocally regulated in PCs

One possibility for the regional difference of CF innervation in PLCβ4 mutant mice would be heterogeneous expressions of PLCβ isoforms in the cerebellum. If PLCβ4 is the major isoform in the rostral cerebellum and other isoforms are highly expressed in the caudal cerebellum, deletion of PLCβ4 would result in defects confined to the rostral cerebellum. Thus, we examined regional differences in expression of these mRNAs along rostrocaudal and mediolateral axes of the cerebellum. Following hybridization and emulsion autoradiography, PCs were identified by counterstaining with toluidine blue, which visualized the perikarya as a metachromatic purple color against the blue perikarya of GCs (Figs 8G–K). The PLCβ1 mRNA was detected at low levels in the PC somata showing no apparent regional differences (Figs 8A, G). Hybridization signals for PLCβ3 mRNA were restricted to PC somata (Figs 8B, E, H, I), exhibiting stronger signals in PCs of the caudal cerebellum than those in the rostral cerebellum (Fig. 8B). The PLCβ4 mRNA was expressed in PCs and GCs (Figs 8C, F, J, K). In contrast to the PLCβ3 mRNA expression pattern, levels of PLCβ4 mRNA in PCs were stronger in the rostral than in the caudal cerebellum (Fig. 8C). The boundary for regions with PCs expressing high levels of PLCβ3 or PLCβ4 mRNA fell in the middle of lobule 6 (Figs 8B, C).

Examination of coronal sections revealed a reciprocal pattern of PLCβ3 and PLCβ4 mRNA expression in the mediolateral cerebellar axis. In sections through the rostral cerebellum, the majority of PCs expressing high levels of PLCβ4 and low levels of PLCβ3 mRNAs formed thick parasagittal zones, which were interspersed periodically by one or two PCs with low PLCβ4 and high PLCβ3 mRNAs (Figs 8E, F). Identical results were obtained using two non-overlapping probes for in situ hybridization. These results indicate that PLCβ3 and PLCβ4 are the major isoforms in PCs, and that the cerebellar regions where PLCβ4 is predominantly expressed (Fig. 8C) correspond to those where a significantly higher percentage of PCs was multiply innervated by CFs in PLCβ4 mutant mice (Fig. 7C).

CF innervation during early postnatal days

We followed the developmental course of CF innervation in the rostral and caudal cerebellum during the first three postnatal weeks. During the first postnatal week (P1–P7), the majority of PCs in the rostral cerebellum were multiply innervated by more than four CFs in both wild-type and PLCβ4 mutant mice (Fig. 9A, upper and middle panels). Majority of PCs in the caudal cerebellum were also multiply innervated, but by generally less number of CFs than those in the rostral cerebellum (Fig. 9A, lower panel). The frequency distribution of the number of CF–EPSC steps per PC in the PLCβ4 mutant mice was not significantly different from that in the wild-type mice in both rostral and caudal cerebellum ($p > 0.05$, χ^2 test).

During the second postnatal week (P8–P14), the percentage of multiply innervated PCs decreased markedly in both wild-type and PLCβ4 mutant mice in the rostral (Fig. 9B, upper and middle panels) and caudal (Fig. 9B, lower panel) cer-

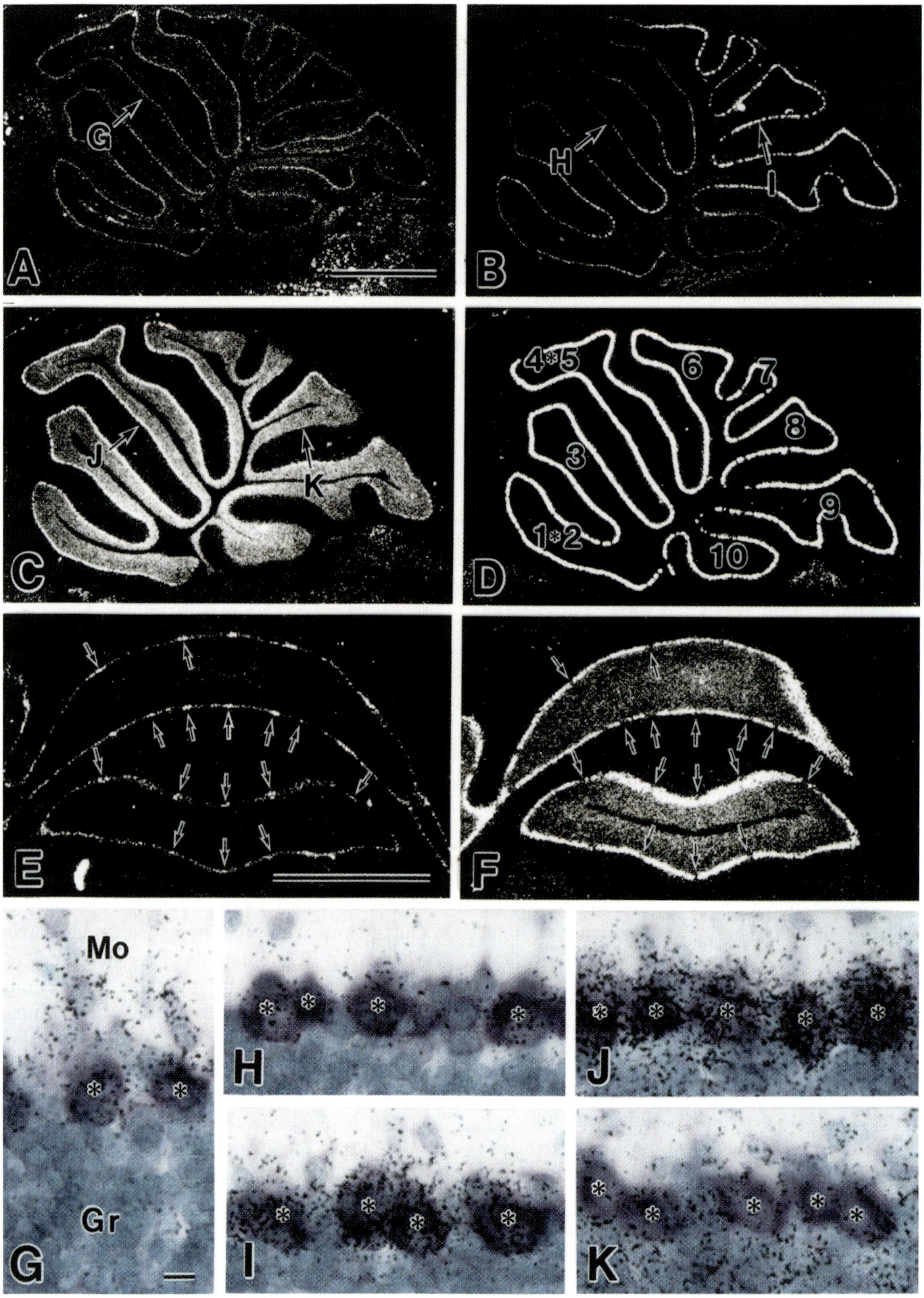
G
A
H
I
B
J
K
C
4*5
6
7
8
3
9
1*2
10
D
E
F
Mo
Gr
G
H
J
I
K

ebellum. However, the distributions were not significantly different between the wild-type and the PLCβ4 mutant mice in both rostral and caudal cerebellum ($p>0.05$, χ^2 test).

During the third postnatal week (P15–P21), clear differences between the two cerebellar regions became obvious. In the rostral cerebellum, the percentage of multiply innervated PCs decreased markedly in the wild-type mice when compared to the value during P8–P14. In contrast, the change in frequency distribution was much less prominent in PLCβ4 mutant mice (Fig. 9C, middle panel). Thus, the frequency distribution histograms were significantly different between the two strains of mice ($p<0.01$, χ^2 test) (Fig. 9C, middle panel). By contrast, in the caudal cerebellum, the reduction of multiply innervated PCs proceeds to the same extent between the wild-type and PLCβ4 mutant mice (Fig. 9C, lower panel).

In the cerebellum of the wild-type mice and also the caudal cerebellum of the PLCβ4 mutant mice, regression of CF innervation appears to continue during the third postnatal week, and the CF innervation pattern observed in adult mice is established by P21. Thus, the distributions of PCs at P15–P21 in the wild-type cerebellum and the caudal region of the PLCβ4 mutant cerebellum (Fig. 9C) are significantly different from those at P35–P55 (Figs 7E, F) ($p<0.01$, χ^2 test). By contrast, PCs at P15–P21 in the rostral cerebellum of PLCβ4 mutant mice (Fig. 9C middle panel) had similar distributions ($p>0.05$, χ^2 test) as those at P35–P55 (Fig. 7E), indicating that no significant change occurred in CF innervation during the third postnatal week.

These results indicate that during early postnatal period (P1–P14) regression of CF innervation appears to occur normally in the rostral cerebellum of PLCβ4 mutant mice (Figs 9A, B). The difference between the two strains became obvious at P15–21 (Fig. 9C), suggesting that the synapse elimination process that occurs during the third postnatal week is specifically impaired in the rostral cerebellum of the PLCβ4 mutant mice. By marked contrast, regression of CFs appears to occur normally throughout postnatal development in the caudal cerebellum of the PLCβ4 mutant mice.

Other electrophysiological parameters of PLCβ4 mutant EPSCs are normal

The 10–90% rise time or the decay time constants were similar in the wild-type and PLCβ4 mutant mice. CF–EPSCs showed prominent paired-pulse depression in both mono innervated and multiply innervated PCs (Konnerth et al., 1990; Aiba et al., 1994; Kano et al., 1995; Kano et al., 1997) derived from the wild-type and PLCβ4 mutant mice at varying inter-pulse intervals. The current-voltage relations were linear in both mono innervated and multiply innervated PCs derived from the two strains of mice. CF–EPSCs in the wild-type and PLCβ4 mutant mice were not affected by AP5 (100 μM), but were totally suppressed by CNQX, (10 μM). These results suggest that, in both the rostral and caudal cerebellum, CF to PC synapses in the PLCβ4 mutant mice are functional and their electrophysiological properties are largely similar to the wild-type mice.

The nature of PF–EPSCs was examined in mice at P27–P50. In the rostral and caudal cerebellum, there was no significant difference between the wild-type and PLCβ4 mutant mice in the kinetics of PF–EPSCs. In both strains of mice, PF–EPSCs displayed paired-pulse facilitation at varying inter-pulse intervals from 10 to 300 ms (Konnerth et al.,

Fig. 8. PLCβ4 and PLCβ3 mRNAs are reciprocally expressed in PCs of the rostral and caudal cerebellum. In situ hybridization showing expressions of the PLCβ1 (A, G), PLCβ3 (B, E, H, I) and PLCβ4 (C, F, J, K) mRNAs in the mature wild-type mouse cerebellum. (A-D) Dark-field micrographs taken from adjacent sagittal cerebellar sections near the midline. As a reference, cell bodies of PCs were visualized by detecting calbindin mRNAs in D. The numbers in D represent the cerebellar lobules in the vermis. (E, F) Dark-field micrographs taken from adjacent coronal sections through the rostral cerebellum. Note that locations of PCs with high PLCβ3 mRNA expression (arrows in E) correspond to those with low PLCβ4 mRNA expression in the adjacent section (arrows in F). (G-K) Bright-field micrographs taken from the sites indicated in cerebellar sections A–C. Asterisks indicate cell bodies of PCs that displayed metachromatic color changes into purple by counterstaining with toluidine blue. Note that PCs only weakly express PLCβ3 mRNA (H) and strongly PLCβ4 mRNA (J) in the rostral cerebellum, and vice versa in the caudal cerebellum (I, K) [Bars = 1 mm (A and E) and 10 μm (G)] (from Kano et al., 1998).

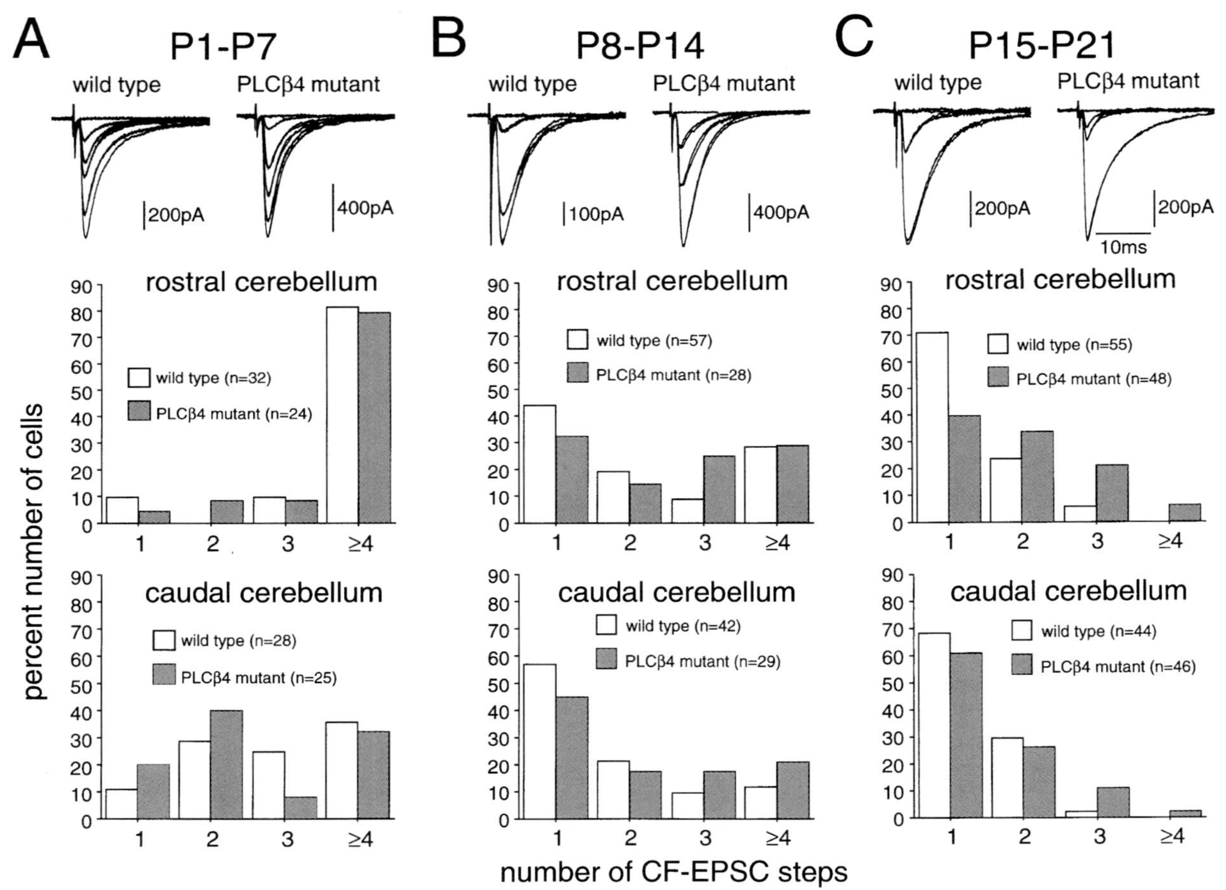

Fig. 9. Early postnatal development of CF innervation. (A) (upper panel) Specimen records of CF–EPSCs of the wild-type (P7, holding potential − 30 mV) and PLCβ4 mutant (P6, holding potential − 60 mV) PCs sampled in the rostral cerebellum. One or two traces were superimposed at each threshold intensity. Stimuli were applied at 0.2 Hz. (middle and lower panels) Frequency distribution histogram for PCs in the rostral (middle panel) and caudal (lower panel) cerebellum from 4 wild-type and 3 PLCβ4 mutant mice at P1–P7. Twenty five out of 32 wild-type PCs in the rostral, and 20 out of 28 in the caudal cerebellum were studied blind to the mouse genotype. As for PLCβ4 mutant mice, all cells in the rostral and caudal cerebellum were studied blind to mouse genotype. (B) Similar to A, but for data from mice at P8–P14. Specimen records were taken from the wild-type (P10, holding potential − 20 mV) and PLCβ4 mutant (P10, holding potential − 60 mV) PCs that were sampled in the rostral cerebellum. Data were obtained from 6 wild-type and 4 PLCβ4 mutant mice. All cells were studied blind to the mouse genotype. (C) Similar to A and B, but for data from mice at P15–P21. Specimen records were taken from the wild-type (P16, holding potential 0 mV) and PLCβ4 mutant (P18, holding potential 0 mV) PCs that were sampled in the rostral cerebellum. Data were obtained from 6 wild-type and 7 PLCβ4 mutant mice. All cells were studied blind to the mouse genotype (from Kano et al., 1998)

1990; Aiba et al., 1994; Kano et al., 1995). Furthermore, PF–EPSCs in the wild-type and PLCβ4 mutant mice were not affected by AP5 (100 μM), but were totally suppressed by CNQX (10 μM). Taken together, these results suggest that PF to PC transmission in the PLCβ4 mutant mice is normal in the rostral and caudal cerebellum.

Discussion

Impaired regression of multiple CFs is correlated with motor discoordination

Our results demonstrate that both Gαq mutant mice and PLCβ4 mutant mice display clear ataxia with typical signs of motor discoordination. Although the causal relationship is not fully understood, motor discoordination and impaired regression of multiple CFs have been reported in several strains of mutant mice deficient in mGluR1 (Aiba et al., 1994; Conquet et al., 1994; Kano et al., 1997; Levenes et al., 1997), PKCγ (Chen et al., 1995; Kano et al., 1995) or the glutamate receptor δ2 subunit (GluRδ2) (Kashiwabuchi et al., 1995). In Gαq mutant mice, multiply innervated PCs were distributed in both the rostral and caudal lobules of the cerebellar vermis (Fig. 3B) with significantly higher percentages when compared to the wild-type mice (Fig. 3A). By contrast, in PLCβ4 mutant mice, multiply innervated PCs were densely localized in the rostral half of the cerebellar vermis (Fig. 7D). Such a regional difference within the cerebellum has not been reported in terms of persistent multiple CF innervation. Thus, the present data on PLCβ4 mutant mice suggest that the rostral half of cerebellar vermis may be important for coordinated locomotion in mice. In other mammals including the cat and rat (Armstrong, 1988; Armstrong, 1990), lesions in the lobules 4 to 6 of the cerebellar vermis resulted in impairment of locomotion. The corresponding region of the cerebellar vermis may be involved in controlling locomotion in mice.

Involvement of Gαq and PLCβ4 in the second phase of CF synapse elimination

Our results indicate that regression of multiple CF innervation in both mutant cerebella occurred normally during P1 to P14 coincident with the first phase of reduction of multiple CF innervation (Figs 4A, 4B, 9A, 9B). The abnormality in the regression process occurs during the third postnatal week in the Gαq mutant cerebellum (Fig. 4C) and the rostral half of the PLCβ4 mutant cerebellum (Fig. 9C). This appears to correspond to the second phase of the CF synapse elimination (Kano et al., 1995; Kano et al., 1997; Offermanns et al., 1997a; Kano et al., 1998)

Previous results indicate that multiple CF innervation persists in adult animals that are devoid of GCs or that lack proper PF–PC synaptogenesis during early postnatal development (Crépel, 1982). In addition, multiple CF innervation was recently reported in the mutant adult mice deficient in GluRδ2 (Kashiwabuchi et al., 1995) in which the number of synaptic contacts from PFs onto PC dendritic spines is less than half of that of the wild type (Kashiwabuchi et al., 1995; Kurihara et al., 1997). These results suggest that the presence of functional PF–PC synapses is essential for the regression of multiple CF innervation. However, the Gαq and PLCβ4 mutant mice have no obvious defects in GCs, in PF synapse formation (Figs 1, 6) or in the basic electrophysiology of PF-mediated excitatory synaptic transmission. In this respect, Gαq and PLCβ4 mutant mice share a common phenotype with previously reported mutant mice with inactivated mGluR1 or PKCγ gene (Kano et al., 1995; Kano et al., 1997).

mGluR1, Gαq and PKCγ are all expressed abundantly and homogeneously in adult wild-type PCs; mGluR1 is localized on the peri-synaptic membrane of PC dendritic spines (Baude et al., 1993; Nusser et al., 1994), Gαq/11 is on the extra-junctional membrane of PC dendritic spines (Tanaka et al., in press), and PKCγ is in the cytoplasm of PC dendrites and somata (Hidaka et al., 1988; Huang et al., 1988; Saito et al., 1988). In the present study, we found that expression of PLCβ4 mRNA is heterogeneous in the rostrocaudal and mediolateral axes (Fig. 3), the expression pattern appears to be similar to that seen when the cerebellum is stained for biochemical markers, such as zebrins (Hawkes and Gravel, 1991). Reflecting this regional heterogeneity, PCs with persistent multiple CF innervation in PLCβ4 mutant mice emerge in cerebellar regions where PLCβ4 mRNA is predominantly expressed. In the vermis, they preferentially occur in the rostral half (lobule 1 to the rostral half of the lobule 6). Based on these and previous results, we suggest that, in the rostral cerebellar vermis, activity at excitatory synapses of PCs modulates the mGluR1 signaling that acts through Gαq, PLCβ4 and PKCγ to facilitate the elimination of multiple CF innervation during the third postnatal week. Furthermore, our present results indicate that PLCβ1 does not compensate for the lack of PLCβ4 function in the rostral cerebellar vermis. This notion is supported by recent results that PLCβ1 deficient mice do not exhibit any signs of motor discoordination (Kim et al., 1997).

In the caudal half of the cerebellar vermis, the developmental elimination normally proceeds in PLCβ4 mutant mice (Figs 7, 9). Here the expression of PLCβ4 mRNA is low and that of PLCβ3 mRNA is high (Fig. 8). Biochemical data indicate that PLCβ3 and PLCβ4 are indistinguishably activated by Gαq and Gα11 (Hepler et al., 1993; Jiang et al., 1994). Thus, it is conceivable that PLCβ3, presumably together with PLCβ1, compensates for the lack of PLCβ4 function in the caudal cerebellum.

Possible molecular mechanisms for CF synapse elimination

The signal transduction cascade from mGluR1 to PKCγ via Gαq and PLCβ4 appears to be activated at PF–PC synapses rather than at CF–PC synapses. This notion is based on the following two findings. First, mGluR1, Gαq/11 and PLCβ4 are colocalized in the PC dendritic spines (Baude et al., 1993; Nusser et al., 1994). Second, the CF elimination process requires activation of NMDA receptors (Rabacchi et al., 1992) which do not contribute to either PF– or CF–EPSCs in PCs (Aiba et al., 1994; Kano et al., 1995), but play significant roles in mossy fiber-GC excitatory transmission (Takahashi et al., 1996). Thus, signals generated in the PC dendritic spines activate PKCγ and then, influence CF to PC synapses either directly or indirectly through further intracellular cascades.

We suggest that signaling proteins including mGluR1, Gαq, PLCβ4 and PKCγ may be clustered in a manner similar to the complex of signaling proteins that are found in the rhabdomeres of the photoreceptor cell in the Drosophila visual system and in the post synaptic density in mammalian neurons (Sheng, 1996; Chevesich et al., 1997). A scaffolding protein interacts with many of the protein components of the signaling pathway, maintains their functional integrity and helps in their localization (Kornau et al., 1995; Dong et al., 1997; Songyang et al., 1997). In the Drosophila visual system the proteins found in the complex include a PLCβ4 homologue, a PKC family member, as well as a receptor and an ion channel protein. Dissociation of any one component from the complex inactivates function (Tsunoda et al., 1997). An analogous assemblage in the dendritic spines of PCs could be involved in processing PF–PC synaptic activity and in coordinating synaptic activity with some of the events involved in the second phase of CF synapse elimination.

Acknowledgements

We thank K. Matsumoto and V. Mancino for expert technical help and Dr.N. Kawai for continuous encouragement. This work has been supported by grants to M.K. from the Japanese Ministry of Education, Science, Sports and Culture, from the Human Frontier Science Program, and by Special Coordination Funds for promoting Science and Technology from STA. This work has also been supported by a grant to H.-S.S. from the Creative Research Initiatives of the Korean Government, and by grants to D.W. and to M.I.S. from the National Institutes of Health, USA. S.O. was a recipient of a fellowship from the Deutsche Forschungsgemeinschaft and the Guenther Foundation.

References

Aiba, A., Kano, M., Chen, C., Stanton, M.E., Fox, G.D., Herrup, K., Zwingman, T.A. and Tonegawa, S. (1994) Deficient cerebellar long-term depression and impaired motor learning in mGluR1 mutant mice. *Cell*, 79: 377–388.

Armstrong, D.M. (1988) The supraspinal control of mammalian locomotion. *J. Physiol. (Lond.)*, 405: 1–37.

Armstrong, D.M. (1990) Topographical localization in the projections from the inferior olive to the paravermal cortex of the anterior lobe and paramedian lobule in the cerebellum of the cat. A brief review. *Arch. Ital. Biol.*, 128: 183–207.

Baude, A., Nusser, Z., Roberts, J.D.B., Mulvihill, E., McIlhinney, R.A.J. and Somogyi, P. (1993) The metabotropic glutamate receptor (mGluR1a) is concentrated at perisynaptic membrane of neuronal subpopulations as detected by immunogold reaction. *Neuron*, 11: 771–787.

Berridge, M.J. (1993) Inositol trisphosphate and calcium signalling. *Nature*, 361: 315–325.

Chen, C., Kano, M., Abeliovich, A., Chen, L., Bao, S., Kim, J.J., Hashimoto, K., Thompson, R.F. and Tonegawa, S. (1995) Impaired motor coordination correlates with persistent multiple climbing fiber innervation in PKCg mutant mice. *Cell*, 83: 1233–1242.

Chevesich, J., Kreuz, A.J. and Montell, C. (1997) Requirement for PDZ domain protein, INAD, for localization of the TRP store-operated channel to a signaling complex. *Neuron*, 18: 95–105.

Conquet, F., Bashir, Z.I., Davies, C.H., Daniel, H., Ferraguti, F., Bordi, F., Franz-Bacon, K., Reggiani, A., Matarese, V., Conde, F., Collingridge, G.L. and Crépel, F. (1994) Motor deficit and impairment of synaptic plasticity in mice lacking mGluR1. *Nature*, 372: 237–243.

Crépel, F. (1982) Regression of functional synapses in the immature mammaliam cerebellum. *Trends Neurosci.*, 5: 266–269.

Dong, H.L., Obrien, R.J., Fung, E.T., Lanahan, A.A., Worley, P.F. and Huganir, R.L. (1997) GRIP: a synaptic PDZ domain-containing protein that interacts wtih AMPA receptors. *Nature*, 386: 279–284.

Edwards, F.A., Konnerth, A., Sakmann, B. and T., T. (1989) A thin slice preparation for patch-clamp recordings from neurons of the mammalian central nervous system. *Pflügers Archiv. (Europ. J. Physiol.)*, 414: 600–612.

Exton, J.H. (1996) Regulation of phosphoinositide phospholipases by hormones, neurotransmitters and other agonists linked to G-proteins. *Annu. Rev. Pharmacol. Toxicol.*, 36: 481–509.

Hawkes, R. and Gravel, C. (1991) The modular cerebellum. *Prog. Neurobiol.*, 36: 309–327.

Hepler, J.R., Kozasa, T., Smrcka, A.V., Simon, M.I., Rhee, S.G., Sternweis, P.C. and Gilman, A.G. (1993) Purification from Sf9 cells and characterization of recommbinant Gqa and G11a. Activation of purified phospholipase C isozymes by G protein a subunit. *J. Biol. Chem.*, 268: 14367–14375.

Hidaka, H., Tanaka, T., Onoda, K., Hagiwara, M., Watanabe, M., Ohta, H., Ito, Y., Tsurudome, M. and Yoshida, T. (1988) Cell type-specific expression of protein kinase C isozymes in the rabbit cerebellum. *J. Biol. Chem.*, 263: 4523–4526.

Huang, F.L., Yoshida, Y., Nakabayashi, H., Young, W.S. and Huang, K.-P. (1988) Immunocytochemical localization of protein kinase C isozymes in rat brain. *J. Neurosci.*, 8: 4734–4744.

Ito, M. (1984) The Cerebellum and Neural Control, New York: Raven Press.

Jiang, H., Wu, D. and Simon, M.I. (1994) Activation of phospholipase C b4 by heterotrimeric GTP-binding proteins. *J. Biol. Chem.*, 269: 7593–7596.

Kano, M., Hashimoto, K., Chen, C., Abeliovich, A., Aiba, A., Kurihara, H., Watanabe, M., Inoue, Y. and Tonegawa, S. (1995) Impaired synapse elimination during cerebellar development in PKCg mutant mice. *Cell*, 83: 1223–1231.

Kano, M., Hashimoto, K., Kurihara, H., Watanabe, M., Inoue, Y., Aiba, A. and Tonegawa, S. (1997) Persistent multiple climbing fiber innervation of cerebellar Purkinje cells in mice lacking mGluR1. *Neuron*, 18: 71–79.

Kano, M., Hashimoto, K., Watanabe, M., Kurihara, H., Offermanns, S., Jiangs, H., Wu, Y., Jun, K., Shin, H.-S., Inoue, Y., Simon, M.I. and Wu, D. (1998) PLCb4 is specifically involved in climbing fiber synapse elimination in the developing cerebellum. *Proc. Natl. Acad. Sci. USA*, 95: 15724–15729.

Kashiwabuchi, N., Ikeda, K., Araki, K., Hirano, T., Shibuki, K., Takayama, C., Inoue, Y., Kutsuwada, T., Yagi, T., Kang, Y., Aizawa, S. and Mishima, M. (1995) Impairment of motor coordination, Purkinje cell synapse formation and cerebellar long-term depression in GiuRd2 mutant mice. *Cell*, 81: 245–252.

Kim, D., Jun, K.S., Lee, S.B., Kang, N.-G., Min, D.S., Kim, Y.-H., Ryu, S.H., Suh, P.G. and Shin, H.-S. (1997) Phospholipase C isozymes selectively couple to specific neurotransmitter. receptors. *Nature*, 389: 290–293.

Konnerth, A., Llano, I. and Armstrong, C.M. (1990) Synaptic currents in cerebellar Prukinje cells. *Proc. Natl. Acad. Sci. USA.*, 87: 2662–2665.

Kornau, H.C., Schenker, L.T., Kennedy, M.B. and Seeburg, P.H. (1995) Domain interaction between NMDA receptor subunits and the postsynaptic density protein PSD–95. *Science*, 269: 1737–1740.

Kurihara, H., Hashimoto, K., Kano, M., Takayama, C., Sakimura, K., Mishina, M., Inoue, Y. and Watanabe, M. (1997) Impaired parallel fiber-Purkinje cell synapse stabilization during cerebellar development of mutant mice lacking the glutamate receptor δ2 subunit (GluRδ2). *J. Neurosci.*, 17: 9613–9623.

Levenes, C., Daniel, H., Jaillard, D., Conquet, F. and Crépel, F. (1997) Incomplete regression of multiple climbing fibre innervation of cerebellar Purkinje cells in mGLuR1 mutant mice. *NeuroReport*, 20: 571–574.

Llano, I., Marty, A., Armstrong, C.M. and Konnerth, A. (1991) Synaptic- and agonist-induced excitatory currents of Purkinje cells in rat cerebellar slice. *J. Physiol. (Lond.)*, 434: 183–213.

Nakanishi, S. (1994) Metabotropic glutamate receptors: synaptic transmission, modulation and plasticity. *Neuron*, 13: 1031–1037.

Napieralski, J.A. and Eisenman, L.M. (1996) Further evidence for a unique developmental compartment in the cerebellum of the meander tail mutant mouse as revealed by the quantitative analysis of Purkinje cells. *J. Comp. Neurol.*, 364: 718–728.

Nishizuka, Y. (1992) Intracellular signaling by hydrolysis of phospholipids and activation of protein kinase C. *Science*, 258: 607–614.

Nusser, Z., Mulvihill, E.P.S. and Somogyi, P. (1994) Subsynaptic segregation of metabotropic and ionotropic glutamate receptors as revealed by immunogold localization. *Neuroscience*, 61: 421–427.

Offermanns, S., Hashimoto, K., Watanabe, M., Sun, W., Kurihara, H., Thompson, R.F., Inoue, Y., Kano, M. and Simon, M.I. (1997a) Impaired motor coordination and persistent multiple climbing fiber innervation of cerebellar Purkinje cells in mice lacking Gαq. *Proc. Natl. Acad. Sci. USA*, 94: 14089–14094.

Offermanns, S., Toombs, C.F., Hu, Y.-H. and Simon, M.I. (1997b) Defective platelet activation in Gαq deficient mice. *Nature*, 389: 183–186.

Pin, J.-P. and Duvoisin, R. (1995) The metabotropic glutamate receptors: structure and functions. *Neuropharmacology*, 34: 1–26.

Rabacchi, S., Bailly, Y., Delhaye-Bouchaud, N. and Mariani, J. (1992) Involvement of the *N*-methyl D-aspartate (NMDA) receptor in synapse elimination during cerebellar development. *Science*, 256: 1823–1825.

Roustan, P., Abitbol, M., Nenini, C., Ribeaudeau, F., Gerard, M., Vekemans, M., Mallet, J. and Dufier, J.-L. (1995) The rat phospholipase Cβ4 gene is expressed at high abundance in cerebellar Purkinje cells. *NeuroReport*, 6: 1837–1841.

Saito, N., Kikkawa, U., Nishizuka, Y. and Tanaka, C. (1988) Distribution of protein kinase C-like immunoreactive neurons in rat brain. *J. Neurosci.*, 8: 369–382.

Sheng, M. (1996) PDZs and receptor/channel clustering: rounding up the latest suspects. *Neuron*, 17: 575–578.

Songyang, Z., Fanning, A.S., Fu, C., Xu, J., Marfatia, S.M., Chishti, A.H., Crompton, A., Chan, A.C., Anderson, J.M. and Cantley, L.C. (1997) Recognition of unique carboxyl-terminal motifs by distinct PDZ domains. *Science*, 275: 73–77.

Strathmann, M.P. and Simon, M.I. (1990) G protein diversity: A distinct class of a subunits is present in vertebrates and invertebrates. *Proc. Natl. Acad. Sci. USA*, 87: 9113–9117.

Takahashi, T., Feldmeyer, D., Suzuki, N., Onodera, K., Cull Candy, S.G., Sakimura, K. and Mishina, M. (1996) Functional correlation of NMDA receptor epsilon subunits expression with the properties of single-channel and synaptic currents in the developing cerebellum. *J. Neurosci.*, 16: 4376–4382.

Tanaka, O. and Kondo, H. (1994) Localization of mRNAs for three novel members (b3, b4 and g2) of phospholipase C family in mature rat brain. *Neurosci. Lett.*, 182: 17–20.

Tanaka, J., Nakagawa, S., Yawasaki, M., Fukaya, M., Iwanaga, T., Inoue, Y., Simon, M.I., Sakimura, K., Kano, M. and Watanabe, M. (2000) Gq protein α subunits Gαq and $G\alpha_{11}$ are localized at postsynaptic extra-junctional membrane of cerebellar Purkinje cells and hippocampal pyramidal cells. *Eur. J. Neurosci.*, in press.

Tsunoda, S., Sierralta, J., Sun, Y., Bodner, R., Suzuki, E., Becker, A., Socolich, M. and Zuker, C.S. (1997) A multivalent PDZ-domain protein assembles signalling complexes in a G-protein-coupled cascade. *Nature*, 388: 243–249.

Watanabe, M., Nakamura, M., Sato, K., Kano, M., Simon, M.I. and Inoue, Y. (1998) Patterns of expression for the mRNA corresponding to the four isoforms of phospholipase Cb in mouse brain. *Eur. J. Neurosci.*, 10: 2016–2025.

Weibel, E.R. (1979) *Stereological Methods – Practical Methods for Biological Morphometry*, London–New York–Sydney–San Francisco: Academic Press.

Wilkie, T.M., Scherle, P.A., Strathman, M.P., Slepak, V.Z. and Simon, M.I. (1991) Characterization of G-protein a subunits in the G_q class: Expression in murine tissue and in stromal and hematopoietic cell lines. *Proc. Natl. Acad. Sci. USA*, 88: 10049–10053.

SECTION II

Interneurons

N.M. Gerrits, T.J.H. Ruigrok and C.I. De Zeeuw (Eds.)
Progress in Brain Research, Vol 124

CHAPTER 4

Lineage, development and morphogenesis of cerebellar interneurons

Karl Schilling*

Anatomisches Institut, Rheinische Friedrich-Wilhelms-Universität, Nussallee 10, D–53115 Bonn, Germany

Introduction

The mammalian cerebellar cortex consists of a rather limited set of neuronal phenotypes, arranged in a strikingly regular, stereotyped way. It arises from a region centered about the mid/hindbrain junction of the neural tube, and the basic histogenetic and morphogenetic events leading to the formation of this unique part of the central nervous system have been described in considerable detail (for reviews see, e.g. Ito, 1984; Hatten and Heintz, 1995; Altman and Bayer, 1997; Oberdick et al., 1998). This has since long nourished the hope that the cerebellum may be utilized as a paradigm to unravel basic principles governing the formation and differentiation of individual nerve cells and their integration in a functional network.

It has been established for quite some time that cerebellar cortical neurons arise from two well separated germinative layers. Purkinje neurons, the sole output (projection) neurons of the cerebellar cortex arise from the neuroepithelium lining the roof of the fourth ventricle, from where, following cell cycle arrest, they migrate radially to reach their final position within the cerebellar anlage. In contrast, cerebellar granule cells, which form excitatory interneurons projecting onto Purkinje cells, arise from a secondary germinal layer located on the surface of the cerebellar anlage, the external granule cell layer. From there, granule cell precursors have to migrate inwardly, past Purkinje neurons, to reach their final position within the (internal) granule cell layer. The origin of inhibitory interneurons of the cerebellar cortex, and basket and stellate cells in particular, has been controversial. The prevailing view was that they share a common ancestry with granule neurons and originate from the external granule cell layer. A series of elegant studies (Otero et al., 1993; Napieralski and Eisenman, 1993; Hallonet and Le Douarin, 1993; Zhang and Goldman, 1996a, 1996b) has recently helped to clarify this long-standing conundrum of cerebellar histogenesis. As it turned out, precursors for inhibitory cerebellar cortical interneurons originate from the neuroepithelium of the fourth ventricle, and migrate outwardly, much like Purkinje neurons, to reach their final positions within the cerebellar cortex. In contrast to Purkinje cells, however, interneuronal precursors continue to proliferate during their transit through the deep cerebellar mass/incipient white matter of the nascent cerebellum. Thus, this region may be perceived as yet another secondary germinal zone, in addition to the external granule cell layer.

These findings have major implications for our understanding of basic histogenetic mechanisms leading to the formation of the cerebellar cortex. In particular, they warrant a reassessment of our views of when the lineages leading to known cerebellar

*Corresponding author. Tel.: +49–228–73–2602;
Fax: +49–228–73–7301;
e-mail: karl.schilling@uni-bonn.de

neuronal phenotypes separate, and how lineage separation and terminal differentiation may be mechanistically realized. Here, I will try to briefly focus on some more recent findings pertinent to our understanding of the origins and development of cerebellar granule cells and cerebellar basket/stellate neurons, the two better understood types of cerebellar cortical interneurons.

The delineation of the early cerebellar anlage – molecules and morphogenesis

It is convenient to view the differentiation of individual neuronal phenotypes as a series of discrete developmental steps, ranging from precursor cell allocation to the synaptic and functional integration of the terminally differentiated nerve cell. Critical steps along this developmental path include the acquisition of regional identity, neurochemical differentiation, and morphological maturation. What do we know about when key developmental decisions are taken by cerebellar interneurons, and how are they realized in terms of molecular genetic regulation?

Prior to the first morphological indications of cerebellar formation, at around embryonic day 8 in the mouse, the mes/metencephalic junction of the neural tube, from which the cerebellar anlage is derived, is molecularly defined by the localized expression of a set of genes, including those coding for the nuclear transcription factors *engrailed*–1 and –2 (*en*–1, *en*–2) and the secreted signaling protein, *wingless* (*wnt*–1) (Joyner, 1996).While it is clear that the *engrailed/wingless* signaling pathway is involved in the regulation of neuronal precursor proliferation (Ikeya et al., 1997), it is not really understood how precisely these genes relate to the well-defined morphogenetic movements which, following their first expression at the mid/hindbrain junction, rapidly lead to the formation of the topographic landmarks characterizing the early cerebellar anlage. Indeed, it should be noted that the expression of these genes is not restricted to the alar (dorsal) plate of the mes/metencephalon, from which the future cerebellum derives, but extends in the ventral part of the neural tube at the mid-hindbrain junction. Consequently, their ablation results in developmental defects beyond the cerebellum that also affect midbrain-derived structures originating ventrally, e.g. midbrain aminergic neurons (Simon et al., 1996). Thus, while the orderly expression of the *engrailed* genes and wingless are absolutely necessary for proper cerebellar development, they are by no means sufficient to molecularly define the origin of the cerebellar anlage. Rather, the *engrailed/wingless* signaling pathway may be perceived as critical to the acquisition of areal identity of prospective mid/hindbrain neurons, including those of the future cerebellum. Notably, initial expression of the constituents of the *engrailed/wingless* signaling cascade is not to be equated with the irrevocable commitment to/determination of a mid/hindbrain fate. Thus, the majority of cells derived from the murine embryonic day 10.5 mid/hindbrain region will lose their mid/hindbrain identity if transplanted, as single cells, into the forebrain (Olsson et al., 1997).

Setting apart lineages for cerebellar cortical neurons

To date, we do not know of any molecular marker (or trigger) that may be associated with the acquisition of a cerebellar, or cerebellar cortical fate. However tempting the idea of a common molecular underpinning of cerebellar fate may seem, it should be clearly spelled out that there is no experimental evidence to suggests its existence. Quite to the contrary, there is ample evidence showing how the diverse neuronal constituents of the cerebellar system are matched to each other and interconnected during histogenesis by regulative mechanisms (e.g. Herrup and Sunter, 1987; Vogel and Herrup, 1989; Smeyne et al., 1995; Baader et al., 1998). Thus, the questions to be answered, if we want to approach cerebellar histogenesis, rather refer to when, where and how the cellular lineages leading to the eventual constituents of the cerebellar cortex emerge from the mid/hindbrain territory.

The first cerebellar neuron for which lineage analysis was achieved was the Purkinje cell. Based on a creative and careful analysis of chimeric mice carrying a definitive Purkinje cell marker, Karl Herrup suggested that Purkinje neurons originate from a small set of founder cells, which are allocated during the early neural tube stage (Wetts and Herrup, 1982; Herrup, 1986; Herrup and Sunter, 1986), at about the same time that the *engrailed/wingless* signaling pathway constituents are first expressed. Later investigations (Baader et al., 1996; Mathis et al., 1997; Hawkes et al., 1998), using different experimental approaches, led to some changes in the numbers originally suggested by the chimera-based analyses, but concur to corroborate the basic conclusions reached by these pioneering studies. However, although it may be reasonably argued that the Purkinje cell lineage is indeed specified at this early stage (see Baader et al., 1996 for discussion and further references), the available experimental evidence is only indirect. There is no marker that allows the cerebellar Purkinje cell lineage to be identified directly, prior to the terminal exit of these cells from the cell cycle, which, in the mouse, occurs between embryonic days 11 to 13 (Miale and Sidman, 1961).

Precursors for cerebellar interneurons can be first recognized following the formation of the trigonum germinativum and the external granule cell layer emerging from it. This highly specialized secondary germinative layer forms, in the mouse, at about embryonic day 13/13.5 (Miale and Sidman, 1961), following a series of conspicuous morphogenetic movements leading to the delineation of the early cerebellar anlage (reviewed in Altman and Bayer, 1997).

As a consequence of the ventral bending of the mes/metencephalic neural tube to form the pontine flexure, the incipient fourth ventricle broadens into its name-giving rhomboid shape. Dorsally, the proliferative neuroepithelium separates, leaving behind a thinned out layer to form the velum medullare (and eventually the choroid plexus). At the sites where the velum medullare is attached to the neural tube, the proliferating neuroepithelium appears thickened, and it is referred to as the rhombic lip. The 'upper lip', or rhombic lip sensu strictu as originally defined by His (1891) extends from the midline rostrally in a latero-caudal direction to join the 'lower lip', which elongates from latero-rostrally to reach the midline again more caudally. Along the line where the (upper) rhombic lip joins the velum medullare, the germinal neuroepithelium undergoes a conspicuous transformation, leading to the formation of the trigonum germinativum, from where dividing neuroblasts emigrate rostrally across the cerebellar anlage to form the external granule cell layer. This process starts laterally, but eventually the trigonum germinativum spans the whole width of the cerebellar anlage.

Molecularly, prospective granule cell neuroblasts are distinguished from dividing cells in the ventricular zone by their down regulation, prior to emigration from the rhombic lip, of the transcription factor genes *cSox2* and *cSox3* (Rex et al., 1998). Conversely, proliferating neuroblasts in and emigrating from the trigonum germinativum are characterized by the expression of the vertebrate homologue of the drosophila proneural gene, *atonal* (mouse: *Math1*; Akazawa et al., 1995; Ben-Arie et al., 1997). This gene is not expressed by proliferating neuroblasts within the neuroepithelium of the fourth ventricle, and its ablation leads to the selective elimination of the external granule cell layer. These data, taken together with recent findings that define the cerebellar granule neuron as the sole descendant of the external granule cell layer (see below; Otero et al., 1993; Napieralski and Eisenman, 1993; Hallonet and Le Douarin, 1993; Zhang and Goldman, 1996a, 1996b) establish *Math1* as a molecular marker of the granule cell lineage. Available evidence suggests that *Math1* is critical to the acquisition of granule cell fate, and/or the proliferative expansion of this lineage (Ben-Arie et al., 1997; Helms and Johnson, 1998).

As already mentioned above, the precursors for cerebellar basket and stellate cells have long been thought to arise, like granule cells, from the external granule cell layer. This view has first been seriously shattered by a series of experiments in which the chick/quail chimera system was utilized to show that cells of the external granule cell layer give rise exclusively to cerebellar granule cells (Otero et al., 1993; Napieralski and Eisenman, 1993; Hallonet and Le Douarin, 1993). Analysis of the murine meander tail mutant indicated that this may also be true for mammals (Napieralski and Eisenman, 1993). Incidentally, *mea,* which abrogates granule cell colonization of the anterior cerebellar anlage, but spares basket/stellate cell generation in this territory, provides evidence for the separation of the basket/stellate and the granule cell lineages no later that embryonic day 16 (the time the *mea* phenotype becomes evident), without requiring any assumptions about the site of origin of these cells. Transplantation experiments using early postnatal cells of the murine external granule cell layer (Gao and Hatten, 1994) concurred with the conclusions reached by the careful analysis of the *mea* mutant. Essentially by exclusion, these studies suggested that basket and stellate cells are derived from the primary proliferative zone of the cerebellar anlage, the neuroepithelium lining the fourth ventricle. This view was vindicated by the elegant studies of Zhang and Goldman (1996a, 1996b), who also added, however, a most essential detail to our understanding of the genesis of cerebellar inhibitory interneurons. Using direct injections of a replication-deficient retroviral marker in the early postnatal cerebellar anlage of rats, these authors showed that precursors for cerebellar inhibitory interneurons, including those for Golgi neurons and basket/stellate cells, proliferate while in transit from the neuroepithelial lining of the fourth ventricle to their final position within the cerebellar cortex. The deep cerebellar mass/white matter of the cerebellar anlage may thus be viewed as a secondary proliferative zone. It is currently not to decide what the lineage relationship between basket/stellate and Golgi neurons might be, although the overlap in the times these neurons are generated (Zhang and Goldman, 1996a) argues against a mode of cell fate commitment analogous to that realized in the retina, where cell fate is decisively influenced by timing of cell cycle withdrawal (Reh and Kljavin, 1989; Belecky-Adams et al., 1996). Finally, as cerebellar neuroglia arises from a population of precursors which also proliferates within the white matter of the developing cerebellum, it is currently not obvious what the lineage relationship between inhibitory cerebellar cortical interneurons and cerebellar neuroglia might be.

In summary, then, these data concur to show that the lineages leading to individual neuronal phenotypes – with the possible but unlikely exception of basket/stellate and Golgi cells – get separated early during cerebellar development, at a phase when essential histogenetic and morphogenetic events shaping the eventual cerebellum are just about to begin. At least at embryonic day 13, and probably considerably earlier, when, in the mouse, the last Purkinje cells become postmitotic and *Math1*-expressing granule cell precursors can be delineated at the rhombic lip, these two lineages are separated from each other, as well as from those cells that will give rise to inhibitory cerebellar cortical interneurons. We do not know what the relationship between these lineages is before this time point. However, the data available suggest that at least the Purkinje cell lineage and granule cell precursors, which are spatially sequestered into the trigonum germinativum, i.e. that area of the incipient cerebellar anlage located along the dorsal midline of the neural tube before its bending into the pontine flexure, are segregated considerably earlier. Given the apparently shared origin of the Purkinje cell lineage and that/those leading to inhibitory cerebellar interneurons in the proliferating neuroepithelium of the roof of the fourth ventricle, it is hard to make, as of this writing, any judicious statement as to the possible lineage relationship between these cells.

An extended view of granule cell specification and lineage restriction

Neurons are terminally postmitotic cells, and cell cycle arrest is generally viewed as one of the earliest, and a most critical event of neuronal differentiation. Besides the significance of proliferation for the generation of an adequate complement of nerve cells, it may also affect cell lineage specification, realized, in all likelihood, through the asymmetric distribution of cytoplasmic constituents among daughter cells (Jan and Jan, 1998). Thus, the susceptibility of forbrain neuroblasts towards cues determining their laminar fate varies in a predictable manner during the mitotic cycle (for a review, see McConnell, 1995; see also Bohner et al., 1997). Similarly, the ultimate fate of retinal neuroblasts is critically dependent on when these cells go through their final mitosis (Reh and Kljavin, 1989; Belecky-Adams et al., 1996).

The realization that the lineages leading to granule cells and inhibitory interneurons of the cerebellar cortex separate early during histogenesis, long before these cells become terminally post-mitotic, opened a new perspective on how cell fate specification and final differentiation may relate to the mitotic history of these cells. Indeed, the maturation of these cells can be followed over a protracted period of time of about three weeks, from the earliest delineation of their precursors to their final migration and synaptic integration. The unique spatial disposition of maturing cerebellar granule cell neuroblasts, which effectively results in the spatial segregation of cells in distinct developmental stages, in particular, provides a unique opportunity to ask when these cells acquire their fate, and possibly how cell fate specification and its realization may be mechanistically explained on a molecular genetic level.

In order to address these issues, Alder et al. (1996) isolated precursor cells from the rhombic lip of embryonic day 14 mice and grew them at high density (i.e. in aggregate cultures). Cultured rhombic lip precursors were found to express the zinc-finger protein RU 49, which, within the cerebellum, is specifically expressed by granule cells and their precursors, but failed to express any other trait characteristic of granule cells (Alder et al., 1996). In order to allow further development of these cells along the path a granule neuron normally takes in situ, they had to be cultured together either with granule cell precursors isolated from the external granule cell layer of 17 days or older embyonic mice, or with granule cell membranes prepared thereof, or they had to be transplanted into neonatal cerebella.. Together, these findings were interpreted as indicating that embryonic day 14 rhombic lip cells are specified to a granule cell neuron identity. It may be noted, however, that the meaning of term specification, as used by these authors, is somewhat unconventional. According to textbook definition, "a group of cells is called specified if, when isolated and cultured in the neutral environment of a simple culture medium away from the embryo, they develop more or less according to their normal fate" (Wolpert et al., 1998, p. 15). Alder et. al. (1996; cf. also Hatten et al., 1997), in contrast, clearly distinguish between cell specification and the competence of a cell to realize the fate it was specified to. So when do granule cell precursors acquire this competence?

A great number of studies have established that cerebellar granule cells obtained from young post-natal mice do indeed develop under reductionistic in vitro conditions into what may be rightfully called bona fide granule cells (e.g. Cox et al., 1990; Gao et al., 1991; Vicario-Abejon et al., 1995; Jankovski et al., 1996; Fletcher et al., 1996). Yet the early postnatal cerebellum contains granule cells in very different developmental stages, from actively proliferating granule cell neuroblasts to post-mitotic, postmigratory granule cells. To test whether granule cell neuroblasts are competent to develop properly if deprived of their normal tissue context prior to their last mitosis, we dissociated late embryonic (embryonic day 17) cerebellar anlage and analyzed granule cell development under controlled in vitro conditions (Baader et al., 1999). At this age, granule cell precursors have not yet become postmitotic in vivo. Upon dissociation and cultivation, granule cell neuroblasts immediately stopped proliferating. Yet although these cells had left the proliferating cell pool at least 3–4, and probably many more cell cycles prematurely, they developed into genuine granule neurons as judged by the expression of specific molecular markers and the acquisition of a typical morphology, including the discrete, cell autonomously regulated, posititioning of their afferent synapses to their (distal) dendrites. Besides differentiating granule cells, dissociated cultures derived from embryonic day 17 murine embryos contained a considerable complement of neuronal precursors that did not differentiate, and indeed died within about 1–2 weeks in vitro. In striking contrast to those granule cells that differentiated in vitro, those that did not incorporated BrdU when exposed to this agent immediately following isolation from the cerebellar anlage (Baader et al., 1999). From these experiments, it may be concluded that the competence of cerebellar granule cells to differentiate appropriately is independent of the precise timing of their final mitosis; however, their sensitivity towards extrinsic developmental signals appears to vary in a cell cycle-dependent manner. Whereas cerebellar neuroblasts isolated before the G1 restriction point, which do not replicate their DNA in vitro, are capable to develop into bona fide cerebellar granule cells independently of the completion of a predetermined number of cell cycles, those that go through S-phase in the reductionistic in vitro environment are not competent to realize their full developmental potential, and eventually die. These latter cells appear to be particularly sensitive towards signals received, or missing, during dissociation or in the reductionistic in vitro environment (for details and further discussion, see Baader et al., 1999).

Together, these data suggest that murine cerebellar granule cell precursors acquire the competence to autonomously realize the cell fate they are specified to while in transit through the external granule cell layer, between embryonic days 14 and 17. However, it appears that, at this age, granule cell precursors are not yet determined (or committed) irrevocably, in the sense that their fate cannot be changed by external signals.

Mitosis and migration

While in most neuronal populations neurons start to migrate to their definitive positions only after their final mitosis, both cerebellar granule cells and inhibitory interneurons of the cerebellar cortex derive from precursors that migrate extensively while actively dividing. The translocation of granule cell precursors from the rhombic lip through the external granule cell layer and into the (internal) granule cell layer may be clearly divided into two distinct phases: While migrating tangentially to along the surface of the cerebellar anlage in transit through the external granule cell layer, granule cell precursors divide; however, they embark on the final leg of their journey only after their terminal mitosis. The transition between these two migratory phases coincides, in *weaver* mutant mice, with the time point at which the phenotype of this mutation becomes apparent, resulting in the death of early postmitotic granule cells in the external granule cell layer (Smeyne and Goldowitz, 1989). Another molecular marker recently associated with postmitotic, premigratory granule cells is the helix-loop-helix gene, *NSCL*–2 (Haire and Chiaramello, 1996).

The mechanisms and structural components subserving tangential migration of granule cell precursors within the external granule cell layer remain largely enigmatic. Recently, the *mea*, *rcm*, and *zic1* genes have been associated with this first phase of granule neuroblast proliferation and/or migration (Napieralski and Eisenman, 1996; Hamre and Goldowitz, 1997; Aruga et al., 1998; Eisenman and Brothers, 1998). The molecular identity of the *mea* gene product is currently not known. *rcm* encodes an UNC–5 like receptor for the netrin ligand (Ackerman et al., 1997), a diffusible factor regulating cell migration and axonal pathfinding (see Tessier-Lavigne and Goodman, 1996 for a review). Zic1 is a zinc finger protein the expression of which is highly restricted to cerebellar granule cells and their precursors (Aruga et al., 1998). However, the functions of these molecules in granule cell neuroblast proliferation and/or migration is currently not known. Ultrastructural studies have shown close contacts of proliferating, migrating granule cell neuroblasts with the basement membrane of the meninges covering the cerebellar anlage, an association that can no longer be demonstrated for granule cells in the deeper parts of the external granule cell layer, which have withdrawn from the cell cycle. This has led to the suggestion that the contact of granule cell neuroblasts with the basement membrane may be causally related to their migratory status and/or their mitotic activity (Hausmann and Sievers, 1985). However, there is currently no direct experimental evidence to support this inference. The interaction of granule cell neuroblasts with components of the extracelluar matrix, as well as direct cell–cell interactions among these cells, however, might be worthwhile further scrutiny, as suggested by some of the riddles still posed by the *weaver* mutant, despite the cloning of this gene (Patil et al., 1995) and the electrophysiological characterization of its protein product (Slesinger et al., 1996; Kofuji et al., 1996; Navarro et al., 1996; Surmeier et al., 1996).

Defining the transition from proliferation to radial, postmitotic migration in granule cells

As mentioned above, the *weaver* mutation manifests itself just as cerebellar neuroblasts become postmitotic (Smeyne and Goldowitz, 1989). The mRNA encoding the Girk2 ion channel defective in this mutant, however, is expressed by granule cell neuroblasts dividing in and migrating through the external granule cell layer well before that time point. Whether this delayed action of the weaver gene is due to deferred mRNA translation or some other posttranscriptional mechanism is currently not known. Cell culture studies (Gao et al., 1992) have shown that the weaver gene does not act cell-autonomously; i.e. weaver granule cells can be rescued from their normally calamitous fate by contact with normal granule cells, or membrane fractions prepared from wild-type cerebellar granule neurons (Gao et al., 1992). Following characterization of the weaver gene product, these observations have been interpreted as evidence that wild-type granule cells can activate a signal transduction pathway that suppresses the action of the Girk2 channel (Kofuji et al., 1996). The molecular structure of this (hypothetical) signal transduction pathway remains elusive. Like granule cell membranes, aprotinin, a Kunitz-type proteinase inhibitor, was found to rescue cultured weaver granule cell neuroblasts (Murtomäki et al., 1995). These data, published before the molecular characterization of the weaver mutant, had been construed as indicating that the increased proteolytic activity observed in cultured weaver granule neurons is causally related to the cell death seen in this mutant (Murtomäki et al., 1995). While proteolysis sensitive to extracellularly applied aprotinin (i.e. proteolysis mediated by secreted serine proteases) may or may not play a role in weaver phenotype, the restoration of the resting membrane potential brought about by aprotinin in weaver granule cell neuroblasts can hardly be explained by its proteinase inhibitory action, given the structural defect of the Girk2 channel in weaver mice. It is currently not to decide to what extent the rescue of weaver granule cell neuroblasts mediated by wild-type granule cell membranes and aprotinin share a common mechanism, i.e. whether and when the signal transduction pathways impinged upon by granule cell membranes and aprotinin converge to assure cell survival. A lead, however may come from the observation that a class of cell surface transmenbrane proteins, the Alzheimer β-amyloid precursor protein (for a short review, see Selkoe, 1996) and its relative, APLP2 (Wasco et al., 1993), which are expressed widely in the central nervous system and also the developing cerebellum (e.g. Löffler and Huber, 1992; Ouimet et al., 1994; Slunt et al., 1994), as well as a growing group of neuron-associated extracellular matrix molecules, including collagens VI and VII (Greenspan, 1993; Du et al., 1996; Kohlfeldt et al., 1996), share a common structural motif with aprotinin, referred to as the Kunitz type proteinase inhibitor (KPI) domain (Fig. 1). Intriguingly, Kunitz type proteinase inhibitors are structurally highly homologous to naturally occurring blockers of potassium channels, such as dendrotoxins (Dufton, 1985; Skarzynski, 1992; Lancelin et al., 1994; Schweitz et al., 1995; Kohlfeldt et al., 1996) and to a blocker of high voltage-activated Ca^{2+}-channels (calciclu-dine; Schweitz et al., 1994). Thus, it might be tempting to speculate whether the interaction of (a) cell surface/extracellular matrix protein(s) containing domains homologous to naturally occurring ion channel blockers with ionic conductances expressed by granule cells and/or (b) their precursors might explain some of the enigmas still inherent to weaver. While this is currently plain speculation, based largely on protein structural resemblances, it offers direct predictions about the weaver cell biology amenable to experimental scrutiny.

Mice in which either SDF–1 (stromal cell-derived factor; a chemokine regulating proliferation and migration of leukocytes and lymphocytes; see Zuo et al., 1998; Ma et al., 1998) and references therein) or its receptor, CXCR4 (Zuo et al., 1998; Ma et al., 1998) are rendered null by homologous recombination have recently shed some more light on the transition from proliferation in the external granule cell layer to postmitotic radial migration. Within the cerebellum, mice lacking SDF–1 or CXCR4 expression show identical phenotypes, pinpointing that the two mutations indeed define a highly specific, non-redundant signaling pathway. In these animals, the inward migration of granule cell precursors from the external granule cell layer starts prematurely, and granule cell neuroblasts continue to proliferate even after leaving the external granule cell layer (Zuo et al., 1998). These findings suggest that the SDF–1/CXCR4-defined signaling pathway is critical for the orderly transition from proliferation to postmitotic differentiation and migration in cerebellar granule cells. Thus, besides Girk2, SDF–1 and CXCR4 may provide a molecular handle to dissect the cell biology of this critical step of granule cell development. Specifically, these mutants suggest that whatever drives that migration of external granule cell layer precursors towards their final location in the (internal) granule cell layer must already be operative in the embryonic cerebellar anlage, well before the first granule cell neuroblasts become postmitotic and start to migrate inwardly past Purkinje cells. The SDF–1/CXCR4-defined signaling pathway seems necessary to counterbalance this driving force and to prevent precocious inward migration of granule cell precursors prior to cell cycle arrest. It is intriguing to ask whether changes in SDF–1/CXCR4 signaling correlate with the stop of proliferation and beginning inward migration of cerebellar granule cells, and whether SDF–1/CXCR4 signaling pathway interacts with the Girk2 channel.

In contrast to tangential migration, radial inward migration of postmitotic granule cells has been extensively studied (for reviews, see Hatten and Heintz, 1995; Goldowitz and Hamre, 1998; Komuro and Rakic, 1998b). It is well established that Bergmann glia provide a rail along which early postmitotic granule cells ride through the molecular layer; what guides these cells through the Purkinje cell layer, and further to their final positions in the (internal) granule cell layer remains less clearly defined. Bergmann glial processes, which extend across the Purkinje cell layer into the (upper)

```
BPTI            RPDFCLEPPYTGPCKARIIR---YFYNAKAGLCQTFVYGGCRAKRNNFKSAEDCMRTCGGA.
IαTI            KEDSCQLDHAQGPCLGMISR---YFYNGTSMACETFQYGGCLGNGNNFASQKECLQTCRTVAA

α-Den           RRKLCILHRNPGRCYDKIPA---FYYNQKKKQCERFDWSGCGGNSNRFKTIEECRRTCIG..
Den K           AAKYCKLPLRIGPCKRKIPS---FYYKWKAKQCLPFDYSGCGGNANRFKTIEECRRTCVG..
Den I           LRKLCILHRNPGRCYQKIPA---FYYNQKKKQCEGFTWSGCGGNSNRFKTIEECRRTCIR..

AsKC1           INKDCLLPMDVGRCRASHPR---YYYNSSSKRCEKFIYGGCRGNANNFHTLEECEKVCGVR
AsKC2           INKDCLLPMDVGRCRARHPR---YYYNSSSRRCEKFIYGGCRGNANNFITKKECEKVCGVR
AsKC3           INGDCELPKVVGRCRARFPR---YYYNLSSRRCEKFIYGGCGGNANNFHTLEECEKVCGVRS

Calci           PPWYCKEPVRIGSCKKQFSS---FYFKWTAKKCLPFLFSGCGGNANRFQTIGECRKKCLGK

APP             VREVCSEQAETGPCRAMISR---WYFDVTEGKCAPFFYGGCGGNRNNFDTEEYCMAVCG...
APLP2              VCSQEAMTGPCRAVMPR---WYFDLSKGKCVRFIYGGCGGNRNNFESEDYCMAVCKAMI

colα3(VI)        TDICKLPKDEGTCRDFILK---WYYDPNTKSCARFWYGGCGGNENKFGSQKECEKVCA
colα1(VII)      SDDPCSLPLDEGSCTAYTLR---WYHRAVTGACHPFVYGGCGGNANRFGTREACERRC...
MEC-9L ku1      NEDICLEDVDPGPCQYYQVQ---WFWDKQVEECKEFHYGGCMGTKNRFSSKQQCVKQCKYKM
MEC-9L ku2      VPDLCLLDADQGHCGDERNGHWWYFFNQESGECEKFFYYGCGGNDNKFYSLHMCRKVCGERL
                    C        C                  C  F   GC GN N F    C   C
```

Fig. 1. Sequence alignment of the Kunitz-type proteinase inhibitor (KPI) domains found in proteinase inhibitors, ion channel blockers, Alzheimer type (cell surface) proteins, and extracellular matrix molecules. Sequences were retrieved from the GeneBank data base. They include those of the prototypical proteinase inhibitors aprotinin (BPTI, bovine pancreatic trypsin inhibitor) and inter-α-trypsin inhibitor (horse; IαTI); those for the dendrotoxins α (α-Den), K (Den K) and I (Den I), highly effective blockers of voltage gated potassium channels; the kalicludines 1, 2 and 3 (AsKC 1, AsKC2, AsKC3), which show both potassium channel blocking activity and trypsin inhibitory action; and the sequence of the calcium channel blocker, calcicludine (Calic). Moreover, the KPI domains of the Alzheimer ß-amyloid precursor protein (APP) and its relative, amyloid precursor like protein 2 (APLP2), collagen α(VI) (colα(VI)) and collagen α(VII) (colα(VII), and the first two out of five KPI domains of the C. elegans neuron-associated collagen, MEC–9L (MEC–9L ku1, MEC–9L-ku2) are shown. The bottom line lists those residues that are 100% conserved by all of the above sequences. For references, see text.

internal granule cell layer during the phase of granule cell migration (Rakic, 1971; see Komuro and Rakic, 1998a for further discussion) might also guide granule cells past Purkinje cell bodies, yet which structure(s) guide the positioning of granule cells within the deeper parts of the granule cell layer remains elusive (Komuro and Rakic, 1998a). Recently, some of the molecular–genetic rules regulating the interplay between granule cells and its guiding glia have begun to emerge. Granule cells, and also proliferating granule cell neuroblasts, synthesize neuregulin, which acts through ErbB4, a tyrosine kinase receptor localized on cerebellar glia, to regulate granule cell migration and also to provide morphogenetic support for cerebellar cortical glia (Rio et al., 1997). Another molecule synthesized by postmitotic granule cells that mediates interaction with and migration along Bergmann glia is astrotactin (Zheng et al., 1996). The receptor of this cell-surface molecule awaits identification; it is noteworthy, however, that astrotactin contains three epidermal growth factor-like (EGF-like) domains, and that the biological action of neuregulin as well as its ErbB4 binding site were localized to the EGF-like domain present in all known splice variants of this protein (Rio et al., 1997; Zhang et al., 1997; Fischbach and Rosen, 1997). Finally, it should be stressed that, rather than simply providing a rail for migrating granule cells, the interaction between Bergmann glia and maturing granule cells constitutes a intricate, reciprocal relationship. That the development of these two cellular phenotypes are tightly linked was known at least from the early descriptions of the weaver mutant, and it has stimulated considerable debate about the primary cellular lieu of action of this mutation.

In an impressive series of experiments, Komuro and Rakic (Komuro and Rakic, 1992, 1993, 1995, 1996, 1998a; for a review see Komuro and Rakic, 1998b) have established that neurotransmitter receptor and ion channel activation, translated intracellularly into patterns of fluctuating levels of ionized calcium, play a significant role in regulating the speed of granule cell migration through the molecular layer. An open issue at this time is what provides the directionality of granule cell translocation from the external to the internal granule cell layer. Clearly, Bergmann glia do not impart polarity to the path taken by granule cells (see Hatten, 1993). A parsimonious explanation would be that granule cells simply move down a concentration gradient of those agents also regulating mobility (speed), which might be provided by mossy fibers (glutamate) of Golgi neurons (GABA), both of which are present in the nascent internal granule cell layer.

Overall, the molecular mechanisms known to be involved in granule cell migration share striking similarities, and indeed overlap, with activity driven/neurotransmitter mediated modulation of neurite outgrowth (Shatz, 1990; Kater and Mills, 1991; Katz and Shatz, 1996). Specifically, granule cell migration may be likened to dendritogenesis: like granule cell translocation, dendritogenesis is highly sensitive to neurotransmitters and ion channel modulation (e.g. Schilling et al., 1991; Metzger et al., 1998). Moreover, the leading process of migrating granule cells, within which the cell nucleus relocates to effect migration (Hager et al., 1995; Komuro and Rakic, 1998a) is molecularly clearly distinct from the trailing axon, as revealed by the polarized expression of cell adhesion molecules (cf. Rivas and Hatten, 1995; Powell et al., 1997 and references cited therein). Intriguingly, granule cell migration also ensues a clear spatial segregation of cellular domains involved in axon formation and cell migration (Ramon y Cajal, 1995; Rivas and Hatten, 1995; Powell et al., 1997). Consequently, it has been argued, (migrating) granule cells provide a unique model to study early events of dendritogenesis and contrast them to those subserving axonal extension (Powell et al., 1997).

Do cerebellar granule cells form a homogenous neuronal population?

The stereotyped morphology and wiring of cerebellar granule cells, their sheer numbers, and their dense packing may all have contributed to the long held feeling that they represent a highly homogeneous class of cells. More recently, neurochemical and molecular characterization of granule cells has somewhat challenged this view.

Besides glutamate, cerebellar granule cells utilize nitric oxide as a signaling molecule. However, based on the levels of expression of the neuronal isoform of nitric oxide synthase (nNOS), granule cells may be grouped in two categories, those expressing nNOS, and those that do not express this enzyme at levels detectable by NADPH-diaphorase histochemistry or immunocytochemistry (Schilling et al., 1994; Hawkes and Turner, 1994). In adult rodents, differentially nNOS-expressing granule cells are grouped in discrete clusters. Patterned expression of nNOS is developmentally regulated (Schilling et al., 1994), and may be correlated to the ingrowth of mossy fibers in the nascent (internal) granule cell layer. In the adult animal, distinct groups of nNOS positive and negative granule cells can be correlated with the distribution discrete spinocerebellar projections (Li et al., 1997). Cell culture experiments established that glutamatergic input, mediated through both NMDA and non-NMDA ionotropic glutamate receptors can down regulate nNOS expression in developing cerebellar granule neurons, and that the sensitivity of nNOS expression towards extrinsic signals, which also comprise neurotrophins and GABA, changes dynamically as granule cells mature (Baader and Schilling, 1996; Baader et al., 1997). Intracellular calcium has been recognized as an important signal transduction molecule in the cell-extrinsically driven regulation of nNOS expression in developing granule cells. Yet how might mossy fibers, all of which utilize glutamate as a transmitter, result in the uneven expression of nNOS in granule cells?

The NO hypothesis of neural development (Gally et al., 1990) posits that NO is involved in the stabilization of neuronal connectivity, such that (retrograde) NO signaling contributes to the translation of patterns, initially manifested by coherent activity of ingrowing afferents, into the final structure of the developing neuronal circuit. In accordance with this hypothesis, one might surmise whether patterned activity of mossy fibers during development gets translated into the differential nNOS expression seen in the adult (cf. Baader and Schilling, 1996; Baader et al., 1997 for further details). However, in order to explain how this patterned expression of nNOS becomes fixed (i.e. insensitive to mossy fiber input) beyond the phase of cerebellar development, one has to postulate a mechanism that uncouples mossy fiber input from nNOS expression. How this mechanism might be realized in terms of molecular constituents is not known.

Alternatively, the patterned expression of nNOS following afferent innervation of granule cells might reflect differences in the internal responsiveness of these cells. Are there any molecular clues to suggest such a difference? In a recent review summarizing molecular aspects of cerebellar histogenesis (Hatten et al., 1997), it has been mentioned in passing that a number of genes are differentially expressed in subpopulations of cerebellar granule cells, yet neither the nature of these genes nor the profile of their expression has so far been published (PubMed review as of January 10, 1999). The *lyn* gene, which codes for a Src-family protein tyrosine kinase, is expressed by discrete subsets of adult cerebellar granule cells, and indeed its pattern of expression is strikingly reminiscent of that reported for nNOS, such that nNOS and Lyn appear to be expressed in complementary subsets of granule neurons (Chen et al., 1996). However, so far a direct, double-labeling based comparison has not been reported. The developmental emergence of patterned Lyn expression during cerebellar development is distinct from the of nNOS: Lyn is already weakly expressed in external granule cell layer neuroblasts (which do not express nNOS; Schilling et al., 1994), and expression remains weak in early postmigratory granule neurons (Chen et al., 1996). It is only during the third postnatal week that patches of strongly Lyn-positive granule cells emerge. The appeal to link nNOS expression to that of *lyn* stems from the fact that AMPA receptors have been found to interact with and signal through Lyn in the cerebellum to regulate gene expression, such as that of the neurotrophin, brain derived neuronotrophic factor (BDNF; Hayashi et al., 1999). Thus, differential expression of *lyn* appears to reflect distinct receptive, signal transduction, and/or signaling capacities of subsets of granule cells.

The shaping of cerebellar granule cells

It has already been mentioned above that granule cell morphogenesis occurs concurrently with its translocation, and that this provides for an intriguing model to separate molecular aspects of axon extension and early dendritogenesis (Powell et al., 1997). A crucial aspect of dendritogenesis concerns the proper arrangement of afferent synapses along the somatodendritic surface and the coordinated sorting of membrane receptors. An outstanding feature of cerebellar granule neurons is that their afferent synapses are strictly localized to the ends of their dendrites, claw-like structures that interact with glutamatergic mossy fibers and inhibitory, GABAergic axons of Golgi neurons in highly organized synaptic complexes, termed glomeruli. Granule cell bodies, which are tightly packed, are notoriously devoid of afferent synapses. In order to begin to understand the basic mechanism which subserve the specific localization of afferent granule cell synapses, we studied the synaptic investment of cerebellar granule cells that had developed under controlled in vitro conditions in primary dissociated cultures (Baader et al., 1999). Cultured granule neurons can be identified by the expression of neuronal nitric oxide syntheses or $GABA_A$ receptor subunit α_6, and they share a number of physiological and pharmacological traits typical for these cells in situ (Baader and Schilling, 1996; Powell et al., 1997; Jones et al., 1997; Baader et al., 1999 and references cited therein), including the expression of NMDA, non-NMDA and $GABA_A$ receptors (Cox et al., 1990; Mathews et al., 1994; Caruncho et al., 1995; Baader and Schilling, 1996; Vallano et al., 1996; Baader et al., 1997). Yet the topographical associations that form between cultured granule cells and other major cerebellar cortical neurons (and glia) present in these cultures (Schilling et al., 1991; Baader and Schilling, 1996) appear strikingly distinct from those in situ. In particular, the regular laminar typical for the cerebellum is not reconstituted in dissociated culture, and granule cells, identified by the expression of neuronal nitric oxide syntheses or $GABA_A$ receptor subunit α_6 come to lie in small groups, or as single cells in such cultures (Schilling et al., 1994; Baader and Schilling, 1996; Jones et al., 1997). Cultured granule cells are integrated in a spontaneously active neuronal network comprising inhibitory, GABAergic and excitatory, glutamatergic neurons (Schilling et al., 1991). Significantly, however, the sources of cognate extracerebellar, excitatory afferents of granule cells, i.e. mossy fibers, are not present in cerebellar cultures. Thus, excitatory input of granule cells in dissociated cultures is heterologous, and actually provided by other granule cells. Golgi neurons, which provide inhibitory input to granule cells in situ, may be supposed to be present in primary dissociated cerebellar cultures; yet Golgi cells are not autonomous in selecting their postsynaptic targets (Sotelo, 1982; Takayama, 1994). Consequently, the finding that, like in situ, afferent synapses of granule cells in culture do not contact cell bodies, but are confined to neurites (Baader et al., 1999), strongly suggest that the specific localization of granule cell afferent synapses is independent of the specific topographic disposition of granule cells in vivo, and likewise independent of the source(s) providing these afferents. Thus, cerebellar granule cells appear to be able to intrinsically and autonomously direct the subcellular arrangement of their afferent synapses. While it may be presumed that this relates to the discrete subcellular distribution of membrane proteins attracting/orchestrating synaptogenesis, the nature of such protein(s) is currently not known.

An intriguing question is how the subcellular localization of granule cell afferent synapses mechanistically relates to the subcellular disposition of neurotransmitter receptors. Analysis of $GABA_A$ receptors in cerebellar granule cells may provide an inroad to approach this issue. Despite the impressive segregation of afferent synapses to the very distal parts of the dendrites of these cells, $GABA_A$ receptors are also found in the membranes of dendritic shafts and granule cell bodies (Nusser et al., 1998). Subsynapically and non-synaptiallly localized $GABA_A$ receptors differ in their precise subunit composition in an intricate and so far, incompetely understood way (e.g. Nusser et al., 1995; Jones et al., 1997; Nusser et al., 1998). It should be noted that the subunit composition of $GABA_A$ receptors defines not only their precise electrophysical properties, but may also directly relate to the subcellular sorting and anchoring of these receptors (Essrich et al., 1998), and hence provide a link to those mechanisms regulating synapse location. While the discrete disposition of functional and molecular variants of $GABA_A$ receptors to synapses and extrasynaptic membranes may play a crucial role in orchestrating the activity of groups of granule cells which send a dendrite to a common glomerulus (Rossi and Hamann, 1998), the functional significance of the discrete subcellular segregation of afferent synapses and individual receptors for individual cells is less well understood. In fact, granule cells have been described as electrically very compact (Gabbiani et al., 1994), suggesting that the segregation of afferents and receptors might serve to compartmentalize granule cells biochemically. The small size of granule neurons and the tangled arrangement of their dendrites in glomeruli has so far precluded to address this issue experimentally. However, there is precedence from other central nervous neurons (e.g. Müller and Connor, 1991), the best studied example of which is, indeed, the postsynaptic target of granule cells, the cerebellar Purkinje neuron (e.g. Denk et al., 1995; Bian et al., 1996; Wanner et al., 1997; Takechi et al., 1998).

Basket/stellate cells en route

Other than that they divide while in transit through the white matter of the cerebellar anlage, we know essentially nothing about the mechanisms that regulate the proliferation of basket/stellate cell precursors. When do these cells leave the neuroepithelium of the fourth ventricle? Is their proliferation related to the cellular microenvironment they migrate through? E.g. do these cells stop to proliferate once they migrate past the white matter in the incipient (internal) granule cell layer? Is the number of basket/stellate cells matched to that of their postsynaptic targets, the cerebellar Purkinje neurons? If so, how?

Our understanding of mechanisms subserving basket/stellate cell precursor migration is also very rudimentary. Understandably, the long lasting uncertainties about the site of origin of these cells did not make them an attractive object of analysis for neuronal migration. The recent findings documenting the radial inside-out migration of these cells (Otero et al., 1993; Napieralski and Eisenman, 1993; Hallonet and Le Douarin, 1993; Zhang and Goldman, 1996a, 1996b), however, raise the intriguing possibility to correlate their migration with the overall histogenesis of the cerebellum, and to contrast it to the radial migration of postmitotic granule cells, which indeed have to translocate in exactly the opposite direction.

Thus, we would like to know which structures provide guideposts for basket and stellate cells as they migrate, subsequently, through the white matter, the (internal) granule cell layer, and across the ganglionic (Purkinje cell) layer into the molecular layer; what is the molecular nature of the entity/entities that regulate(s) translocation of these cells? Purkinje cells are located to provide a thoroughfare through most of these layers, from deep cerebellar nuclei in the white matter to the molecular layer. From the (internal) granule cell layer onward, granule cell axons might serve as a substratum; and beyond the Purkinje cell layer, Bergmann glia have to be considered, to mention only those cellular elements whose major longitudinal axis runs parallel to the expected migratory path of basket and stellate neurons. In weaver, basket/stellate cells migrate towards the cerebellar surface in the absence of granule cell axons. Cultured cerebellar GABAergic interneurons (Hekmat et al., 1989; Magyar-Lehmann et al., 1995), including molecularly identified basket/stellate cells (T. Kosckeck and KS, unpublished observations) have been observed to migrate along granule cell axons and/or astroglial cells, independently of contact with Purkinje cells. These findings do not rule out a role for these structures in basket/stellate cell migration during regular cerebellar development, but they seem to suggest a certain redundancy of the migratory substratum/substrata offered by various cells in the cerebellum, or a degree of promiscuity of basket/stellate cells as far as migratory substrates are concerned.

In the weaver mutant, basket/stellate cells are found intermingled with granule cell precursors which fail to migrate from the external granule cell layer (Rakic and Sidman, 1973). Similarly, in rats in which external granule cell neuroblast migration to the (internal) granule cell layer has been experimentally disrupted and in which granule neurons develop at the surface of the cerebellum or even in the subarachnoid space, basket stellate cells can be found amid these ectopically located (yet morphologically mature appearing) granule cells (Sievers et al., 1983). While this has traditionally been considered an argument for the external granular cell layer origin of molecular layer interneurons (e.g. Rakic and Sidman, 1973; Sievers et al., 1983; Hausmann and Sievers, 1985), it may also point to the fact that granule neurons provide an attractive signal towards which basket/stellate cells migrate. The notion that such a signal might emanate from cerebellar meningeal cells seems less likely, as basket/stellate cells still migrate vs. the cerebellar surface even in animals in which the meninges have been pharmacologically ablated (Sievers et al., 1983; Hausmann and Sievers, 1985), but apparently fail to do so properly in the anterior part of the cerebellum of the *meander tail* mutant (Napieralski and Eisenman, 1993; and L. Eisenman, personal communication). As mentioned above, the anterior part of the cerebellar anlage in this strain of mice is not, or only very sparsely, colonized by external granule cell neuroblasts. If granule cells provide signals(s) that directs basket/stellate cell migration into the molecular layer, it should be sorted preferentially to granule cell axons; otherwise, it would be hard to explain why basket/stellate cells bypass granule cell bodies and dendrites in the (internal) granule cell layer. Moreover, the localization of basket/stellate cells in the external granule cell layer of the weaver mouse indicates that such a signal must already be produced by relatively immature granule cells, clearly before they emanate an axon.

However, it should be kept in mind that a simple attractive force can hardly explain the even distribution of basket/stellate cells within the molecular layer. Intelligibly, basket/stellate cell migration to and subsequent proper disposition within the molecular layer requires more that one single factor.

Morphogenesis of basket/stellate cells

To date, what we know about the morphological development of basket and stellate neurons is based on primarily descriptive analyses, including that of murine cerebellar mutants (e.g. Lemkey-Johnston and Larramendi, 1968; Rakic, 1972; Rakic and Sidman, 1973; Sotelo, 1975). Clearly, dendritic morphology and axonal target areas of these interneurons varies predictably with their vertical position within the molecular layer. Yet presently there is no indication that molecular layer interneurons can indeed be unambiguously classified in two groups, as might be suggested by their names; rather, these names refer to two morphological traits (one describing axonal, the other dendritic morphology) that both appear to vary continuously. If there is a correlation between position and neurite morphology, is there also a causal relationship to it? As the ultimate vertical position of basket/stellate cells within the molecular layer closely correlates with the time point at which these cells underwent their last mitosis ('birth date'), one might wonder whether position and axonal projection of basket/ stellate cells is realized via mechanism(s) similar to the one operative in cortical neurons, where birth date determines laminar fate (McConnell and Kaznowski, 1991; McConnell, 1995). Alternatively, one might envision that the sequentially generated basket/stellate cells, once having passed the Purkinje cell layer, successively occupy those areas of the molecular layer not yet colonized by their siblings. Likewise, later arriving basket/stellate cells would contact those areas of Purkinje cell bodies and smooth dendritic surfaces not yet occupied. A prediction of such a scenario would be that the affinity of basket/stellate cell axons towards the axon initial segment of Purkinje cells should be greater that that towards Purkinje cell bodies, and least towards the (smooth surface) of the Purkinje cell dendritic tree. Indeed, there is evidence that cell adhesion molecules are distributed along the Purkinje cell membrane following such a gradient (Davis et al., 1996; Zhou et al., 1998).

The morphology of basket/stellate cells in weaver mice (Sotelo, 1975) strongly suggests that the development of both dendrites and axons of molecular layer interneurons is dependent on network integrity, and determined at least in part by non-cell autonomous mechanisms. Schachner and collaborators (Hekmat et al., 1989; Magyar-Lehmann et al., 1995) showed that, in vitro, local cues and cell adhesion molecules are involved in the perpendicular orientation of the dendrites of GABAergic cerebellar interneurons, supposedly basket and stellate cells, relative to granule cell axons. Yet still, how the cell-extrinsic regulation of basket/stellate cell morphogenesis and differentiation is realized on a molecular genetic level remains largely to be understood. In particular the intriguing synaptic/axonal projection specificity realized at the pinceau synapse located at the Purkinje cell axon segment, in conjunction with the ready genetic accessability of the cerebellum–Purkinje cells may be targeted via the L7/*pcp*–2 promoter (Oberdick et al., 1990; Vandaele et al., 1991; Oberdick et al., 1993), granule cells using the gene coding the α_6 subunit of the $GABA_A$ receptor (Bahn et al., 1997), and Golgi cells via their rather specific expression of type 2 metabotropic glutamate receptors (Watanabe et al., 1998) – clearly recommend this neuronal circuit as a prolific paradigm to study basic mechanism by which the precise location of individual synapses are realized.

Concluding remarks

Thus despite, or maybe because of, long years of meticulous and fruitful investigations into the cell biology, development, and function of the cerebellum, cerebellar interneurons still confront us with many unsolved and probing question; yet they also offer unique inroads and opportunities to approach issues fundamental not only to our understanding of cerebellar development, but to developmental neurobiology in general and neuronal regeneration alike.

Acknowledgements

I am indebted to numerous colleagues for stimulating discussions and encouraging comments. Work in my lab has been supported by grants from the Deutsche Forschungsgemeinschaft and NATO.

References

Ackerman, S.L., Kozak, L.P., Przyborsky, S.A., Rund, L.A., Boyer, B.B. and Knowles, B.B. (1997) The mouse rostral cerebellar malformation gene encodes an UNC–5-like protein. *Nature*, 386: 838–842.

Akazawa, C., Ishibashi, M., Shimizu, C., Nakanishi, S. and Kageyama, R. (1995) A mammalian helix-loop-helix factor structurally related to the product of drosophila atonal is a positive transcriptional regulator expressed in the developing nervous system. *J. Biol. Chem.*, 270: 8730–8738.

Alder, J., Cho, N.K. and Hatten, M.E. (1996) Embryonic precursor cells from the rhombic lip are specified to a cerebellar granule cell identity. *Neuron*, 17: 389–399.

Altman, J. and Bayer, S.A. (1997) *Development of the Cerebellar System in Relation to its Evolution, Structure, and Functions*, Boca Raton: CRC Press.

Aruga, J., Minowa, O., Yaginuma, H., Kuno, J., Nagai, T., Noda, T. and Mikoshiba, K. (1998) Mouse Zic1 is involved in cerebellar development. *J. Neurosci.*, 18: 284–293.

Baader, S.L., Bergmann, M., Mertz, K., Fox, P.A., Gerdes, J., Oberdick, J. and Schilling, K. (1999) The differentiation of cerebellar interneurons is independent of their mitotic history. *Neuroscience*, 90: 1243–1254.

Baader, S.L., Gruchmann, S. and Schilling, K. (1997) The developmental expression of neuronal nitric oxide synthase in cerebellar granule cells is sensitive to GABA and neurotrophins. *Dev. Neurosci.*, 12: 283–290.

Baader, S.L., Sanlioglu, S., Berrebi, A.S. and Oberdick, J. (1998) Ectopic overexpression of engrailed–2 in cerebellar Purkinje cells: restricted cell loss and retarded external germinal layer development at lobule junctions. *J. Neurosci.*, 18: 1763–1773.

Baader, S.L. and Schilling, K. (1996) Glutamate receptors mediate dynamic regulation of nitric oxide synthase expression in cerebellar granule cells. *J. Neurosci.*, 16: 1440–1449.

Baader, S.L., Schilling, M.L., Rosengarten, B., Pretsch, W., Teutsch, H.F., Oberdick, J. and Schilling, K. (1996) Purkinje cell lineage and the topographic organization of the cerebellar cortex: a view from X inactivation mosaics. *Dev. Biol.*, 174: 393–406.

Bahn, S., Jones, A. and Wisden, W. (1997) Directing gene expression to cerebellar granule cells using γ-aminobutyric acid type A receptor α_6 subunit transgenes. *Proc. Natl. Acad. Sci. USA*, 94: 9417–9421.

Belecky-Adams, T., Cook, B. and Adler, R. (1996) Correlations between terminal mitosis and differential fate of retinal precursor cells in vivo and in vitro: analysis with the 'window-labelling' technique. *Dev. Biol.*, 178: 304–315.

Ben-Arie, N., Bellen, H.J., Armstrong, D.L., McCall, A.E., Gordatze, P.R., Guo, Q., Matzuk, M.M. and Zoghbi, H.Y. (1997) Math1 is essential for genesis of cerebellar granule neurons. *Nature*, 390: 169–172.

Bian, F., Chu, T., Schilling, K. and Oberdick, J. (1996) Differential mRNA transport and the regulation of protein synthesis: selective sensitivity of Purkinje cell dendritic mRNAs to translational inhibition. *Molec. Cell. Neurosci.*, 7: 116–133.

Bohner, A.P., Akers, R.M. and McConnell, S.K. (1997) Induction of deep layer cortical neurons in vitro. *Development*, 124: 915–923.

Caruncho, H.J., Puia, G., Mohler, H. and Costa, E. (1995) The density and distribution of six $GABA_A$ receptor subunits in primary cultures of rat cerebellar granule cells. *Neuroscience*, 67: 583–593.

Chen, S., Ren, Y.Q. and Hillman, D.E. (1996) Transient expression of lyn gene in Purkinje cells during cerebellar development. *Dev. Brain Res.*, 92: 140–146.

Cox, J.A., Felder, C.C. and Henneberry, R.C. (1990) Differential expression of excitatory amino acid receptor subtypes in cultured cerebellar neurons. *Neuron*, 4: 941–947.

Davis, J.Q., Lambert, S. and Bennett, V. (1996) Molecular composition of the node of Ranvier: identification of ankyrin-binding cell adhesion molecules neurofascin (mucin + /third FNIII domain-) and NrCAM at nodal axon segments. *J. Cell Biol.*, 135: 1355–1367.

Denk, W., Sugimori, M. and Llinás, R. (1995) Two types of calcium response limited to single spines in cerebellar Purkinje cells. *Proc. Natl. Acad. Sci. USA*, 92: 8279–8282.

Du, H., Gu, G., William, C.M. and Chalfie, M. (1996) Extracellular proteins needed for C.elegans mechanosensation. *Neuron*, 16: 183–194.

Dufton, M.J. (1985) Proteinase inhibitors and dendrotoxins. Sequence classification, structural prediction and structure/activity. *Eur. J. Biochem.*, 153: 647–654.

Eisenman, L.M. and Brothers, R. (1998) Rostral cerebellar malformation (rcm/rcm): a murine mutant to study regionalization of the cerebellum. *J. Comp. Neurol.*, 394: 106–117.

Essrich, C., Lorenz, M., Benson, J.A., Fritschy, J.-M. and Lüscher, B. (1998) Postsynaptic clustering of major $GABA_A$ receptor subtypes requires the γ2 subunit and gephyrin. *Nature Neurosci.*, 1: 563–571.

Fischbach, G.D. and Rosen, K.M. (1997) ARIA: a neuromuskular junction neuregulin. *Ann. Rev. Neurosci.*, 20: 429–458.

Fletcher, C.F., Lutz, C.M., O'Sullivan, T.N., Shaughnessy, J.D.Jr., Hawkes, R., Frankel, W.N., Copeland, N.G. and Jenkins, N.A. (1996) Absence epilepsy in tottering mutant mice is associated with calcium channel defects. *Cell*, 87: 607–617.

Gabbiani, F., Midtgaard, J. and Knöpfel, T. (1994) Synaptic integration in a model of cerebellar granule cells. *J. Neurophysiol.*, 72: 999–1009.

Gally, J.A., Montague, P.R., Reeke, G.N., Jr. and Edelman, G.M. (1990) The NO hypothesis: possible effects of a short-lived, rapidely diffusible signal in the development and function of the nervous system. *Proc. Natl. Acad. Sci. USA*, 87: 3547–3551.

Gao, W.-Q. and Hatten, M.E. (1994) Immortalizing oncogenes subvert the establishment of granule cell identity in developing cerebellum. *Development*, 120: 1059–1070.

Gao, W.-Q., Heintz, N. and Hatten, M.E. (1991) Cerebellar granule cell neurogenesis is regulated by cell-cell interactions in vitro. *Neuron*, 6: 705–715.

Gao, W.-Q., Liu, X.-L. and Hatten, M.E. (1992) The weaver gene encodes a nonautonomous signal for CNS neuronal differentiation. *Cell*, 68: 841–854.

Goldowitz, D. and Hamre, K. (1998) The cells and molecules that make a cerebellum. *TINS*, 21: 375–382.

Greenspan, D.S. (1993) The carboxy-terminal half of type VII collagen, including the non-collagenous NC–2 domain and intron/exon organization of the corresponding region of the COL7A1 gene. *Hum. Mol. Genet.*, 2: 273–278.

Hager, G., Dodt, H.-U., Ziegelgänsberger, W. and Liesi, P. (1995) Novel forms of neuronal migration in the rat cerebellum. *J. Neurosci. Res.*, 40: 207–219.

Haire, M.F. and Chiaramello, A. (1996) Transient expression of the basic helix-loop-helix protein NSCL–2 in the mouse cerebellum during postnatal development. *Mol. Brain Res.*, 36: 174–178.

Hallonet, M.E.R. and Le Douarin, N.M. (1993) Tracing neuroepithelial cells of the mesencephalic and metencephalic alar plates during cerebellar ontogeny in quil-chick chimaeras. *Eur. J. Neurosci.*, 5: 1145–1155.

Hamre, K.M. and Goldowitz, D. (1997) Meander tail acts intrinsic to granule cell precursors to disrupt cerebellar development: analysis of meander tail chimeric mice. *Development*, 124: 4201–4212.

Hatten, M.E. (1993) The role of migration in central nervous system neuronal development. *Curr. Op. Neurobiol.*, 3: 38–44.

Hatten, M.E., Alder, J., Zimmerman, K. and Heintz, N. (1997) Genes involved in cerebellar cell specification and differentiation. *Curr. Op. Neurobiol.*, 7: 40–47.

Hatten, M.E. and Heintz, N. (1995) Mechanisms of neural patterning and specification in the developing cerebellum. *Ann. Rev. Neurosci.*, 18: 385–408.

Hausmann, B. and Sievers, J. (1985) Cerebellar external granule cells are attached to the basal lamina from the onset of migration up to the end of their proliferative activity. *J. Comp. Neurol.*, 241: 50–62.

Hawkes, R., Faulkner-Jones, B., Tam, P. and Tan, S.S. (1998) Pattern formation in the cerebellum of murine embryonic stem cell chimeras. *Eur. J. Neurosci.*, 10: 790–793.

Hawkes, R. and Turner, R.W. (1994) Compartmentation of NADPH-diaphorase activity in the mouse cerebellar cortex. *J. Comp. Neurol.*, 346: 499–516.

Hayashi, T., Umemori, H., Mishina, M. and Yamamoto, T. (1999) The AMPA receptor interacts with and signals through the protein tyrosine kinase Lyn. *Nature*, 397: 72–76.

Hekmat, A., Künemund, V., Fischer, G. and Schachner, M. (1989) Small inhibitory cerebellar interneurons grow in a perpendicular orientation to granule cell neurites in culture. *Neuron*, 2: 1113–1122.

Helms, A.W. and Johnson, J.E. (1998) Progenitors of dorsal commissural interneurons are defined by MATH1 expression. *Development*, 125: 919–928.

Herrup, K. (1986) Cell lineage relationships in the development of the mammalian CNS: role of lineage in control of cerebellar Purkinje cell number. *Dev. Biol.*, 115: 148–154.

Herrup, K. and Sunter, K. (1986) Cell lineage dependent and independent control of Purkinje cell number in the mammalian CNS: further quantitative studies of lurcher chimeric mice. *Dev. Biol.*, 117: 417–427.

Herrup, K. and Sunter, K. (1987) Numerical matching during cerebellar development: quantitative analysis of granule cell death in staggerer mouse chimeras. *J. Neurosci.*, 7: 829–836.

His, W. (1891) Die Entwicklung des menschlichen Rautenhirns vom Ende des ersten bis zum Beginn des dritten Monats.

Abhandlungen der königlich sächsischen Gesellschaft der Wissenschaften, Mathematische-physikalische Klasse, 29: 1–74.

Ikeya, M., Lee, S.M.K., Johnson, J.E., McMahon, A.P. and Takada, S. (1997) Wnt signalling required for expansion of neural crest and CNS precursors. *Nature*, 389: 966–970.

Ito, M. (1984) *The Cerebellum and Neural Control*, 1st edn. New York: Raven Press.

Jan, Y.N. and Jan, L.Y. (1998) Asymmetric cell division. *Nature*, 392: 775–778.

Jankovski, A., Rossi, F. and Sotelo, C. (1996) Neuronal precursors in the postnatal mouse cerebellum are fully committed cells: evidence from heterochronic transplantations. *Eur. J. Neurosci.*, 8: 2308–2319.

Jones, A., Korpi, E.R., McKernan, R.M., Pelz, R., Nusser, Z., Mäkelä, R., Mellor, J.R., Pollard, S., Bahn, S., Stephenson, F.A., Randall, A.D., Sieghart, W., Somogyi, P., Smith, A.J.H. and Wisden, W. (1997) Ligand-gated ion channel subunit partnerships: $GABA_A$ receptor α_6 subunit gene inactivation inhibits δ subunit expression. *J. Neurosci.*, 17: 1350–1362.

Joyner, A.L. (1996) Engrailed, Wnt and Pax genes regulate midbrain-hindbrain development. *TIG*, 12: 15–20.

Kater, S.B. and Mills, L.R. (1991) Regulation of growth cone behavior by calcium. *J. Neurosci.*, 11: 891–899.

Katz, L.C. and Shatz, C.J. (1996) Synaptic activity and the construction of cortical circuits. *Science*, 274: 1133–1138.

Kofuji, P., Hofer, M., Millen, K.J., Millonig, J.H., Davidson, N., Lester, H.A. and Hatten, M.E. (1996) Functional analysis of the weaver mutant GIRK2 K^+ channel and rescue of weaver granule cells. *Neuron*, 16: 941–952.

Kohlfeldt, E., Göhring, W., Mayer, U., Zweckstetter, M., Holak, T.A., Chu, M.-L. and Timpl, R. (1996) Conversion of the Kunitz-type module of collagen VI into a highly active trypsin inhibitor by site-directed mutagenesis. *Eur. J. Biochem.*, 238: 333–340.

Komuro, H. and Rakic, P. (1992) Selective role for calcium channels in neuronal migration. *Science*, 257: 806–809.

Komuro, H. and Rakic, P. (1993) Modulation of neuronal migration by NMDA receptors. *Science*, 260 , 95–97.

Komuro, H. and Rakic, P. (1995) Dynamics of granule cell migration: a confocal microscopy study in acute cerebellar slice preparations. *J. Neurosci.*, 15: 1110–1120.

Komuro, H. and Rakic, P. (1996) Intracellular Ca^{2+} fluctuations modulate the rate of neuronal migration. *Neuron*, 17: 275–285.

Komuro, H. and Rakic, P. (1998a) Distinct modes of neuronal migration in different domains of developing cerebellar cortex. *J. Neurosci.*, 18: 1478–1490.

Komuro, H. and Rakic, P. (1998b) Orchestration of neuronal migration by activity of ion channels, neurotransmitter receptors, and intracellular Ca^{2+} fluctuations. *J. Neurobiol.*, 37: 110–130.

Lancelin, J.-M., Foray, M.-F., Poncin, M., Hollecker, M. and Marion, D. (1994) Proteinase inhibitor homologues as potassium channel blockers. *Nature Struct. Biol.*, 1: 246–250.

Lemkey-Johnston, N. and Larramendi, L.M.H. (1968) Morphological characteristics of mouse stellate and basket cells and their neuroglial envelope: an electron microscopic study. *J. Comp. Neurol.*, 134: 39–72.

Li, J., Chen, S., Lin, R.C.S. and Smith, S.S. (1997) Cerebellar nitric oxide synthase is expressed within granule cell patches innervated by specific mossy fiber terminals: a developmental profile. *Dev. Neurosci.*, 19: 274–282.

Löffler, J. and Huber, G. (1992) β-amyloid precursor protein isoforms in various rat brain regions and during brain development. *J. Neurochem.*, 59: 1316–1324.

Ma, Q., Jones, D., Borghesani, P.R., Segal, R.A., Nagasawa, T., Kishimoto, T., Bronson, R.T. and Springer, T.A. (1998) Impaired B-lymphopoiesis, myelopoiesis, and derailed cerebellar neuron migration in CXCR4- and SDF–1-deficient mice. *Proc. Natl. Acad. Sci. USA*, 95: 9448–9453.

Magyar-Lehmann, S., Suter, C.S., Stahel, W. and Schachner, M. (1995) Behaviour of small inhibitory interneurons in early postnatal mouse cerebellar microexplant cultures. *Eur. J. Neurosci.*, 7: 1449–1459.

Mathews, G.C., Bolos-Sy, A.M., Holland, K.D., Isenberg, K.E., Covey, D.F., Ferrendelli, J.A. and Rothman, S.M. (1994) Developmental alteration in $GABA_A$ receptor structure and physiological properties in cultured cerebellar granule neurons. *Neuron*, 13: 149–158.

Mathis, L., Bonnerot, C., Puelles, L. and Nicolas, J.F. (1997) Retrospective clonal analysis of the cerebellum using genetic laacZ/lacZ mouse mosaics. *Development*, 124: 4089–4104.

McConnell, S.K. (1995) Constructing the cerebral cortex: neurogenesis and fate determination. *Neuron*, 15: 761–768.

McConnell, S.K. and Kaznowski, C.E. (1991) Cell cycle dependence of laminar determination in developing neocortex. *Science*, 254: 282–285.

Metzger, F., Wiese, S. and Sendtner, M. (1998) Effect of glutamate on dendritic growth in embryonic rat motoneurons. *J. Neurosci.*, 18: 1735–1742.

Miale, I.R. and Sidman, R.L. (1961) An autoradiographic analysis of histogenesis in the mouse cerebellum. *Exp. Neurol.*, 4: 277–296.

Murtomäki, S., Trenkner, E., Wright, J.M., Saksela, O. and Liesi, P. (1995) Increased proteolytic activity of the granule neurons may contribute to neuronal death in the weaver mouse cerebellum. *Dev. Biol.*, 168: 635–648.

Müller, W. and Connor, J.A. (1991) Dendritic spines as individual neuronal compartments for synaptic Ca^{2+} responses. *Nature*, 354: 73–76.

Napieralski, J.A. and Eisenman, L.M. (1993) Developmental analysis of the external granule cell layer in the meander tail mutant mouse: do cerebellar microneurons have independent progenitors? *Dev. Dynamics*, 197: 244–254.

Napieralski, J.A. and Eisenman, L.M. (1996) Further evidence for a unique developmental compartment in the cerebellum of the meander tail mutant mouse as revealed by the quantitative analysis of Purkinje cells. *J. Comp. Neurol.*, 364: 718–728.

Navarro, B., Kennedy, M.E., Velimirovic, B., Bhat, D., Peterson, A.S. and Clapham, D.E. (1996) Nonselective and

Gβgamma-insensitive weaver K^+ channels. *Science*, 272: 1950–1953.

Nusser, Z., Roberts, J.D.B., Baude, A., Richards, J.G. and Somogyi, P. (1995) Relative densities of synaptic and extrasynaptic $GABA_A$ receptors on cerebellar granule cells as determined by a quantitative immunogold method. *J. Neurosci.*, 15: 2948–2960.

Nusser, Z., Sieghart, W. and Somogyi, P. (1998) Segregation of different $GABA_A$ receptors to synaptic and extrasynaptic membranes of cerebellar granule cells. *J. Neurosci.*, 18: 1693–1703.

Oberdick, J., Baader, S.L. and Schilling, K. (1998) From zebra stripes to postal zones: deciphering patterns of gene expression in the cerebellum. *TINS*, 21: 383–391.

Oberdick, J., Schilling, K., Smeyne, R.J., Corbin, J.G. and Morgan, J.I. (1993) Control of segment-like patterns of gene expression in the mouse cerebellum. *Neuron*, 10: 1007–1018.

Oberdick, J., Smeyne, R.J., Mann, J.R., Zackson, S. and Morgan, J.I. (1990) A promoter that drives transgene expression in cerebellar Purkinje and retinal bipolar neurons. *Science*, 248: 223–226.

Olsson, M., Campbell, K. and Turnbull, D.H. (1997) Specification of mouse telencephalic and mid-hindbrain progenitors following heterotopic ultrasound-guided embryonic transplantation. *Neuron*, 19: 761–772.

Otero, R.A., Sotelo, C. and Alvarado-Mallart, R.-M. (1993) Chick/quail chimeras with partial cerebellar grafts: an analysis of the origin and migration of cerebellar cells. *J. Comp. Neurol.*, 333: 597–615.

Ouimet, C.C., Baerwald, K.D., Gandy, S.E. and Greengard, P. (1994) Immunocytochemical localization of amyloid precursor protein in rat brain. *J. Comp. Neurol.*, 348: 244–260.

Patil, N., Cox, D.R., Bhat, D., Faham, M., Myers, R.M. and Peterson, A.S. (1995) A potassium channel mutation in weaver mice implicates membrane excitability in granule cell differentiation. *Nature Genet.*, 11: 126–129.

Powell, S.K., Rivas, R.J., Rodriguez-Boulan, E. and Hatten, M.E. (1997) Development of polarity in cerebellar granule neurons. *J. Neurobiol.*, 32: 223–236.

Rakic, P. (1971) Neuron-glia relationship during granule cell migration in developing cerebellar cortex. A Golgi and electronmicroscopic study in maccacus rhesus. *J. Comp. Neurol.*, 141: 283–312.

Rakic, P. (1972) Extrinsic cytological determinants of basket and stellate cell dendritic pattern in the cerebellar molecular layer. *J. Comp. Neurol.*, 146: 335–354.

Rakic, P. and Sidman, R.L. (1973) Organization of cerebellar cortex secondary to deficit of granule cells in weaver mutant mice. *J. Comp. Neurol.*, 152: 133–161.

Ramon y Cajal, S. (1995) *Histology of the Nervous System. Volume I*, New York: Oxford University Press.

Reh, T.A. and Kljavin, I.J. (1989) Age of differentiation determines rat retinal germinal cell phenotype: induction of differentiation by dissociation. *J. Neurosci.*, 9: 4179–4189.

Rex, M., Church, R., Tointon, K., Ichihashi, R.M.A., Mokhtar, S., Uwanogho, D., Sharpe, P.T. and Scotting, P.J. (1998) Granule cell development in the cerebellum is punctuated by changes in Sox gene expression. *Mol. Brain Res.*, 55: 28–34.

Rio, C., Rieff, H.I., Qi, P. and Corfas, G. (1997) Neuregulin and erbB receptors play a critical role in neuronal migration. *Neuron*, 19: 39–50.

Rivas, R.J. and Hatten, M.E. (1995) Motility and cytoskeletal organization of migrating cerebellar granule neurons. *J. Neurosci.*, 15: 981–989.

Rossi, D.J. and Hamann, M. (1998) Spillover-mediated transmisstion at inhibitory synapses promoted by high affinity α_6 subunit $GABA_A$ receptors and glomerular geometry. *Neuron*, 20: 783–795.

Schilling, K., Dickinson, M.H., Connor, J.A. and Morgan, J.I. (1991) Electrical activity in cerebellar cultures determines Purkinje cell dendritic growth patterns. *Neuron*, 7: 891–902.

Schilling, K., Schmidt, H.H.H.W. and Baader, S.L. (1994) Nitric oxide synthase expression reveals compartments of cerebellar granule cells and suggests a role for mossy fibers in their development. *Neuroscience*, 59: 893–903.

Schweitz, H., Bruhn, T., Guillemare, E., Moinier, D., Lancelin, J.-M., Beress, L. and Lazdunski, M. (1995) Kalicludines and kaliseptine. Two different classes of sea anemone toxins for voltage-sensitive K^+ channels. *J. Biol. Chem.*, 270: 25121–25126.

Schweitz, H., Heurteaux, C., Bois, P., Moinier, D., Romey, G. and Lazdunski, M. (1994) Calcicludine, a venom peptide of the Kunitz-type protease inhibitor family, is a potent blocker of high threshold Ca^{2+} channels with a high affinity for L-type channels in cerebellar granule neurons. *Proc. Natl. Acad. Sci. USA*, 91: 878–882.

Selkoe, D.J. (1996) Amyloid β-protein and the genetics of Alzheimer's disease. *J. Biol. Chem.*, 271: 18295–18298.

Shatz, C.J. (1990) Impulse activity and the patterning of connections during CNS development. *Neuron*, 5: 745–756.

Sievers, J., Mangold, U. and Berry, M. (1983) 6-OHDA-induced ectopia of external granule cells in the subarachnoid space covering the cerebellum. *Cell Tissue Res.*, 230: 309–336.

Simon, H.H., Wurst, W. and O'Leary, D.D.M. (1996) En–1 and En–2 control midbrain development in a gene dose-dependent manner. *Soc. Neurosci. Abstr.*, 22: 528–528.(Abstract)

Skarzynski, T. (1992) Crystal structure of α-dendrotoxin from the green mamba venom and its comparison with the structure of the bovine pancreatic trypsin inhibitor. *J. Mol. Biol.*, 224: 671–683.

Slesinger, P.A., Patil, N., Liao, Y.J., Jan, Y.N., Jan, L.Y. and Cox, D.R. (1996) Functional effects of the mouse weaver mutation on G protein-gated inwardly rectifying K^+ channels. *Neuron*, 16: 321–331.

Slunt, H.H., Thinakaran, G., von Koch, C., Lo, A.C.Y., Tanzi, R.E. and Sisodia, S.S. (1994) Expression of a ubiquitous, cross-reactive homologue of the mouse β-amyloid precursor protein (APP). *J. Biol. Chem.*, 269: 2637–2644.

Smeyne, R.J., Chu, T., Lewin, A., Bian, F., Crisman, F., Kunsch, C., Lira, S.A. and Oberdick, J. (1995) Local control of granule cell generation by cerebellar Purkinje cells. *Molec. Cell Neurosci.*, 6: 230–251.

Smeyne, R.J. and Goldowitz, D. (1989) Development and death of external granular layer cells in the weaver mouse cerebellum: a quantitative study. *J. Neurosci.*, 9: 1608–1620.

Sotelo, C. (1975) Anatomical, physiological and biochemical studies of the cerebellum from mutant mice. II. Morphological study of cerebellar cortical neurons and circuits in the weaver mouse. *Brain Res.*, 94: 19–44.

Sotelo, C. (1982) Synaptic remodeling in agranular cerebella. In: S.L. Palay and V. Chan-Palay (Eds) *The Cerebellum – New Vistas*, Berlin Heidelberg New York: Springer Verlag, pp. 50–68.

Surmeier, D.J., Mermelstein, P.G. and Goldowitz, D. (1996) The weaver mutation of GIRK2 results in a loss of inwardly rectifying K^+ current in cerebellar granule clls. *Proc. Natl. Acad. Sci. USA*, 93: 11191–11195.

Takayama, C. (1994) Altered distribution of inhibitory synaptic terminals in reeler cerebellum with special reference to malposition of GABAergic neurons. *Neurosci. Res.*, 20: 239–250.

Takechi, H., Eilers, J. and Konnerth, A. (1998) A new class of synaptic response involving calcium release in dendritic spines. *Nature*, 396: 757–760.

Tessier-Lavigne, M. and Goodman, C.S. (1996) The molecular biology of axon guidance. *Science*, 274: 1123–1133.

Vallano, M.L., Lambolez, B., Audinat, E. and Rossier, J. (1996) Neuronal activation differentially regulates NMDA receptor subunit expression in cerebellar granule cells. *J. Neurosci.*, 16: 631–639.

Vandaele, S., Nordquist, D.T., Feddersen, R.M., Tretjakoff, I., Peterson, A.C. and Orr, H.T. (1991) Purkinje cell protein–2 regulatory regions and transgene expression in cerebellar compartments. *Genes Develop.*, 5: 1136–1148.

Vicario-Abejon, C., Cunningham, M.G. and McKay, R.D.G. (1995) Cerebellar precursors transplanted to the neonatal dentate gyrus express features characteristic of hippocampal neurons. *J. Neurosci.*, 15: 6351–6363.

Vogel, M.W. and Herrup, K. (1989) Numerical matching in the mammalian CNS: lack of a competitive advantage of early over late-generated cerebellar granule cells. *J. Comp. Neurol.*, 283: 118–128.

Wanner, I., Baader, S.L., Brich, M., Oberdick, J. and Schilling, K. (1997) Subcellular localization of specific mRNAs and their protein products in Purkinje cells by combined fluorescent in situ hybridization and immunocytochemistry. *Histochem. Cell Biol.*, 108: 345–357.

Wasco, W., Gurubhagavatula, S., paradis, M.d., Romano, D.M., Sissodia, S.S., Hyman, B.T., Neve, R.L. and Tanzi, R.E. (1993) Isolation and characterization of APLP2 encoding a homologue of the Alzheimer's associated amyloid β protein precursor. *Nature Genet.*, 5: 95–99.

Watanabe, D., Inokawa, H., Hashimoto, K., Suzuki, N., Kano, M., Shigemoto, R., Hirano, T., Toyama, K., Kaneko, S., Yokoi, T., Moriyoshi, K., Suzuki, M., Kobayashi, K., Nagatsu, T., Kreitman, R.J., Pastan, I. and Nakanishi, S. (1998) Ablation of cerebellar Golgi cells disrupts synaptic integration involving GABA inhibition and NMDA receptor activation in motor coordination. *Cell*, 95: 17–27.

Wetts, R. and Herrup, K. (1982) Cerebellar Purkinje cells are descendent from a small number of progenitors committed during early development: quantitative analysis of lurcher chimeric mice. *J. Neurosci.*, 2: 1494–1498.

Wolpert, L., Beddington, R., Brockes, J., Jessell, T., Lawrence, P. and Meyerowitz, E. (1998) *Principles of Development*, Oxford: Oxford University Press.

Zhang, D., Sliwkowski, M.X., Mark, M., Frantz, G., Akita, R., Sun, Y., Hillan, K., Crowley, C., Brush, J. and Godowski, P.J. (1997) Neuregulin–3 (NRG3): a novel neural tissue-enriched protein that binds and activated Erb4. *Proc. Natl. Acad. Sci. USA*, 94: 9562–9567.

Zhang, L. and Goldman, J.E. (1996a) Developmental fates and migratory pathways of dividing progenitors in the postnatal rat cerebellum. *J. Comp. Neurol.*, 370: 536–550.

Zhang, L. and Goldman, J.E. (1996b) Generation of cerebellar interneurons from dividing progenitors in white matter. *Neuron*, 16: 47–54.

Zheng, C., Heintz, N. and Hatten, M.E. (1996) CNS gene encoding astrotactin, which supports neuronal migration along glial fibers. *Science*, 272: 417–419.

Zhou, D., Lambert, S., Malen, P.L., Carpenter, S., Boland, L.M. and Bennett, V. (1998) Ankyrin G is required for clustering of voltage-gated channels at axon initial segments and for normal action potential firing. *J. Cell Biol.*, 143: 1295–1304.

Zuo, Y.-R., Kottmann, A.H., Kuroda, M., Taniuchi, I. and Littman, D.R. (1998) Function of the chemokine receptor CXCR4 in haematopoiesis and in cerebellar development. *Nature*, 393: 595–599.

N.M. Gerrits, T.J.H. Ruigrok and C.I. De Zeeuw (Eds.)
Progress in Brain Research, Vol 124

CHAPTER 5

Transgenic methods for directing gene expression to specific neuronal types: cerebellar granule cells

A. Jones, M. Paterlini, W. Wisden* and D. Merlo

MRC Laboratory of Molecular Biology, MRC Centre, Cambridge CB2 2QH, UK

Introduction

Cerebellar granule cells are the glutamatergic interneurons innervating Purkinje cells. Granule cells have two notable features: they are tiny (5–8 μM in diameter), and they are the most numerous cell type in the brain, being densely packed in the cerebellar granule cell layer (Llinás and Walton, 1998). Their neuropharmacological and electrophysiological properties are well known (Traynelis et al., 1993; Gabbiani et al., 1994; D'Angelo et al., 1995; Kaneda et al., 1995; Brickley et al., 1996; Wisden et al., 1996; Bahn and Wisden, 1997; Mäkelä et al., 1997; Maex and De Schutter, 1998). In transgenic mouse lines where adult granule cells selectively die (Pennacchio et al., 1998; Shmerling et al., 1998), or their GABAergic input is removed (Watanabe et al., 1998), severe cerebellar impairment results. Nevertheless, the *specific* function of granule cells in the circuit is uncertain. Network models have been proposed (Marr, 1969; Gabbiani et al., 1994; Bower, 1997; Maex and De Schutter, 1998). To elaborate these models, it will be useful to genetically modify granule cell properties, as done successfully for Purkinje cells (De Zeeuw et al., 1998; Oberdick et al., 1998). Ultimately, a reversible form of gene modulation may be possible: genes could be turned on and off specifically in granule cells.

*Corresponding author. Tel.: 0044–1223–402306; Fax: 0044–1223–402310; e-mail: ww1@mrc-lmb.cam.ac.uk

This review will illustrate a way of directing gene expression precisely to adult cerebellar granule cells using gene targeting by homologous recombination in mouse embryonic stem cells (Jasin et al., 1996); as an example, we describe expression of the reporter gene β-galactosidase (lacZ) to image the developing granule cell layer (Jones et al., 1997; Mellor et al., 1998). This gene 'knock-in' method could be adapted to deliver any gene product (e.g. green fluorescent protein, Cre recombinase, kinase inhibitors) to the adult granule cells (see Fig. 2). The Δα6lacZ mice described here provide a valuable resource of marked granule cells. Genetically tagged granule cells can be used, for example, in transplantation (Bahn et al., 1999), electrophysiological or developmental studies (Mellor et al., 1998).

The 'knock-in' strategy is general: it can be used for any gene (Hanks et al 1995; Jasin et al., 1996). Targeting genes in their natural chromosomal/chromatin context means that the expression is already optimised; there are no position effects or variability in expression levels associated with pronuclear injection of transgenes into mouse eggs (Jasin et al., 1996; Grant and Wisden, 1997). Developing useful transgenes for pronuclear injection (i.e. identifying the regulatory elements of a gene) can be a long and unfruitful process: in fact, few transgenes give *reliable* expression to specific neuronal populations (although there are exceptions e.g. Oberdick et al., 1990, 1993). Even if they contain suitable elements, injected transgenes integrate randomly into chromosomes, and may be silenced or give ectopic expression, depending on their integration position (Palmiter and Brinster, 1986). To sort out which transgenic mice are suitable often requires the analysis of many founder lines. In contrast, mice with gene knock-ins usually express the 'knocked-in' reading frame with the same pattern and intensity as the native gene.

Granule cell development

Cerebellar cell fate is specified in embryogenesis (Hatten and Heintz, 1995; Alder et al., 1996; Jankovski et al., 1996; Joyner, 1996; Hatten et al., 1997; Goldowitz and Hamre, 1998), but in rodents the final building phase happens postnatally (Altman and Bayer, 1996). Granule cell precursors are present at birth on the cerebellum's exterior as a thin layer of cells – the external granule cell layer. In the first postnatal weeks, these precursors divide vigorously (Miale and Sidman, 1961; Fujita et al., 1966; Altman, 1969). After mitosis, the cells migrate transversally in the deeper part of the external layer (Ryder and Cepko, 1994), then migrate through the molecular layer, and enter the internal granule cell layer (Komuro and Rakic, 1998); there, the deepest granule cells are the oldest, and the layer develops in an 'inside-out' order (Altman, 1972a, b). The end product – the adult granule cell layer – is a convoluted sheet of densely packed cells. The midline portion is termed the vermis, whereas the lateral sheets form the hemispheres (reviewed in Marani and Voogd, 1979; Voogd et al., 1996; Voogd and Glickstein, 1998). Different areas of the external sheet cease dividing at different times, e.g. granule cells contributing to vermis lobules IX and X are born first (Altman, 1969).

α6-lacZ expression is a selective marker for adult granule cells

A granule cell's history – mitosis, premigration, migration, maturation – is marked by distinct phases of gene expression (Kuhar et al., 1993; Goldowitz and Hamre, 1998): for example, the $GABA_A$ receptor α6 subunit gene, a unique marker for the granule cell-lineage, is switched on only

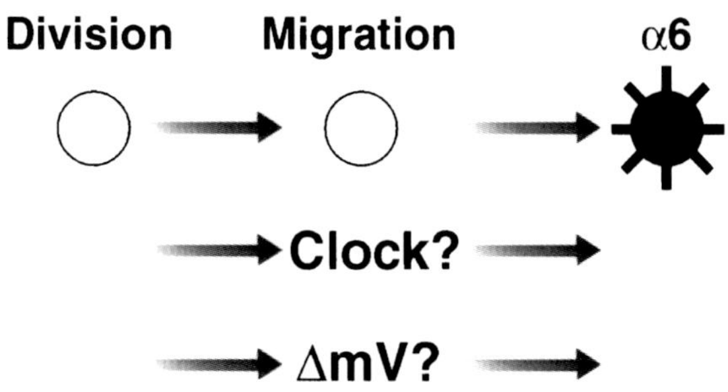

Fig. 1. $GABA_A$ receptor α6 gene expression commences in post-migratory granule cells. Determining factors for initiating this expression might be an intrinsic clock combined with changes in membrane potential (ΔmV).

when granule cells reach the internal layer (Fig. 1; Laurie et al., 1992; Zheng et al., 1993; Varecka et al., 1994; Mellor et al., 1998). To study α6 subunit gene regulation and to provide a convenient histochemical marker for adult granule cells, we targeted the α6 gene by homologous recombination, inserting a β-galactosidase (lacZ) expression cassette into exon 8 to make the Δα6lacZ mouse line (Jones et al., 1997) (Fig. 2). An internal ribosome entry site (IRES) causes cap-independent translation of β-galactosidase (β-gal) from within the α6 mRNA, and so the production of β-gal in the same pattern as the native α6 gene (Fig. 2, Jones et al., 1997; Mellor et al., 1998; the use of the IRES is reviewed by Mountford and Smith, 1995; homologous recombination is reviewed by Hasty and Bradley, 1993). The Δα6lacZ mice are healthy (Korpi et al., 1999), allowing the use of lacZ expression to image the final phases of cerebellar maturation. The adult cerebellar granule cell and dorsal cochlear nuclei layers are intensely lacZ-positive (Jones et al., 1997).

α6-lacZ gene expression spreads from the vermis to the hemispheres

The developmental expression from the α6-lacZ hybrid gene was followed by staining staged postnatal brains for lacZ activity, using whole mount preparations (Fig. 3). At postnatal day 5 (P5), no expression is detected. By P7, clear expression is found in the midline dorsal vermis, with the strongest activity present in lobules IV–V, and IX (other expressing lobules can not be seen from this view – see Fig. 4). At this age, no expression is found in the hemispheres e.g. the simple lobule or crusI and crusII (ansiform lobule). One day later (P8), expression has moved outwards from the midline (Fig. 3 arrows) – cells in certain hemispheric divisions (simple lobule and crusII) and the adjoining lobules have turned on the gene, whereas others (crusI) are still non-expressing. Two lacZ-negative stripes are seen bordering the midline vermis (lobule VIa) at P8. This striped pattern was obtained with all animals. By P9, most regions have been filled in with α6 expression – the anterior part of crusI is now expressing, but expression is still weak in the posterior part of crusI; lobule VIb has less expression, and residual stripes are still visible in VIa; expression is developing in the paraflocculus. At P10, the spatial expression is fully developed (Fig. 3), although the lacZ stain is more intense in older animals (data not shown). Expression in the inferior colliculi develops between P6 and P10 (Fig. 3) (this expression is not seen by in situ hybridization or immunocytochemistry to the α6 subunit, and could reflect the high sensitivity of the lacZ assay).

The developing α6 subunit gene expression pattern was traced in sagittal sections along the rostral-caudal midline of the vermis (marked R—C in Fig. 4). Expression starts in the most posterior part, lobule X (see also Varecka et al., 1994). At P5, a few lacZ-positive cells are found in the deepest part of the internal granule cell layer of lobule X (Fig. 4A). Other lobules, at this time point, are not expressing detectable β-gal. A day later (P6), more cells are expressing in the deep layers of X, and scattered lacZ-positive cells are found in the dorsal aspect of lobule IX (Fig. 4B). By P7, expression in lobules X and IX has increased, but is still confined

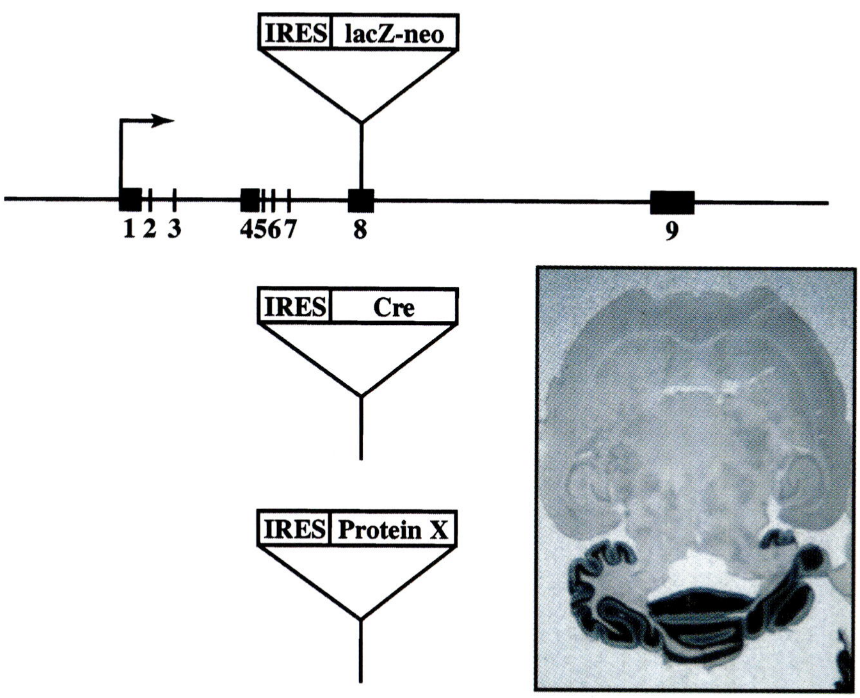

Fig. 2. Gene delivery to granule cells by targeting an IRES-expression cassette into the native $GABA_A$ receptor α6 subunit gene by homologous recombination. The top line shows the structure of the Δα6lacZ mouse $GABA_A$ receptor α6 subunit gene. The IRES-lacZ cassette is inserted into exon 8. The arrow marks the transcriptional start sites; shaded boxes are exons. Other reading frames, such as those encoding Cre recombinase, or any other protein ('X') could replace the lacZ; such proteins would then be delivered specifically to granule cells. The inset picture shows the production of β-galactosidase specifically in granule cells of an adult Δα6lacZ mouse brain (Jones et al., 1997); horizontal section. IRES, internal ribosome entry site.

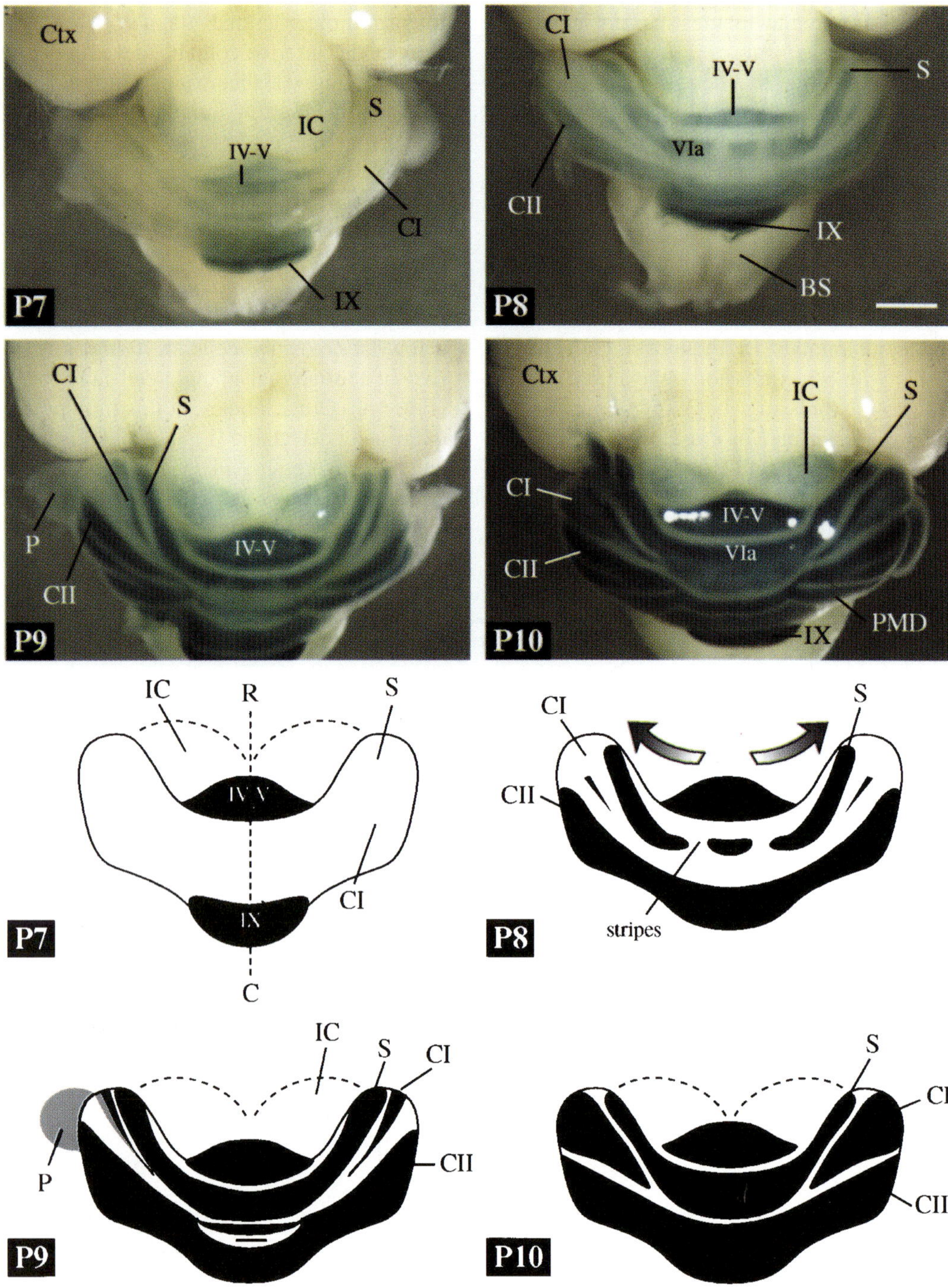

Fig. 3. Dorsal views of whole brains showing the development of α6-lacZ gene expression (blue) in Δα6lacZ mice over a four day period between postnatal day seven (P7) and P10. Expression is also seen in the inferior colliculi (IC). The schematics highlight features of the evolving pattern; black shading corresponds to prominent lacZ staining. An, ansiformis; BS, brain stem; C, caudal; CI, crus 1 of the ansiform lobule; CII, crus 2 of the ansiform lobule; Ctx, neocortex; P, paraflocculus; PMD, paramedian lobule; R, rostral; S, simple lobule; Roman numerals identify vermis lobules. Scale bar, 1.3 mm.

to the deeper layers (Fig. 4C). At this stage, expression is beginning to appear in the sulci (stems) of the anterior lobules I–V (Fig. 4C). As with lobules IX and X, these α6 gene expressing cells in lobules I–V are in the deeper part of the internal layer closest to the white matter tracts.

By P9, α6 expression is established in all lobules except the gyri (outer folds) of VI and VII (Fig. 4D). In the lobules (X and IX, I–V), where expression first started to appear (Fig. 4B), β-gal activity is becoming apparent in the internal part of the molecular layer. This presumably corresponds

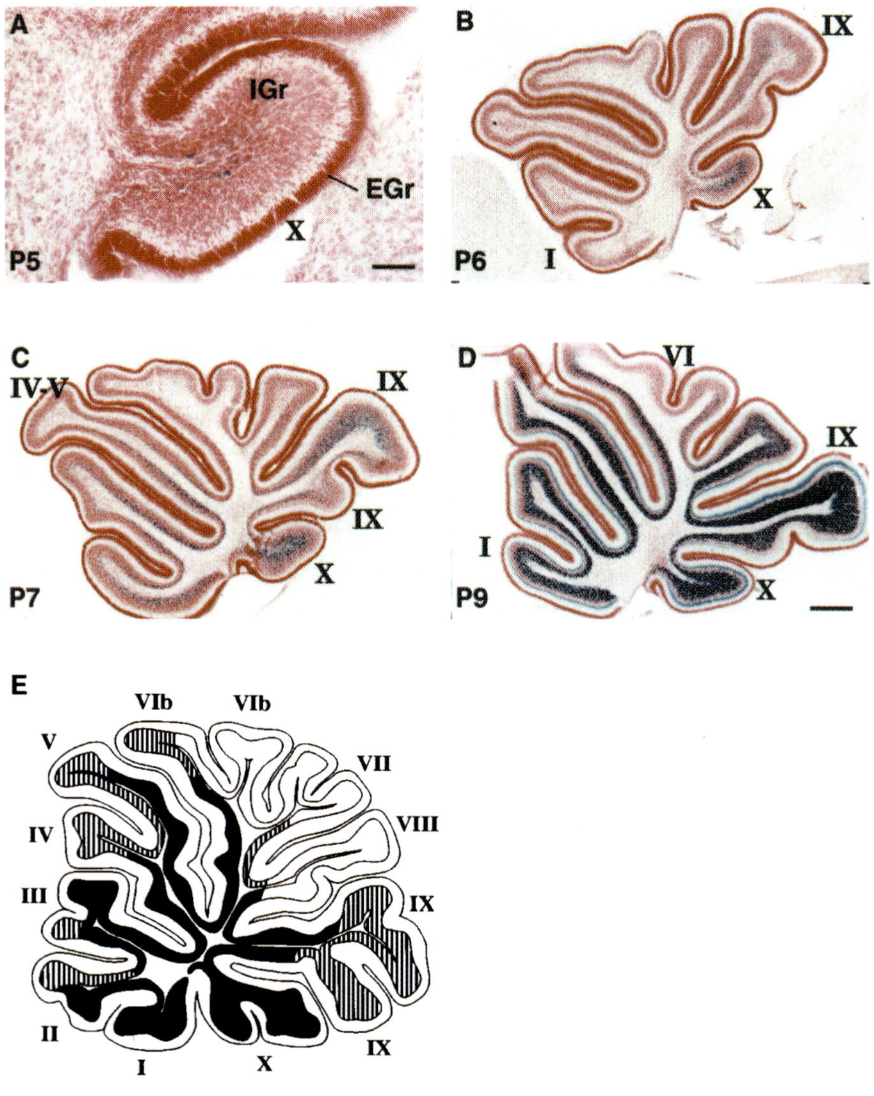

Fig. 4. Sagittal sections of developing Δα6lacZ mouse cerebella taken through the rostral-caudal midline marked in Fig. 3, and stained for lacZ activity (blue). The rostral direction is on the left-hand side. A, Lobule X, postnatal day five (P5). B–D, sections from P6 through to P9. Sections are counterstained with neutral red. E, Altman's [^{3}H]thymidine birth dating autoradiographic study on the age of cerebellar granule cells and showing their contribution to different folia. Early maturing (black), intermediate (cross-hatched) and late-maturing granule cells (white) are marked in an adult rat cerebellar mid-line sagittal section roughly comparable to the mouse sections shown in panels B–D (modified from Altman, 1969). EGr, external granule cell layer; IGr, internal granule cell layer; Roman numerals identify the vermis lobules. Scale bar A, 80 μm; Scale bar D, 0.3 mm.

to transport of β-gal into the granule cell axons – the parallel fibres (Figs 4D, 5A & B). By P10–11, the entire internal granule cell layer in all lobules is lacZ-positive (data not shown); over the following week, as more granule cells migrate into the internal layer and switch on the gene, the intensity of expression continues to increase (data not shown). N.B. these pictures are from sections through the midline: by taking more lateral sagittal sections through the hemispheres, a different sequence of regional staining would be seen (see Fig. 3).

Purkinje cells are compartmentalized, as defined by alternating sagittal bands of marker gene expression e.g. L7/pcp–2 and Zebrin–2/aldolase C (Eisenman and Hawkes, 1993; Oberdick et al., 1993; Voogd et al., 1996; Hawkes, 1997; Lin and Cepko, 1998; Oberdick et al., 1998). Adult granule cells and the granule cell layer are also heterogeneous by some biochemical markers and physical features (Ozol and Hawkes, 1997) e.g. acetylcholinesterase staining (Boegman et al., 1988), NO synthase staining (Hawkes and Turner, 1994; Schilling et al., 1994), expression of the calcium-binding protein VILIP (Lenz et al., 1996) and wrinkling or blebs (Hawkes, 1997; Hawkes et al., 1998); however, developing α6 expression does not resemble either the Purkinje longitudinal stripes or the adult NOS-stained or wrinkled granule cell layer patterns. In the adult granule cell layer, α6 expression is uniform in all granule cells (Laurie et al., 1992; Jones et al., 1997). Nevertheless, at P9, α6 gene expression does define three clear domains: expression in vermal lobules I–V, no expression in the central VI lobe, and expression in the posterior (IX) and inferior (X) lobes (Fig. 4D). This corresponds to functional subdivisions of the vermis (Altman and Bayer, 1996), and resembles the combined expression patterns of the Otx1 and Otx2 transcription factors (Frantz et al., 1994). The postnatal development of transcription factor *Engrailed*–2 (*En–2*) expression is similar (see Fig. 7 in Millen et al., 1995), but precedes the majority of α6 expression by 1.5 weeks. At P0, *En–2* expression starts off in the external granule cells of the mid-line vermis, then spreads to external granule cells in parts of the hemispheres by P5. In midline sagittal sections, *En–2* expression initiates in lobules I to V and inferior lobule X (Millen et al., 1995). The *En–2* pattern could reflect the same underlying cerebellar maturation process responsible for α6 gene induction, shifted forward by a week. *En–2* may therefore contribute to the specification of α6 subunit gene expression.

The best predictor of α6 gene expression is cell age: α6-lacZ transgene expression *correlates* with [^{3}H]thymidine birth-dating experiments with respect to both: (i) the rat granule cell regional fate map (a summary is shown in Fig. 4E) and (ii) an inner to outer gradient of maturing granule cells in the internal granule cell layer (Altman, 1972a, b). From [^{3}H]thymidine pulse-chase designs, a map of cell birthday versus cell position can be built up – shown for an adult rat cerebellar vermis in Fig. 4E (modified from Altman, 1969). The oldest cells are in lobules IX and X. A correlation with α6-lacZ expression stands out. In particular, granule cells are generated earlier near the depth of fissures (lacZ-positive cells appear there first), the outer loops tend to be born later (same as the sequence of lacZ-positive maturation), and the central lobe (VII) is the youngest (becomes lacZ-positive last) (see Fig. 4). The external granule cells of the hemispheres are the last to incorporate [^{3}H] thymidine (Altman, 1969), matching with the development of the α6 pattern (Fig. 3). This could be directly tested by bromodeoxyuridine incorporation into cerebella of developing Δα6lacZ mice, combined with lacZ staining.

Two zones of cells in the developing internal granule cell layer

The contrast in maturation stages between vermis lobules, as reflected by α6 gene expression, is shown in detail in Fig. 5. At P9, nearly all the granule cells in the internal layer of lobule IX are lacZ-positive; in the internal layer of VIII, however, expression is mosaic – a clear proportion of cells are still not lacZ positive (Fig. 5A). By the criterion of α6 expression, the internal granule cell layer of a developing lobule divides into two shells or zones: an outer zone of non-α6 expressing cells; and a deeper inner zone of expression (Fig. 5A & B, marked as *). This outer shell of lacZ-negative cells is made largely of granule cells, since it is positive for *En–2*, a selective granule cell marker at this age (Fig. 5C, asterisks) (Millen et al., 1995); the cells in this layer are therefore likely to be newly arrived postmigratory granule cells which have not yet turned on the α6 gene. This *En–2* -positive/α6-negative zone is transient. α6-lacZ gene expression is never found in the external granule cell layer (which contains dividing progenitors and premigratory granule cells - Fig. 5A, EGr) or in the migrating granule cells crossing the molecular layer (Figs 3 & 4).

Altman (1972b) concluded that the deeper cells in the internal granule cell layer are the oldest: "Cells that differentiate in the first week after birth have a tendency to settle in the lower half of the internal granular layer. The cells formed after the 8–15 day period have a tendency to be located in the upper half. However, a strict zonal segregation was not observed. Presumably, there are forces operating in the organisation of the granular layer which interfere with such a strict segregation based on time of origin". (Altman, 1972b). Indeed, Komoru and Rakic (1998) and Bahn et al., (1999) have shown that migrating granule cells can push deep into the internal granule cell layer. Nevertheless, the tendency for the deeper cells to become α6-lacZ-positive first (e.g. see Fig. 5A), correlates with the [^{3}H]thymidine data, suggesting α6 is expressed only in the oldest cells.

Corresponding to the inside-out gradient of cells with age in the internal granule cell layer, there is an equivalent stacking of parallel fibres in the molecular layer: the axons closest to the granule cell layer are the oldest (Altman, 1972a; Altman and Bayer, 1996); the parallel fibres coming from the younger internal granule cells are located more superficially in the molecular layer (Altman, 1972a). This would explain the blue coloration in the developing molecular layer of Δα6lacZ mice: the blue stain starts off in the region closest to the internal granule cell layer (Fig. 5A &B, black arrowheads), and presumably comes from the deepest and oldest granule cells. As the lobules mature, and more granule cells migrate in and turn on the α6 gene, the blue in the molecular layer spreads out. In Fig. 5A, lobule IX (left) is more mature than lobule VIII (right), and the blue strip in the molecular layer (black arrowhead, Fig. 5A) is correspondingly stronger and deeper.

Our description of maturing α6-lacZ gene expression splitting the internal granule cell layer into two transient zones supports other observations of a transient subclass of granule cell in the developing internal granule cell layer (Zdilar et al., 1991; Kuhar et al., 1993; Duncan et al., 1997): at P6–14, the liver 10-formyltetrahydrofolate dehydrogenase, heart fatty acid binding protein and transcription factor Id–1 genes are transcribed only in granule-like cells that have just migrated past the Purkinje cell, and their expression in the deeper cells is reduced (Kuhar et al., 1993; Duncan et al., 1997). Conversely, in P8–P10 mouse cerebellum, $GABA_A$ receptor subunit α1 mRNA is almost absent from a narrow strip beneath the Purkinje cell layer (Zdilar et al., 1991), although α1 mRNA is in all adult granule cells (Laurie et al., 1992). Thus, in the maturing cerebellum, 'adult-like' granule cells are found in the deeper parts of the layer, fitting with Altman's 'inside-out' order of development.

What initiates and controls α6 gene expression?

Knowing how α6 gene expression is specified may help in understanding other types of neuron-specific gene expression (Grant and Wisden, 1997). Information for specifying α6 expression to the cerebellar granule cells is contained within the 5′ half (7 kb) of the gene (Jones et al., 1996; Bahn et al., 1997); in fact, 450 bp of proximal 5′ sequence is enough to specify granule cell-specific expression, depending on the integration position of the transgene (Jones et al., 1996). The relevant transcription factors are still undefined. In principle, transcription could be activated by external cues e.g. from glutamatergic mossy fibres or GABAergic Golgi innervation as cells reach the internal layer; however, ionotropic or metabotropic GABA and glutamate receptor antagonists fail to block α6 induction in dissociated cells (Mellor et al., 1998).

We believe that α6 transcription is regulated by an internal clock, e.g. timing from the last cell division. Granule cell precursors are committed to their fate, including α6 expression, by the time they reach the early postnatal period (Gao and Hatten, 1994; Alder et al., 1996; Jankovski et al., 1996; Bahn et al., 1999). Transplantation studies show that P1 granule cells are committed to express the

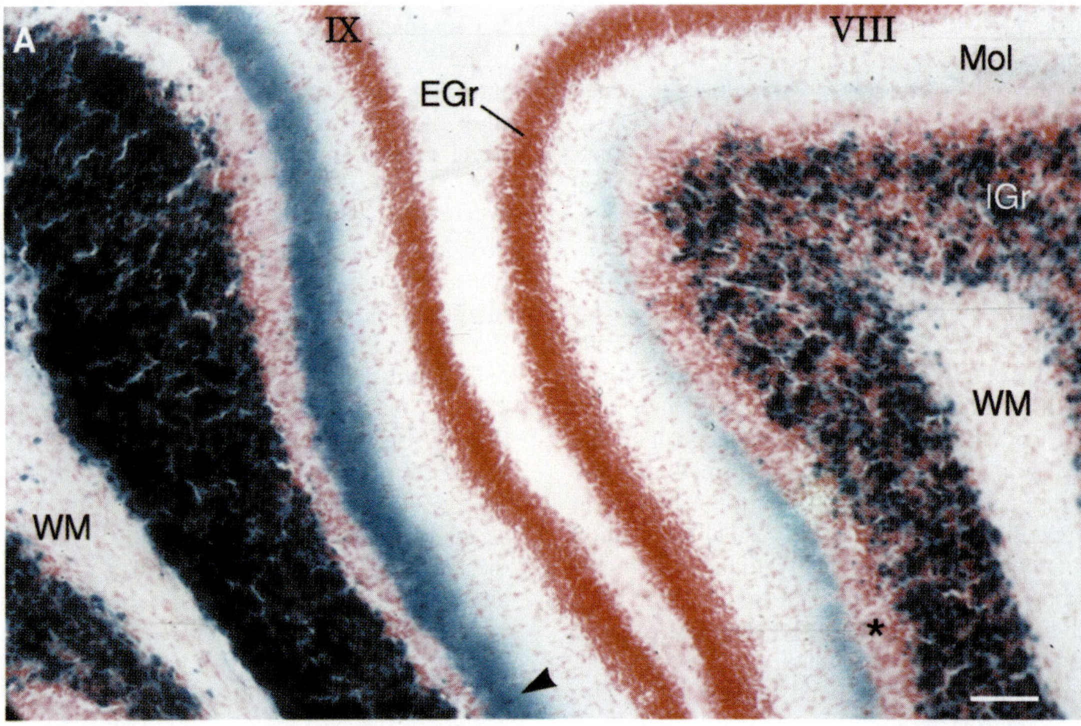

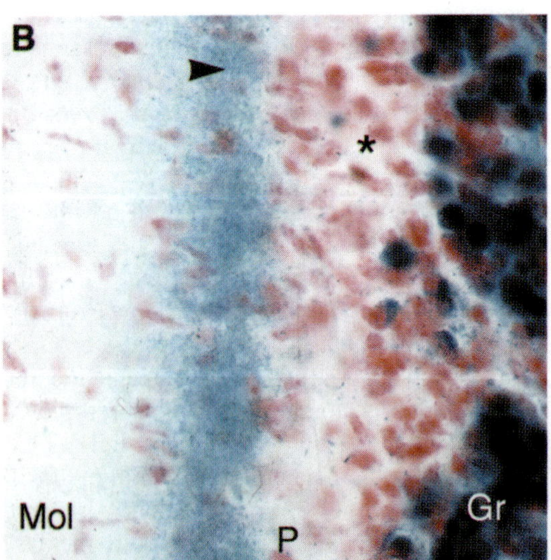

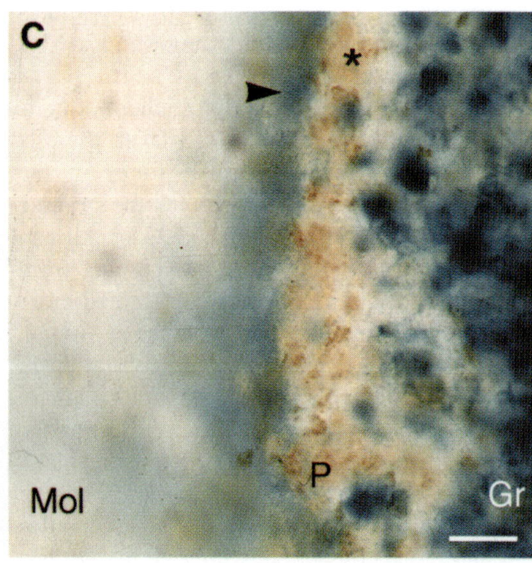

Fig. 5. A, P9 sagittal slice with neighbouring cerebellar granule cell layers – loops of lobules VIII (right) and IX (left) – at differing stages of maturation as seen by lacZ staining (blue); non-expressing cells are red. B, Higher power view of lobule VIII, showing that the upper part of the internal granule cell layer (*) has no α6-lacZ expression; arrowhead marks β-gal enzyme transported into granule cell axons – the parallel fibres – in the molecular layer. C, LacZ (blue) plus brown anti-engrailed stain (*En–2*) showing granule-like cells in the upper-part of the internal granule cell layer which are *En–2* positive, α6-lacZ negative. EGr, external granule cell layer; IGr, internal granule cell layer; Mol, molecular layer; P, Purkinje cell layer; WM, white matter. Scale bars: A, 40 μM; C, 20 μM.

α6 gene regardless of their environment (Bahn et al., 1999). $GABA_A$ receptor α6 subunit gene expression can be established by dissociated granule cell precursors when they are placed into different adult brain environments. Establishing and maintaining α6 gene expression is unlikely to need specific patterns of innervation, or an environment unique to the developing brain. For example, the place within the cerebellum (surface, molecular layer, granule cell layer) for allowing the establishment of α6 gene transcription is irrelevant; expression can be initiated and maintained in different locations in the hippocampus, in the amygdala, and even on the outside brain surface (Bahn et al., 1999). Electrical activity, brain-derived-neurotrophic factor and cAMP all affect the α6 gene expression levels (Thompson et al., 1996a, b; Ghose et al., 1997; Lin et al., 1998; Mellor et al., 1998), but no agent fundamentally changes the α6 gene expression pattern. This remains intrinsic to granule cells.

An intrinsic clock might cause α6 gene induction by allowing a granule cell-specific transcription factor(s) to accumulate. This factor might interact with proteins such as *En*–2 (Millen et al., 1995), RU–49 (Yang et al., 1996), members of the Zic family (Aruga et al., 1994), or *w*idely-*i*nterspersed-*z*inc finger (wiz) proteins (Matsumoto et al., 1998); all these factors are highly expressed in adult cerebellar granule cells, but are also found in other neuronal types and other body organs.

Conclusions

We have shown that homologous recombination at the $GABA_A$ receptor α6 gene locus can be used to precisely deliver other gene products to adult cerebellar granule cells (Jones et al 1997; Mellor et al., 1998). As an example, we used α6-lacZ expression to trace the maturation of the granule cell layer. This strategy can be used for any protein, or be used for any other gene with the desired expression pattern (Hanks et al., 1995; Jasin et al., 1996). It would also be possible to make induction conditional, e.g. by driving the tet-activator protein from the α6 gene locus (Furth et al., 1994). By so doing, the functional properties of granule cells could be genetically manipulated.

Acknowledgements

We thank: A. Joyner (Skirball Institute, New York) for the gift of α-Enhb–1 serum; A. Lenton and J. Westmorland (MRC Visual Aids) for help with figure preparation. This work was supported by Human Frontier Science Program Organization Grant RG0017 to W.W.

References

Alder, J., Cho, N.K. and Hatten, M.E. (1996) Embryonic precursor cells from the rhombic lip are specified to a cerebellar granule neuron identity. *Neuron*, 17: 389–399.

Altman, J. (1969) Autoradiographic and histological studies of postnatal neurogenisis III. Dating the time of production and onset of differentiation of cerebellar microneurons in rats. *J. Comp. Neurol.*, 136: 269–294.

Altman, J. (1972a) Postnatal development of the cerebellar cortex in the rat I. The external germinal layer and the transitional molecular layer. *J. Comp. Neurol.*, 145: 353–398.

Altman, J. (1972b) Postnatal development of the cerebellar cortex in the rat III. Maturation of the components of the granular layer. *J. Comp. Neurol.*, 145: 465–514.

Altman, J. and Bayer, S.A. (1996) *Development of the Cerebellar System in Relation to its Evolution, Structure and Functions*. CRC Press, Boca Raton, Florida .

Aruga, J., Yokota, N., Hashimoto, M., Furuichi, T., Fukuda, M. and Mikoshiba, K. (1994) A novel zinc finger protein, Zic, is involved in neurogenesis, especially in the cell lineage of cerebellar granule cells. *J. Neurochem.*, 63: 1880–1890.

Bahn, S. and Wisden, W. (1997). A map of non-NMDA receptor subunit expression in the vertebrate brain derived from *in situ* hybridization histochemistry. In: D.T. Monaghan and R.J. Wenthold, (Eds) *The Ionotropic Glutamate Receptors*. Humana Press Inc, Totowa, NJ, pp. 149–187.

Bahn, S., Jones, A. and Wisden, W. (1997) Directing gene expression to cerebellar granule cells using γ-aminobutyric acid type A receptor α6 subunit transgenes. *Proc. Natl. Acad. Sci. USA*, 94: 9417–9421.

Bahn, S., Wisden, W., Dunnett, S.B. and Svendsen, C. (1999) The intrinsic specification of γ-aminobutyric acid type A receptor α6 subunit gene expression in cerebellar granule cells. *Eur.J. Neurosci.*, 11: 2194–2198.

Boegman, R.J., Parent, A. and Hawkes, R. (1988) Zonation in the rat cerebellar cortex: patches of high acetylcholinesterase activity in the granular layer are congruent with Purkinje cell compartments. *Brain Res.*, 448: 237–251.

Bower, J.M. (1997) Is the cerebellum sensory for motor's sake, or motor for sensory's sake: the view from the whiskers of a rat? In: C.I. de Zeeuw, P. Strata and J. Voogd (Eds), *Progress in Brain Research, Vol. 114*, Elsevier Science BV, pp. 463–496.

Brickley, S.G., Cull-Candy, S.G. and Farrant, M. (1996) Development of a tonic form of synaptic inhibition in rat cerebellar granule cells resulting from persistent activation of $GABA_A$ receptors. *J. Physiol.*, 497: 753–759.

D'Angelo, E., De Filippi, G., Rossi, P. and Taglietti, V. (1995) Synaptic excitation of individual rat cerebellar granule cells in situ: evidence for the role of NMDA receptors. *J. Physiol.*, 484: 397–413.

De Zeeuw, C.I., Hansel, C., Bian, F., Koekkoek, S.K., van Alphen, A.M., Linden, D.J. and Oberdick, J. (1998) Expression of a protein kinase C inhibitor in Purkinje cells blocks cerebellar LTD and adaptation of the vestibulo-ocular reflex. *Neuron*, 20: 495–508.

Duncan, M.K., Bordas, L., Dicicco-Bloom, E. and Chada, K.K. (1997) Expression of the helix-loop-helix genes Id–1 and NSCL–1 during cerebellar development. *Dev. Dynamics*, 208: 107–114.

Eisenman, L.M. and Hawkes, R. (1993) Antigenic compartmentation in the mouse cerebellar cortex: zebrin and HNK–1 reveal a complex, overlapping molecular topography. *J. Comp. Neurol.*, 335: 586–605.

Frantz, G.D., Weimann, J.M., Levin, M.E. and McConnell, S.K. (1994) Otx1 and Otx2 define layers and regions in developing cerebral cortex and cerebellum. *J. Neurosci.*, 14: 5725–5740.

Fujita, S., Shimada, M. and Nakamura, T. (1966) H^3-Thymidine autoradiographic studies on the cell proliferation and differentiation in the external and the internal granular layers of the mouse cerebellum. *J. Comp. Neurol.*, 128: 191–208.

Furth, P.A., St Onge, L., Boger, H., Gruss, P., Gossen, M., Kistner, A., Bujard, H. and Hennighausen, L. (1994) Temporal control of gene expression in transgenic mice by a tetracycline-responsive promoter. *Proc. Natl. Acad. Sci. USA*, 91: 9302–9306.

Gabbiani, F., Midtgaard, J. and Knopfel, T. (1994) Synaptic integration in a model of cerebellar granule cells. *J. Neurophysiol.*, 72: 999–1009

Gao, W.Q and Hatten, M.E. (1994) Immortalizing oncogenes subvert the control of granule cell lineage in developing cerebellum. *Development*, 120: 1059–1070.

Ghose, S., Wroblewska, B., Corsi, L., Grayson, D.R., De Blas, A.L., Vicini, S. and Neale, J.H. (1997) *N*-acetylaspartylglutamate stimulates metabotropic glutamate receptor 3 to regulate expression of the $GABA_A$ α6 subunit in cerebellar granule cells. *J. Neurochem.*, 69: 2326–2335.

Goldowitz, D. and Hamre, K. (1998) The cells and molecules that make a cerebellum. *Trends Neurosci.*, 21: 375–382.

Grant, A.L. and Wisden, W. (1997) Neuron-specific gene expression. In: R.W. Davies and B.J.Morris (Eds), *Molecular Biology of the Neuron*, Bios Scientific Publishers, Oxford, pp. 67–93.

Hanks, M., Wurst. ,W., Anson-Cartwright, L., Auerbach, A.B. and Joyner, A.L. (1995) Rescue of the *En*–1 mutant phenotype by replacement of *En*–1 with *En*–2. *Science*, 269: 679–682.

Hasty, P. and Bradley, A. (1993) Gene targeting vectors for mammalian cells. In A.L. Joyner (Ed.), *Gene Targeting: A Practical Approach*, IRL Press at Oxford University Press, Oxford, pp. 1–31.

Hatten, M.E., Alder, J., Zimmerman, K. and Heintz, N. (1997) Genes involved in cerebellar cell specification and differentiation. *Curr. Opin. Neurobiol.*, 7: 40–47.

Hatten, M.E. and Heintz, N. (1995) Mechanisms of neural patterning and specification in the developing cerebellum. *A. Rev. Neurosci.*, 18: 385–408.

Hawkes, R. (1997) An anatomical model of cerebellar modules. In: C.I. de Zeeuw, P. Strata and J. Voogd (Eds), *Progress in Brain Research, Vol.114*, Elsevier Science BV, pp. 39–52.

Hawkes.R. and Turner, R.W. (1994) Compartmentation of NADPH-diaphorase activity in the mouse cerebellar cortex. *J. Comp. Neurol.*, 346: 499–516.

Hawkes, R., Gallagher, E. and Ozol, K. (1998) Blebs in the mouse cerebellar granular layer as a sign of structural in homogeneity. 2. Posterior lobe vermis. *Acta Anat.*, 163: 47–55.

Jankovski, A., Rossi, F. and Sotelo, C. (1996) Neuronal precursors in the postnatal mouse cerebellum are fully committed cells: evidence from heterochronic transplantations. *Eur.J. Neurosci.*, 8:2308–2319.

Jasin, M., Moynahan, M.E. and Richardson, C. (1996) Targeted transgenesis. *Proc. Natl. Acad. Sci. USA*, 93: 8804–8808.

Jones, A., Bahn, S, Grant, A.L., Köhler, M. and Wisden, W. (1996) Characterization of a cerebellar granule cell-specific gene encoding the γ-aminobutyric acid type A receptor α6 subunit. *J. Neurochem.*, 67: 907–916.

Jones, A., Korpi, E.R., McKernan, R.M., Pelz, R., Nusser, Z., Mäkelä, R., Mellor, J.R., Pollard, S., Bahn, S., Stephenson, F.A., Randall, A.D., Sieghart, W., Somogyi, P., Smith, A.J.H. and Wisden, W. (1997) Ligand-gated ion channel subunit partnerships: $GABA_A$ receptor α6 subunit gene inactivation inhibits δ subunit expression. *J. Neurosci.*, 17: 1350–1362.

Joyner, A.L. (1996) Engrailed, Wnt and Pax genes regulate midbrain-hindbrain development. *Trends Genet.*, 12: 15–20.

Kaneda, M., Farrant, M. and Cull-Candy, S.G. (1995) Whole-cell and single-channel currents activated by GABA and Glycine in granule cells of the rat cerebellum. *J. Physiol.*, 485: 419–435.

Komuro, H. and Rakic, P. (1998) Distinct modes of neuronal migration in different domains of developing cerebellar cortex. *J. Neurosci.*, 18: 1478–1490.

Korpi, E.R., Koikkaliainen, P., Vekovischeva, O.Y., Mäkelä, R., Kleinz, R., Uusi-Oukari, M. and Wisden, W. (1999) Cerebellar granule cell-specific $GABA_A$ receptors attenuate benzodiazepine-induced ataxia: evidence from α6 subunit-deficient mice. *Eur. J. Neurosci.*, 11: 233–240.

Kuhar, S.G., Feng, L., Vidan, S., Ross, M.E., Hatten, M.E. and Heintz, N. (1993) Changing patterns of gene expression define four stages of cerebellar granule neuron differentiation. *Development*, 117:97–104.

Laurie, D.J., Wisden, W. and Seeburg, P.H. (1992) The distribution of thirteen $GABA_A$ receptor subunit mRNAs in the rat brain. III. Embryonic and postnatal development. *J. Neurosci.*, 12: 4151–4172.

Lenz, S.E., Jiang, S., Braun, K. and Gundelfinger, E.D. (1996) Localization of the neural calcium-binding protein VILIP (visinin-like protein) in neurons of the chick visual system and cerebellum. *Cell Tissue Res.*, 283: 413–424.

Lin, J.C. and Cepko, C.L. (1998) Granule cell raphes and parasagittal domains of Purkinje cells: complementary patterns in the developing chick cerebellum. *J. Neurosci.*, 18: 9342–9353.

Lin, X., Cui, H. and Bulleit, R.F. (1998) BDNF accelerates gene expression in cultured cerebellar granule cells. *Dev. Brain Res.*, 105: 277–286.

Llinás, R.R. and Walton K.D. (1998) Cerebellum. In: G.M. Shepherd (Ed.), *The Synaptic Organization of the Brain*, Oxford University Press, New York. pp. 255–288.

Maex, R. and De Schutter, E. (1998) Synchronization of Golgi and granule cell firing in a detailed network model of the cerebellar granule cell layer. *J. Neurophysiol.*, 80: 2521–2537.

Mäkelä, R., Uusi-Oukari, M., Homanics, G.E., Quinlan, J.J., Firestone, L.L., Wisden, W. and Korpi, E.R. (1997) Cerebellar γ-aminobutyric acid type A receptors: pharmacological subtypes as revealed by mutant mouse lines. *Mol. Pharmacol.*, 52: 380–388.

Marani, E. and Voogd, J. (1979) The morphology of the mouse cerebellum. *Acta Morphol. Neerl-Scand.*, 17: 33–52.

Marr, D. (1969) A theory of cerebellar cortex. *J. Physiol.*, 202: 437–470.

Matsumoto, K., Ishii, N., Yoshida, S., Shiosaka, S., Wanaka, A.and Tohyama, M. (1998) Molecular cloning and distinct developmental expression pattern of spliced forms of a novel zinc finger gene *wiz* in the mouse cerebellum. *Mol. Brain Res.*, 61: 179–189.

Mellor, J.R., Merlo, D., Jones, A., Wisden, W. and Randall, A.D. (1998) Mouse cerebellar granule cell differentiation: electrical activity regulates the $GABA_A$ receptor α6 subunit gene. *J. Neurosci.*, 18: 2822–2833.

Miale, I.L. and Sidman, R.L. (1961) An autoradiographic analysis of histogenesis in the mouse cerebellum. *Exp. Neurol.*, 4: 277–296.

Millen, K.J., Hui, C-C. and Joyner A.L. (1995) A role for *En–2* and other murine homologues of *Drosophila* segment polarity genes in regulating positional information in the developing cerebellum. *Development*, 121: 3935–3945.

Mountford, P.S. and Smith, A.G. (1995). Internal ribosome entry sites and dicistronic RNAs in mammalian transgenesis. *Trends Genet.*, 11: 179–184.

Oberdick, J., Smeyne, R.J., Mann, J.R,, Zackson, S. and Morgan, J.I. (1990) A promoter that drives transgene expression in cerebellar Purkinje and retinal bipolar neurons. *Science*, 248: 223–226.

Oberdick, J., Schilling, K., Smeyne, R.J., Corbin, J.G., Bocchiaro, C., and Morgan, J.I. (1993) Control of segment-like patterns of gene expression in the mouse cerebellum. *Neuron*, 10: 1007–1018.

Oberdick, J., Baader, S.L. and Schilling, K. (1998) From zebra stripes to postal zones: deciphering patterns of gene expression in the cerebellum. *Trends Neurosci.*, 21: 383–390.

Ozol, K.O. and Hawkes, R. (1997) Compartmentation of the granular layer of the cerebellum. *Histol. Histopathol.*, 12: 171–184.

Palmiter, R.D. and Brinster, R.L. (1986) Germ-line transformation of mice. *Ann. Rev. Gen.*, 20: 465–499.

Pennacchio, L.A., Bouley, D.M., Higgins, K.M., Scott, M.P., Noebels, J.L. and Myers, R.M. (1998) Progressive ataxia, myoclonic epilepsy and cerebellar apoptosis in cystatin B-deficient mice. *Nature Genet.*, 20: 251–258.

Ryder, E.F. and Cepko, C.l. (1994) Migration patterns of clonally related granule cells and their progenitors in the developing chick cerebellum. *Neuron*, 12: 1011–1029.

Schilling, K., Schmidt, H.H. and Baader, S.L. (1994) Nitric oxide synthase expression reveals compartments of cerebellar granule cells and suggest a role for mossy fibers in their development. *Neuroscience*, 59: 893–903.

Shmerling, D., Hegyi, I., Fischer, M., Blattler, T., Brandner, S., Gotz, J., Rulicke, T., Flechsig, E., Cozzio, A., von Mering, C., Hangartner, C., Aguzzi, A. and Weissmann, C. (1998) Expression of amino-terminally truncated PrP in the mouse leading to ataxia and specific cerebellar lesions. *Cell*, 93: 203–214.

Thompson, C.L., Pollard, S. and Stephenson, F.A. (1996a) Bi-directional regulation of $GABA_A$ receptor α1 and α6 subunit expression by acyclic AMP-mediated signalling mechanism in cerebellar granule cells in primary culture. *J. Neurochem.*, 67: 434–437.

Thompson, C.L., Pollard, S. and Stephenson, F.A. (1996b) Developmental regulation of expression of $GABA_A$ receptor α1 and α6 subunits in cultured rat cerebellar granule cells. *Neuropharmacology*, 35: 1337–1346.

Traynelis, S.F., Silver, R.A. and Cull-Candy, S.G. (1993) Estimated conductance of glutamate receptor channels activated during EPSCs at the cerebellar mossy fiber-granule cell synapse. *Neuron*, 11: 279–289.

Varecka, L., Wu, C-H., Rotter, A. and Frostholm, A. (1994) $GABA_A$/benzodiazepine receptor α6 subunit mRNA in granule cells of the cerebellar cortex and cochlear nuclei: expression in developing and mutant mice. *J. Comp. Neurol.*, 339: 341–352.

Voogd, J., Jaarsma, D. and Marani, E. (1996) The cerebellum: chemoarchitecture and anatomy. In: L.W. Swanson, A. Björkland and T. Hökfelt (Eds), *Handbook of Chemical Neuroanatomy. Integrated Systems of the CNS,* part III: *Cerebellum, Basal Ganglia and Olfactory System. Vol. 12,* Elsevier, Amsterdam, pp. 1–369.

Voogd, J. and Glickstein, M. (1998) The anatomy of the cerebellum. *Trends Neurosci.*, 21: 37–375.

Watanabe, D., Inokawa, H., Hashimoto, K., Suzuki, N., Kano, M., Shigemoto, R., Hirano, T., Toyama, K., Kaneko, S., Yokoi, M., Moriyoshi, K., Suzuki, M., Kobayashi, K., Nagatsu, T., Kreitman, R.J., Pastan, I. and Nakanishi, S. (1998) Ablation of cerebellar Golgi cells disrupts synaptic integration involving GABA inhibition and NMDA receptor activation in motor coordination. *Cell*, 95: 17–27.

Wisden, W., Korpi, F.R. and Bahn, S. (1996) The cerebellum: a model system for studying $GABA_A$ receptor diversity. *Neuropharmacology*, 35: 1139–1160.

Yang, X.W., Zhong, R. and Heintz, N. (1996) Granule cell specification in the developing mouse brain as defined by expression of the zinc finger transcription factor RU–49. *Development*, 122: 555–566.

Zdilar, D., Rotter, A. and Frostholm, A. (1991) Expression of $GABA_A$/benzodiazepine receptor α1-subunit mRNA and $[^3H]$flunitrazepam binding sites during postnatal development of the mouse cerebellum. *Dev. Brain Res.*, 61: 63–71.

Zheng, T., Santi, M-R., Bovolin, P., Marlier, LNJ-L. and Grayson, D.R. (1993) Developmental expression of the α6 $GABA_A$ receptor subunit mRNA occurs only after cerebellar granule cell migration. *Dev. Brain Res.*, 75: 91–103.

N.M. Gerrits, T.J.H. Ruigrok and C.I. De Zeeuw (Eds.)
Progress in Brain Research, Vol 124

CHAPTER 6

The function of cerebellar Golgi cells revisited

Erik De Schutter*, Bart Vos and Reinoud Maex

Laboratory for Theoretical Neurobiology, Born-Bunge Foundation, University of Antwerp, B2610 Antwerp, Belgium

Introduction

The inhibitory interneurons of the cerebellar cortex have received little attention compared to the granule and Purkinje cells, and Golgi cells are no exception. Theoretical considerations of Golgi cell function have evolved little since the late sixties and experimental studies were sparse until the last few years. Recent modeling and in vivo experimental studies by our group, combined with in vitro experimental studies by others, have provided new insights into the properties of these cells which necessitate a revisiting of their function.

The connectivity of the Golgi cell

The anatomical facts are rather simple. The numerically most important input to the cerebellum is the mossy fiber system (Murphy and Sabah, 1971; Brodal and Bjaalie, 1997). If we limit ourselves to this input, the anatomy of cerebellar cortex can be described as a two-layered network. The input layer, corresponding to the granular layer, processes the incoming mossy fiber signals and transmits them by the parallel fiber system to the output layer, consisting mainly of the Purkinje cells. In both layers activity is controlled by inhibitory neurons, the Golgi cells in the input layer, and the basket and stellate cells in the output layer.

Mossy fibers activate both the excitatory granule cells and the inhibitory Golgi cells (Fig. 1A). The granule cell axon forms the parallel fibers, which not only transmit information to the output layer, but also provide additional excitatory input to Golgi cells. Each Golgi cell in turns inhibits the many granule cells present within the range of its axonal arbor (Eccles et al., 1966) with probably some overlap between adjacent Golgi cells. The combination of parallel fiber excitation of Golgi cells with their inhibition of granule cells constitutes a feedback inhibition circuit (Fig. 1C). The direct

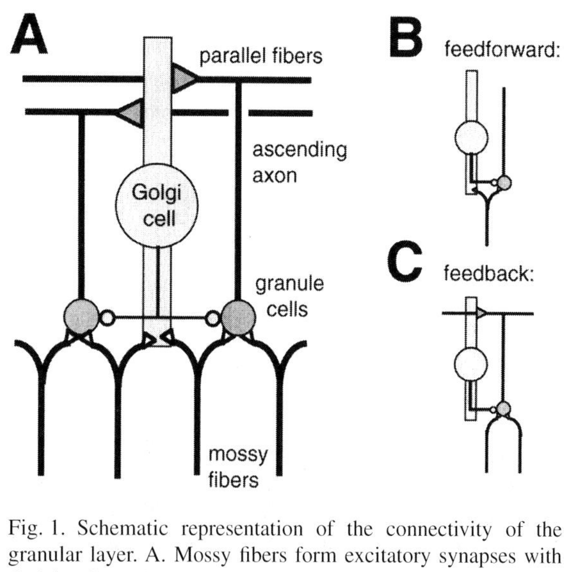

Fig. 1. Schematic representation of the connectivity of the granular layer. A. Mossy fibers form excitatory synapses with Golgi cells and granule cells (triangles), which in turn excite Golgi cells by the parallel fibers. Golgi cells make inhibitory contacts onto granule cells (circles). B. Representation of the components forming a feed-forward inhibition loop. C. Representation of the components forming a feedback inhibition loop.

*Corresponding author. Fax: +32–3–8202669;
e-mail: erik@bbf.uia.ac.be

excitation of Golgi cells by mossy fibers (Fig. 1B) provides a feed-forward connection. It should be noted, however, that the existence of mossy fiber contacts onto Golgi cells could not be confirmed in electron microscopal reconstructions of cerebellar glomeruli (Jakab and Hámori, 1988).

The classic view: feedback performs gain control

An early proposal on the function of Golgi cell inhibition was formulated by Eccles et al. (1967). They suggested that "this Golgi cell inhibition would suppress the discharges from all weakly excited granule cells, and thus would serve to focus the response to those granule cells strongly excited by the mf input".

Marr (1969), and most theoreticians since, formulated the function of Golgi cells differently. He saw them subserving the need of keeping "the numbers of active parallel fibres ... reasonably small over quite large variation in the number of active mossy fibres" and therefore introduced the concept of Golgi cells setting "the threshold of granule cells". In Marr's model Golgi cells regulate the codon size wich is the number of mossy fiber inputs onto a granule cell that need to be co-activated to make it cross threshold. In this theory it is important that "the codon size set for a particular mossy fibre input must depend only on that input; so that the same input is always translated into the same parallel fibres". Albus (1971), having an engineering background, used the word 'automatic gain control' to describe the feedback inhibition by Golgi cells. In both these theories gain control is assumed to keep the total output of the granular layer, i.e. the number of active parallel fibers, relatively constant over time. Marr suggests a range of 500–9,000 active granule cells contacting a single Purkinje cell, while Albus proposes a temporally and spatially constant activity level of about 1% of the parallel fibers.

Marr was the first to distinguish between the role of the 'descending' (or basal) dendrites of the Golgi cell which probably receive the mossy fiber contacts and its 'ascending' (or apical) dendrites on which parallel fiber synapses are located. He provided the first with a local and fast sampling role and the latter with a more delayed and global sampling function. "One cannot say that the local or global sample will always give the best solution". He proposed that the Golgi cell would perform an exclusive-OR function: only the strongest activated dendrite should determine Golgi cell output. This means that the mossy fiber and parallel fiber excitatory inputs are not allowed to summate. Though Marr did not specify his reasons in detail it is likely that synaptic summation would make the Golgi cell response too non-linear, leading to an unpredictable codon size. He notes that "refutation of this would be awkward but not fatal". Modern insights into the passive properties of dendrites and into the synaptic control over spike initiation make this part of his proposal unlikely, unless it is implemented by a very specific form of inhibition (Koch, 1998). Moreover, intracellular recordings from Golgi cells in rat cerebellar slice show that mossy fiber synapse responses are stronger than parallel fiber ones (Dieudonné, 1998a), suggesting that the first input would dominate if an exclusive-OR arrangement was implemented.

Later modeling studies by Pellionisz and Szentagothai (1973) demonstrated that a simple network model implementing either a pure mossy fiber activation of Golgi cells or a binary OR function between mossy and parallel fiber excitation performed good gain control over the parallel fiber activation during steady mossy fiber input, which largely confirmed Marr's view. In contrast, a network model with pure parallel fiber activation of Golgi cells caused a focusing effect similar to that proposed by Eccles et al. (1967). But while these modeling results were innovative at the time they were produced (Pellionisz and Szentagothai, 1973), they are not up to modern standards as no parameter space evaluation was performed to evaluate their robustness. Nevertheless, the latter results contradict Albus' intuition. He considered the difference in time delays between the mossy fiber feed-forward and parallel fiber feedback paths more important. For him the gain control function of the feedback pathway dominated and "the mossy fiber inputs to Golgi cells probably serve to stabilize parallel fiber rates under transient conditions". Keeler (1990) returned to the distinction between the effect of local compared to long-range sampling on threshold setting and suggested that the Golgi cell would operate as an "arbitrator in a bidding scheme".

Important features of a gain control circuit

At this point it is useful to consider the properties of a gain control function in more detail. First, because of the presence of the feedback inhibitory loop, it is inevitable that Golgi cells provide some degree of gain control over granule cell output. As has been pointed out by many authors before, such gain control could be quite useful in cerebellar cortex. Indeed, because of the massive excitatory projection to Purkinje cells, receiving more than 150,000 parallel inputs in the rat (Harvey and Napper, 1991), it may be important to prevent granule cells from collectively firing at high rates in response to strong mossy fiber input. If not in the context of a coding scheme as proposed by Marr (1969), Albus (1971) and (Ito, 1984), then maybe just to prevent excessive excitation of Purkinje cells up to neurotoxic levels (Brorson et al., 1995; Strahlendorf et al., 1998). But, considering the strong stimulus-evoked modulations in Purkinje cell firing rate that can often be observed (Simpson et al., 1989; Fu et al., 1997), it seems unlikely that the total parallel fiber activity is kept relatively constant like Albus proposed. A more realistic statement may be that a perfect gain controller should constrain the total parallel fiber activity impinging onto a Purkinje cell to a physiological range.

An important consideration in this context is the time frame over which the Golgi cell is presumed to exert its gain control function. In essence the theories by Marr and Albus are based on a rate-coding assumption (Rieke et al., 1997). The cerebellum is assumed to code information by the average firing rate of units, e.g. Marr (1969) formulates it as "a signal in a mossy fibre is represented by a burst of impulses lasting many tens of milliseconds; and that a signal from a Purkinje cell is represented by a prolonged increase in its firing rate". In later paragraphs we will argue that typical Golgi cell responses to punctate stimuli are so accurately timed and short that a rate coding scheme is unlikely. While not everybody may agree to such a statement, most would concur that changes in spike frequencies over time windows of only a few tens of msecs are relevant to cerebellar function in sensorimotor coordination and in timing in general (Welsh et al., 1995; Raymond et al., 1996; Ivry, 1997; Thach, 1998). In other words, a perfect gain controller will have to operate on this tens of msecs time scale to be useful for cerebellar processing. If the gain controller is only concerned with preventing neurotoxicity it might still be effective if it operates over much longer time scales.

Before considering the possible gain control function of Golgi cells in more detail, recent modeling and experimental results from our laboratory will be reviewed. More information about these results can be found at http://www.cerebellum.org

A network model predicts synchronous dynamics

Previous modeling studies of cerebellar cortex have always been based on the mean firing rates of cerebellar neurons (e.g. by using logical units (Moore et al., 1989) or rate coding units (Chauvet and Chauvet, 1995; Schweighofer et al., 1998)). Two previous modeling studies have considered to some degree the temporal dynamics specific to spiking neurons. Fujita (1982) predicted that Golgi cells would cause phase differences between co-activated granule cells. But this model made a large number of simplifying assumptions (e.g. Golgi cells behave as leaky integrator units) and did not actually simulate the spiking dynamics. The model by Pellionisz and Szentagothai (1973) computed spike times of logical units. While the authors did not recognize the importance of this observation, their model produced oscillatory patterns of parallel fiber activity when parallel fiber input to Golgi cells dominated (their Fig. 8).

Such oscillatory dynamics are a consequence of well-known, but unique, property of the granule cell–Golgi cell circuit. In fact, the feedback inhibition loop is absolutely pure as no synaptic connections exist between granule cells or between Golgi cells (Ito, 1984; Voogd and Glickstein, 1998), which is in contrast to circuits in cerebral cortex. Based on theoretical considerations Freeman

(1972) already recognized that such a pure feedback circuit would be driven to oscillatory dynamics, but he did not realize that it exists in the cerebellar cortex.

Therefore, retrospectively, it should have come as no surprise that we found our spiking neuron network model of the cerebellar granular layer to be highly prone to synchronous oscillations (Maex and De Schutter, 1998b). A typical example of such a simulation is shown in the raster plots of Fig. 2A. Initially no input is provided (the mossy fibers are silent). Granule cells do not fire spontaneously (D'Angelo et al., 1995), but Golgi cells do

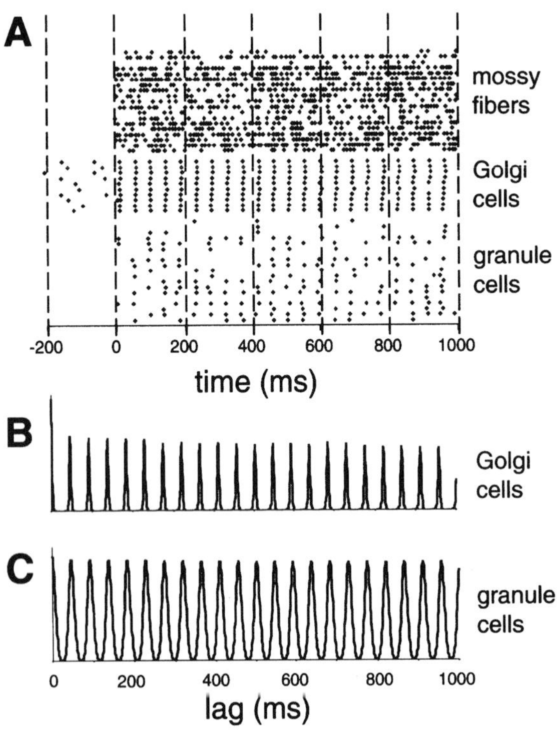

Fig. 2. Simulation of network dynamics in the granular layer. A. Raster plots of activity in a subset of mossy fibers, Golgi cells and granule cells. B. Autocorrelogram of the Golgi cell population. C. Autocorrelogram of the granule cell population. The mossy fiber input is initially silent, after 200 msec mossy fibers start firing randomly with average firing rates of 5 to 75 Hz. Total network size was 30 Golgi cells, 5355 granule cells and 540 mossy fibers, with an average 602 parallel fiber synapses on each Golgi cell. Golgi and granule neurons were modeled as single compartments containing several voltage and calcium-gated channels. See Maex and De Schutter (1998b) for more details.

(Dieudonné, 1998b) and in the network this results in a desynchronized firing (Fig. 2A left). Immediately upon activation of the mossy fibers, however, Golgi cells synchronize and start firing rhythmically. The granule cells are entrained in this synchronous oscillation, but they fire less precisely (as evidenced by the broader peaks in their autocorrelogram in Fig. 2C) and often skip cycles (Fig. 2A bottom).

The generation of these synchronous oscillations is easily explained by the intrinsic dynamics of the feedback inhibition circuit. Lets first consider a single Golgi cell and its population of postsynaptic granule cells. Inhibitory neurons can exert a strong influence on the exact timing of action potentials in their target neurons (Lytton and Sejnowski, 1991; Cobb et al., 1995). The simulated granule cells tend to fire when inhibition is at its lowest, which is just before the next Golgi cell spike. Because a single Golgi cell inhibits many granule cells (Palkovits et al., 1971), this means that a large population of granule cells shall fire at about the same time. The loosely synchronous granule cell activity excites the Golgi cell and causes it to fire immediately, leading to the establishment of a synchronized oscillation within this small circuit (comparable to 'the comparator circuit' of Kammen et al. (1989)). The unique structure of the cerebellar cortex couples many of these oscillatory circuits together through the common parallel fiber input to the Golgi cells. Consequently the long parallel fibers (Pichitpornchai et al., 1994) cause Golgi cells and granule cells located along the same transverse axis to fire synchronously.

It is important to realize that this is a *dynamic* property of the circuitry: once the granular layer is activated beyond a minimum level the most stable form of spiking is a synchronous oscillation (Maex and De Schutter, 1998b). This has several consequences. First, the synchronization is immediate upon activation (Fig. 2), there is no delay due to the slow parallel fiber conduction velocity (Maex and De Schutter, 1999). Second, the accuracy of synchronization will increase with increased network activity, which corresponds to an increased Golgi cell firing rate (Maex and De Schutter, 1998b). This leads to a fire-rate dependency of the synchronization (Maex et al., 1998). Third, the synchronization requires a minimal number of weak parallel fiber synapses onto each Golgi cell (Maex and De Schutter, 1998a). The strength of parallel fiber synapses onto Golgi cells is indeed rather small (Dieudonné, 1998b). Fourth, synchronization occurs despite the slow conduction velocities of the parallel fibers (Vranesic et al., 1994) and entrains Golgi cells over the complete extent of the transverse axis where mossy fibers are activated, even if this is much longer than the mean parallel fiber length.

These observations emphasize the importance of the parallel fiber excitation of Golgi cells. Direct mossy fiber input to Golgi cells can desynchronize the network as it may cause Golgi cell spikes which are not coupled to the oscillatory pattern. In simulations we found that this depended on the relative strength of mossy fiber compared to parallel fiber synapses. As long as the total synaptic conductance of all mossy fiber inputs onto a Golgi cell did not exceed that of all parallel fiber inputs, which are individually weaker (Dieudonné, 1998a, b) but probably much more numerous (e.g. Pellionisz and Szentagothai, 1973), the network dynamics were unaffected (Maex and De Schutter, 1998b).

Parallel fibers synchronize Golgi cells

The network model predicted that Golgi cells receiving common parallel fiber input should synchronize. As the dynamics described in the previous paragraph depend on common parallel fiber excitation one expects that Golgi cells which are separated along the parasagittal axis will not fire synchronously, unless they are so close to each other that their dendritic trees may overlap (Dieudonné, 1998b).

These predictions were confirmed using multi single unit recordings of spontaneous Golgi cell activity in the cerebellar hemisphere of the anesthetized rat (Vos et al., 1999a). Two or three electrodes were placed in one of two orientations: transverse or sagittal. Golgi cells were identified based on their characteristic spiking patterns (Edgley and Lidierth, 1987; VanKan et al., 1993; Atkins et al., 1997) and their positions in the granular layer were confirmed histologically. The results were independent of the kind of anesthesia used.

A total of 42 Golgi cell pairs in 38 ketamine-xylazine anesthetized rats were recorded. Of these 26 pairs were positioned along the transverse axis, with the other pairs along the sagittal axis. Synchronization was measured as the height of the central peak in the normalized cross-correlogram. All transverse pairs except one showed a highly significant coherence. An example of a pair of Golgi cells along the tranverse axis is shown in Fig. 3. The synchrony between these cells was highly significant. The central peak had a z-score of 6.4 (range for all: 3.0–9.3), but was rather wide (half-width of 21 ms, range: 11.5–39.5 ms). Conversely, in 12 out of 16 sagittal pairs no synchrony could be found. The remaining four sagittal pairs showed low levels of coherence (z-score <4), but in each of these four pairs the Golgi cells were located within 200 μm from each other.

These findings confirm the first prediction generated by our network simulations. Additionally, as also predicted by the model, the accuracy of synchronization, evidenced by increasing z-scores and decreasing widths, increased with Golgi cell firing rate (Fig. 1 of Vos et al., 1999a). This has the important implication that synchronization may be much more accurate in awake animals, compared to the loose synchrony observed in the anesthetized rat. The only data available from awake animals to date are field potential recordings (Pellerin and Lamarre, 1997; Hartmann and Bower, 1998), which are more suitable for measuring oscillatory activity than our single unit recordings. The studies in awake animals have indeed demonstrated the presence of oscillations in the granular layer that may correspond to those predicted by the model.

Stimulus-evoked responses of Golgi cells

We will summarize the results described in the companion paper (Vos et al., 2000) and elsewhere (Vos et al., 1999b) because they are relevant to the discussion presented next. Golgi cell receptive fields for punctate tactile stimuli are very large and often bilateral which is in contrast to the fractured somatotopy observed in field potential recordings (presumably mostly reflecting activation of mossy fiber synapses; e.g. Shambes et al., 1978; Bower et al., 1981; Bower and Kassel, 1990). Within these receptive fields one can distinguish different response patterns which consist usually of two components, an early one reflecting trigeminal inputs and a later one reflecting corticopontine activation, as was also found in granular layer field potential recordings (Bower et al., 1981; Morissette and Bower, 1996). Especially the early component of the response is highly accurate in timing and a subclass of responses with two or more sharp peaks in this component can be distinguished. This subclass can be evoked from only a restricted area of the face and has a very short latency. It probably reflects direct mossy fiber activation of the Golgi cell. Because of their large receptive fields the other response types are presumably caused by parallel fiber activation. Finally, most of the Golgi cell responses are distinguished by a long silent period following the excitatory response.

Is gain control the main function of Golgi cells?

Because of the feedback connectivity it is inevitable that Golgi cells perform some level of gain control over the output of granule cells. This does not imply, however, that gain control is their only, or even main, function. In this section we will argue, based on the response characteristics described earlier, that Golgi cells are bad gain controllers. To put these arguments in perspective it is useful to repeat our definition of an ideal gain control circuit first. It should keep parallel fiber input to a Purkinje cell within a range which is much smaller than the range of possible mossy fiber input activities (Marr, 1969), and it should operate on the order of tens of msec.

A first question to consider is whether the large receptive fields observed in Golgi cell responses are compatible with a gain control function. In view of the large receptive fields a tactile stimulus on a small part of the face will lead to inhibition of most of the granule cells in crus I/II, even though many of them actually receive mossy fiber input from other parts of the face. The usefulness of such a diffuse inhibition for a gain control circuit will

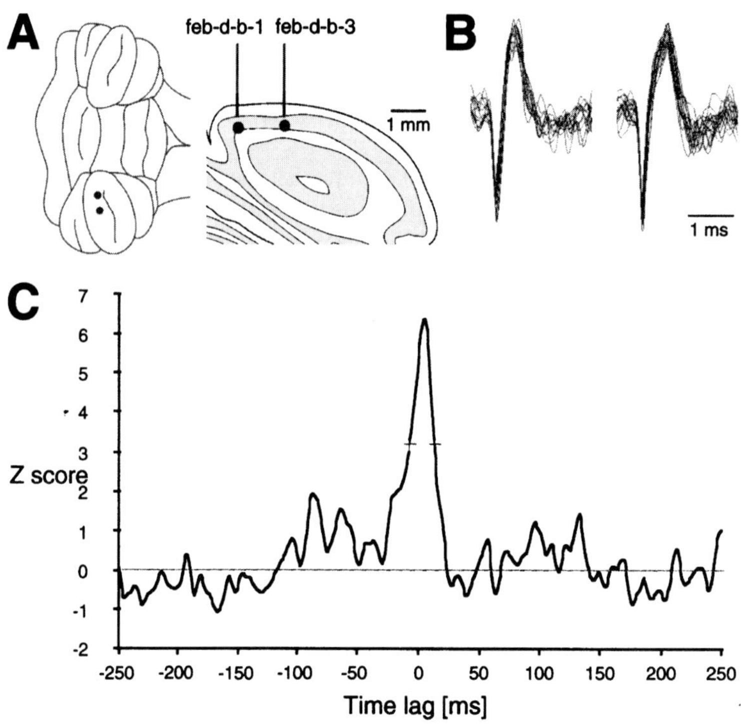

Fig. 3. Synchronous activity of two Golgi cells simultaneously recorded in anesthetized rat. A. Location of recording sites in crus II. B. Examples of extracellular waveforms of each cell. C. Normalized cross-correlogram. See Vos et al. (1999a) for more detail on experimental methods.

depend on how Purkinje cells are excited by granule cell activity. If one believes that Purkinje cells are activated by parallel fiber beams (Eccles et al., 1967), without distinction of the origin of the parallel fiber signal, then the observed arrangement would seem best. The large receptive fields of Golgi cells are probably due to parallel fiber activation (Vos et al., 1999b, c), which would ensure a good global sampling of parallel fiber activity for the gain controller. But if Purkinje cells are mainly activated by underlying granule cell activity, as demonstrated in electrophysiological recordings (Bower and Woolston, 1983) and in recent optical imaging (Cohen and Yarom, 1998, 2000), then local sampling is needed. Indeed, it is presumed that the restricted Purkinje cell responses are due to activation by synapses from the ascending granule cell axon (Gundappa-Sulur et al., 1999) while parallel fiber synapses seem much less effective in exciting the cell (Cohen and Yarom, 1998). This means that the large receptive fields of Golgi cells would be rather unsuitable for controlling the gain of Purkinje cell excitation.

The large receptive fields fit better with the "arbitrator in a bidding scheme" function suggested by Keeler (1990). A more intuitive way of describing this effect is that the granular layer circuitry performs lateral inhibition along the parallel fiber axis. Note that while the effects are similar to those of classic lateral inhibition, with a potential for implementation of a "first area activated wins all" scheme, the mechanisms are not. The parallel fibers provide lateral *excitation* to the *inhibitory* Golgi cells so that a highly activated patch of granule cells can lead to inhibition of nearby patches. More recent recordings suggest that the size of the stimulus is not an important parameter in this lateral inhibition (Volny-Luraghi et al., 1999).

While the importance of the large receptive fields will depend on how Purkinje cells are excited in vivo, the fine and accurate timing of Golgi cell responses is difficult to reconcile with fast gain control. These responses do not fit well with a rate coding scheme as the average firing rate is a poor measure of the true Golgi cell response, consisting of a few accurately timed spikes within 30 msec followed by a silent period of 100 msec or more (e.g. see Fig. 5). In fact, the average firing rate of Golgi cells changes only very little when their responses to tactile input are averaged over 300 ms (Fig. 8 of Vos et al., 1999b). As a consequence, it is unlikely that Golgi cells can provide smooth inhibiton of granule cells. While it is not possible to isolate granule cells responses in vivo, one can estimate their firing rates to some degree from the simple spike patterns in Purkinje cells. Figure 4 shows an example of a cross-correlogram between a Golgi cell and a nearby Purkinje cell during spontaneous activity. A pronounced dip is observed which peaks just past the zero time point. This indicates that on average a single Golgi cell spike is sufficient to diminish granule cell activity so much that the Purkinje cell is less likely to spike for more than 100 msec. In other words, Golgi cell inhibition does not maintain a steady parallel fiber activation of Purkinje cells during the tens of msec following each Golgi cell spike.

This clearly violates our definition of an ideal gain controller. In fact, it can be interpreted as an overcompensation of the signal by the gain controller as Golgi cell spiking seems to lead to below average activity of granule cells. Notice also that it would have been perfectly possible to design the circuitry so that this problem does not arise. If Golgi cells were much faster firing neurons and if they didn't show silent periods their rate could vary much more smoothly, which would allow for a continuously varying inhibition of granule cells as opposed to the discrete inhibition observed in Figs 4 and 5. The fact that this is not the case for Golgi

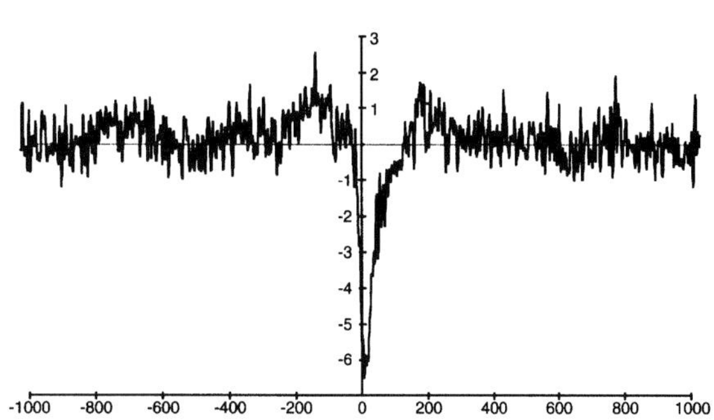

Fig. 4. Cross-correlogram between spontaneous activity of a Golgi cell and of a Purkinje cell recorded in crus IIA of the anesthetized rat. Notice the large dip in Purkinje cell activity starting at the time of the Golgi cell spike (zero time point) and during more than 100 msec.

cells implies that they have not been optimized by evolution to perform an ideal gain control function.

The significance of the long silent period that follows the initial Golgi cell response evoked by tactile stimulation (Fig. 5) is even more difficult to explain in the context of a gain control function. Silent periods are also present in other parts of the somatosensory system (Mountcastle et al., 1957; Mihailoff et al., 1992; Nicolelis and Chapin, 1994), where they are assumed to be the consequence of local feedforward inhibition loops (Dykes et al., 1984). Note, however, that in somatosensory cortex these silent periods are observed in excitatory neurons while inhibitory neurons remain active (Brumberg et al., 1996). This is clearly not the case in the granular layer where a silent period is observed in the inhibitory neuron.

It is not known what the excitatory granule cells do during this silent period. Theoretical analysis shows that this depends on what causes the silent periods observed in Golgi cell responses. Four potential sources can be distinguished which may very well operate together. First, it is likely that the diminished mossy fiber input due to the silent periods observed elsewhere in the somatosensory system plays a role. But this cannot be the only explanation as simultaneously recorded Golgi cells can show different duration silent periods in response to the same stimulus (Fig. 5), indicating that factors intrinsic to Golgi cells (Dieudonné, 1998b) or local circuit effects contribute. One such local circuit effect may be the strong tonic component of granule cell inhibition which has recently been described in cerebellar slice preparations (Brickley et al., 1996; Wall and Usowicz, 1997). This tonic inhibition is probably caused by Golgi cell activity resulting in a spillover of GABA within the glomerulus (Rossi and Hamann, 1998), which then activates high affinity receptors on granule cells (Nusser et al., 1998). It is therefore expected to correlate with the amplitude of Golgi cell responses, similar to the correlation between the duration of the silent period and the excitatory response amplitude (Fig. 8 of Vos et al., 1999b). The tonic inhibition of granule cells will cause an

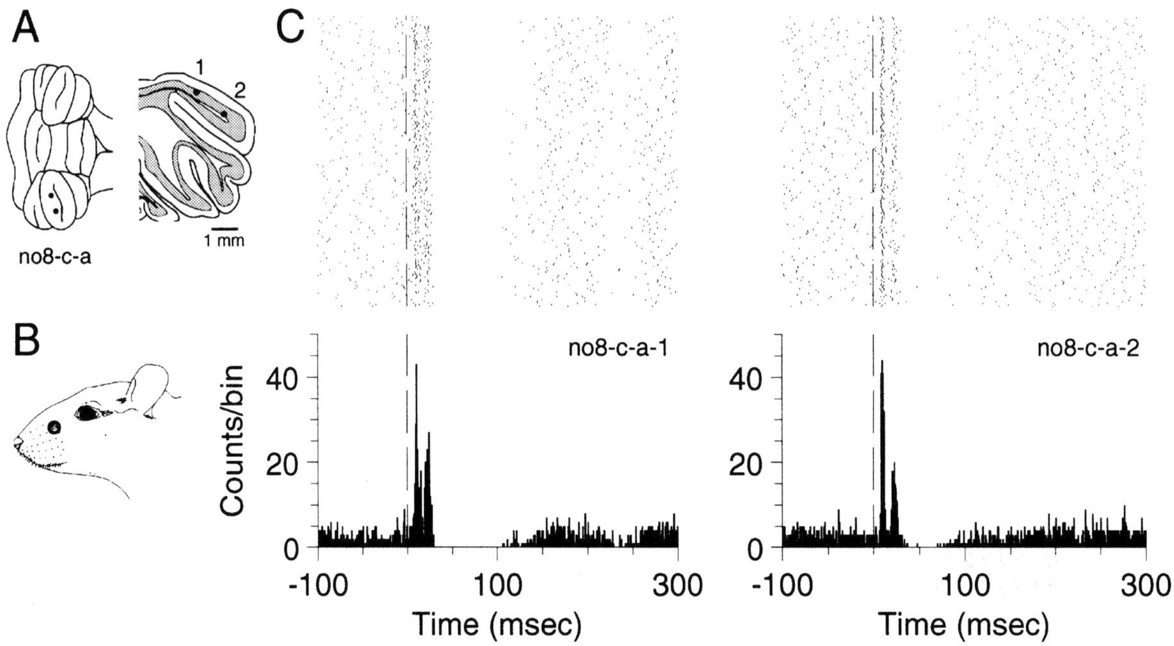

Fig. 5. Responses to tactile stimulation around the A1 vibrissa of two Golgi cells simultaneously recorded in crus IIA of the anesthetized rat. Notice large difference in duration of the silent period after the excitatory response. See Vos et al. (1999b) for more detail on experimental methods.

interruption of parallel fiber excitation to the Golgi cell, enhancing the effect of diminished mossy fiber input during the silent period. Another local circuit effect could be inhibition of the Golgi cell by molecular layer interneurons (Dieudonné, 1995). Compared to tonic inhibition, which may cause a complete silencing of local granular layer activity, molecular layer inhibition of Golgi cells and/or an afterhyperpolarisation intrinsic to Golgi cells (Dieudonné, 1998b) will cause a disinhibition of local granule cells. While these effects are opposite, both seem unsuitable for a gain control circuit. As both effects may have important implications for how the cerebellum operates it will be important to elucidate the cause of the silent periods.

A beam of activated Golgi cells

In the previous section we raised serious doubts about the role of Golgi cells as gain controllers. In this and the following section we will speculate on alternative functions of Golgi neurons.

An overall conclusion of the results reviewed here is that the parallel fiber input to Golgi cells does not only provide for a feedback inhibition of granule cells but also conveys a spatial organization onto Golgi cells. Indeed, while both old (Bower and Woolston, 1983) and new (Cohen and Yarom, 1998, 2000) experimental data demonstrate the absence of parallel fiber beams at the level of Purkinje cell activity as proposed by Eccles and colleagues in the sixties, our multi-unit recordings indicate that Golgi cells are organized into such beams (Vos et al., 1999a). Moreover, the stimulus-evoked responses can be explained only by assuming an important contribution of the parallel fibers which transfer signals from distant mossy fiber receptive fields (Vos et al., 1999b, c).

The functional importance of this particular spatial organization may be twofold. First, as already mentioned above, due to their large receptive fields Golgi cells can provide a form of lateral inhibition along the parallel fiber axis. Second, by synchronizing the firing of granule cell spikes (e.g. Fig. 2), they may organize granule cells into functional assemblies (König and Schillen, 1991; Nicolelis et al., 1997) along the parallel fiber axis.

The control of timing of granule cell spiking

A consistent finding in our modeling and experimental work is that Golgi cells exert a strong control over the timing of granule cell spikes. The network model predicts that granule cells will preferentially spike just before their afferent Golgi cell (Fig. 2) and the example in Fig. 4 suggests indeed a strong effect of Golgi cell activation on the timing of granule cell spikes. This control over timing of granule spikes could have functional consequences at many levels.

At the cellular level Golgi cells entrain granule cells into an oscillatory rhythm which may make the latter more prone to burst (i.e. fire a few spikes at high frequency). Such behavior can be observed in the network model for certain ranges of parameter values (unpublished observations) and burst firing has been observed in granule cells in slice (D'Angelo et al., 1998). Similar to what has been described in many invertebrate systems (Calabrese and De Schutter, 1992; Marder and Calabrese, 1996), an oscillatory pattern of inhibition may thus turn regular spiking granule cells into bursting neurons. This would have important functional consequences as the parallel fiber synapse onto Purkinje cells is quite sensitive to high frequency activation, which preferentially activates the metabotropic glutamate receptor (Batchelor and Garthwaite, 1997; Tempia et al., 1998) leading to calcium release from intracellular stores (Finch and Augustine, 1998).

At the network level Golgi cells impose a loose synchronization onto granule cells lying along the same parallel fiber axis (Figs. 2 and 3). Such synchronization may contribute to the transformation of spatial patterns present in the mossy fiber input into temporal patterns which are transmitted along the parallel fiber system. Hopfield (1995) hypothesized that the nervous system might use the relative phase lags between spikes to encode information. In the context of pattern recognition such a temporal code has the advantage of being much less sensitive to stimulus amplitude than standard rate codes. Because the mossy fiber input to the cerebellar cortex has a patchy, fractured somatotopy (Shambes et al., 1978; Bower et al., 1981; Bower and Kassel, 1990) tactile stimuli are expected to co-activate several patches in the granular layer (see also Peeters et al., 1999). In other words, mossy fiber inputs caused by different stimuli may be distinguished, at least partially, by the complex spatial pattern of activation of patches. But how can such a pattern be recognized by Purkinje cells, especially if one assumes a population coding of the signal? Because of the extended length of the parallel fibers (Mugnaini, 1983; Pichitpornchai et al., 1994) signals originating from granule cells in multiple patches will become mixed together, making their origin and their relation to each other obscure unless one can rely on the relative timing of the spikes. By synchronizing granule cell spiking along the parallel fiber axis it becomes possible to convert the spatial code implicit in the patchy pattern of mossy fiber activity into a reliable temporal code of phase differences between parallel fiber spikes originating in distinct patches. Even though the variable conduction velocities of parallel fibers (Bernard and Axelrad, 1991; Vranesic et al., 1994) will cause small phase differences between spikes from identical patches, each temporal pattern of phase differences recorded at a particular location along the transverse axis of the folium will correspond to a unique spatial pattern of mossy fiber activation. In support of this hypothesis we have found that the only stimulus aspect that determines the fine temporal shape of Golgi cell responses is the receptive field being activated (which determines the spatial pattern of input), while the stimulus has little or no effect (Vos et al., 1999b, 2000).

But many questions remain to be answered before the temporal code hypothesis can be taken for granted. First, in anesthetized rats the synchronization of Golgi cells is rather loose, but it is predicted to be much more accurate in awake animals (Maex et al., 1998; Vos et al., 1999a). Nevertheless it remains to be proven that sufficiently precise synchronization can be achieved to support temporal coding. Next, it remains unclear if Purkinje cells can decode this temporal information. Our Purkinje cell model (De Schutter and Bower, 1994a, b) is a very poor coincidence detector (De Schutter, 1998; Santamaria et al., 2000). This is not surprising as the active dendrite of Purkinje cells does not contain fast sodium channels (Stuart and Häusser, 1994), but, instead, much slower activating calcium channels (Regan, 1991; Usowicz et al., 1992) which will determine its window of temporal integration. So while Braitenberg et al. (1997) and Meek (1992) have proposed that Purkinje cells would act as coincidence detectors of externally generated temporal patterns (they do not assume any processing by the granular layer), we have suggested instead that these temporal patterns specifically change the Purkinje cell responsiveness to activation of its ascending granule cell axon synapses, with an effect similar to a complex filtering function (De Schutter, 1995, 1998). An alternative hypothesis is that Purkinje cells may learn to recognize particular temporal patterns (Steuber and Willshaw, 1999).

Conclusions

We have presented experimental evidence that Golgi cells perform poorly as gain controllers at the time scales of interest for cerebellar motor control. While their true function in cerebellar cortex remains to be proven, our modeling and experimental data suggest the presence of beams of Golgi cells being synchronized by common parallel fiber input. This may result in lateral inhibition by Golgi cells, but also in a tight control over the timing of granule spikes which could implement a temporal code. Additional modeling and experimental work is needed to clarify these issues.

Acknowledgements

Our research is funded by the University of Antwerp (GOA 99/2/33), FWO Flanders (G.0113.96), IUAP Belgium (P4/22) and the European Commission (BIO4-CT98–0182).

References

Albus, J.S. (1971) A theory of cerebellar function. *Math. Biosci.*, 10: 25–61.

Atkins, M.J., VanAlphen, A.M. and Simpson, J.I. (1997) Characteristics of putative Golgi cells in the rabbit cerebellar flocculus. *Abstr. Soc. Neurosci.*, 23: 1287.

Batchelor, A.M. and Garthwaite, J. (1997) Frequency detection and temporally dispersed synaptic signal association through a metabotropic glutamate receptor pathway. *Nature*, 385: 74–78.

Bernard, C. and Axelrad, H. (1991) Propagation of parallel fiber volleys in the cerebellar cortex: a computer simulation. *Brain Res.*, 565: 195–208.

Bower, J.M., Beerman, D.H., Gibson, J.M., Shambes, G.M. and Welker, W. (1981) Principles of organization of a cerebro-cerebellar circuit. Micromapping the projections from cerebral (SI) to cerebellar (granule cell layer) tactile areas of rats. *Brain Behav. Evol.*, 18: 1–18.

Bower, J.M. and Kassel, J. (1990) Variability in tactile projection patterns to cerebellar folia crus IIA of the Norway rat. *J. Comp. Neurol.*, 302: 768–778.

Bower, J.M. and Woolston, D.C. (1983) Congruence of spatial organization of tactile projections to granule cell and Purkinje cell layers of cerebellar hemispheres of the albino rat: vertical organization of cerebellar cortex. *J. Neurophysiol.*, 49: 745–766.

Braitenberg, V., Heck, D. and Sultan, F. (1997) The detection and generation of sequences as a key to cerebellar function. Experiments and theory. *Behav. Brain Sci.*, 20: 229–245.

Brickley, S.G., Cull-Candy, S.G. and Farrant, M. (1996) Development of a tonic form of synaptic inhibition in rat cerebellar granule cells resulting from presisitent activation of $GABA_A$ receptors. *J. Physiol.*, 497: 753–759.

Brodal, P. and Bjaalie, J.G. (1997) Salient anatomic features of the cortico-ponto-cerebellar pathway. *Prog. Brain Res.*, 14: 227–249.

Brorson, J.R., Manzolillo, P.A., Gibbons, S.J. and Miller, R.J. (1995) AMPA receptor desensitization predicts the selective vulnerability of cerebellar Purkinje cells to excitotoxicity. *J. Neurosci.*, 15: 4515–4524.

Brumberg, J.C., Pinto, D.J. and Simons, D.J. (1996) Spatial gradients and inhibitory summation in the rat whisker barrel system. *J. Neurophysiol.*, 76: 130–140.

Calabrese, R.L. and De Schutter, E. (1992) Motor pattern generating networks in invertebrates: modeling our way toward understanding. *Trends Neurosci.*, 15: 439–445.

Chauvet, P. and Chauvet, G.A. (1995) Mathematical conditions for adaptive control in Marr's model of the sensorimotor system. *Neur. Networks*, 8: 693–706.

Cobb, S.R., Buhl, E.H., Halasy, K., Paulsen, O. and Somogyi, P. (1995) Synchronization of neuronal activity in hippocampus by individual GABAergic interneurons. *Nature*, 378: 75–78.

Cohen, D. and Yarom, Y. (1998) Patches of synchronized activity in the cerebellar cortex evoked by mossy-fiber stimulation: questioning the role of parallel fibers. *Proc. Natl. Acad. Sci. USA*, 98: 15032–15036.

Cohen, D. and Yarom, Y. (2000) Unravelling cerebellar circuitry: an optical imaging study. *Prog. Brain Res.*, 124: 107–114.

D'Angelo, E., DeFilippi, G., Rossi, P. and Taglietti, V. (1995) Synaptic excitation of individual rat cerebellar granule cells *in situ*: evidence for the role of NMDA receptors. *J. Physiol.*, 482: 397–413.

D'Angelo, E., Filippi, G.D., Rossi, P. and Taglietti, V. (1998) Ionic mechanism of electroresponsiveness in cerebellar granule cells implicates the action of a persistent sodium current. *J. Neurophysiol.*, 80: 493–503.

De Schutter, E. (1995) Cerebellar long-term depression might normalize excitation of Purkinje cells: a hypothesis. *Trends Neurosci.*, 18: 291–295.

De Schutter, E. (1998) Dendritic voltage and calcium-gated channels amplify the variability of postsynaptic responses in a Purkinje cell model. *J. Neurophysiol.*, 80: 504–519.

De Schutter, E. and Bower, J.M. (1994a) An active membrane model of the cerebellar Purkinje cell.I. Simulation of current clamps in slice. *J. Neurophysiol.*, 71: 375–400.

De Schutter, E. and Bower, J.M. (1994b) An active membrane model of the cerebellar Purkinje cell: II. Simulation of synaptic responses. *J. Neurophysiol.*, 71: 401–419.

Dieudonné, S. (1995) Glycinergic synaptic currents in Golgi cells of the rat cerebellum. *Proc. Natl. Acad. Sci. USA*, 92: 1441–1445.

Dieudonné, S. (1998a). *Etude fonctionelle de deux interneurones inhibiteurs du cortex cérébelleux: les cellules de Lugaro et de Golgi*. Unpublished doctoral thesis, Université Pierre et Marie Curie, Paris VI.

Dieudonné, S. (1998b) Submillisecond kinetics and low efficacy of parallel fibre-Golgi cell synaptic currents in the rat cerebellum. *J. Physiol.*, 510: 845–866.

Dykes, R.W., Landry, P., Metherate, R. and Hicks, T.P. (1984) Functional role of GABA in cat primary somatosensory cortex: shaping receptive fields of cortical neurons. *J. Neurophysiol.*, 52: 1066–1093.

Eccles, J.C., Ito, M. and Szentagothai, J. (1967) *The Cerebellum as a Neuronal Machine*. Springer-Verlag, Berlin.

Eccles, J.C., Llinás, R.R. and Sasaki, K. (1966) The mossy fibre-granule cell relay of the cerebellum and its inhibitory control by Golgi cells. *Exp. Brain Res.*, 1: 82–101.

Edgley, S.A. and Lidierth, M. (1987) The discharges of cerebellar Golgi cells during locomotion in the cat. *J. Physiol.*, 392: 315–332.

Finch, E.A. and Augustine, G.J. (1998) Local calcium signalling by inositol-1,4,5-trisphosphate in Purkinje cell dendrites. *Nature*, 396: 753–756.

Freeman, W.J. (1972) Waves, pulses, and the theory of neural masses. In: R. Rosen and F.M. Snell (Eds), *Progress in Theoretical Biology*. (Vol. 2). Academic Press, New York, pp. 87–165.

Fu, Q.-G., Flament, D., Coltz, J.D. and Ebner, T.J. (1997) Relationship of cerebellar Purkinje cell simple spike discharge to movement kinetics in the monkey. *J. Neurophysiol.*, 78: 478–491.

Fujita, M. (1982) Adaptive filter model of the cerebellum. *Biol. Cybern.*, 45: 195–206.

Gundappa-Sulur, G., De Schutter, E. and Bower, J.M. (1999) The ascending granule cell axon: an important component of the cerebellar cortical circuitry. *J. Comp. Neurol.*, 408: 580–596.

Hartmann, M.J. and Bower, J.M. (1998) Oscillatory activity in the cerebellar hemispheres of unrestrained rats. *J. Neurophysiol.*, 80: 1598–1604.

Harvey, R.J. and Napper, R.M.A. (1991) Quantitative studies of the mammalian cerebellum. *Prog. Neurobiol.*, 36: 437–463.

Hopfield, J.J. (1995) Pattern recognition computation using action potential timing for stimulus representation. *Nature*, 376: 33–36.

Ito, M. (1984) *The Cerebellum and Neural Control*. Raven Press, New York.

Ivry, R. (1997) Cerebellar timing systems. In: J.D. Schmahmann (Ed.), *The Cerebellum and Cognition*. (Vol. 41). Academic Press, San Diego, pp. 556–573.

Jakab, R.I. and Hámori, J. (1988) Quantitative morphology and synaptology of cerebellar glomeruli in the rat. *Anat. Embryol.*, 179: 81–88.

Kammen, D.M., Holmes, P.J. and Koch, C. (1989) Cortical architecture and oscillations in neuronal networks: Feedback versus local coupling. In: R.M.J. Cotterill (Ed.), *Models of Brain Function*. Cambridge University Press, Cambridge, pp. 273–284.

Keeler, J.D. (1990) A dynamical system view of cerebellar function. *Physica D*, 42: 396–410.

Koch, C. (1998) *Biophysics of Computation: Information Processing in Single Neurons*. Oxford University Press.

König, P. and Schillen, T.B. (1991) Stimulus-dependent assembly formation of oscillatory responses: I. Synchronization. *Neural Comput.*, 3: 155–166.

Lytton, W.W. and Sejnowski, T.J. (1991) Simulations of cortical pyramidal neurons synchronized by inhibitory interneurons. *J. Neurophysiol.*, 66: 1059–1079.

Maex, R. and De Schutter, E. (1998a). The critical synaptic number for rhythmogenesis and synchronization in a network model of the cerebellar granular layer. In: L. Niklasson, M. Bodén and T. Ziemke (Eds), *ICANN 98*. Springer-Verlag, London, pp. 361–366.

Maex, R. and De Schutter, E. (1998b) Synchronization of Golgi and granule cell firing in a detailed network model of the cerebellar granule cell layer. *J. Neurophysiol.*, 80: 2521–2537.

Maex, R. and De Schutter, E. (1999). An optimal connection radius for long-range synchronization. In: D. Willshaw (Ed.), *ICANN 99*, IEE Proceedings No. 470: 557–562.

Maex, R., Vos, B.P. and De Schutter, E. (1998) The height and width of central peaks on spike train cross-correlograms. *Eur. J. Neurosci.*, Suppl. 10. 303.

Marder, E. and Calabrese, R.L. (1996) Principles of rhythmic motor pattern generation. *Physiol. Rev.*, 76: 687–717.

Marr, D.A. (1969) A theory of cerebellar cortex. *J. Physiol.*, 202: 437–470.

Meek, J. (1992) Why run parallel fibers parallel? Teleostean Purkinje cells as possible coincidence detectors, in a timing device subserving spatial coding of temporal differences. *Neuroscience*, 48: 249–283.

Mihailoff, G.A., Kosinski, R.J., Azizi, S.A., Lee, H.S. and Border, B.G. (1992) The expanding role of the basilar pontine nuclei as a source of cerebellar afferents. In: R.R. Llinás, and C. Sotelo (Eds), *The Cerebellum Revisited*. Springer-Verlag, Berlin, pp. 135–164.

Moore, J.W., Desmond, J.E. and Berthier, N.E. (1989) Adaptively timed conditioned responses and the cerebellum: a neural network approach. *Biol. Cybern.*, 62: 17–28.

Morissette, J. and Bower, J.M. (1996) Contribution of somatosensory cortex to responses in the rat cerebellar cortex granule cell layer following peripheral tactile stimulation. *Exp Brain Res*, 109, 240–250.

Mountcastle, V.B., Davies, P.W. and Berman, A.L. (1957) Response properties of neurons of cat's somatic sensory cortex to peripheral stimuli. *J Neurophysiol*, 20, 374–407.

Mugnaini, E. (1983) The length of cerebellar parallel fibers in chicken and rhesus monkey. *J. Comp. Neurol.*, 220: 7–15.

Murphy, J.T. and Sabah, N.H. (1971) Cerebellar Purkinje cell responses to afferent inputs.II. Mossy fiber activation. *Brain Res.*, 25: 469–482.

Nicolelis, M.A.L. and Chapin, J.K. (1994) Spatiotemporal structure of somatosensory responses of many-neuron ensembles in the rat ventral posterio medial nucleus of the thalamus. *J. Neurosci.*, 14: 3511–3532.

Nicolelis, M.A.L., Ghazanfar, A.A., Faggin, B.M., Votaw, S. and Oliveira, L.M.O. (1997) Reconstructing the engram: simultaneous, multisite, many single neuron recordings. *Neuron*, 18: 529–537.

Nusser, Z., Sieghart, W. and Somogyi, P. (1998) Segregation of different $GABA_A$ receptors to synaptic and extrasynaptic membranes of cerebellar granule cells. *J. Neurosci.*, 18: 1693–1703.

Palkovits, M., Magyar, P. and Szentagothai, J. (1971) Quantitative histological analysis of the cerebellar cortex in the cat. II. Cell numbers and densities in the granular layer. *Brain Res.*, 32: 15–30.

Peeters, R.R., Verhoye, M., Vos, B.P., Van Dyck, D., Van Der Linden, A. and De Schutter, E. (1999) A patchy horizontal organization of the somatosensory activation of the rat cerebellum demonstrated by functional MRI. *Eur. J. Neurosci.*, 11: 2720–2730.

Pellerin, J.-P. and Lamarre, Y. (1997) Local field potential oscillations in primate cerebellar cortex during voluntary movement. *J. Neurophysiol.*, 78: 3502–3507.

Pellionisz, A. and Szentagothai, J. (1973) Dynamic single unit simulation of a realistic cerebellar network model. *Brain Res.*, 49: 83–99.

Pichitpornchai, C., Rawson, J.A. and Rees, S. (1994) Morphology of parallel fibers in the cerebellar cortex of the rat: an experimental light and electron microscopic study with biocytin. *J. Comp. Neurol.*, 342: 206–220.

Raymond, J.L., Lisberger, S.G. and Mauk, M.D. (1996) The cerebellum: a neuronal learning machine? *Science*, 272: 1126–1131.

Regan, L.J. (1991) Voltage-dependent calcium currents in Purkinje cells from rat cerebellar vermis. *J. Neurosci.*, 11: 2259–2269.

Rieke, F., Warland, D., de Ruyter van Steveninck, R.R. and Bialek, W. (1997) *Spikes. Exploring the Neural Code*. The MIT Press, Cambridge, MA.

Rossi, D.J. and Hamann, M. (1998) Spillover-mediated transmission at inhibitory synapses promoted by high affinity α6 subunit $GABA_A$ receptors and glomerular geometry. *Neuron*, 20: 783–795.

Santamaria, F., Jaeger, D., Bower, J.M. and De Schutter, E. (2000) Dendritic temporal integration properties of a Purkinje cell are modulated by background activity: A modeling study. in preparation.

Schweighofer, N., Spoelstra, J., Arbib, M.A. and Kawato, M. (1998) Role of the cerebellum in reaching movements in humans. II. A neural model of the intermediate cerebellum. *Eur. J. Neurosci.*, 10: 95–105.

Shambes, G.M., Gibson, J.M. and Welker, W. (1978) Fractured somatotopy in granule cell tactile areas of rat cerebellar hemispheres revealed by micromapping. *Brain Behav. Evol.*, 15: 94–140.

Simpson, J.I., Vandersteen, J., Tan, J., Graf, W. and Leonard, C.S. (1989) Representations of ocular rotations in the cerebellar flocculus of the rabbit. *Prog. Brain Res.*, 80: 213–223.

Steuber, V. and Willshaw, D.J. (1999) Adaptive leaky integrator models of cerebellar Purkinje cells can learn the clustering of temporal patterns. *Neurocomputing*, 26: 271–276.

Strahlendorf, J.C., Brandon, T., Miles, R. and Strahlendorf, H.K. (1998) AMPA receptor-mediated alterations of intracellular calcium homeostasis in rat cerebellar Purkinje cells in vitro: correlates to dark cell degeneration. *Neurochem. Res.*, 23: 1355–1362.

Stuart, G. and Häusser, M. (1994) Initiation and spread of sodium action potentials in cerebellar Purkinje cells. *Neuron*, 13: 703–712.

Tempia, F., Miniaci, M.C., Anchisi, D. and Strata, P. (1998) Postsynaptic current mediated by metabotropic glutamate receptors in cerebellar Purkinje cells. *J. Neurophysiol.*, 80: 520–528.

Thach, W.T. (1998) What is the role of the cerebellum in motor learning and cognition. *Trends Cogn. Sci.*, 2: 331–337.

Usowicz, M.M., Sugimori, M., Cherksey, B. and Llinás, R.R. (1992) Characterization of P-type calcium channels in cerebellar Purkinje cells. *Abstr. Soc. Neurosci.*, 18: 974–974.

Van Kan, P.L.E., Gibson, A.R. and Houk, J.C. (1993) Movement-related inputs to intermediate cerebellum of the monkey. *J. Neurophysiol.*, 69: 74–94.

Volny-Luraghi, A., De Schutter, E. and Vos, B.P. (1999) Responses of cerebellar Golgi cells to tactile stimuli of different sizes. *Abstr. Soc. Neurosci.*, 25: 941.

Voogd, J. and Glickstein, M. (1998) The anatomy of the cerebellum. *Trends Neurosci.*, 21: 370–375.

Vos, B.P., Maex, R., Volny-Luraghi, A. and De Schutter, E. (1999a) Parallel fibers synchronize spontaneous activity in cerebellar Golgi cells. *J. Neurosci.*, 19: RC6: 1–5.

Vos, B.P., Volny-Luraghi, A. and De Schutter, E. (1999b) Cerebellar Golgi cells in the rat: receptive fields and timing of responses to facial stimulation. *Eur. J. Neurosci.*, 11: 2621–2634.

Vos, B.P., Volny-Luraghi, A., Maex, R. and De Schutter, E. (2000) Precise spike timing of tactile-evoked cerebellar Golgi cell responses: a reflection of combined mossy fiber and parallel fiber activation? *Prog. Brain Res.*, 124: 95–105.

Vranesic, I., Iijima, T., Ichikawa, M., Matsumoto, G. and Knöpfel, T. (1994) Signal transmission in the parallel fiber Purkinje cell system visualized by high-resolution imaging. *Proc. Natl. Acad. Sci. USA*, 91: 13014–13017.

Wall, M.J. and Usowicz, M.M. (1997) Development of action potential-dependent and independent spontaneous $GABA_A$ receptor-mediated currents in granule cells of postnatal rat cerebellum. *Eur. J. Neurosci.*, 9: 533–548.

Welsh, J.P., Lang, E.J., Sugihara, I. and Llinás, R. (1995) Dynamic organization of motor control within the olivocerebellar system. *Nature*, 374: 453–457.

N.M. Gerrits, T.J.H. Ruigrok and C.I. De Zeeuw (Eds.)
Progress in Brain Research, Vol 124

CHAPTER 7

Precise spike timing of tactile-evoked cerebellar Golgi cell responses: a reflection of combined mossy fiber and parallel fiber activation?

Bart P. Vos, Antonia Volny-Luraghi, Reinoud Maex and Erik De Schutter*

Laboratory Theoretical Neurobiology, Born-Bunge Foundation, University of Antwerp, B–2610 Antwerp, Belgium

Introduction

In recent years a number of papers emerged with tales of ventures in what J.I. Simpson (this Volume) named "the *terra incognita* in the cerebellar cortex": the gray zone (granular layer) between input (mossy fibers) and output (Purkinje cells). This gray zone holds feedforward and feedback circuits which pre-process afferent information before it reaches the Purkinje cells.

Some of these recent publications specifically focused on the Golgi cell, the primary inhibitory interneuron of the granular layer circuitry (Dieudonné 1998a, b; Maex and De Schutter, 1998a, b; Watanabe et al., 1998; Vos et al., 1999a, b). They unequivocally concluded that Golgi cells do more than control the gain of parallel fiber input to Purkinje cells (see also De Schutter et al., Chapter 6, this Volume), which was suggested to be their only function in the classic cerebellar theories of Marr (1969), Albus (1971) and Ito (1984).

Figure 1 shows a schematic drawing of the anatomical connections of Golgi cells. Golgi cells receive direct excitatory input from mossy fibers (Hámori and Szentágothai, 1966). They are also contacted by parallel fibers (Fox et al., 1967), which excite Golgi cells after activation of granule cells by mossy fibers. In addition, Golgi cells do not contact each other (nor do granule cells) (Palay and Chan-Palay, 1974). Being inhibitory, Golgi cells exert both a feedback (Eccles et al., 1966b) and a feed-forward (Precht and Llinás, 1969) inhibition of granule cell activity. The classic view of Golgi cells as local controllers of granule cell

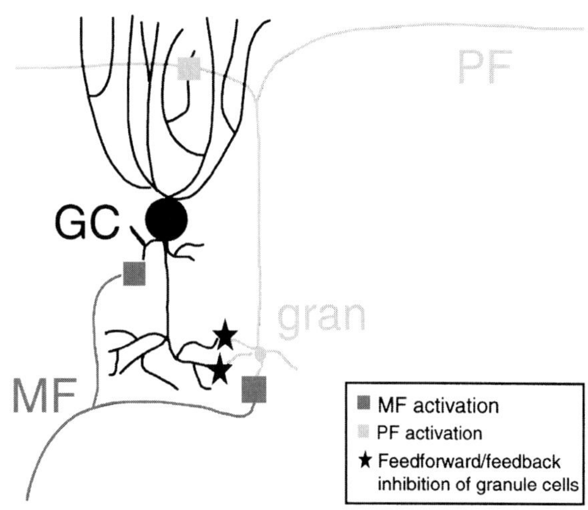

Fig. 1. Connectivity in the granular layer. Granule cells (gran; light gray) receive excitation from mossy fibers (MF; darker gray) and send their output along their parallel fiber (PF). Golgi cells (GC; black) receive both MF (dark squares) and PF (light squares) excitation and reciprocally inhibit (stars) granule cells.

*Corresponding author. Fax: +32-3-8202669; e-mail: erik@bbf.uia.ac.be activity was inspired by descriptions of these anatomical connections.

Despite the rather straightforward afferent 'wiring' scheme, the characterization of the response properties of Golgi cells appeared rather difficult in paradigms of vestibulo-ocular reflex adaptation (Miles et al., 1980; Atkins et al., 1997), locomotion (Edgley and Lidierth, 1987) or limb movement (Van Kan et al., 1993). We recently conducted a series of experiments to further assess firing and response properties of Golgi cells in anesthetized rats (Vos et al., 1999a, b). In the present chapter we give a brief account of our observations on discharge patterns of cerebellar Golgi cells in Crus I–II, evoked by punctate, tactile stimulation of the face.

Methods

Multi-electrode extracellular recordings

Recordings (for full description of the recording procedure see Vos et al., 1999a, b) were made in the cerebellar cortex (Crus I–II) of ketamine-xylazine anesthetized, male Sprague-Dawley rats (350–500g, $n = 38$) with tungsten microelectrodes (2 MOhm).

Signals were filtered and amplified (bandpass = 400–20,000 Hz; gain = 5,000–15,000) using a Multi-channel Neuronal Acquisition Processor (Plexon Inc., Austin, TX, USA). Spike waveforms were discriminated with a real-time hardware-implemented combined time/voltage window discriminator (Nicolelis and Chapin, 1994). Electrolytic lesions (15 μA, 12 s, cathodal DC current) were made to mark the location of the electrode tips.

Identification of Golgi cells

Typically, putative Golgi cells were recognized by the distinctive rhythm of their activity at rest (Atkins et al., 1997): spikes appeared as pronounced 'pops' at a slow cadence with appreciable intervals (no bursting). A total of 87 units were categorized as Golgi cells based on the same criteria as those used by others (Eccles et al., 1966b; Miles et al., 1980; Edgley and Lidierth, 1987; Van Kan et al., 1993; Atkins et al., 1997): low discharge rates at rest (interspike intervals >20 ms), long duration (>0.8 ms) di-phasic (negative-positive or positive-negative) wave shapes, long tuning distances, no complex-spike discharges and location of the electrolytic lesion in the granular layer (e.g. Fig. 2A, B, C).

In addition, Golgi cells could also be quantitatively differentiated from other cells with a non-parametric coefficient of variation (nCV) that was calculated based on the ratio between the median interspike interval and its median absolute deviation (MAD) (see Fig. 2 of Vos et al., 1999b). Golgi cell firing at rest (average 8.4 spikes/s) appeared not more regular than simple-spike firing of Purkinje cells ($n = 38$; average 27.9 spikes/s), but more regular than the bursting firing pattern of presumed mossy fibers ($n = 30$; 8.75 spikes/s).

Tactile stimulation

Once a unit was categorized as a putative Golgi cell, and spontaneous activity was recorded,

Fig. 2. Firing and response characteristics of two simultaneously recorded Golgi cells in Crus II of a ketamine-xylazine anesthetized rat. A. Schematic representation of the position of the electrodes in Crus IIa and line drawing of the transverse section with the location of the electrolytic lesions in the granular layer. B. Superimposed records of 100 spikes from both units (jn8-b-a–1, jn8-b-a–2). C. Interspike interval histograms (1 ms bins) based on 2040 spikes (jn8-b-a–1) and 4540 spikes (jn8-b-a–2) captured during a 400 s recording of activity at rest. Average firing rates of the cells were respectively 5.08 and 11.32 spikes/s. Median ISIs were 61 ms and 48 ms. D. The rhinarium was mechanically stimulated at 1 Hz (1 mm probe, 10 ms duration). For both units, rasterized traces of recorded spikes captured on 450 successive trials are shown ($-100/+300$ ms from stimulus onset, indicated by dashed line; hatched bars on left of rasters indicate two periods of 100 successive trials with stimulation). Below the rasters the cumulative frequency distributions are shown for the 200 trials with stimulation (cumulative sum of spikes/bin, 1 ms bins, pre-stimulus spike counts are used as zero-offset). The lower panels show the peri-event histograms for both units ($-100/+300$ ms from stimulus onset, 1 ms bins; dashed line indicates stimulus onset) with the number of spikes/bin over 200 trials with stimulation. Firing rates at rest (30 ms epoch before stimulation) were respectively 5.16 and 8.51 spikes/s. Post stimulus firing rates were 58.6 and 63.6 spikes/s (30 ms epoch). The spike pattern evoked by the same stimulus is different in both simultaneously recorded units. Both units show a silent period.

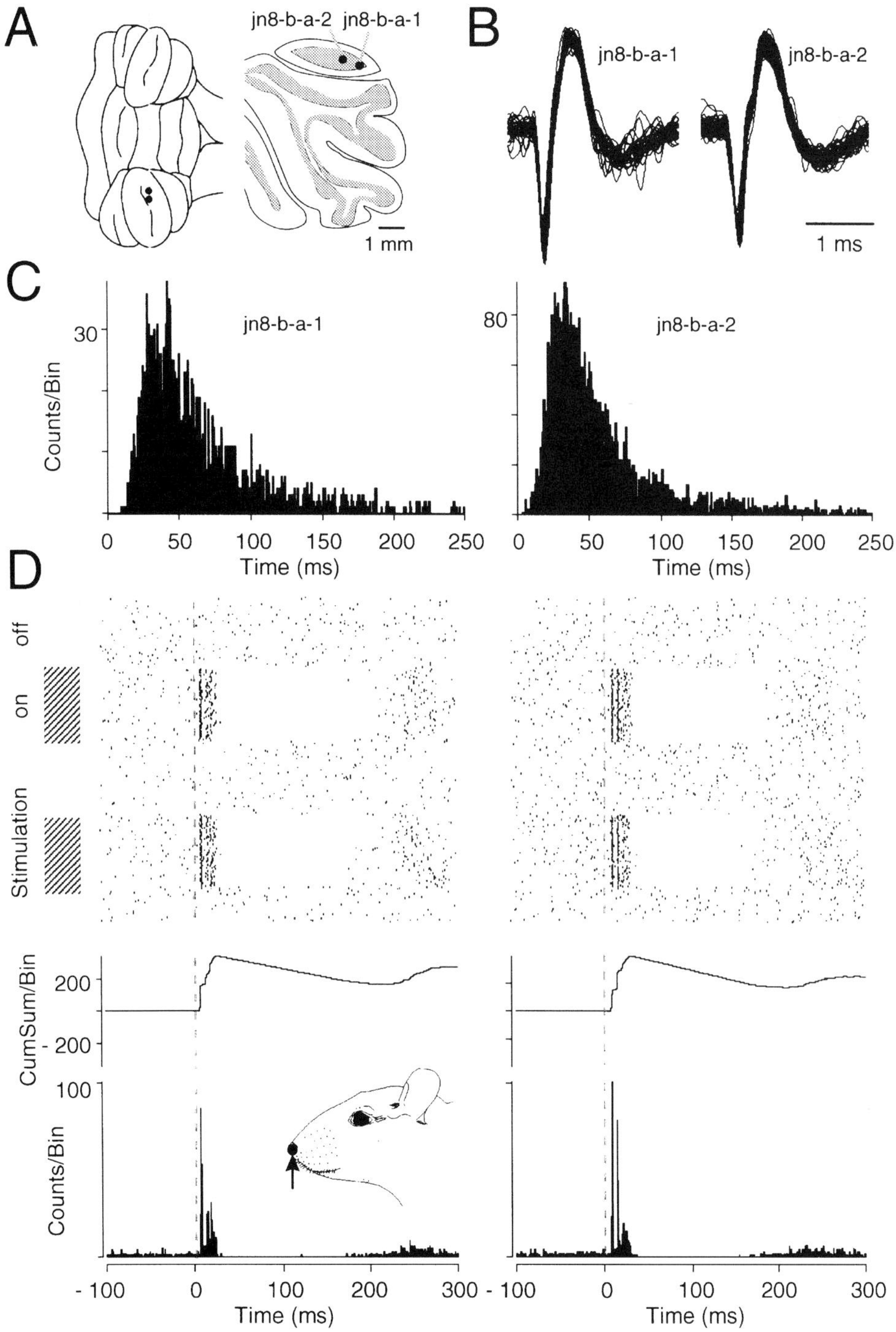
A
jn8-b-a-2
jn8-b-a-1
1 mm
B
jn8-b-a-1
jn8-b-a-2
1 ms
C
30
80
Counts/Bin
jn8-b-a-1
jn8-b-a-2
0 50 100 150 200 250
Time (ms)
D
off
on
Stimulation
200
- 200
CumSum/Bin
100
Counts/Bin
- 100 0 100 200 300
Time (ms)

mechanical taps with a hand-held cotton-tipped wooden probe were used to explore its receptive field (RF). In addition, somatic responses were quantitatively tested in half of the rats, with controlled mechanical stimuli. A stainless steel probe (1 mm, cylindrical, flat surface, maximal excursion 2.5 mm) mounted on an electromagnetic activator (12V solenoid) and driven by a digital stimulator was precisely positioned at (at least four) different facial loci ipsi-and/or contralateral to the recording site to deliver innocuous mechanical stimuli (< 10ms). In some rats one specific locus was stimulated at different frequencies (0.5–6Hz). For each stimulus configuration the same stimulation paradigm consisting of (at least) 100 trials without stimulation, followed by (at least) 100 trials with stimulation, was repeated twice (Fig. 2D).

Golgi cell response properties

Golgi cell receptive fields are large

Of all Golgi cells tested ($n = 87$), most responded to tactile stimulation (84%). Responsive Golgi cells had facial receptive fields, which is consistent with previously described trigeminal-evoked responses of granule and Purkinje cells in Crus I and II (e.g. Shambes et al., 1978; Bower et al., 1981). In contrast to the patchy organization of mossy fiber projections to Crus I and II (Glickstein et al., 1994), and the fractured somatotopic maps found for granular layer field potentials (Shambes et al., 1978; Bower et al., 1981) or Purkinje cells (Bower and Woolston, 1983) in these lobules, most Golgi cells recorded in our study (Vos et al., 1999b) had rather large RFs. In 67% of the Golgi cells tested, the RF covered the entire territory of the maxillary branch of the trigeminal nerve (including vibrissal pad, furry buccal pad, rhinarium and upper lip) ipsilateral to the recording site. Sometimes the RF extended to territories of the other trigeminal nerve branches (14%), or included contralateral parts (35%) (e.g. Fig. 4).

Extended Golgi cell RFs have also been reported by others. Edgley and Lidierth (1987) found 85/87 Golgi cells to respond to tactile stimulation in the awake cat, the majority of which had large and bilateral RFs. Also in halothane-anesthetized rats, an unspecified proportion of Golgi cells responded to peripheral stimulation (Schulman and Bloom, 1981). In barbiturate-anesthetized animals (7 monkeys, 11 cats) only few Golgi cells (3/18) responded to somatosensory stimulation (Van Kan et al., 1993). In awake monkeys however, all of the Golgi cells tested (5/5) responded to passive joint manipulation, though they lacked directional specificity (van Kan et al., 1993). For Golgi cells recorded in the cerebellar flocculus of the awake rabbit and the awake monkey a great diversity of modulation patterns was found (Miles et al., 1980; Atkins et al., 1997). The latter two findings are, like the presence of large tactile RFs observed by others and in the present study, an indication of an enormous convergence of excitatory inputs onto single Golgi cells.

Excitatory responses are followed by a silent period

All tactile-evoked responses were typified by a long silent period (duration: 13–200ms) that followed the initial 'excitatory' response (latency: 28–69 ms; e.g. Fig. 2D, Fig 5D). This silent period was quite noticeable because all cells were spontaneously active before stimulus onset. The duration of this silent period appeared to be related to the amplitude of the excitatory component (Vos et al., 1999b, Fig. 8). Sometimes a 'rebound' increase in activity was observed at the end of the silent period (e.g. Fig. 3D).

Similar silent periods have been observed in other Golgi cell preparations, e.g. in anesthetized cats in response to parallel and mossy fiber stimulation but not to inferior olive stimulation (Eccles et al., 1966b), in awake cats in response to peripheral stimulation (Armstrong and Drew, 1980), and in anesthetized rats in response to inferior olive stimulation (Schulman and Bloom, 1981).

The cause and function of this silent period remains unclear. Silent periods have been observed in visual and somatosensory pathways and this suggests that a lack of mossy fiber activation underlies the silent period. In fact, Eccles et al. (1971) reported that discharges of presumed mossy fibers could be silenced for almost 100 ms, even in the absence of any initial excitation.

However, factors intrinsic to Golgi cells or to the cerebellar circuitry may contribute to the appearance of the silent period. In a study of cerebellar interneurons, Eccles et al. (1966a) observed that parallel fiber stimulation initially facilitates responses of presumed inhibitory interneurons to subsequent parallel fiber excitation. The initial facilitation gives way to a marked depression of the responses to new parallel fiber input that arrives 50–500 ms after the conditioning stimulus. Such a post-activation depression could also underlie the observed silencing of the Golgi cell, although we did not test whether the cells were actually non-responsive during the silent period.

In cerebellar slice preparations, a strong tonic form of granule cell inhibition was recently

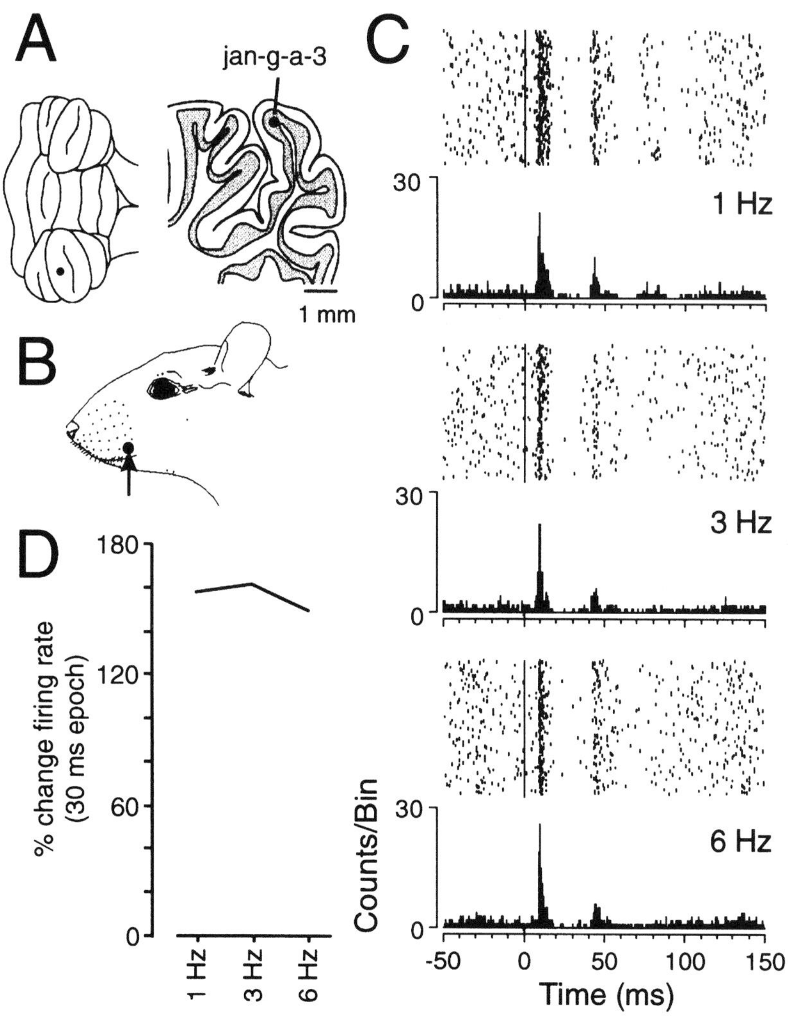

Fig. 3. Effect of stimulation frequency on a Golgi cell response. A. Schematic representation of the position of the electrode in Crus IIa and line drawing of the sagittal section with the location of the electrolytic lesion in the granular layer. B. The same stimulus (1 mm probe, 10 ms duration) was applied to the hairy skin between the gamma vibrissa and vibrissa E1 at 1, 3 and 6 Hz. C. Rasterized traces and PSTHs (− 50/ + 150 ms from stimulus onset; 1 ms bins) with the responses (average spikes/s) to each stimulus frequency (200 trials each). D. Line graph representing the percent increase in firing rate (30 ms epochs) evoked by each stimulus frequency. Response amplitudes did not differ between the stimulation frequencies.

observed (Brickley et al., 1996; Wall and Usowicz, 1997). This tonic inhibition was postulated to be caused by Golgi cell activity resulting in spillover of GABA which activates high affinity receptors (Rossi and Hamann, 1998), and its duration should thus be correlated with the amplitude of the Golgi cell response. Since tonic granule cell inhibition causes an interruption of parallel fiber excitation to the Golgi cell, the effect of diminished or absent mossy fiber input could be enhanced. Other evidence that supports this hypothetical scenario is the strong on-beam inhibition evoked by short bursts of parallel fiber excitation, and observed in an isolated guinea pig cerebellum (Cohen and Yarom, 1998).

A last possibility is that the stimulus-induced mossy fiber activation evokes a (damped) oscillation in the granular circuit, as predicted by Maex and De Schutter (1998b). In the raster plots shown in Figure 3C (especially for 1 Hz stimulation),

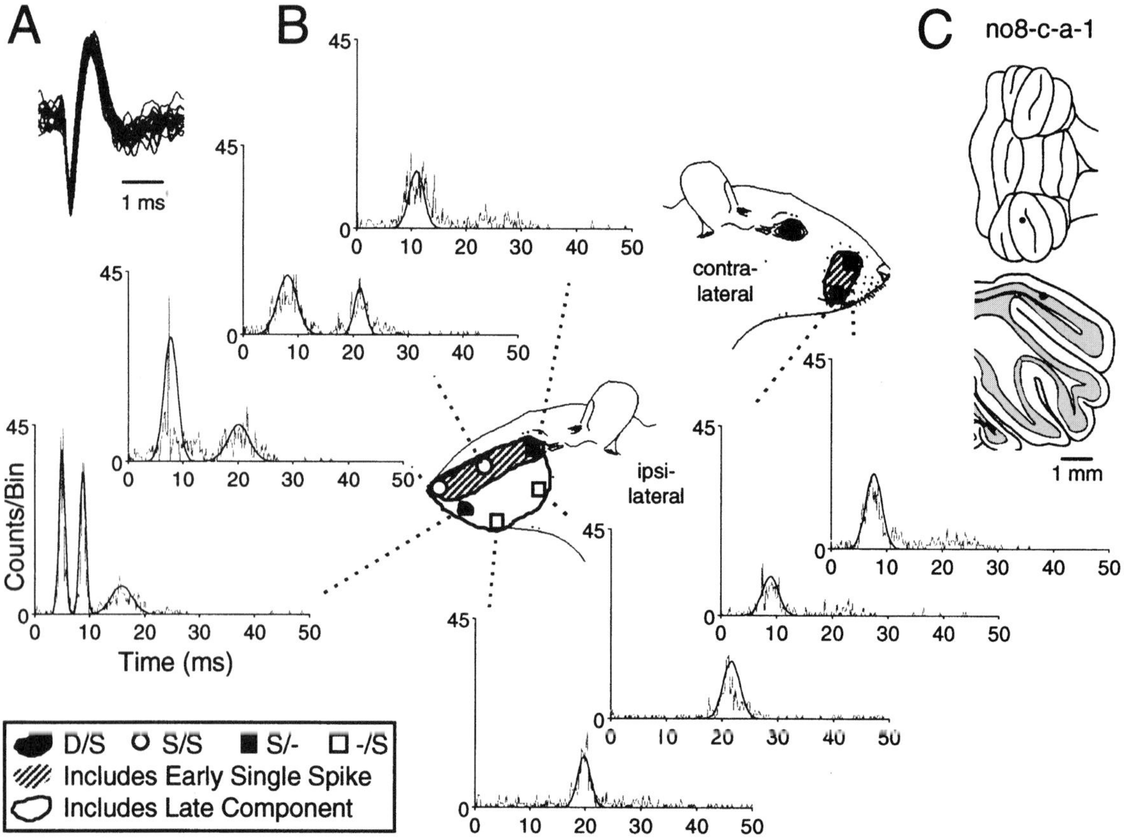

Fig. 4. Large receptive fields of Golgi cells. A. Superimposed records of 100 spikes recorded from unit no8-c-a–1. B. The same stimulus (1 mm probe, 1 Hz) was applied to nine different locations. High resolution PSTHs (0/50 ms from stimulus onset, 0.25 ms bins) based on stimulation of each of the locations (200 trials each) including the gaussian fit for each of the peaks are shown. The exact location from which each of the response profiles were evoked is indicated on graphical representations of the rat's face, contra- and ipsilateral to the recording site. The relation between the receptive field size and spike timing is illustrated: the facial field from which responses with an early double PSTH-peak response (black area), responses with an early single PSTH-peak response (shaded area), and responses that include a late component (white area) are drawn on the face-drawings. C. Schematic representation of the position of the electrode in Crus IIa and line drawing of the transverse section with the location of the electrolytic lesion in the granular layer.

oscillatory 'ripples' in the Golgi cell response can be discerned.

Timing of tactile-evoked spikes is precise and location-dependent

Discharge patterns of cerebellar Golgi cells, evoked by punctate, tactile stimuli, showed a strict temporal patterning. The timing of the evoked spikes appeared very robust and precise over trials (e.g. Fig. 2D, 3D, 5D). In order to study the temporal characteristics of the evoked responses in more detail, distinct peaks on 'high resolution' peristimulus time histograms (0.25 ms bins, 0–55 ms) were fitted with Gaussians (see Vos et al., 1999b; e.g. Fig. 4B). Each PSTH-peak represents the fact that in a substantial number of stimulus trials, a spike was fired at a specific, fixed latency from stimulus onset. The fitted PSTH-peaks were used to quantify the evoked response profiles. For each PSTH the number of 'significant' peaks was determined, and for each peak, the latency (mean of the Gaussian), the accuracy of timing (standard deviation of the Gaussian) and the robustness (per cent of trials contributing a spike to the peak) were deduced.

In agreement with known trigemino-cerebellar (e.g. Watson and Switser, 1978; Woolston et al., 1981; Jacquin et al., 1982; Huerta et al., 1983) and cortico-ponto-cerebellar (e.g. Welker, 1987; Brodal and Bjaalie, 1997) projections, Golgi cell responses showed an early (5–10 ms) and a late (13–26 ms) 'excitatory' component (Fig.4B). In most instances, both the early and the late component appeared as a single peak on the PSTH (Fig. 4B). Sometimes, the first component consisted of two distinct PSTH-peaks with a short (<4 ms) interval (e.g. Fig 2D: unit jn8-b-a–2; Fig 4B: stimulation between vibrissae ED–56). Other responses consisted of the early or the late component alone (Fig 4B: only late component: stimulation around the genal vibrissae; only early component: contralateral stimulus locations). Figures 3D and 4B demonstrate that the stimulus location – not the stimulus frequency – determined the temporal pattern of the spikes (see also Vos et al., 1999b). The use of another anesthetic (α-chloralose) did not affect these response characteristics (Fig. 5D), although it slowed down the firing rate at rest considerably (Fig. 5E).

Golgi cell responses reflect trigemal and cortico-pontine input

The latencies of the early and the late component of the Golgi cell response, match closely those of the early and late response component of granular layer field potential responses which were recorded in Crus II under similar experimental conditions (e.g. Morissette and Bower, 1996). This close match strongly suggests that the first component of the Golgi cell responses results from direct trigemino-cerebellar projections (e.g. Woolston et al., 1982), and the second component from cortico-pontine projections (e.g. Morissette and Bower, 1996) (Fig. 6 lower part).

Although both projections arrive in the cerebellar cortex as mossy fibers, it is not very likely that the whole of each Golgi cell response was caused by direct mossy fiber input. Mossy fiber input to Crus I-II is fractured (Shambes et al., 1978; Bower et al., 1981; Bower and Kassel, 1990) and trigeminal and cortico-pontine mossy fiber projections largely overlap (e.g. Voogd and Glickstein, 1998). Thus if Golgi cell responses would result from pure mossy fiber excitation alone they should have had small RFs corresponding to one or, at most, a few patches (see Welker, 1987 for review). In contrast, Golgi cells had very large RFs, which probably reflected an important excitation by parallel fibers.

Parallel fibers can, because of their length (Pichitpornchai et al., 1994) originate in many different patches. As such, they carry input from several parts of the rat's face outside the mossy fiber projection field, onto the same Golgi cell. Some patches in Crus II have contralateral or bilateral RFs (Welker, 1987) which could explain the bilateral RFs we observed in 41% of the Golgi cells.

Double peak responses may result from 'monosynaptic' mossy fiber activation

In view of the observations that responses with an early double peak had (a) significantly shorter delays, (b) a significantly higher spike timing accuracy (i.e. less jitter), (c) a larger amplitude, and (d) a more restricted RF (e.g. Fig 4B), compared to other response components, we hypothesize that they resulted from a strong, direct mossy fiber input (see Vos et al., 1999b). For most 'double peaks', the second peak was not much smaller than the first peak (e.g. Fig. 4B). This indicated that spikes were fired at both latencies in about the same number of trials. These 'double spikes' may be a reflection of the presence of a doublet in the afferent mossy fiber input. It is also possible that the first spike results from mossy fiber activation and the second spike from subsequent parallel fiber activation. Least likely is the possibility that Golgi cells respond with 'doublets' to single spike mossy fiber input.

We further hypothesize that early single peak components (Fig. 6) were produced by parallel fiber excitation which was delayed and less precise in timing because of the slow and variable parallel fiber conduction velocities (Bernard and Axelrad, 1991; Vranesic et al., 1994). It should be stressed that PSTH-profiles with and those without an early double peak were observed in the same individual Golgi cell (Fig. 4B). This rules out the possibility

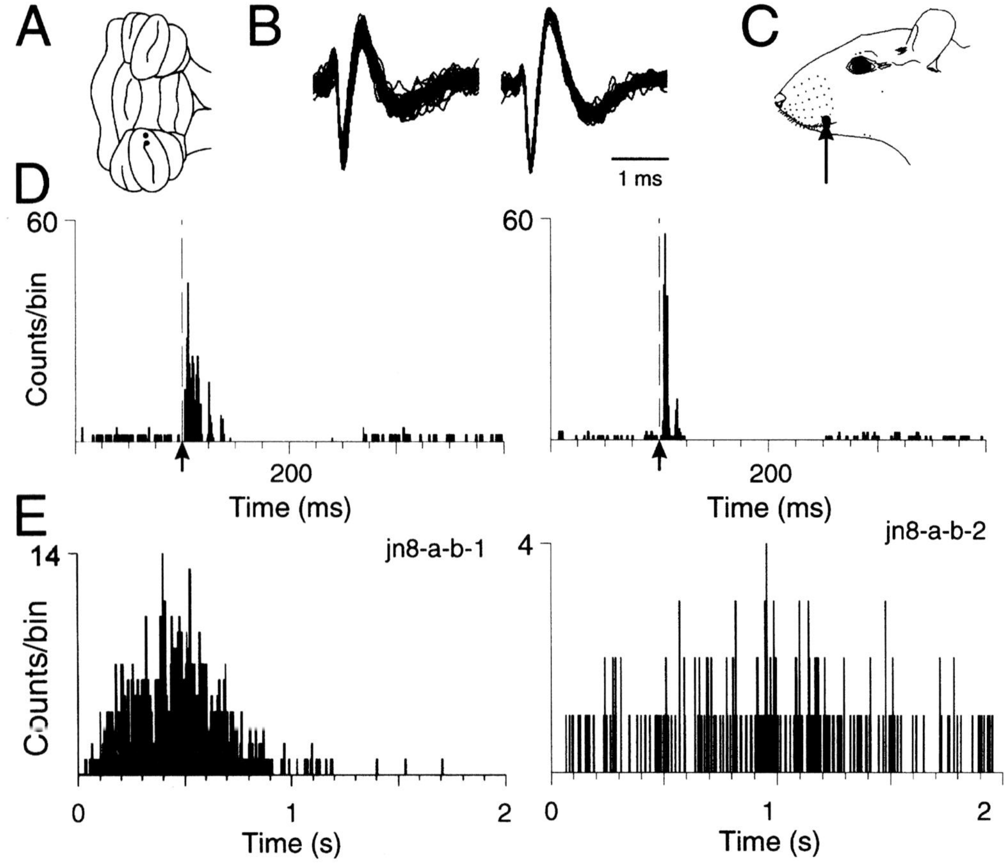

Fig. 5. Responses of Golgi cells in α-chloralose-anesthetized rats. A. Schematic representation of the position of the electrodes in Crus II. B. Superimposed records of 100 spikes from both units. C. The hairy skin between around the E1-vibrissa was mechanically stimulated at 1 Hz (1 mm probe, 10 ms duration). D. PSTHs for both units (− 200/ + 600 ms from stimulus onset, 2 ms bins; arrows indicate stimulus onset) with the total number of spikes/bin over 200 stimulation trials. Post stimulus firing rates were 42 and 32 spikes/s (30 ms epoch). Like in ketamine-anesthetized rats, the spike pattern evoked by the same stimulus is different in both simultaneously recorded units. Both units show a silent period. E. ISIs (4 ms bins) based on 797 spikes (jn8-a-b–1) and 261 spikes (jn8-a-b–2) captured during a 360 s recording of activity at rest. Average firing rates of the cells were respectively 2.17 and 0.71 spikes/s. Median ISIs were 457 ms and 1106 ms. nCVs were 0.32 and 0.38.

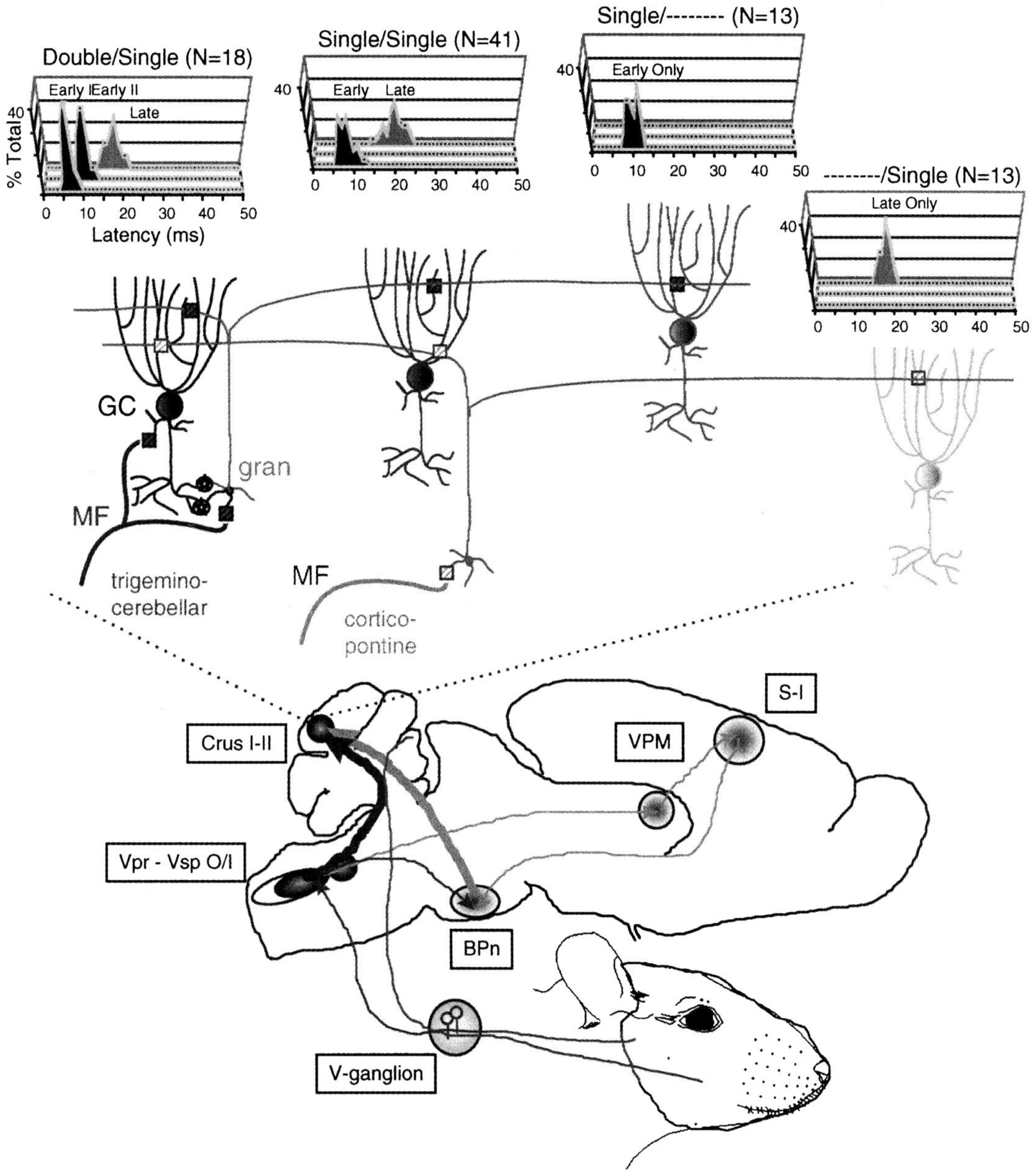

Fig. 6. Hypothesis: temporal pattern of Golgi cell responses reflects combined mossy fiber and parallel fiber activation. Lower part: Trigeminal information reaches the cerebellum directly via trigemino-cerebellar projections (dark arrows) and indirectly via cortico-pontine projections (light arrows) (trigeminal pathways to cerebellum are indicated: V-ganglion, trigeminal ganglion; BPn, Basilar Pontine nuclei; VPM, Ventral Posteromedial Thalamic nucleus; SI, somatosensory cortex; Vpr, Trigeminal nucleus Principalis; Vsp O/I, pars Oralis and Interpolaris of Trigeminal subnucleus Spinalis). Upper part: In the cerebellar cortex, trigeminal activation of mossy fibers will excite Golgi cells either 'mono-synaptically' or via the parallel fibers. In view of the shorter latency (see frequency histograms showing the number of 'peaks' per 1 ms bin, on top), and other properties (see text) of responses with an early double peak, we hypothezise that these responses are the result of excitation by direct trigemino-cerebellar mossy fiber input. Single peaks are hypothesized to result from parallel fiber activation. The late response component is hypothesized to result from corticopontine input.

that all early double peak responses were in fact recorded directly from mossy fibers (and not from Golgi cells).

Since responses with only one component also had large RFs (e.g. Fig. 4B), it is most likely that they reflect mostly a parallel fiber evoked event: either only the trigemino-cerebellar input (only early component), or only the cortico-pontine input (only late component). Single peaks also had significantly larger standard deviations (i.e. were less accurately timed; see Vos et al., 1999b). The latter is consistent with the assumption that the transmission over the parallel fibers weakens with increasing distance from the original mossy fiber input.

Conclusions

The somatic receptive fields of Golgi cells recorded in Crus I and II of anesthetized rats were larger than expected based on known mossy fiber projection patterns in this region. The timing of the evoked Golgi cell discharges appeared remarkably precise and robust, and depended on the location of the stimulus within the RF. The presence of an early and a late component in the evoked responses probably reflected converging trigemino-cerebellar and cortico-ponto-cerebellar inputs. We also hypothesized that the early double-spike components resulted from direct mossy fiber activation, and that single spike components were evoked by parallel fiber inputs (Fig. 6).

The temporal and topographical characteristics of the somatic responses do not completely fit the traditional theories of cerebellar function which assume a pure gain control function of Golgi cells (Marr, 1969; Albus, 1971). Our results may necessitate a re-evaluation of their role in cerebellar function.

Acknowledgements

The authors wish to thank Inge Bats for the photography, and Evelyne De Leenheir and Ursula Lubke for the histology. We gratefully acknowledge the technical wizardry of Mike Wijnants. We would also like to thank the reviewers for their comments on an earlier version. This research was funded by EC contract BIO4-CT98–0182, by IPA Belgium (P4/22) and by the Fund for Scientific Research – Flanders (FWO-Vl) (# 1.5.504.98). BPV and EDS are supported by the FWO-Vl.

References

Albus, J.S. (1971) A theory of cerebellar function. *Math. Biosci.*, 10: 25–61.

Armstrong, D.M. and Drew, T. (1980) Responses in the posterior lobe of the rat cerebellum to electrical stimulation of cutaneous afferents to the snout. *J. Physiol.*, 309: 357–374.

Atkins, M.J., VanAlphen, A.M. and Simpson, J.I. (1997) Characteristics of putative Golgi cells in the rabbit cerebellar flocculus. *Soc. Neurosci. Abstr.*, 23: 1287.

Bernard, C. and Axelrad, H. (1991) Propagation of parallel fiber volleys in the cerebellar cortex: a computer simulation. *Brain Res.*, 565: 195–208 .

Bower, J.M., Beerman, D.H., Gibson, J.M., Shambes, G.M. and Welker, W. (1981) Principles of organization of a cerebro-cerebellar circuit. Micromapping the projections from cerebral (SI) to cerebellar (granule cell layer) tactile areas of rats. *Brain Behav. Evol.*, 18: 1–18.

Bower, J.M. and Kassel, J. (1990) Variability in tactile projection patterns to cerebellar folia crus IIA of the Norway rat. *J. Comp. Neurol.*, 302: 768–778.

Bower, J.M. and Woolston, D.C. (1983) Congruence of spatial organization of tactile projections to granule cell and Purkinje cell layers of cerebellar hemispheres of the albino rat: vertical organization of cerebellar cortex. *J. Neurophysiol.*, 49: 745–766.

Brickley, S.G., Cull-Candy, S.G. and Farrant, M. (1996) Development of a tonic form of synaptic inhibition in rat cerebellar granule cells resulting from persistent activation of $GABA_A$ receptors. *J. Physiol.*, 497: 753–759.

Brodal, P. and Bjaalie, J.G. (1997) Salient anatomic features of the cortico-ponto-cerebellar pathway. In: C.I. De Zeeuw, P. Strata and J. Voogd (Eds), *The Cerebellum: From Structure to Control*, Progress in Brain Research, Vol. 114, Elsevier, Amsterdam, pp. 227–349.

Cohen, D., and Yarom, Y. (1998) Patches of synchronized activity in the cerebellar cortex evoked by mossy-fiber stimulation: questioning the role of parallel fibers. *Proc. Natl Acad. Sci. USA*, 95: 15032–15036.

Dieudonné S (1998a). *Etude fonctionelle de deux interneurones inhibiteurs du cortex cérébelleux: les cellules de Lugaro et de Golgi.* Unpublished doctoral thesis, Université Pierre et Marie Curie, Paris VI.

Dieudonné S (1998b) Submillisecond kinetics and low efficacy of parallel fibre-Golgi cell synaptic currents in the rat cerebellum. *J. Physiol.*, 510: 845–866.

Eccles, J.C., Faber, D.S., Murphy, J.T., Sabah, N.H., and Táborková, H. (1971) Afferent volleys in limb nerves influencing inpulse discharges in cerebellar cortex. *Exp. Brain Res.*, 13: 15–35.

Eccles, J.C., Llinás, R.R. and Sasaki, K. (1966a) The inhibitory interneurones within the cerebellar cortex. *Exp. Brain Res.*, 1: 1–16.

Eccles, J.C., Llinás, R.R. and Sasaki, K. (1966b) The mossy fibre-granule cell relay of the cerebellum and its inhibitory control by Golgi cells. *Exp. Brain Res.*, 1: 82–101.

Edgley, S.A. and Lidierth, M. (1987) The discharges of cerebellar Golgi cells during locomotion in the cat. *J. Physiol.*, 392: 315–332.

Fox, C.A., Hillman, D.E., Sigesmund, K.A., and Dutta, C.R. (1967) The primate cerebellar cortex: a Golgi and electron microscopic study. *Progr. Brain Res.*, 25: 174–225.

Glickstein, M., Gerrits, N., Kralj-Hans, I., Mercier, B., Stein, J., and Voogd, J. (1994) Visual pontocerebellar projections in the macaque. *J. Comp. Neurol.*, 349: 51–72.

Hámori, J. and Szentágothai, J. (1966) Participation of Golgi neurone processes in the cerebellar glomeruli an electron microscope study. *Exp. Brain Res.*, 2: 35–48.

Huerta, M.F., Frankfurter, A. and Harting, J.K. (1983) Studies of the principal sensory and spinal trigeminal nuclei of the rat: projections to the superior colliculus, inferior olive, and cerebellum. *J. Comp. Neurol.*, 220: 147–167.

Ito, M. (1984) The Cerebellum and Neural Control. Raven Press, New York.

Jacquin, M.F., Semba, K., Rhoades, R.W. and Egger, M.D. (1982) Trigeminal primary afferents project bilaterally to dorsal horn and ipsilaterally to cerebellum, reticular formation, and cuneate, solitary, supratrigeminal and vagal nuclei. *Brain Res.*, 246: 285–291.

Maex, R. and DeSchutter, E. (1998a) The critical synaptic number for rhythmogenesis and synchronization in a network model of the cerebellar granular layer. In: L. Niklasson, M. Bodén and T. Ziemke (Eds), *ICANN 98*. Springer-Verlag, London, pp. 361–366.

Maex, R. and DeSchutter, E. (1998b) Synchronization of Golgi and granule cell firing in a detailed network model of the cerebellar granule cell layer. *J. Neurophysiol.*, 80: 2521–2537.

Marr, D.A. (1969) A theory of cerebellar cortex. *J. Physiol.*, 202: 437–470.

Miles, F.A., Fuller, J.H., Braitman, D.J. and Dow, B.M. (1980) Long-term adaptive changes in primate vestibuloocular reflex. III. Electrophysiological observations in flocculus of normal monkeys. *J. Neurophysiol.*, 43: 1437–1476.

Morissette, J. and Bower, J.M. (1996) Contribution of somatosensory cortex to responses in the rat cerebellar cortex granule cell layer following peripheral tactile stimulation. *Exp. Brain Res.*, 109: 240–250.

Nicolelis, M.A.L. and Chapin, J.K. (1994) Spatiotemporal structure of somatosensory responses of many-neuron ensembles in the rat ventral posterio medial nucleus of the thalamus. *J. Neurosci.*, 14: 3511–3532.

Palay, S.L. and Chan-Palay, V. (1974) *Cerebellar Cortex*, Springer-Verlag, New York.

Pichitpornchai, C., Rawson, J.A. and Rees, S. (1994) Morphology of parallel fibers in the cerebellar cortex of the rat:.an experimental light and electron microscopic study with biocytin. *J. Comp. Neurol.*, 342: 206–220.

Precht, W. and Llinás, R. (1969) Functional organization of the cestibular afferents to the cerebellar cortex of frog and cat. *Exp. Brain Res.*, 9: 30–52.

Rossi, D.J. and Hamann, M. (1998) Spillover-mediated transmission at inhibitory synapses promoted by high affinity α–6 subunit $GABA_A$ receptors and glomerular geometry. *Neuron*, 20: 783–795.

Schulman, J.A. and Bloom, F.E. (1981) Golgi cells of the cerebellum are inhibited by inferior olive activity. *Brain Res.*, 210: 350–355.

Shambes, G.M., Gibson, J.M. and Welker, W. (1978) Fractured somatotopy in granule cell tactile areas of rat cerebellar hemispheres revealed by micromapping. *Brain Behav. Evol.*, 15: 94–140.

Van Kan, P.L.E., Gibson, A.R. and Houk, J.C. (1993) Movement-related inputs to intermediate cerebellum of the monkey. *J. Neurophysiol.*, 69: 74–94.

Vranesic, I., Iijima, T., Ichikawa, M., Matsumoto, G. and Knöpfel, T. (1994) Signal transmission in the parallel fiber Purkinje cell system visualized by high-resolution imaging. *Proc. Natl Acad. Sci. USA*, 91: 13014–13017.

Voogd, J. and Glickstein, M. (1998) The anatomy of the cerebellum. *Trends Neurosci.*, 21: 370–375.

Vos, B.P., Maex, R., Volny-Luraghi, A., and De Schutter, E. (1999a) Parallel fibers synchronize spontaneous activity in cerebellar Golgi cells. *J. Neurosci.*, 19: RC1–5.

Vos, B.P., Volny-Luraghi, A., and De Schutter, E. (1999b) Spike timings and receptive fields for trigeminal-evoked responses of rat cerebellar Golgi cells. *Eur. J. Neurosci.*, 11: 2621–2634.

Vos, B.P., Volny-Luraghi, A., Maex, R. and De Schutter, E. (2000) Precise spike timing of tactile-evoked cerebellar Golgi cell responses: a reflection of combined mossy fiber and parallel fiber activation? *Prog. Brain Res.*, 124: 95–105.

Wall, M.J. and Usowicz, M.M. (1997) Development of action potential-dependent and independent spontaneous $GABA_A$ receptor-mediated currents in granule cells of postnatal rat cerebellum. *Eur. J. Neurosci.*, 9: 533–548.

Watanabe, D., Inokawa, H., Hashimoto, K., Suzuki, N., Kano, M., Shigemoto, R., Hirano, T., Toyama, K., Kaneko, S., Yokoi, M., Moriyoshi, K., Suzuki, M., Kobayashi, K., Nagatsu, T., Kreitman, R.J., Pastan, I. and Nakanishi, S. (1998) Ablation of cerebellar Golgi cells disrupts synaptic integration involving GABA inhibition and NMDA receptor activation in motor coordination. *Cell*, 95: 17–27.

Watson, C.R.R. and Switzer, R.C., III (1978) Trigeminal projections to cerebellar tactile areas in the rat: origin from n. interpolaris and n. principalis. *Neurosci. Lett.*, 10: 77–82.

Welker, W. (1987) Spatial organization of somatosensory projections to granule cell cerebellar cortex: functional and connectional implications of fractured somatotopy (summary of Wisconsin studies). In: J.S. King (Ed.), *New Concepts in Cerebellar Neurobiology*. Alan R. Liss, Inc., New York, pp. 239–280.

Woolston, D.C., LaLonde, J.R. and Gibson, J.M. (1982) Comparison of response properties of cerebellar and thalamic-projecting interpolaris neurons. *J. Neurophysiol.*, 48: 160–173.

N.M. Gerrits, T.J.H. Ruigrok and C.I. De Zeeuw (Eds.)
Progress in Brain Research, Vol 124

CHAPTER 8

Unravelling cerebellar circuitry: an optical imaging study

Dana Cohen and Yosef Yarom*

Department of Neurobiology, Life Sciences Institute and Center for Neural Computation, Hebrew University, Jerusalem 91904, Israel

Introduction

The anatomical organization of the mammalian cerebellar cortex, its circuitry and the intrinsic properties of its neuronal elements have been studied intensively (Cajal, 1955; Eccles, et al., 1967; Palay and Chan-Palay, 1974; Llinas and Simpson, 1981; Ito, 1984), but few attempts have been made to explore its functional organization in time and space (Konnerth, et al., 1987; Kim, et al., 1989; Elias, et al., 1993; Welsh, et al., 1995). In order to unravel the functional circuitry of the cerebellum, we have developed over the past five years, a unique recording system for optical imaging of voltage-sensitive dyes in an isolated cerebellar preparation. This system reveals the temporal and spatial flow of activity within the cerebellar cortex, providing detailed information on its functional organization.

The spatial organization of the cerebellar cortex shows several dominant features that have greatly influenced theories of cerebellar function. For example, the parallel fiber system, organized along the folium in a medio-lateral direction, plays a prominent role in cerebellar theories on timing function (Braitenberg, 1961); coordination of movements (Thach, et al., 1992) and learning theories (Marr, 1969; Albus, 1971). Yet, it has been difficult to demonstrate the spread of information along this system in-situ. These difficulties gave rise to the hypothesis that the ascending part of the granule cell axons is the main drive of Purkinje cells as opposed to the parallel fiber system (Llinas, 1982). This hypothesis could account for the mosaic-like distribution of the receptive fields to tactile inputs (Bower and Woolstone, 1983).

The climbing fiber input to the cerebellum imposes a parasagittal organization on the cerebellar cortex (Groenwegen and Voogd, 1977; Groenwegen, et al., 1979; Voogd and Glickstein, 1998). This organization, which has been visualized with a number of staining techniques (Hawkes et al., 1992), overlaps the cortico-nuclear projections (Gravel et al., 1987). The morphological sagittal organization suggests a functional organization. Indeed, physiological measurements have shown that climbing fiber responses are organized within sagittal compartments (Llinas and Sasaki, 1989).

Anatomical studies of the inhibitory interneurons in the cerebellar cortex reveal another organized system. The axons of the molecular layer interneurons (stellate and basket cells) are either confined to the immediate vicinity of the cell body or project up to 500μ perpendicularly to the parallel fibers (Palay and Chan-Palay, 1974). It has been suggested that the short axons of stellate cells provide each Purkinje cell with inhibition proportional to its excitation (Marr, 1969; De Schutter and Bower, 1994). The long-range axons of the molecular layer interneurons induce lateral inhibition in the sagittal direction (Eccles, et al., 1967). In contrast, the Golgi cells axons ramify within the granular layer in a limited space close to their cell

*Corresponding author. Tel.: 972–2–6585172; Fax: 972–2–6586296; e-mail: yarom@vms.huji.ac.il

body. These neurons, which are activated by the parallel fibers, inhibit the granule cells, preventing their activation by mossy fiber. Thus, the Golgi cell inhibitory responses are limited to the area of active parallel fibers. Purkinje cell collaterals inhibit other cortical interneurons in their vicinity (Llinas and Precht, 1969; O'Donoghue and Bishop, 1990). Difficulties in extracellular recording of inhibitory responses have hampered physiological conformation of this morphological organization. Therefore, the role of the inhibitory interneurons of the cerebellar cortex remains largely unknown.

Here we demonstrate that optical imaging reveals essential organizational principles in the cerebellar cortex. The major contribution of the ascending branches of the granule cells has been confirmed and the spatio-temporal recordings of inhibitory responses shed new light on their function.

The system

The experimental system uses an isolated guinea pig cerebellum preparation maintained in vitro and an optical system based on a photodiode array, schematized in Fig. 1. The isolated Guinea pig cerebellar preparation has been described in detail previously (Cohen and Yarom, 1999). Briefly, the intact cerebellum and the brain stem were isolated under barbital anesthesia and a cannula was inserted into one of the vertebral arteries. Physiological solution was perfused via the vertebral artery at a rate of 1 ml/min using a peristaltic pump. The intravascular solution consisted of (in mM) 124 NaCl; 5 KCl; 1.3 $MgSO_4$; 1.2 KH_2PO_4; 26 $NaHCO_3$; 10 Glucose; 2.4 $CaCl_2$; 5% dextran. A similar solution without dextran was used as the external solution. The preparation was maintained at 28°C.

Voltage-sensitive dye (RH–414 or RH–795) was injected into one of the cerebellar folia using a glass micropipette filled with 2 μg/μl of the dye. Optical signals were recorded from 128 photodiodes organized in a 12 × 12 array. With 20 × or 40 × objectives, each element of the array detects light from a surface of about 100 μm × 100 μm or 50 μm × 50 μm respectively. The photodiode signals were amplified in two AC-coupled stages with a time constant of 200 msec and were sampled and digitized with 12-bit accuracy at maximal resolution of 50 μsec. The data were analyzed and displayed as traces of absolute change in fluorescence as a function of time at each location (Fig. 1, middle) or as the false color spatial distribution of the voltage at a given time (Fig. 1, top).

In analyzing the signals recorded with this system, we have found that it can detect inhibitory and excitatory signals from as deep as the granule cell layer. One unexpected advantage is that the optical signal is composed mainly of postsynaptic potentials; that is, the parallel fiber action potentials do not contribute to the optical signals. The response thus reflects activity only in cell somata and dendrites, but not axons (Cohen and Yarom, 1999).

Beams and patches in the cerebellar cortex

The longitudinal morphology of the cerebellar cortex suggests that stimulating the cerebellar

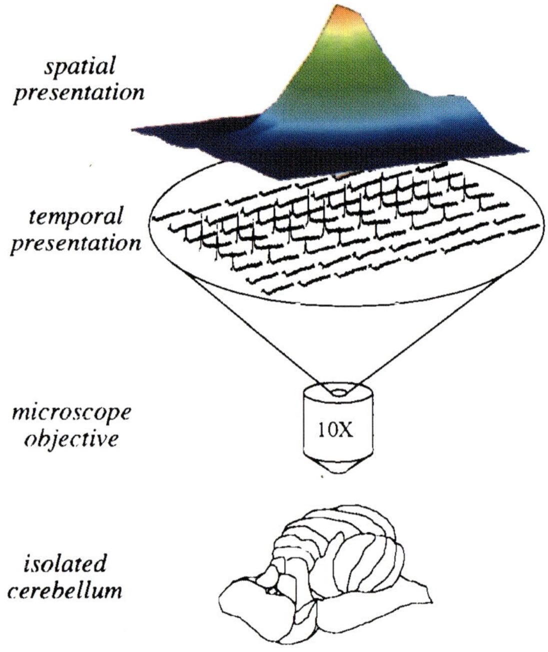

Fig. 1. Schematic representation of the experimental system. The isolated cerebellum is viewed through a 20 × (or 40 ×) objective by a 12 × 12 photodiode array. Data are presented as traces of absolute change in fluorescence as a function of time at each location (temporal presentation) or as the false color distribution of voltage over the area monitored by the photodiode array, at a given time (spatial presentation).

surface activates a beam of parallel fibers. Indeed, the classical work of Eccles and his colleagues (Eccles, et al., 1967) demonstrated that a local surface stimulus elicits a beam of activity that propagates along the cerebellar folium at an almost constant velocity. The field potential of this activity derives from action potentials in the parallel fibers as well as the post-synaptic responses of Purkinje cells (Llinas, 1981). Our optical measurements have confirmed these classic observations; stimulation of the cerebellar cortex elicits a narrow beam of activity (Fig. 2). The optical response is characterized by a fast positive wave (time to peak 2 msec, Fig. 3) followed by a slow positive decay lasting up to 50 msec. The amplitude of the signal and the width of the beam increased with stimulus intensity. The amplitude of the initial peak decreases and its duration increases as the signal propagates along the beam (at a velocity of 0.3 m/s, Fig. 3). Negative signals, reflecting inhibition, can be observed on both sides of the beam (Figs 2 and 3, see Figs 6 and 7). Such lateral inhibition would be predicted from the morphological organization of the axons of the molecular layer interneurons as described above.

A completely different response is obtained when the mossy fibers were stimulated. Stimulation of the cerebellar white matter elicits a circular patch of activity of 100 to 500 μm in diameter, centered above the stimulating electrode (Fig. 4). The size of the patch and amplitude of the response increase with stimulus intensity. Contrary to surface stimulation, the activity arises simultaneously throughout the patch with a delay of 2–4 msec (Fig. 4). The calcium sensitivity of this response, coupled with its complete blockade by the inhibitory system (Cohen and Yarom, 1998), strongly suggest that the patch of activity represents the activation of Purkinje cells and interneurons via the mossy fiber input.

These different patterns of activity most likely result from the morphological organization of the parallel fiber system. The granule cells, the source of the parallel fiber system, each send a single axon ascending through the molecular layer. The axon bifurcates at different levels within this layer, dividing into two branches that run in opposite directions and perpendicularly to Purkinje cell dendrites (Cajal, 1955). En passant synaptic contacts are regularly distributed along its ascending and parallel branches (Pichitpornchai, et al., 1994). As the ascending branch travels along a Purkinje cell all the way up to its point of bifurcation, it can form 10 to 30 synapses with this Purkinje cell, depending on the level at which bifurcation occurs. In contrast the parallel branches cross the thin Purkinje cell dendritic tree and, therefore, form only one, or rarely two, synapses with each Purkinje cell. Stimulating a focal group of granule cells thus activates a group of ascending axons that collectively excite a focal group of Purkinje cells. However, the parallel branches from this group of granule cells fail to significantly activate the Purkinje cells located along their path.

The importance of the ascending branch was initially proposed by Llinas (Llinas, 1982). He demonstrated that stimulating the granule cells in slice preparation evoked substantial synaptic

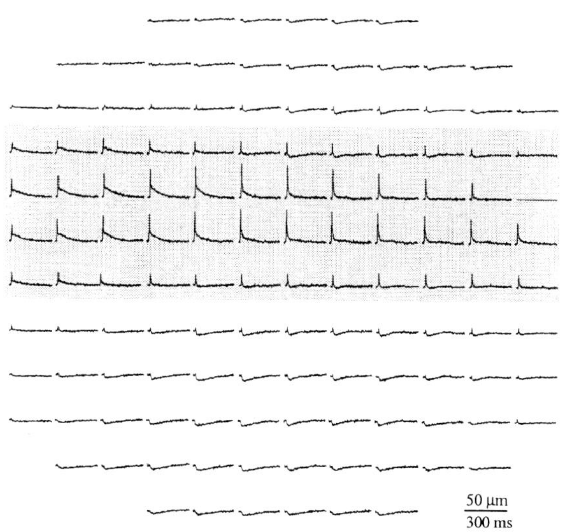

Fig. 2. A single surface stimulus elicits a beam-like response. The data are displayed as absolute change in fluorescence as a function of time at 128 sites. Each of the 128 traces is an average response of three identical stimuli repeated at a frequency of 0.3 Hz. The stimulating electrode was placed on the surface of the cerebellar cortex to the left of the recorded area. Each stimulus elicited a wave of positive responses that propagated in a beam (shaded area) in a medio-lateral direction with a conduction velocity of 0.3 m/s. Depending on stimulus intensity, the length of this beam could reach 4 mm. Inhibitory signals can be seen laterally to the excitatory beam. The calibration bar gives the scales for both time and space.

responses in Purkinje cells located above the point of stimulation. Bower and his colleagues (Bower and Woolstone, 1983) have argued that the fractured mapping of the cerebellar cortex to tactile stimulation is a manifestation of the ascending axon input.

The cerebellar inhibitory systems

The optical imaging system allowed us to explore the temporal properties and spatial distribution of the inhibitory systems of the cerebellar cortex. We found that, apart from the excitatory beam of activity, surface stimulation of the cerebellar cortex activates two types of inhibitory responses: lateral inhibition mentioned above, and on-beam inhibition. Lateral inhibition is evoked by a single surface stimulation and appears on both sides of the activated beam (Fig. 2). A train of stimuli is required to evoke on-beam inhibition and the response is limited to the area of the excitatory beam (Fig. 5). Although both inhibitory responses are blocked by bicuculline (Cohen and Yarom, 1998; Cohen and Yarom, 1999), they differ in their mode of activation, their spatial distribution (Figs 2 and 5), their dynamic range and the kinetics of their temporal integration.

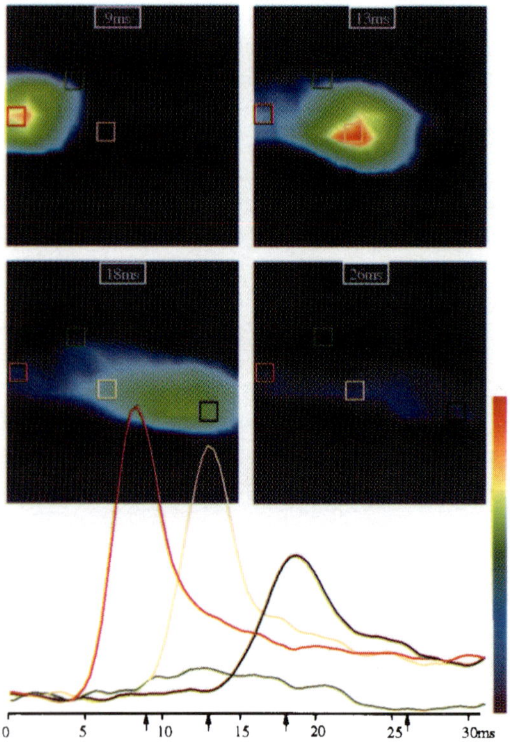

Fig. 3. The optical signal propagated along the parallel fiber axis. Top four panels are a color-coded representation of the responses measured at the times given. The color of each pixel was set according to the color bar on the right. The stimulus first evoked activity close to the stimulating electrode (red square), which propagated in a beam away from the electrode (yellow square). After about 26 msec, the positive peak had become very small and a negative wave could be observed along the edges of the beam (black square). The four superimposed traces show the change in florescence as a function of time measured at the locations of the colored squares. Arrows mark the time that the colored pictures were taken. The conduction velocity was calculated to be about 0.3 m/s, which fits that reported for parallel fibers.

The dynamic range

As shown in Fig. 2, the lateral inhibition that appears on both sides of the excitatory beam,

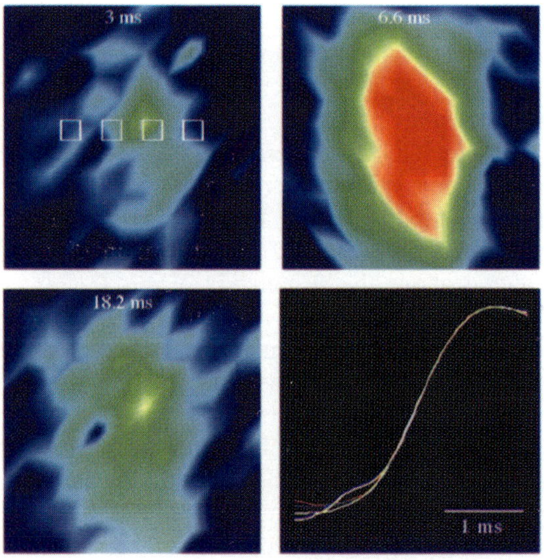

Fig. 4. White matter stimulation evokes a patch of activity. Representation as in Fig. 3. The time after stimulation at which each picture was taken is given at the top. The four superimposed traces in the lower right panel show the change in florescence as a function of time measured at the locations of the white squares (top left panel). These diodes were located along the parallel fiber axis. The amplitude of the optical responses was matched and only the rise time of the responses is shown. The striking overlap of all the signals shows that this evoked activity occurred simultaneously throughout the activated area.

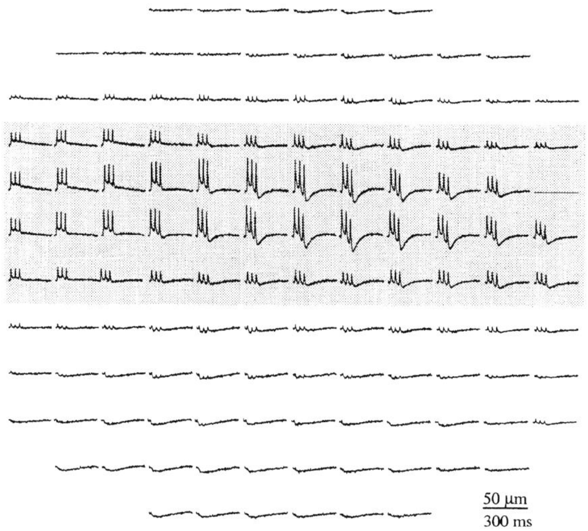

Fig. 5. A train of surface stimulation elicits on-beam inhibition. Data presented as in Fig. 2. The stimulating electrode was placed on the surface of the cerebellar cortex to the left of the recorded area. A train of stimuli elicited a train of positive responses that propagated as a beam along the parallel fiber axis (shaded area). The train of excitatory responses was followed by a prolonged inhibitory response only within this beam area.

spreads up to 700 μm from its center. Proximally to the beam, the inhibitory signals are preceded by a positive wavelet that decreases in amplitude with distance from the beam (Figs 6A and 7). The amplitude of the inhibitory response depends on stimulus intensity (Figs 6A and 6B) but, unlike the excitatory response at the center of the beam, it reaches saturation at a relatively low stimulus intensities (Fig. 6B). In addition, a negative response may be recorded along the center of the beam at very low stimulus intensities. The amplitude of the lateral inhibition appears related to the amplitude of the excitatory response at the center of the beam and not to the amplitude of the positive wavelet that preceded it (Fig. 6C).

The amplitude of on-beam inhibition depends on stimulus train duration, inter-pulse frequency and stimulus intensity. Both the peak of the positive response and the peak of the prolonged inhibition increase with stimulus intensity (Figs 6D and 6E). Quantitative analysis reveals a linear relationship between the positive and negative responses recorded at each location, at each stimulus intensity (Fig. 6F).

The different properties of the two inhibitory responses indicate that they arise from different origins. The linear relationship between the excitation and inhibition of the on-beam inhibition suggests either that both the excitatory and inhibi-

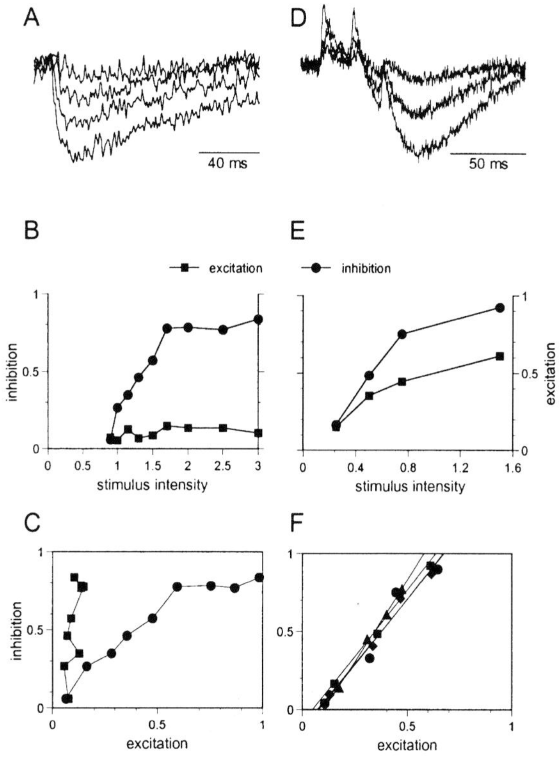

Fig. 6. The amplitudes of lateral and on-beam inhibition increase with stimulus intensity. A: Four superimposed traces of lateral inhibition recorded at a distal location and activated by four different stimulus intensities. B: Peak amplitude of the inhibitory response (circles) and the preceding excitatory wavelet (squares) as a function of stimulus intensity. C: Inhibition as a function of the excitation. Note the lack of dependence of the inhibition on the amplitude of excitation at the same location (squares). A linear relation was observed between the lateral inhibitory response and excitation recorded at the center of the beam (circles). D–F: As in A–C for the on-beam inhibition evoked by a train of three stimuli at time intervals of 20 msec. In F, the inhibition was plotted as a function of the excitation in four diodes (different symbols) located along the activated beam. Note the linear relation between these two parameters measured at the same location.

tory signals are driven by the parallel fibers or that mainly the excitatory signals drive the inhibitory signals. The lateral inhibition has the lowest threshold for activation, suggesting that it is either driven directly by the stimulus or by neurons closer to the center of the beam. These observations all match the morphological organization of the cerebellar cortex. If the lateral inhibition is activated by the molecular layer interneurons, their small size would account for the low threshold of activation. Indeed, a relatively small number of parallel fibers are sufficient to activate the stellate cells compared with Purkinje cells (Barbour, 1993). The recent finding that the molecular layer interneurons form local inhibitory networks interconnected via electrotonic coupling (Mann-Metzer and Yarom, 1999), further explains the relatively fast saturation of this response. Additionally, the axons of the molecular layer interneurons are either restricted to the vicinity of the cell body or extend to different lengths in a plane perpendicular to the parallel fiber axis. This morphological feature may account for the observed spatial distribution of the inhibitory responses.

The properties of the on-beam inhibition can readily be explained if we assume that it is driven by the activation of Golgi cell axons or Purkinje cell collaterals. First, the spatial organization of these axons and collaterals fits the spatial pattern of the on-beam inhibition. Second, both cell types are relatively large neurons and are activated by the same beam of parallel fibers. Hence, their excitation is linearly related to the number of activated parallel fibers. The recorded hyperpolarization reflects the inhibition exerted either by Purkinje cells collaterals on Golgi cells, or by Golgi cells on granule cells. Therefore, both possibilities can account for the observed linear relation between the excitatory and inhibitory responses (Fig. 6F). It should be noted that although one may expect that the optical inhibitory response generated by the granule cells will be smaller than that of Golgi cells, it is likely that their large number will compensate for their size. Therefore, it is rather difficult to distinguish between the two possibilities. A recent detailed description of the Purkinje cells collaterals (O'Donoghue and Bishop, 1990) showed their tendency to saggital arborization. This finding favors the Golgi cell as the main contributor to the on-beam inhibition.

Temporal integration of the inhibitory responses

The difference between lateral and on-beam inhibition, with the greatest functional significance, is revealed by analyzing the temporal properties of the evoked responses. Unlike on-beam inhibition, lateral inhibition is completely blocked by a preceding stimulus. This difference is depicted in Fig. 7 where responses, at the side of the beam, to two consecutive stimuli delivered to the cerebellar surface are shown. The second stimulus failed to elicit lateral inhibitory response when the interval between stimuli was 20 msec (Fig. 7A). This blockade of the second inhibitory response lasts for up to 50 msec, followed by a gradual recovery at longer intervals. In addition, there was an increase in the depolarizing wavelet that preceded the

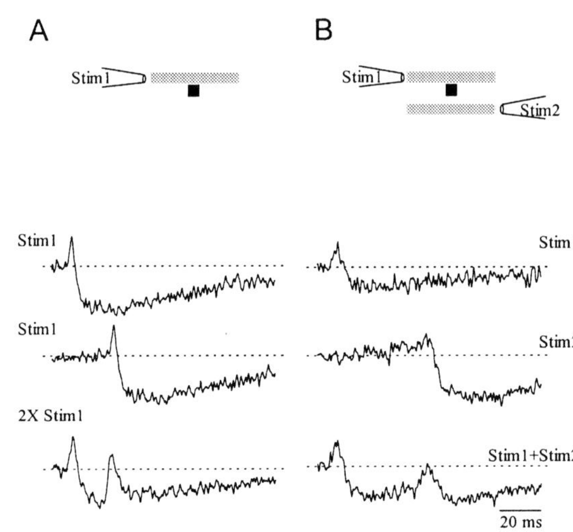

Fig. 7. Lateral inhibition is blocked by a preceding lateral inhibition. The insets in A and B are schematic representations of the experimental arrangement. The thick gray line represents the activated beam and location of the recording sites relative to the beam is marked by a black square. Only one beam was activated in A. The responses to each of the stimuli are shown in the top two traces, the response to both of them is shown in the third trace. Note the complete blockade of the second inhibitory response and the increase in the second excitatory wavelet. B: As in A, for activation of two separate beams. As with activation of a single beam, the second inhibitory response was completely blocked and the preceding excitatory wavelet increased.

inhibitory response. This result indicates that despite the bisynaptic pathway of the inhibitory response, it is fast and strong enough to cut short the excitatory response. Therefore, the excitatory response increases when the inhibitory response is blocked.

The complete blockade of the second lateral inhibitory response can be attributed either to synaptic depression or to presynaptic inhibition. To distinguish between these two possibilities, we studied the temporal interaction between the lateral inhibition elicited by different sources. Two separate parallel beams were evoked using two stimulating electrodes, so that inhibitory responses were obtained in the area between the beams by either beam (inset Fig. 7B). Regardless of the stimulation sequence, the second inhibitory response is always blocked by the first stimulus (Fig. 7B). There is a complete blockade of the inhibitory response for about 50 msec, during which the depolarizing wavelet increases. This result strongly suggests that the blockade of the lateral inhibition is due to presynaptic inhibition rather than synaptic depression.

The robust and efficient presynaptic inhibition of the lateral inhibition is particularly interesting. It suggests that the role of the lateral inhibition may be disinhibition rather than inhibition per se. Thus, the lateral spread of inhibition caused by the molecular layer interneurons may increase the responsiveness of lateral areas to incoming information over a long time.

A different behavior was observed at the center of the activated beam (Fig. 5). At intervals shorter than 50 msec the second stimulus elicits a pronounced inhibitory response that increases in amplitude with decrease in the interval between the stimuli. This inhibitory response has a much faster time course of recovery than the lateral inhibition (Fig. 6).

These results strongly suggest that a facilitatory process governs the on-beam inhibition, whereas, the lateral inhibition undergoes a prolonged depression. The lateral inhibition arises from the activity of molecular layer interneurons, whereas the on-beam inhibition most likely reflects Golgi cell inhibition. The facilitatory process occurs at the parallel fiber-Golgi cell connection, at the Golgi cell-granular cell synapse or at the Purkinje cell-Golgi cell synapse.

Concluding remarks

Two major new concepts emerge from our study of the evoked activity in the cerebellar cortex. First, white matter stimulation elicits a patch of activity, throughout which activation is simultaneous. In contrast, as would be expected from the morphology, surface stimulation generates a beam of activity that propagates along the parallel fiber system. Second, our results reveal two inhibitory systems, the lateral and on-beam inhibition, which modulate the activity in the cerebellar cortex. The two systems differ in their origin as well as in their spatial and temporal properties. Lateral inhibition occurs throughout an activated area, whereas on-beam inhibition is limited to the activated beam of parallel fibers. On-beam inhibition facilitates with repetitive stimulation, whereas a single stimulus produces prolonged depression of the lateral inhibition.

The patchy organization of the response to mossy fiber input and the powerful depression of the lateral inhibition have significant consequences for our view of cerebellar function. They imply that the basic, independent modules of the cerebellar cortex consist of a localized group of granule cells and the Purkinje cells above them. The feedback inhibition of the Golgi cells and the feed-forward inhibition of the stellate cells regulate the activity of each module. Mutual interactions between different modules occur via the parallel fibers in the medio-lateral axis and via the long range axons of the molecular layer interneurons in the sagittal axis. The medio-lateral interaction via the parallel fibers is rather weak and is efficient only within a narrow time window, whereas the sagittal interaction, due to disinhibition, is far more significant and efficient over a much broader time window.

Acknowledgements

This study was supported by the US-Israel Binational Science Foundation, the Israel Academy of

Sciences and the European Commission.D.C. was the recipient of a Clore Felloweship.

References

Albus, J.S. (1971) A theory of cerebellar function. *Math. Biosci.*, 10: 25–61.

Barbour, B. (1993) Synaptic currents evoked in Purkinje cells by stimulating individual granule cells. *Neuron*, 11: 759–769.

Bower, J.M. and Woolstone, D. (1983) Congruence of spatial organization of tactile projections to granule cell and Purkinje cell layers of cerebellar hemispheres of the albino rat: vertical organization of cerebellar cortex. *J. Neurophysiol.*, 49: 745–766.

Braitenberg, V. (1961) Functional interpretation of cerebellar histology. *Nature*, 190: 539–540.

Cajal, R.Y. (1955) Histology du Systeme Nerveux de l'Homme et des Vertebres. *C.S.I.C*, Madrid.

Cohen, D. and Yarom, Y. (1998) Patches of synchronized activity in the cerebellar cortex evoked by mossy fiber stimulation: questioning the role of parallel fibers. *Proc. Natl Acad. Sci.*, 95: 15032–15036.

Cohen, D. and Yarom, Y. (1999) Optical measurements of synchronized activity in mammalian isolated cerebellum. *Neuroscience*, 94: 859–866.

De Schutter, E. and Bower, J.M. (1994) An active membrane model of the cerebellar Purkinje cell II. simulation of synaptic responses. *J. Neurophysiol.*, 71: 401–419.

Eccles, J.C., Ito, M. and Szentagothai, J. (1967) *The Cerebellum as a Neuronal Machine*, Springer-Verlag, Berlin.

Elias, S.A., Yae, H. and Ebner, T.J. (1993) Optical imaging of parallel fiber activation in the rat cerebellar cortex: spatial effects of excitatory amino acids. *Neuroscience*, 52: 771–786.

Gravel, C., Eisenman, L.M., Sasseville, R. and Hawkes, R. (1987) Parasagittal organization of the rat cerebellar cortex: direct correlation between antigenic Purkinje cell bands revealed by mabQ113 and the organization of the olivocerebellar projection. *J. Comp. Neurol.*, 265: 295–310.

Groenwegen, H.J. and Voogd, J. (1977) The parasagittal zonation within the olivocerebellar projections.I. Climbing fiber distribution in the vermis of cat cerebellum. *J. Comp. Neurol.*, 174: 417–488.

Groenwegen, H.J., Voogd, J. and Freedman, S.L. (1979) The parasagittal zonation within the olivocerebellar projections. II. Climbing fiber distrubution in the intermediate and hemispheric parts of cat cerebellum. *J. Comp. Neurol.*, 183: 551–601.

Hawkes, R., Brochu, G., Dore, L., Gravel, C. and Leclerc, N. (1992) Zebrins: molecular markers of compartmentation in the cerebellum. In: R. Llinas and C. Sotelo (Eds), *The Cerebellum Revisited*, Springer-Verlag, New York.

Ito, M. (1984) *The Cerebellum and Neural Control*, Raven Press, New York.

Kim, J.H., Dunn, M.B., Hua, Y., Rydberg, J., Yae, H., Elias, S.A. and Ebner, T.J. (1989) Imaging of cerebellar surface activation in vivo using voltage sensitive dyes. *Neuroscience*, 31: 613–623.

Konnerth, A., Obaid, A.L. and Salzberg, B.M. (1987) Optical recording of electrical activity from parallel fibers and other cell types in skate cerebellar slices. *J. Physiol.*, 393: 681–702.

Llinas, R. (1981) Electrophysiology of the cerebellar network. In: V.M. Brooks (Ed.), *Handbook of Physiology – the Nervous System*, American Physiological Society, Bethesda, pp. 831–876.

Llinas, R. (1982) Radial connectivity in the cerebellar cortex: a novel view regarding the functional organization of the molecular layer. In: S.L. Palay and V. Chan-Palay (Eds), *The Cerebellum: New Vistas*, Springer, New York, pp. 189–192.

Llinas, R. and Precht, W. (1969) Recurrent facilitation by disinhibition in Purkinje cells of the cat cerebellum. In R. Llinas (Ed.), *Neurobiology of Cerebellar Evolution and Development*, American Medical Association, Chicago, pp. 619–627.

Llinas, R. and Sasaki, K. (1989) The functional organization of the olivocerebellar system as examined by mutliple Purkinje cell recordings. *Eur. J. Neurosci.*, 1: 587–602.

Llinas, R.R. and Simpson, J.I. (1981) Cerebellar control of movement. In: A.L. Towe and E.S. Luschei (Eds), *Handbook of Behavioral Neurobiology, Vol. 5*, Plenum, pp. 231–302.

Mann-Metzer, P. and Yarom, Y. (1999) Electrotonic coupling interacts with intrinsic properties to generate synchronized activity in the cerebellar inhibitory interneuron networks. *J. Neurosci.*, 19: 3298–3306.

Marr, D. (1969) A theory of cerebellar cortex. *J. Physiol.*, 202: 437–470.

O'Donoghue, D.L. and Bishop, G.A. (1990) A quantitative analysis of the distribution of Purkinje cells axonal collaterals in different zones of the cat's cerebellum: an intracellular HRP study. *Exp. Brain Res.*, 80: 63–71.

Palay, S.L. and Chan-Palay, V. (1974) *Cerebellar Cortex*, Springer-Verlag, New York.

Pichitpornchai, C., Rawson, J.A. and Rees, S. (1994) Morphology of parallel fibers in the cerebellar cortex of the rat: an experimental light and electron microscopic study with biocytin. *J. Comp. Neurol.*, 342: 206–220.

Thach, W.T., Goodkin, H.P. and Keating, J.G. (1992) The cerebellum and the adaptive coordination of movement. *Annu. Rev. Neurosci.*, 15: 403–442.

Voogd, J. and Glickstein, M. (1998) The anatomy of the cerebellum. *Trends Neurosci.*, 21: 370–375.

Welsh, J.P., Lang, E.J., Sugihara, I. and Llinas, R. (1995) Dynamic organization of motor control within the olivocerebellar system. *Nature*, 374: 453–457.

N.M. Gerrits, T.J.H. Ruigrok and C.I. De Zeeuw (Eds.)
Progress in Brain Research, Vol 124

CHAPTER 9

Electrotonic coupling synchronizes interneuron activity in the cerebellar cortex

Puah Mann-Metzer and Yosef Yarom*

Department of Neurobiology, Life Sciences Institute and Center for Neural Computation, Hebrew University, Givat Ram, Jerusalem 91904, Israel

Introduction

Anatomical studies of the cerebellar cortex revealed a modular structure. Each module is activated by two inputs, the mossy and climbing fibers. A single output, the Purkinje cells (PC) axons, transmits the processed information back into the deep cerebellar nuclei. These modules perform a computational process that may participate in the various functions attributed to the cerebellum. Inhibitory interneurons play a prominent role in modulating the activity within and between modules. In particular the molecular layer interneurons (MLIs) have been implicated in modulating PC firing or shaping the flow of information along the cortex.

Here, we describe further electrical characteristics of the MLIs based on in vitro electrophysiological experiments in guinea pig cerebellar slices. We show that these inhibitory interneurons of the cerebellar cortex form small local networks in which activity is synchronized by electrical coupling (Mann-Metzer and Yarom, 1999).

The MLIs are classically divided into basket and stellate cells. Basket cells are usually located in the inner third of the molecular layer and they are characterized by the basket-like structures that their axons form around the PC bodies, close to the axon. Their soma diameter is about 20 μm (Eccles et al., 1967). Stellate cells are usually found in the outer two thirds of the molecular layer and their cell bodies are smaller than those of the basket cells. They innervate the dendrites of the PC. Although these two cell types have been further classified into several subtypes according to their exact axonal shape (Eccles et al., 1967; Palay and Chan-Palay, 1974), a recent study used principal component analysis (Sultan and Bower, 1998) to show that these cells vary continuously in their morphology depending on the location of the soma in the molecular layer. The study concluded, as did Ramón y Cajal in 1911, that MLIs constitute one population of cells rather than distinct subclasses.

The dendritic and axonal branching of the MLIs is confined to the molecular layer, yet the axonal field is usually larger than the dendritic field (Eccles et al., 1967; Sultan and Bower, 1998). Both dendrites and axons exhibit a highly planar structure, lying in the same plane as the PC's dendritic trees. The axons may also extend for some length in the coronal plane, in the same direction as parallel fibers, before branching parasagittally (Eccles et al., 1967; Midtgaard, 1992a; Sultan and Bower, 1998).

The primary excitatory input to the MLIs comes from the parallel fibers (Eccles et al., 1967). Barbour (1993) has shown that an action potential via a single parallel fiber is sufficient to fire a MLI.

*Corresponding author. Tel.: 972–2–6585172; Fax: 972–2–6586296; e-mail: yarom@vms.huji.ac.il

Because the MLIs form a mutually connected network, their main source of inhibition derives from neighboring MLIs. The PC axon collaterals, which ramify in the vicinity of their cell of origin (O'Donoghue and Bishop, 1988), provide a further inhibitory input.

The main output of the MLIs is fed to the soma and dendrites of the PC. MLIs outnumber the PCs tenfold (Korbo et al., 1993), and it has been estimated that each PC receives about 1500 inhibitory synapses (Jaeger et al., 1997). The synaptic connections are powerful, so that one MLI can prevent a PC from firing (Midtgaard, 1992b; Häusser and Clark, 1997) provided that its output is appropriately timed.

Several hypotheses have been proposed for the role of the molecular layer inhibitory system in cerebellar function. According to the lateral inhibition hypothesis, proposed by Eccles and his colleagues (1967), when a beam of parallel fibers is activated, the long axons of the basket and some of the stellate cells evoke lateral inhibition on both sides of the activated beam, thereby confining the activity to a narrow strip. A recent alternative hypothesis is that the MLIs limit the spread of activity along the beam (Santamaria and Bower, 1997). This may account for observations that mossy fiber input to the cerebellar cortex evokes a patch of activity (Bower and Woolstone, 1983; Cohen and Yarom, 1998), rather than the expected activity flow along beams of parallel fibers.

Other studies have suggested that the role of the inhibitory neurons is to shape the pattern of PCs firing. Indeed, using a detailed PC model, Jaeger et al. (1997) suggested that rather than the excitation it is the inhibition, interacting with intrinsic voltage dependent currents, which determines the firing pattern of the PCs. This fits well with Häusser and Clark's (1997) findings that the inhibitory input inflicts irregularity on the firing pattern of the PCs and the MLIs. Alternatively, Ross and colleagues (Callaway et al., 1995) have suggested that the role of the outer MLIs is to suppress the PC dendritic calcium response evoked by climbing fiber activity. They demonstrated that there is a critical time window in which MLI activity can completely block the dendritic calcium response, whereas the somatic voltage response is unaffected.

The electrical properties of the MLIs

From an electrophysiological viewpoint, the MLI seems a homogenous population, with all neurons displaying similar properties. The MLIs fire spontaneously and show extensive spontaneous subthreshold activity (Fig. 1A). Most of the synaptic potentials are inhibitory (only inhibitory neurons remain intact in the parasagittal slice) and most likely derive from the spontaneous activity of other MLIs (Häusser and Clark 1997; Kondo and Marty 1998). The marked decrease in synaptic activity in conditions that block action potentials (Fig. 1A) supports this possibility. Blocking the inhibitory synaptic transmission by synaptic blockers, such as bicuculline or picrotoxin, increases the firing rate of the spontaneous activity, and alters its pattern (Häusser and Clark, 1997). It is therefore likely that the resting potential of the MLIs lies close or above the firing threshold, enabling spontaneous activity.

The action potentials of the MLIs are brief and show only moderate accommodation. Firing frequencies of up to hundreds of action potentials per second can easily be induced by intracellular current injection (Figs 1B and C), giving almost linear current – frequency curves for low intensity current injections (Fig. 1C). Frequency within the train either decreased or increased as a function of time (Figs 1B and 1C), the latter being more common when the train was triggered from more negative membrane potentials.

The inter-spike trajectory (Fig. 1D) shows that slow currents modulate the membrane potential around the threshold. Indeed, a substantial, inactivating potassium current reminiscent of the A-type current can be demonstrated (Fig. 1E). This current causes the increase in firing rate during a prolonged train of action potentials evoked from negative membrane potentials. Non-inactivating Na^+ current and Ca^{++} currents were also found to participate in modulating neuronal excitability (not shown).

A prominent electrical property of the MLIs is a strong and abrupt outward rectification that dominates the membrane current–voltage relationship (Fig. 1F). This rectification occurred at a membrane potential of about -30 mV, i.e. above the firing threshold. The resting input resistance, which is of the order of 220 MΩ at membrane potentials of up to − 30 mV, dropped to a level of 20 MΩ at − 25 mV, indicating that the rectifying current has an extremely sharp steady state activation curve. In addition, the activation time of this current is less than one millisecond (Fig. 1F). This rapid rate of activation and the steep steady-state activation curve endow this rectifying current with threshold-like properties. As the threshold of this current is higher than the membrane resting potential, it does

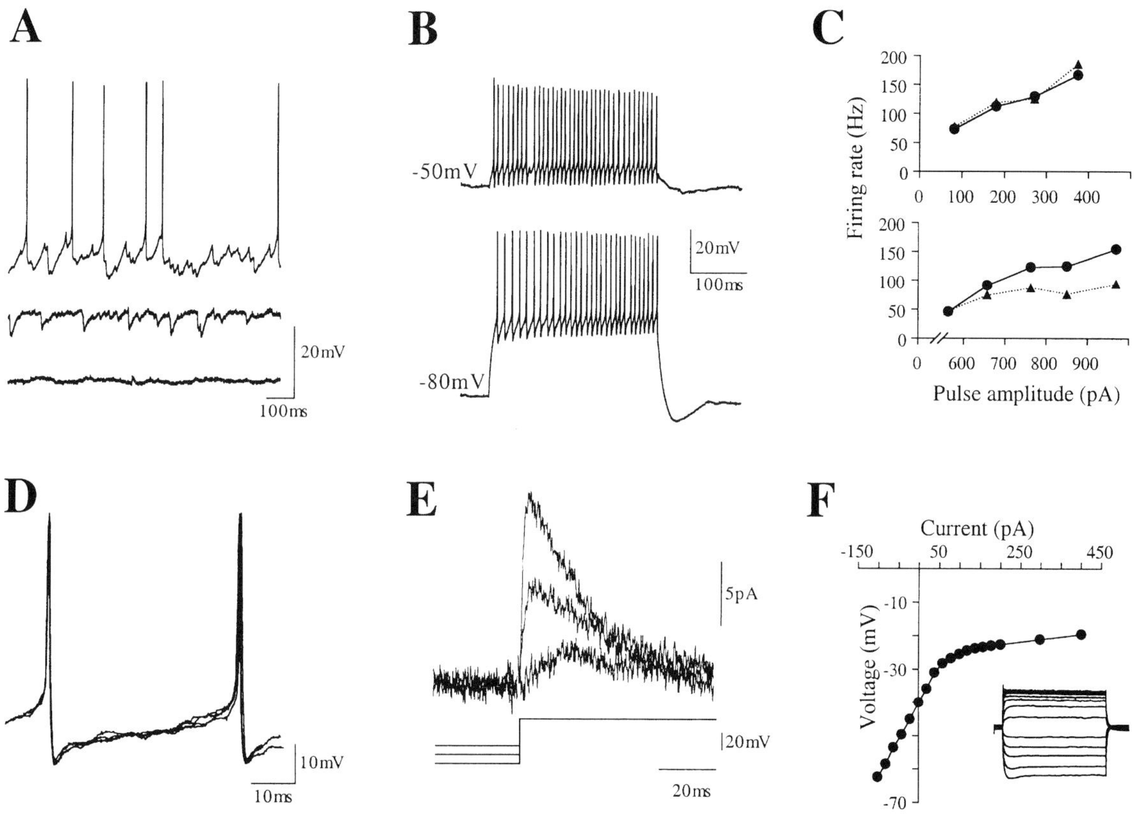

Fig. 1. The electrical properties of the MLIs. A. Intracellular recordings from a stellate cell in a sagittal cerebellar slice. Top trace: Spontaneous activity of a MLI. Middle trace: Spontaneous synaptic potentials. The membrane potential was slightly hyperpolarized to prevent action potentials. Bottom trace: Adding TTX (0.1 μM) to the extracellular medium blocked most of the synaptic activity. B. A depolarizing current step given at a membrane potential of − 50 mV induced a train of action potentials with a constant frequency, while depolarization from a hyperpolarized membrane potential (− 80 mV) induced a train with increasing frequency. C. Frequency – current curve of a MLI. Trains of action potentials were evoked by prolonged current injections (as in B). The average firing rates during the train (solid line) and the instantaneous firing rate at the beginning of the train (dashed line) were plotted against the amplitude of the depolarizing pulse, given from a − 50 mV (Top panel) or − 80 mV (Bottom panel) membrane potential. Note the lower firing rate at the beginning of the train when depolarization is induced from a hyperpolarized membrane potential (bottom panel, dashed line). D. Inter-spike trajectory. Four traces with similar inter-spike intervals were aligned according to the first spike to demonstrate the recurrent characteristic of the spontaneous firing. Note the strong hyperpolarization at the end of the spike, followed by a slow depolarization, which eventually crosses threshold. E. Demonstration of an A-like current. The membrane potential was clamped to − 30 mV from different resting levels (− 30, − 60, − 70 and − 80 mV bottom traces). The resultant transient current (top traces) became larger as the holding membrane potential was more hyperpolarized, indicating a hyperpolarization-dependent deinactivation of this current. This current was blocked by 4-AP 2mM, not shown). This experiment was carried out in the presence of TTX (0.1 μM). Leak current was subtracted from the current traces. F. Typical current – voltage relation of a MLI, with a large and abrupt rectification at potentials above − 30 mV. Inset: the voltage responses to current pulses at different, equally spaced amplitudes. Note the delayed anomalous rectification at hyperpolarizing pulses.

not contribute to the modulation of the membrane voltage at resting level. However its rapid activation and termination suggest that it modulates the action potential time course, most probably causing the brevity of the action potentials and the rapid resetting of the membrane potential, both of which allow the high firing rates.

These findings agree with the current picture of MLIs which are known to fire repetitively after a single stimulus (Eccles et al., 1966), as expected, given the low threshold of activation and the high resting input resistance (see also Midtgaard, 1992a). The rapid resetting of the membrane potential by the fast and powerful rectifying current enables a single input to generate a train of action potentials.

Of particular interest is the spontaneous firing of these neurons, which continues in the presence of synaptic blockers. This firing is generated by slow currents activated at membrane potentials between resting potential and firing threshold. These currents generate slow creeping of the membrane potential towards threshold and thus set the neuron's firing frequency. If this type of activity also occurs in vivo, then these MLIs will continuously regulate the PCs, their main target cells.

It is interesting to note that spontaneous activity and high frequency trains of brief duration action potentials are a common feature among inhibitory interneurons in the CNS. A recent study on cortical inhibitory interneurons demonstrated the prominent role of K^+ channels in shaping some of these properties (Erisir et al., 1999).

Electrical coupling between MLI

Simultaneous recordings from pairs of adjacent MLIs revealed a significant tendency for synchronized firing (Fig. 2A). Aligning recordings from two cells by the spikes in one of them shows the tendency of the spikes in the second cell to cluster around the time of the spike in the first cell. However the time window during which the second cell shows a higher probability for firing is rather wide. The significant peak of the normalized cross-correlation histogram of the spike trains quantifies this tendency for synchronous firing (Fig. 2B). Synchronized firing was observed in about 40% of the recordings, in which a time window of 15 to 20 msec was calculated.

Synchronized firing is particularly important in neuronal systems, especially when the outputs of the system converge to the same target. In this case, synchronization limits the output to a given time window in which its effectiveness is increased. The MLIs innervate PCs with a well-described spatial distribution. There is a large convergence of several MLIs into a single PC, as well as divergence of a single MLI to several PCs. Thus, synchronous activity of MLIs can have a profound effect on the activity of the cerebellar cortex. The time window of the synchronized firing can prolong the inhibitory effect of the MLIs and set it to the desired duration.

Synchronized firing may be due to chemical synaptic transmission, electrical synapses or non-synaptic interactions. Since synchronization was not affected by specific synaptic blockers nor by low Ca^{++} concentration, we tested for electrical connections. Simultaneous recording from two MLIs revealed that current injection into either cell induces voltage responses in both cells (Fig. 2C). The local response in the presynaptic cell (the cell into which the current was injected) was an order of magnitude larger than that recorded in the postsynaptic cell. This bi-directional, direct current flow results from electrical coupling between the two cells. Like the synchronized firing, this coupling was observed in about 40% of the recordings.

Average coupling coefficient i.e. the ratio between pre- and postsynaptic voltages elicited by prolonged current injection, was 0.12. The coupling coefficient evidently depends on the postsynaptic input resistance and the coupling resistance. Interestingly, it also depends on the membrane potential of the postsynaptic cell; uncoupling between the neurons was observed when the membrane potential of the postsynaptic cell rose above -30 mV (Fig. 2D). This apparent voltage-dependence of the electrotonic coupling reflects the voltage-dependence of the membrane resistance (see Fig. 1G). Given the input resistance of the cells, we calculated coupling resistance to be of the order of 5 GΩ with a rather narrow distribution among the different pairs.

Dye-coupling, despite its controversial significance, commonly serves to provide supporting evidence for electrical coupling. Dye-coupling was found in 47% of dye injections to single cells, with staining spreading to up to nine neighboring cells and including stellate as well as basket cells (Fig. 3). However, the distribution of dye-coupling was non-uniform. It occurred more frequently, and larger groups of cells were stained, when dye was injected to superficial stellate cells, compared with deep stellate and basket cells. The superficial neurons appear to innervate the distal dendrites of the PCs, whereas the deeper interneurons, particularly the basket cells, innervate PC somata. Effective inhibition of the distal PC dendrites may require the collaborative effort of a group of neurons. In contrast, the deeper interneurons, which densely innervate the soma and initial segment of the PC axons, provide a very efficient inhibition that can be executed by a single neuron.

Coupling enhancement by intrinsic currents

The demonstration of a direct current flow between neurons does not necessarily mean that the current

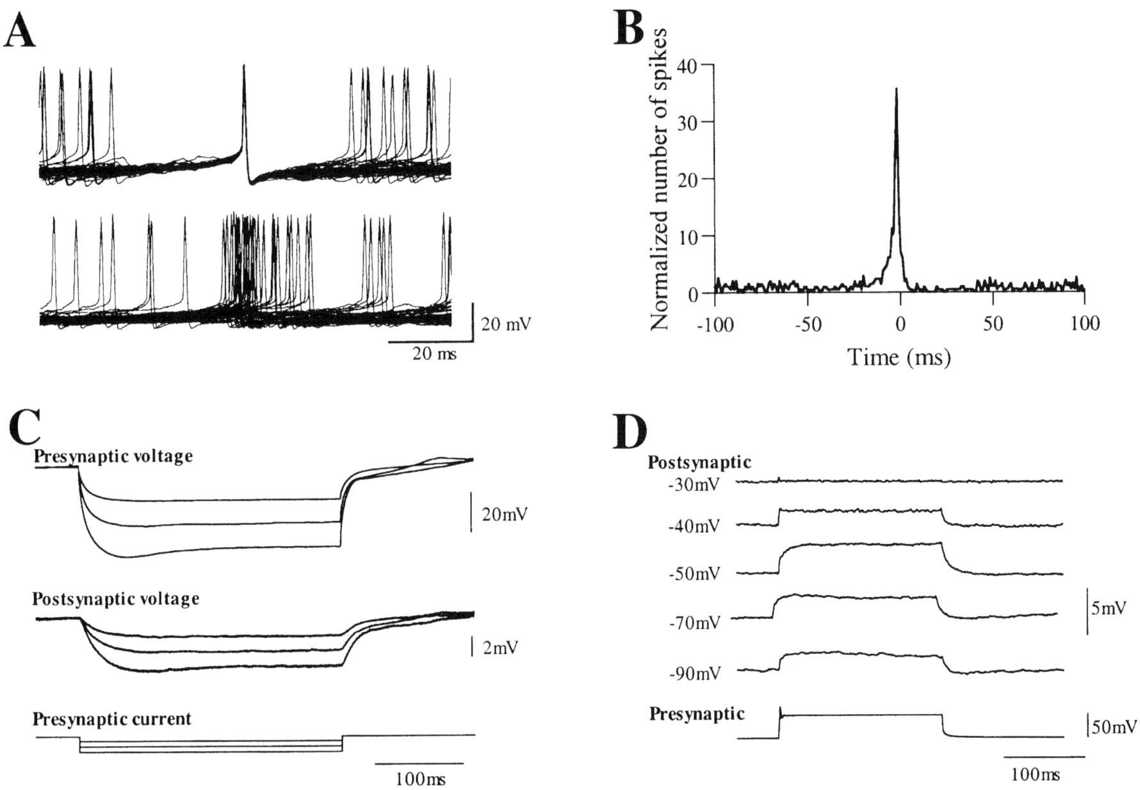

Fig. 2. Electrical coupling between MLIs. A. Fifty consecutive spontaneous activity traces of two simultaneously recorded interneurons. The traces were aligned according to the spikes in the cell in the top panel. Note the clustering of the spikes in the second cell (bottom panel). B. Normalized spike cross-correlogram of the spontaneous activity of the neuron pair in A. The histogram was normalized by dividing the number of observed spikes by the expected number of spikes considering the firing rate and number of traces and assuming a random firing of the two neurons. The width of the peak is 14 ms. C. Direct current flow between a pair of simultaneously recorded MLIs. Current pulses of different amplitudes (bottom traces) were injected to one cell, defined as the presynaptic cell. Voltage responses were measured in this cell (top traces) and in the postsynaptic cell (middle traces). Note difference in scale. Similar responses were recorded when current was injected in what had previously been the postsynaptic cell (not shown). D. The postsynaptic response is voltage-dependent. Current pulses of constant amplitude were injected into the presynaptic cell. The voltage response in the presynaptic cell is shown in the bottom trace. The voltage response in the postsynaptic cell was measured at different membrane potentials, as set by DC current injection. Coupling is greatly reduced at depolarizing membrane potentials.

generated by the brief action potential is sufficient to induce synchronized firing. Furthermore, the 15–20 msec time window of synchronization is longer than would be expected from the brief current flow generated by a presynaptic action potential.

Analysis of the postsynaptic potential during spontaneous firing of the presynaptic cell revealed two components in the postsynaptic response: a prolonged depolarizing potential and a short attenuated spikelet (Fig. 4A). The spikelet reflects the current flow evoked by the presynaptic action potential, while the prolonged depolarizing component reflects the slow depolarizing process leading to the presynaptic action potential. However, it continues beyond the duration of the spike, which may be due to voltage-dependent currents in the postsynaptic cell contributing to this slow depolarizing component. Indeed, hyperpolarizing the postsynaptic cell decreases the prolonged depolarization. Furthermore, the contribution of postsynaptic currents to the compound response was confirmed by blocking postsynaptic Na^+ channels with intracellular QX–314; the postsynaptic action potentials and the prolonged depolarizing component were blocked almost immediately after breaking the neuron membrane and establishing the whole cell recording. Since, under normal conditions, the membrane potential lies within the voltage range where these currents are active, the most common subthreshold postsynaptic response will have the two components.

The following scenario explains how the participation of intrinsic currents can induce a relatively wide time window of synchronization. It starts with the presynaptic cell, where the intrinsic current

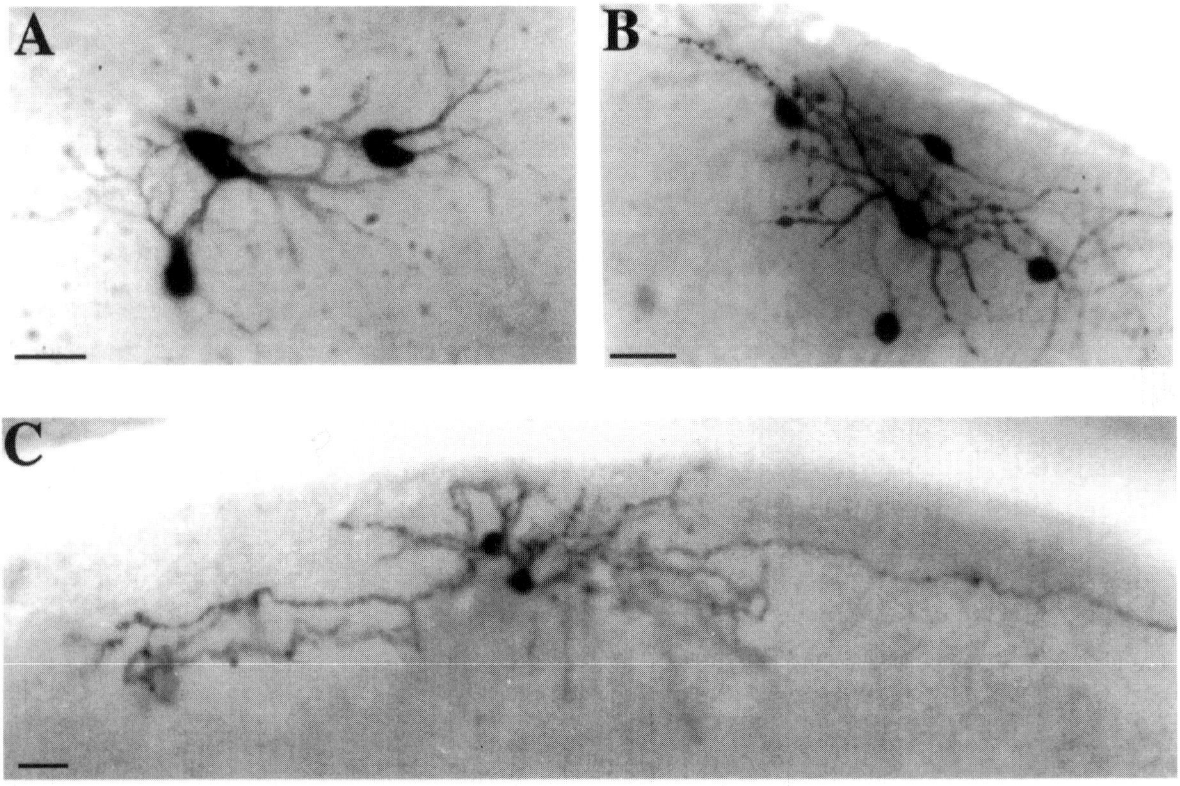

Fig. 3. Three examples of dye-coupling between MLIs, demonstrated by injecting a single neuron with biocytin (0.5%). The stained cell bodies typically lie close together in the same sagittal plane, and the dendrites intermingle. A. A group of three cells in the middle of the molecular layer. B and C. Two groups located in the upper part of the molecular layer which contain 5 and 2 stained neurons respectively. Note the sagittal extention of the axons in C. All scale bars denote 20 μm.

slowly depolarizes the membrane potential. This slow depolarization propagates via the gap junctions to activate the same type of currents in postsynaptic cells. The slow process of depolarization proceeds in the coupled cells until threshold is reached, and an action potential is elicited in one of them. The current generated by the action potential also propagates to the postsynaptic cells and, if the combined depolarization reaches threshold, the postsynaptic cell fires within a short time window of a few milliseconds. If the spikelet fails to activate the postsynaptic cell, the slow depolarizing process continues until it eventually leads to a postsynaptic action potential. Thus, although the postsynaptic action potential is synchronized with the presynaptic action potential, the time window is relatively wide.

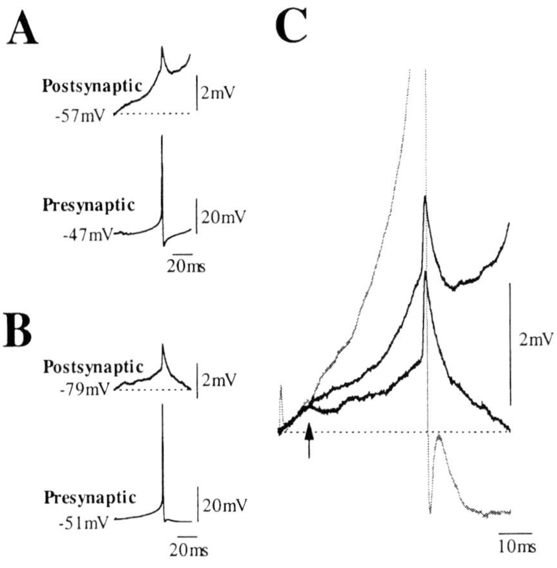

Fig. 4. Intrinsic currents amplify electrotonic coupling between MLIs. A. Spike triggered average of the subthreshold postsynaptic response to a spike in the presynaptic cell. Traces were selected according to the criteria that the postsynaptic cell did not respond with an action potential before and at least 20 ms after the spike in the presynaptic cell. The horizontal dashed line represents the average postsynaptic resting membrane potential measured 40 ms before the action potential in the presynaptic cell. Note the prolonged depolarization and the spikelet that characterize the postsynaptic response. B. The subthreshold postsynaptic response measured as in A during hyperpolarization of the postsynaptic cell to − 79 mV by DC current injection. The prolonged depolarizing response is attenuated and the membrane potential declines to resting level after the spikelet. C. Three superimposed traces of the presynaptic action potential (gray line) and the postsynaptic spikelets under normal and hyperpolarizing conditions.

The increase in apparent coupling by intrinsic currents is particularly interesting as it contributes significantly to what we regard as the physiological role of the coupling. Specifically, a weakly coupled network of spontaneously active neurons can generate synchronized firing within a time window that is larger than would be predicted from the time course of the presynaptic action potential.

Closing remarks

The main and novel finding of this in vitro investigation is that 40% of the interneurons in the molecular layer of the cerebellar cortex are electrically coupled, suggesting that these interneurons form local networks that give rise to synchronized activity. We have demonstrated here that electrical synapses and intrinsic currents interact to form a highly modifiable communication pathway. We propose that a small modulation of membrane potential or membrane resistivity, either by neuromodulators or by spontaneous synaptic activity, can strongly modulate the strength of electrotonic coupling as well as its interaction with intrinsic currents.

Acknowledgements

This study was supported by The Israel Science Foundation

References

Barbour, B. (1993) Synaptic currents evoked in Purkinje cells by stimulating individual granule cells. *Neuron*, 11: 759–769.

Bower, J.M. and Woolstone, D. (1983) Congruence of spatial organization of tactile projections to granule cell and Purkinje cell layers of cerebellar hemispheres of the albino rat: vertical organization of cerebellar cortex. *J. Neurophysiol.*, 49: 745–766.

Callaway, J.C., Lasser-Ross, N. and Ross, W.N. (1995) IPSPs strongly inhibit climbing fiber-activated $[Ca^{2+}]_i$ increases in the dendrites of cerebellar Purkinje neurons. *J. Neurosci.*, 15: 2777–2787.

Cohen, D. and Yarom, Y. (1998) Patches of synchronized activity in the cerebellar cortex evoked by mossy fiber stimulation: questioning the role of parallel fibers. *Proc. Natl Acad. Sci.*, 95: 15032–15036.

Eccles, J.C., Ito, M. and Szentágothai, J. (1967) *The Cerebellum as a Neuronal Machine*, Springer-Verlag, Berlin.

Eccles, J.C., Llinas, R. and Sasaki, K. (1966) The inhibitory interneurons within the cerebellar cortex. *Exp. Brain Res.*, 1: 1–16.

Erisir, A., Lau, D., Rudy, D.B. and Leonard, C.S. (1999) Function of specific K^+ channels in sustained high-frequency firing of fast-spiking neocortical interneurons. *J. Neurophysiol.*, 82: 2476–2489.

Häusser, M. and Clark, B.A. (1997) Tonic synaptic inhibition modulates neuronal output pattern and spatiotemporal synaptic integration. *Neuron*, 19: 665–678.

Jaeger, D., De Schutter, E. and Bower, J.M. (1997) The role of synaptic and voltage-gated currents in the control of Purkinje cell spiking: a modeling study. *J. Neurosci.*, 17: 91–106.

Kondo, S. and Marty, A. (1998) Differential effects of noradrenaline on evoked, spontaneous and miniature IPSCs in rat cerebellar stellate cells. *J. Physiol.*, 509: 233–243.

Korbo, L., Bo-Anderson, B., Ladefoged, O. and Møller, A. (1993) Total numbers of various cell types in rat cerebellar cortex estimated using an unbiased stereological method. *Brain Res.*, 609: 262–268.

Mann-Metzer, P. and Yarom, Y. (1999) Electrotonic coupling interacts with intrinsic properties to generate synchronized activity in the cerebellar inhibitory interneuron networks. *J. Neurosci.*, 19: 8298–3306.

Midtgaard, J. (1992a) Membrane properties and synaptic responses of golgi cells and stellate cells in the turtle cerebellum in vitro. *J. Physiol.*, 457: 329–354.

Midtgaard, J. (1992b) Stellate cell inhibition of Purkinje cells in the turtle cerebellum in vitro. *J. Physiol.*, 457: 355–367.

O'Donoghue, D.L. and Bishop, G.A. (1990) A quantitative analysis of the distribution of Purkinje cells axonal collaterals in different zones of the cat's cerebellum: an intracellular HRP study. *Exp. Brain Res.*, 80: 63–71.

Palay, S.L. and Chan-Palay, V. (1974) *Cerebellar Cortex, Cytology and Organization*, Springer-Verlag, Berlin, pp. 180–233

Ramón y Cajal, S. (1911) Histologie du Système Nerveux de l'Homme et des Vertébrés, *C.S.I.C.*, Paris.

Santamaria, F. and Bower, J.M. (1997) A realistic cerebellar network: temporal, spatial, direct and indirect influences of granule cell axons on Purkinje cells. *Neuroscience Abstract*, 2: 2364.

Sultan, F. and Bower, J.M. (1998) Quantitative Golgi study of the rat cerebellar molecular layer interneurons using principal component analysis. *J. Comp. Neurol.*, 393: 353–373.

N.M. Gerrits, T.J.H. Ruigrok and C.I. De Zeeuw (Eds.)
Progress in Brain Research, Vol 124

CHAPTER 10

Unipolar brush cells of the vestibulocerebellum: afferents and targets*

Maria R. Diño, Maria Grazia Nunzi, Roberta Anelli and Enrico Mugnaini**

Northwestern University Institute for Neuroscience, 320 E. Superior Street, Chicago, IL 60611, USA

Introduction

Unipolar brush cells (UBCs) are putative interneurons of the cerebellar granular layer. They are densely concentrated in folia of the vestibulocerebellum and possess unique morphological and synaptic properties (reviewed in Mugnaini et al., 1997; Slater et al., 1997). The term UBC describes the main features of these small neurons. They usually have a single dendrite of varying length (5–50 μm), which ends in a brush-like tuft of dendrioles. The brush may also arise directly from the cell body. The brush has a mean diameter of 10–15 μm (Mugnaini and Floris, 1994; Takács et al., 1999), and a single mossy fiber (MF) sits in its center forming synapses with the UBC dendrioles. The configuration of MF–UBC synapses varies, ranging from continuous, giant synapses, each measuring up to 3.0 μm, to a series of smaller, fragmented synapses, each measuring 0.3–1.0 μm (Mugnaini et al., 1994). Although the total synaptic area of individual UBCs has not been precisely determined, it has been estimated to measure 12–40 μm^2. Ionotropic glutamate receptors (AMPA, KA, and NMDA) have been localized in the postsynaptic membrane, (Jaarsma et al., 1995), while metabotropic glutamate receptors (mGluRs) lie perisynaptically on special appendages (Jaarsma et al., 1998). White matter stimulation evokes a long-lasting, biphasic excitatory postsynaptic current (EPSC) that has been attributed to the entrapment of glutamate in the unusually extensive area of synaptic apposition (Rossi et al., 1995; Kinney et al., 1997). The round or ovoid cell body (8–15 μm in diameter) of the UBC emits a single axon that gives off collaterals and terminates as large swellings resembling the mossy rosettes of cerebellar afferent fibers (Berthié and Axelrad, 1994; Rossi et al., 1995). During the last few years much has been learned on the anatomy UBCs of the vestibulocerebellum, a cerebellar region of considerable interest in the context of this special volume of Progress in Brain Research.

Mapping the UBC distribution

Joys and perils of using cell class-specific markers

Golgi impregnation of the UBCs was first noted in the Laboratory of Neuromorphology at the University of Connecticut during the fall/winter of 1991, in cerebellar sections of adult animals prepared with a variation of the Golgi–Kopsch protocol. The Golgi–Kopsch protocol stained an unusually large number of cells in the same cerebellar folium, although it usually revealed only the initial axonal portions, by constrast to the widely favored, classical rapid method, which impregnates only a small percentage of neurons.

*This paper is dedicated to Jan Voogd a friend, colleague, and collaborator of long standing, on the occasion of his retirement from administrative duties.

**Corresponding author. Tel.: 312-503-4300; Fax: 312-503-7345; e-mail: e-mugnaini@nwu.edu

This might explain why UBCs had passed unnoticed in Golgi studies for over a century (Mugnaini and Floris, 1994). Nevertheless, these cells had previously been observed by several authors using electron microscopy, standard histology, and immunocytochemistry. In the absence of a consistent and clear characterization with the Golgi method, however, the UBCs had been interpreted as special forms of Golgi cells or granule cells. In the Golgi–Kopsch sections, UBCs were not encountered at random throughout the cerebellum. Rather, they were more frequently observed in the vestibulocerebellum than in other folia. This suggested a non-homogeneous distribution and emphasized their resemblance to pale cells, rat–302 positive cells, and cells positive for chromogranin peptides, which were also enriched in the caudal cerebellum. Because the Golgi technique is notorious for inconsistent impregnation, we embarked on a search for cell class-specific markers that would be more reliable and useful for quantitative estimates. In rat, UBCs were distinctly stained by antisera to the heavy molecular weight neurofilament protein, NFH (Harris et al., 1993), which permitted their electron microscopic characterization. Unfortunately, this marker protein was not equally useful in several other species, including mouse. Furthermore, the calcium binding protein calretinin (CR) appeared to be an early and promising candidate marker, as members of our laboratory recognized densely labeled UBCs in calretinin stained cerebellar sections shown at poster presentations by other researchers, who were promptly notified of this fact. Using the excellent calretinin antiserum from the laboratory of David Jacobowitz, a systematic light and electron microscopic study verified the preferential distribution of UBCs in the vestibulocerebellum, confirmed their presence in the dorsal cochlear nucleus, and helped highlight the ultrastructural features unique to UBCs, such as the non-synaptic appendages along the plasmalemma of the somatodendritic compartment and the peculiar cytoplasmic inclusion consisting of ringlet subunits (Floris et al., 1994). These results encouraged us to analyze the UBC distribution in different mammalian species using CR as a marker (see below; Floris et al., 1994; Mugnaini et al., 1997; Diño et al., 1999).

However, in subsequent studies, specifically those mapping glutamate receptors (GluRs), we became aware that antisera to GluR2/3 and mGluR1α stained UBCs more distinctly than antisera to CR or mGluR2/3 (Jaarsma et al., 1995, 1998). Spatz (1999) recently reported that in marmosets, but not in rodents, there is an abundance of CR-immunoreactive (CR-ir) UBCs in the cerebellum and a scarcity in the cochlear nucleus (CN). The reverse pattern, i.e. an abundance in the CN but a paucity in the cerebellum, was observed using mGluR2/3 as a marker. Moreover, in chickens, UBCs display mGluR1α (Takács et al., 1999), but not CR immunoreactivity (Rogers, 1989; our unpublished observations). Taken together, these studies suggest that combinations of immunocytochemical markers, whose expressions may be differentially regulated in various species, define subclasses of UBCs participating in distinct sensory or sensori-motor functions.

Studies utilizing double and triple labeling with CR, mGluR1α, mGluR2/3 and neurofilament protein antisera are currently in progress in our laboratory. Also, ongoing attempts to identify cohorts of genes specifically expressed by UBCs might uncover markers consistently expressed in all species.

Lessons learned from comparative mapping studies

Bolk was the first to relate relative sizes of different parts of the cerebellum to differences in the kind of movements which an animal characteristically makes (Glickstein and Voogd, 1995). Although Bolk's theory proved incorrect in its details, it has since become apparent that as an animal's locomotor and/or manual dexterity advanced, its cerebellum shows dramatic lateral expansion and a massive increase in foliation (Nieuwenhuys, 1967). Using a similar rationale, we mapped the distribution of UBCs using CR as a marker in a variety of mammals whose repertoire of movements vary greatly (opossum, mouse, rat, guinea pig, rabbit, cat, and monkey) in an effort to relate these morphological data to their putative functional/behavioral significance. The recently published quantitative study by Takács et al. (1999), where they used mGluR1α to map UBCs in chicken, rat, guinea pig, cat, and monkey, arrived at conclusions similar to ours. Surprisingly, the demonstration of mGluR1α immunoreactive UBCs in the chicken revealed interesting features of chicken UBCs and 'inverse' patterns in UBC distribution between mammalian and avian cerebella.

Aside from noting minor variations in dendritic length and somal size among the mammalian species, both studies (Diño et al., 1999; Takács et al., 1999) report that in all species studied there is an uneven distribution of UBCs in the different vermal folia but a consistent enrichment in the vestibulocerebellum (Figs 1 and 2). Expressing the density of mGluR1α-positive UBCs as UBCs/Purkinje cells (UBC/Pu), Takács and coworkers found that the nodulus and lingula were consistently in the list of the top three lobules with the highest UBC/Pu ratio; the uvula also belonged to

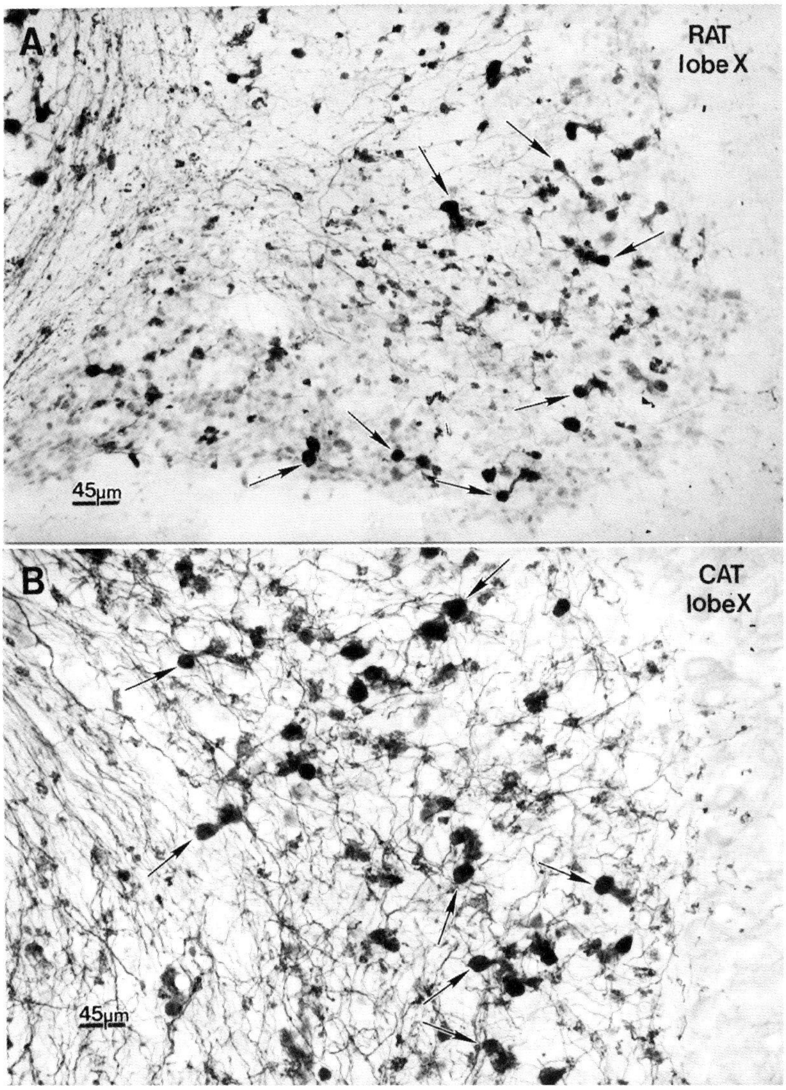

Fig. 1. Calretinin immunostaining in the rat (A) and cat (B) nodulus (lobulus X) illustrates the typical morphology of UBCs (arrows) and the increased density in cats.

this short list in rats, guinea pigs, and cats, but not in chickens and monkeys where it was replaced by lobule II. Hot spots of CR and mGluR1α-positive UBCs were noted in crus I in rats and guinea pigs, and in lobule VII in the cat and monkey. Both groups were intrigued by the fact that these areas of UBC hot spots correspond to the 'oculomotor vermis' (reviewed by Noda, 1991). This is further supported by our findings that the 'flocculus-like' region of the ventral paraflocculus (as described by Gerrits and Voogd, 1982 and Voogd et al., 1996) of the different species enlarges and becomes increasingly filled with CR-positive UBCs in direct correlation to the sophistication of the animal's eye movements.

Lest we give the impression that UBCs participate exclusively in circuits pertaining to vestibular modalities, distribution patterns in several species

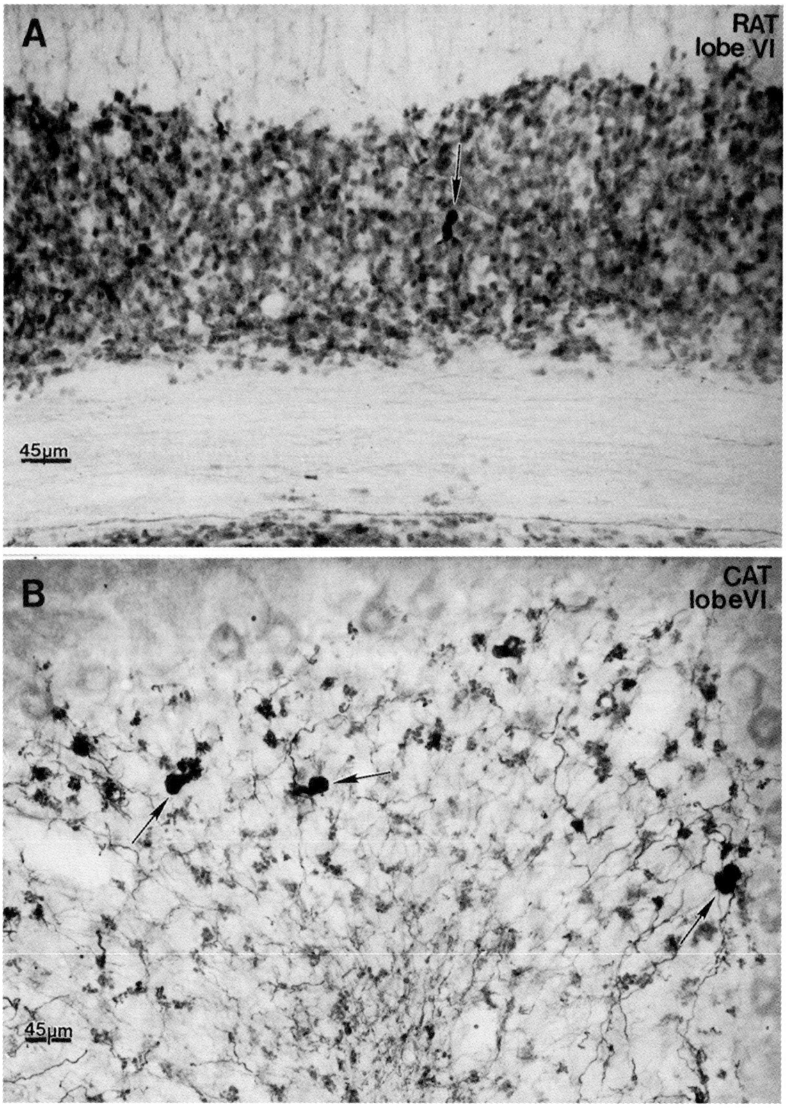

Fig. 2. Calretinin immunostaining in the rat (A) and cat (B) lobulus VI. In the cat vermis, lobulus VI has the lowest density of calretinin-positive UBCs. Note, however, that the density of CR-positive UBCs (arrows) in cat lobulus VI is higher than that of the rat. Granule cell staining is dense in rat lobulus VI, but minimal in the cat.

suggest otherwise. The expansion of CR-positive UBCs into the intermediate and lateral cerebellar hemispheres of carnivores and primates (Braak and Braak, 1993 and Diño et al, 1999) and their absence from regions receiving a predominant pontine input, indicate that they may participate in other spinal reflexes and sensori-motor transformations. Furthermore, the wider distribution of CR-positive UBCs over the corpus cerebelli of Brazilian opossum than in that of the mouse and rat, suggests that some of these sensori-motor transformations may be related to species-specific modes of cerebellar operations. Thus, the findings of Takács et al. (1999) that the occurrence of mGluR1α-immunoreactive UBCs in non-vestibular cerebellar lobules of chickens is higher than in mammals is especially intriguing. It is also noteworthy that the density of mGluR1α-positive UBCs in chickens is comparable to that in guinea pigs, is higher than in rat, but is lower than either in cat or monkey. Aside from having relatively smaller cell bodies, chicken UBCs also had the smallest dendritic brush among the species they examined. Among mammals, mean diameter of the brush was largest in guinea pigs, followed by monkeys, cats, and rats. In addition they described several 'inverse' distribution patterns between avian and mammalian mGluR1α-positive UBCs. For example, in mammals, UBCs are prevalent in the deep fissures whereas in chickens they abound in the superficial portion of folia. Avian UBCs frequently lie at the border of the granule cell layer and the white matter, whereas CR-positive UBCs, especially in monkeys (Yan and Garey, 1996; Diño et al., 1999) and humans (our unpublished observations; Braak and Braak, 1993), lie mainly in the superficial portion of the granular layer. It would be interesting to know if avian and mammalian mGluR1α-immunoreactive UBCs are distributed into more or less distinct parasagittal bands, similar to those described for CR-positive UBCs (Diño et al., 1999; Yan and Garey, 1996). To what extent CR-positive UBC bands correspond to the chemoarchitectonic zones extensively described by Voogd (Hess and Voogd, 1986; Jaarsma et al., 1995), is unknown. Furthermore, Takács et al. (1999) reported preliminary data indicating that the cerebellar nuclei may receive synaptic inputs from UBC axons.

Factors affecting the density of UBCs and their differential distribution to specific cerebellar lobules

The variations in density of UBCs in different species and the unequal distribution of UBCs in different cerebellar lobules may be determined by putative growth factors and by guidance and/or repulsive signals that interact with multipotent neuronal precursors and committed, migrating precursors. As a first step to explore these issues, dissociated cell cultures were prepared from embryonic and postnatal rodent cerebella (Anelli et al., 1996, 1997). Both of these cultures contained a low density of UBC-like cells (Fig. 3A, B). However, the density of UBCs was higher in embryonic cultures than in postnatal cultures (Anelli and Mugnaini, 1998), in accordance with the notion that the UBCs originate from the subventricular zone in the roof of the fourth ventricle mostly between E15 and P0 (Sekerková and Mugnaini, 1997). In a pilot study, the addition of individual growth factors to dissociated cerebellar cultures affected the density of UBCs moderately (BDNF) or insignificantly (CNTF, NT3, and NGF). The density of UBC-like cells in postnatal cultures after 12 days in vitro was much higher when dissociated cells were grown in media containing 5mM KCl (Fig. 3C) than in 25 mM KCl, while the density of viable granule cells was the opposite (Anelli and Mugnaini, 1998). In a pilot study of dissociated cerebellar neurons co-cultured with or without pontine nuclei neurons the UBC density was not affected. Co-cultures of cerebellar dissociated cells with other pre-cerebellar neurons, such as vestibular ganglion and vestibular nuclei neurons are in progress. In embryonic cultures, some of the UBC-like cells were provided with a conspicuous brush and formed synapses with large mossy fiber-like rosettes (Anelli and Mugnaini, 1998). A possible source of these large endings may have been the cerebellar nuclei neurons, which were present in embryonic cultures (Dunn et al., 1998), but not in postnatal cultures. In postnatal cultures, which consist mostly of granule cells, UBC-like cells provided with a dendritic brush were rare and mossy endings synapsing on UBC-like cells were not observed (Anelli and Mugnaini, 1999).

UBC inputs

Primary vestibular afferents

The results from comparative studies beg the questions: do UBCs receive vestibular inputs, and if so, which ones? To determine whether UBCs receive inputs from primary vestibular afferents tract-tracers were injected into vestibular end-organs of gerbils and CR was used as a UBC marker in double labeling studies (Diño et al., 1997). As illustrated in Fig. 4, the primary, Biocytin-labeled fibers reaching lobules IX and X form large mossy fiber rosettes, some of which terminate in close apposition to CR-positive UBC brushes (UBC glomeruli; Fig. 4B), while others terminate in UBC-free glomeruli (granule cell glomeruli; Fig. 4A). Ultrathin sections of this material confirm that primary vestibular fibers do form synapses with granule cells and UBCs (Fig. 5).

Secondary vestibular afferents

Previous work of Barmack et al. (1992) using a combination of tract tracing and cholineacetyl-transferase (ChAT) immunohistochemistry had shown that cholinergic mossy fibers in the nodulus and uvula originate from neurons in the medial vestibular nucleus (MVN), and to a lesser extent the nucleus prepositus hypoglossi (NPH). Building on this work we explored the possibility that secondary cholinergic vestibular fibers synapse with UBCs (Jaarsma et al., 1996). Light and electron microscopic ChAT-immunohistochemistry revealed that about one fifth of UBCs situated in the nodulus are innervated by ChAT-positive mossy fiber rosettes. These synapses came in the form of

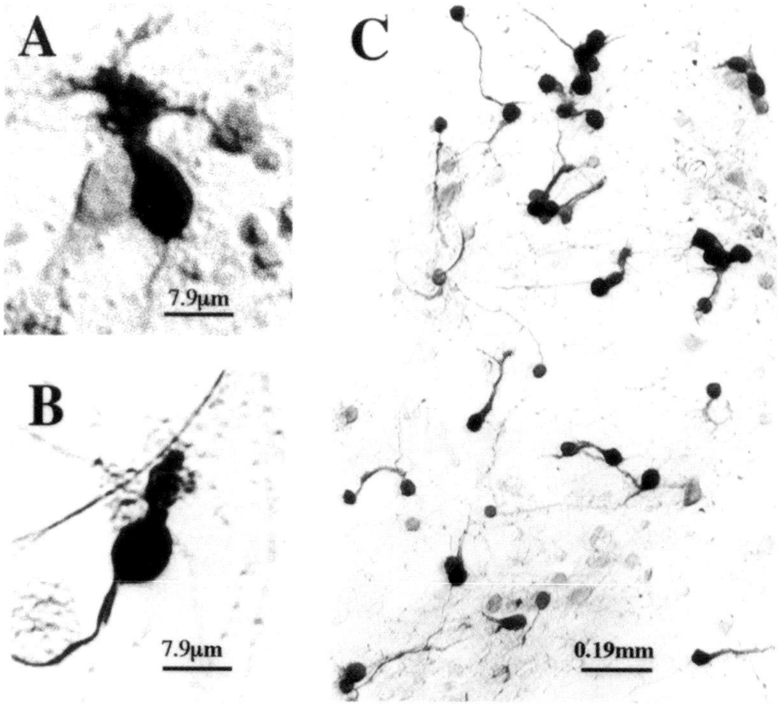

Fig. 3. Calretinin immunostained UBC-like cells in dissociated cultures from embryonic (A, C) and postnatal (B) rodent cerebella. A, UBC-like cell from a dissociated culture of rat E18 cerebellum, after 12 DIV. B, UBC-like cell from a dissociated culture of rat P2 cerebellum, after 12 DIV. C, enrichment of UBC-like cells from a dissociated culture of rat E19 cerebellum grown for 12 DIV in 5 mM KCl medium.

both continuous and segmented varieties, indicating that the configurations of MF–UBC synapses do not correlate with the mossy fiber's transmitter content.

Inhibitory input

Electron microscopy revealed that UBCs receive inhibitory input, usually in the form of axodendritic boutons provided with pleomorphic synaptic vesicles (Mugnaini et al., 1994). UBC dendrioles usually formed symmetric synaptic junctions with these boutons, although asymmetric synaptic junctions paired with presynaptic pleomorphic vesicles were also observed on the peripheral portions of the dendrioles. These putative GABAergic inhibitory boutons presumably belong to the axonal plexus of Golgi cells, although their nature remains to be established with appropriate methods.

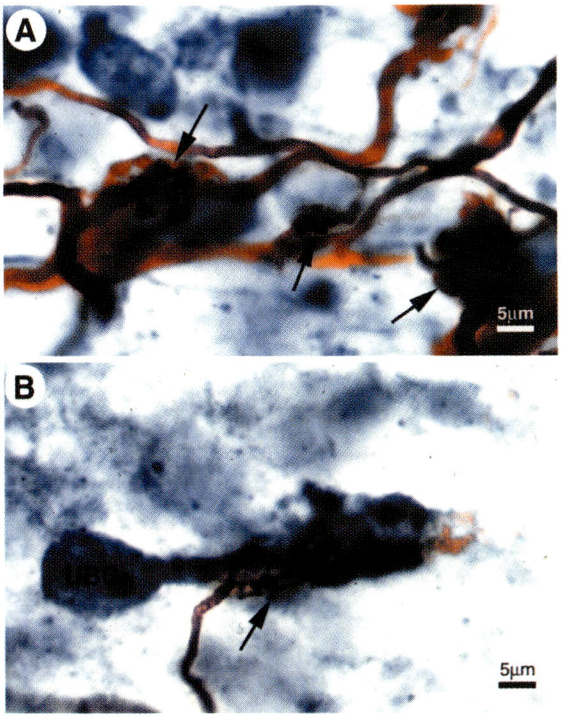

Fig. 4. Parasagittal sections through the cerebellar nodulus of the gerbil double-labeled for anterogradely transported Biocytin (brown) and calretinin (blue). Biocytin-labeled primary vestibular afferents terminate in the nodulus as mossy rosettes (arrows). Some rosettes comprise the central component of glomeruli free of UBC brush formations (A), while others end in close apposition to UBC brushes (B).

Targets of the UBC axon

Prolonged white matter stimulation evokes long lasting EPSCs and a burst of action potentials from UBCs (Rossi et al., 1995). To understand how UBCs transmit this information in the cerebellar microcircuits, it is necessary to identify the targets of UBC axons. Three approaches were used for this purpose: serial ultrathin sectioning of UBCs intracellularly labeled in rat cerebellar slices to trace their axons; surgical isolation of folia of nodulus and uvula in adult rats followed by ultrastructural study of non-degenerating (i.e. cortex-intrinsic) mossy rosettes; and slice-cultures of the isolated mouse nodulus. Taken together, the results of these three approaches prove with certainty that UBC axons form a cortex-intrinsic fiber system that provides a substantial number of bona fide, albeit unorthodox, mossy rosettes in the vestibulocerebellum, and provide glutamatergic innervation to granule cells, other UBCs, and presumably also Golgi cells.

Intracellular fills

After electrophysiological identification and labeling of UBCs with patch-clamp methods in fresh cerebellar slices, Lucifer Yellow- and/or Biocytin-filled UBCs were processed for light or electron microscopy. Light microscopic results were similar to those previously described by Rossi et al. (1995): a single thin axon emanated from the UBC soma and gave rise to 2–3 axon collaterals that terminated as mossy fiber-like rosettes within a few hundred microns in the granular layer. It is possible that each of these axons had collaterals terminating outside the individual slices. We observed some cases where the axon collaterals crossed the white matter and then terminated in the adjacent granular layer. Electron microscopy revealed that the UBC axon is unmyelinated and its mossy rosettes serve as the central components of glomeruli where they form asymmetric synaptic junctions with dendrites of granule cells and/or UBCs (Fig. 6). Synaptic contacts with presumptive Golgi cell dendrites were also observed.

Surgical isolation of lobules IX and X

Because serial sectioning of intracellularly labeled UBCs one at a time is labor intensive and does not afford a large sample, we used other methods to

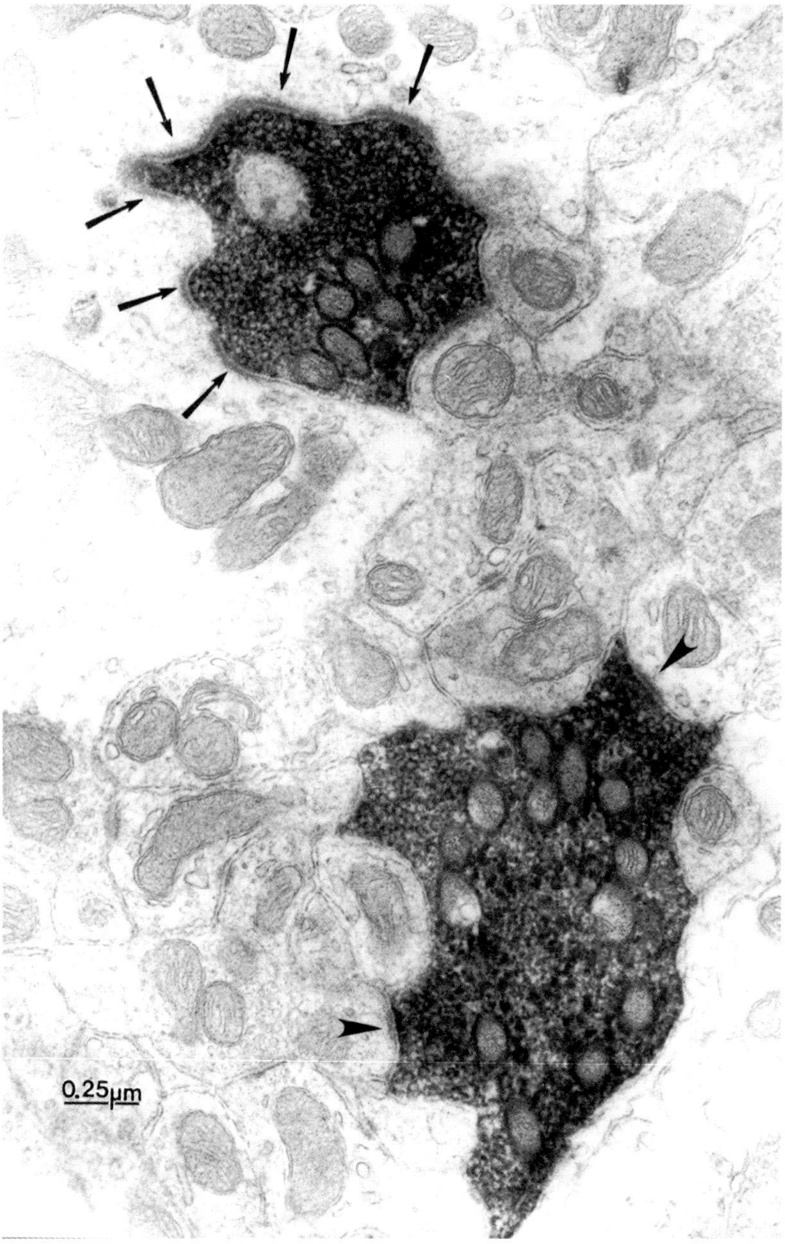

Fig. 5. Electron microscopy reveals that the Biocytin-labeled primary vesitibular afferents form synapses with both UBC (arrows) and granule cell (arrowheads) dendrites.

analyze the system of intrinsic mossy fibers formed by UBCs. Folia of lobules IX and X in the rat were isolated by cutting the medial cerebellum parasagittally on each side of the vermis, followed by a section in the coronal plane approximately 3 mm rostral to the caudal pole of lobulus IX. Rats were perfused with buffered aldehydes seven days after surgery. Rats in which the nodulo-uvular folia were incompletely isolated or appeared necrotic were discarded. After exposure of the cerebellum, three rats, in which the nodulo-uvular folia appeared macroscopically severed from the rest of the cerebellum and possessed the color typical of healthy tissue, were analyzed further by electron microscopy. In this material the majority of glomeruli in the severed folia were devoid of mossy

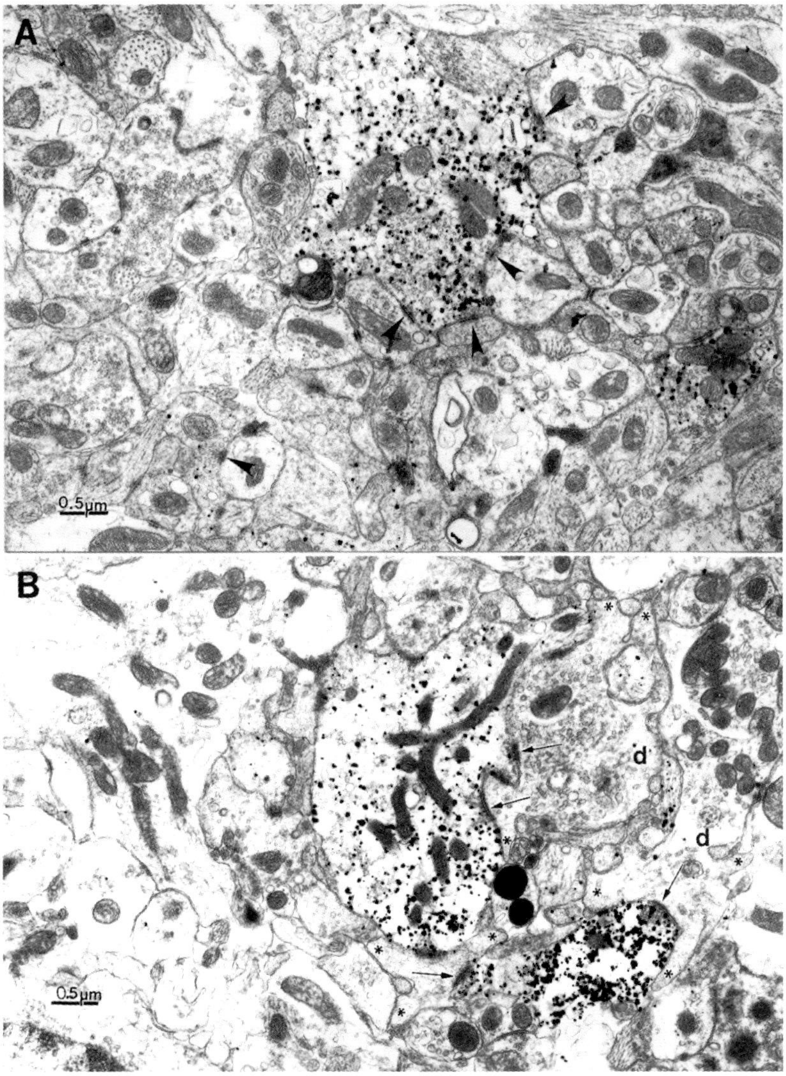

Fig. 6. Biocytin labeled axon terminals of intracellularly filled UBCs visualized with gold-conjugated/silver-intensified Avidin (*electron dense granular deposits*). (A) The UBC rosette forms short asymmetric synaptic contacts (arrowheads) with granule cells. (B) Arrows point to postsynaptic densities of junctions formed by the UBC terminal with unlabeled UBC dendrioles (d), which are provided with non-synaptic appendages (asterisks).

rosettes, while a minority of glomeruli contained normal mossy rosettes (Fig. 7). Degenerating rosettes still identifiable within glial sheets were rare, while degenerating myelinated axons were frequently encountered, suggesting that all cortex-extrinsic afferents had degenerated in the isolated folia. In accordance with this assumption, climbing fiber terminals in the molecular layer were not observed. Consequently, all of the normal mossy rosettes should pertain to axons of the UBCs. These rosettes formed synapses with both granule cell and UBC dendrites (Fig. 7), thereby confirming the results from fresh slices. However, complete surgical severance of all extrinsic afferents, including the nucleo-cortical mossy fibers, could not be incontrovertibly established. Moreover, preservation of ultrastructure in the small isolated folia was far from optimal.

For these reasons, quantitative analysis and immunocytochemistry of nodular mossy fibers were not done on this material. Similar studies on larger mammals, in which the posterior portion of the nodulus can be more consistently isolated from the rest of the cerebellum, are under evaluation.

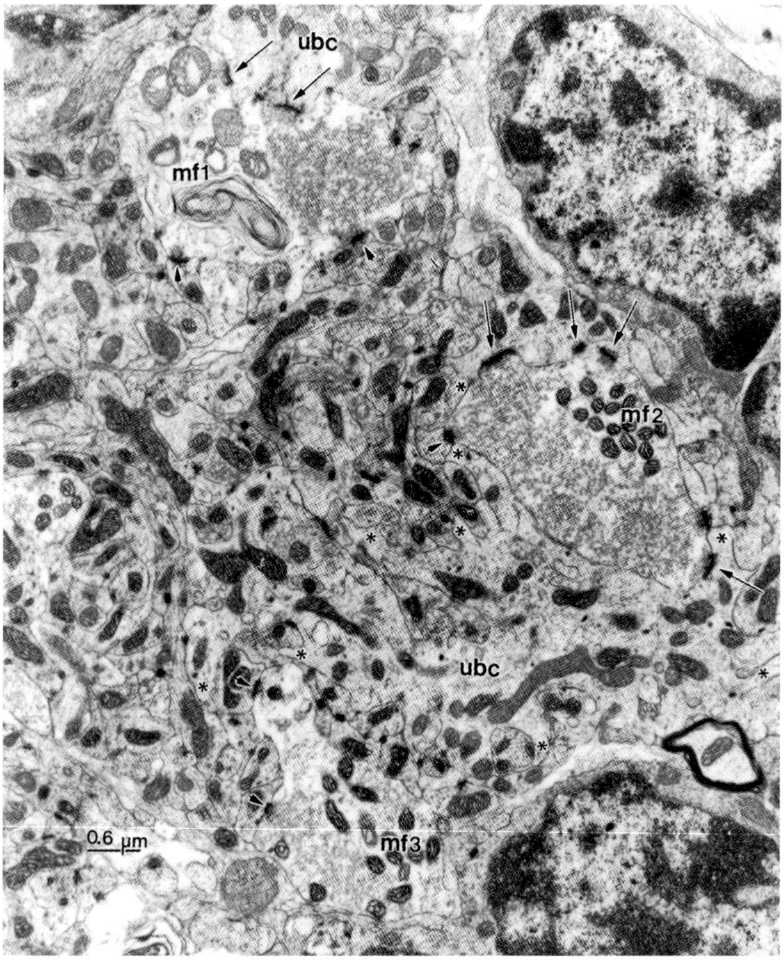

Fig. 7. Electron microscopy of surgically isolated rat nodulus. This micrograph shows three mossy fiber rosette profiles: *mf1*, which forms synapses with a UBC (arrows); *mf2*, which forms synapses with both UBC and granule cell (short arrows) dendrites; and *mf3*, which forms synapses with granule cells only. Asterisks indicate non-synaptic appendages of the UBC dendrioles. The quality of the ultrastructure is low, due to the reduced vascularization of the isolated small folia.

Short-term slice-culture approach

To estimate the relative contribution of the UBC axons to the MF innervation of the vestibulocerebellum, the nodulus was isolated from vermal slices of postnatal day (P) eight mice and cultured for 2–4 days in vitro (DIV) (Nunzi and Mugnaini, 1999). By electron microscopy, the peak of degeneration of terminal axons severed from the parent cell bodies was observed at two DIV. Quantification of degenerating and non-degenerating MF terminals indicated that about one-half of the MF terminals were provided by local UBC axons synapsing on dendrites of granule cells and other

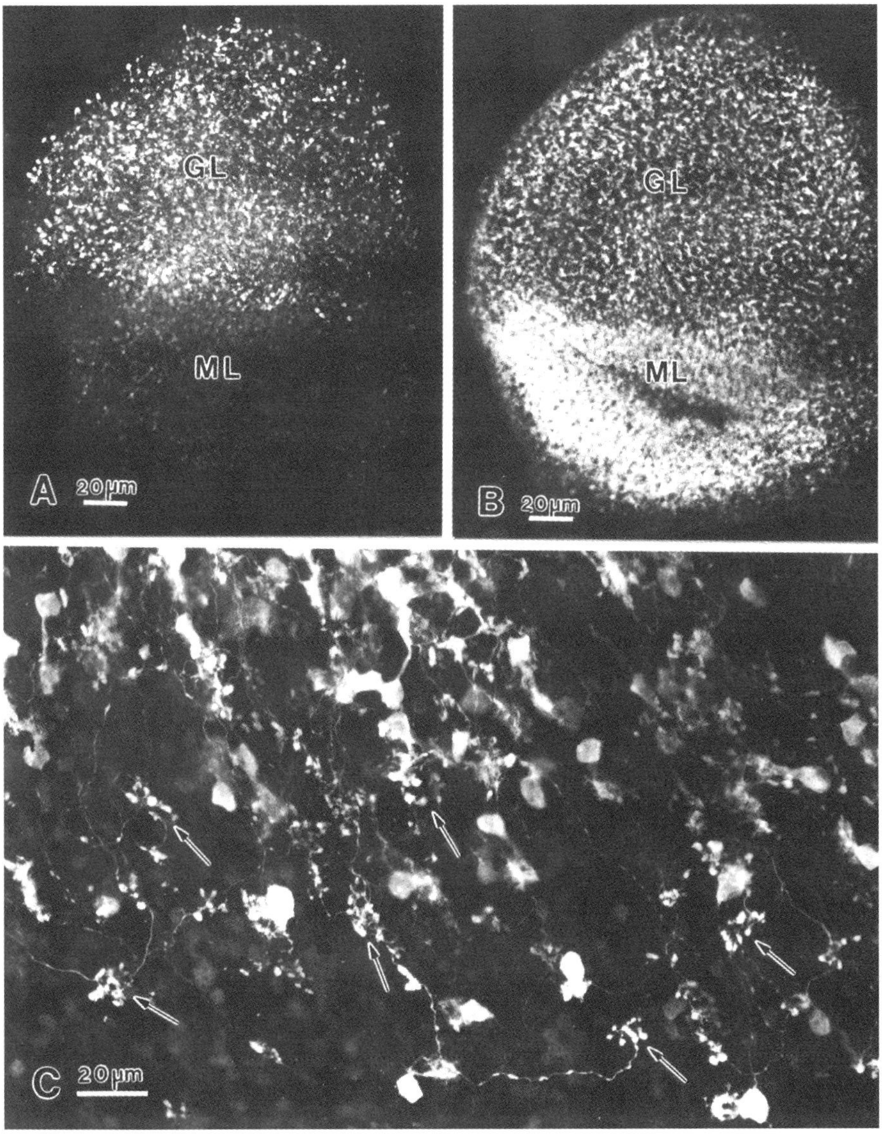

Fig. 8. Organotypic slice cultures from a P8 mouse nodulus grown for 25 DIV. Dual labeling for calretinin (A) and synaptophysin (B) highlights the richness in UBCs and their mossy fiber terminals in the granular layer (GL). The molecular layer (ML) is marked by the high density of small synaptic terminals visualized by synaptophysin immunostaining. C shows calretinin immunostained UBCs and their axons. The axons form mossy terminals (arrows).

UBCs. The proportion of degenerating vs. non-degenerating MF terminals indicated that approximately one-third of the UBCs received MF inputs of extrinsic origin. Thus, at P8 the system of cortex-intrinsic mossy fiber rosettes formed by the UBC axons is indeed substantial.

Long term slice-culture approach: the efferent UBC rosettes

To analyze normal axon terminals undoubtedly belonging to differentiated UBCs, slices of the isolated mouse nodulus were cultured for 15–30 DIV, and prepared for light microscopic immunocytochemistry and standard electron microscopy. In these cultures the granular layer differentiated organotypically (Fig. 8A, B). Although extrinsic mossy fibers had undergone rapid degeneration and removal during the first few days in culture, UBCs formed a rich network of fibers generating numerous MF rosettes, as distinctly revealed by CR immunostaining (Fig. 8C). Electron microscopy showed typical cerebellar glomeruli centered on rosettes of the UBC axon that were surrounded by profiles of granule cells and other UBCs. The UBCs in slices-cultures sustained a compact and healthy granular layer network for more than four weeks in vitro. The rosettes were bulbous or elaborate in outline. They contained round synaptic vesicles with interspersed dense core vesicles and formed asymmetric synaptic junctions. The junctions formed with granule dendrites were smaller than those with UBC dendrioles. Varicosities of the Golgi cell axonal plexus marked the glomerular periphery. They contained pleomorphic synaptic vesicles and formed symmetric and occasionally also asymmetric synaptic junctions with granule cell dendrites and UBC dendrioles. We also observed occasional axosomatic organotypic arrays (Fig. 9), i.e. mossy rosettes synapsing with the UBC perikarya, which in vivo have been referred to as 'en marron' synapses (Palay and Chan-Palay, 1974; Monteiro, 1986; Mugnaini et al., 1994).

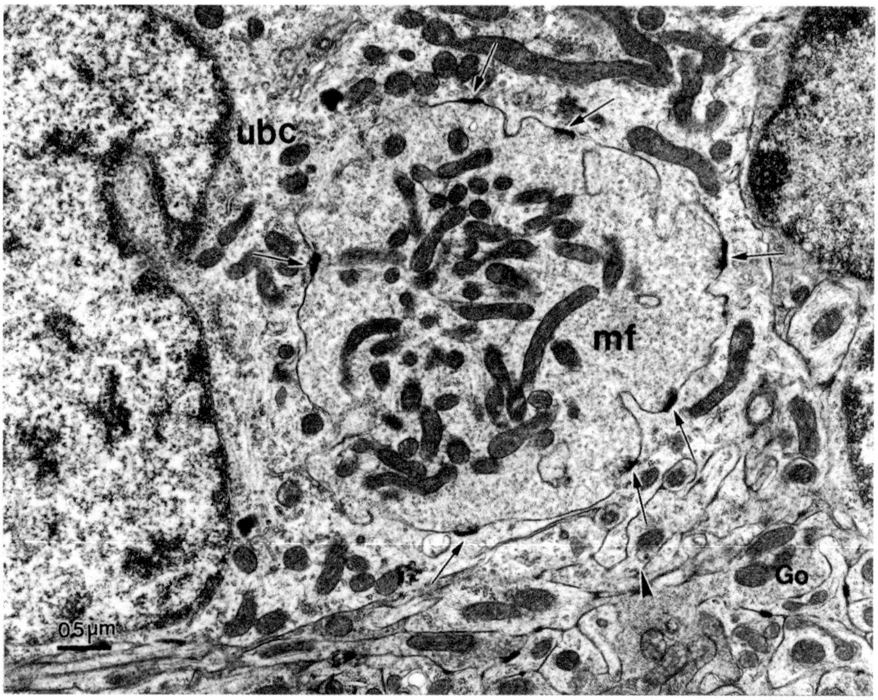

Fig. 9. Organotypic slice culture (25 DIV) of the nodulus from a P8 mouse cerebellum. The electron micrograph shows the mossy terminal of an UBC axon (mf) synapsing with a swallow indentantion of an UBC perikaryon (ubc). Arrows point to synaptic contacts. Golgi terminal (Go).

Long term cultures of the isolated nodulus were then used to identify the neurotransmitter and receptors at the UBC/UBC synapse. The slice cultures grown for 15–30 DIV were processed for either postembedding immunogold or light microscopic immunocytochemistry. Quantification of glutamate immunogold labeling showed that UBCs and their MF terminals were highly reactive for glutamate (Fig. 10). Furthermore, UBCs expressed postsynaptic ionotropic and metabotropic glutamate receptors, as shown by double fluorescence immunostaining for the cell class marker CR and the glutamate receptors GluR2/3, GluR5, NMDAR1, and mGluR2/3 (not shown).

Conclusions

UBCs occur at high density in the vestibulocerebellum of mammals and birds (Yan and Garey, 1996; Diño et al., 1999; Takács et al., 1999). With expansion of the cerebellar cortex in large mammals the density of UBCs increases substantially throughout the vermis and intermediate cortex (Diño et al., 1999; Takács et al., 1999).

Factors affecting the proliferation of UBC precursors and their differential distribution in various cerebellar lobules remain to be identified. In vitro studies suggest that differentiation of the UBC brush requires an appropriate excitatory synaptic input and that this input cannot be provided by granule cells (Anelli and Mugnaini, 1998).

In vivo, UBCs situated in the vestibulocerebellum receive extrinsic mossy fiber inputs from at least two sources: primary vestibular fibers from otolyth and canal organs (Diño et al., 1997), and secondary vestibular ChAT-positive fibers from the vestibular nuclei/prepositus hypoglossi complex (Jaarsma et al., 1996). Inhibitory input to the UBCs is provided by the GABAergic Golgi cell axons (Mugnaini et al., 1994; Nunzi and Mugnaini, 1998; Diño et al., 1999).

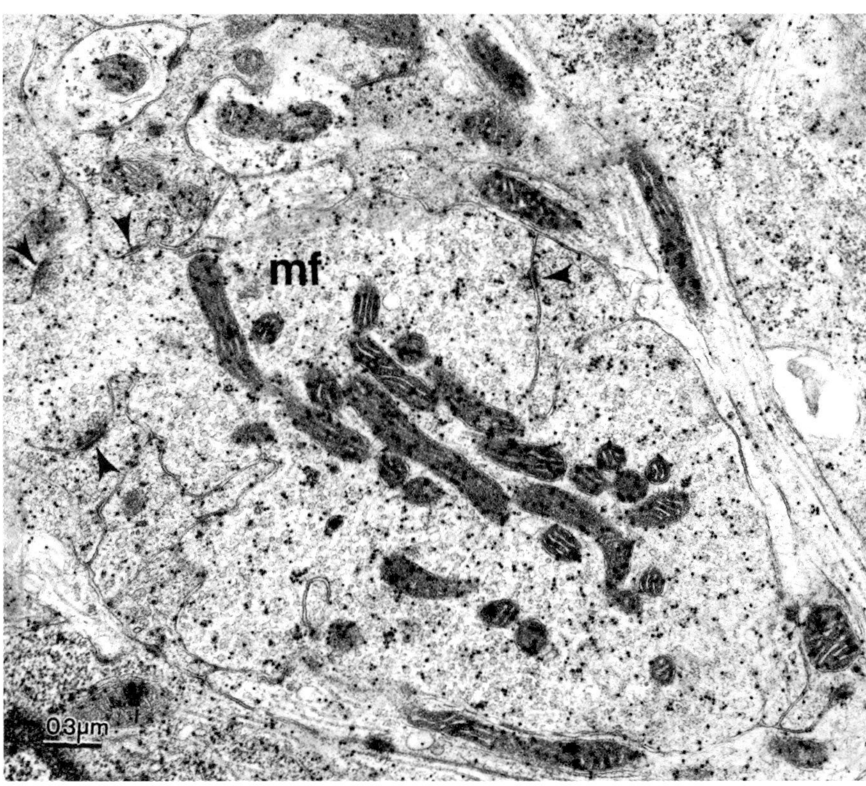

Fig. 10. Postembedding immunogold labeling for glutamate. Gold particles heavily decorate the mossy terminal (mf) of an UBC axon synapsing with granule cell dendrites (arrows).

The targets of the UBC axons are granule cells and other UBCs, with the possible addition of Golgi cells (Diño et al., 1999; Nunzi and Mugnaini, 1999). UBC axons form a substantial system of cortex-intrinsic mossy fibers (Nunzi and Mugnaini, 1999).

Several observations indicate that UBC are excitatory neurons. They are glutamate-positive (Nunzi and Mugnaini, 1998) and GABA- and glycine-negative (Aoki et al., 1986; Mugnaini et al., 1994). Their dendrioles (Mugnaini et al., 1994) and axons (Diño et al., 1999; Nunzi and Mugnaini, 1998) contain round synaptic vesicles and form asymmetric synapses with their target cells.

Through their extensive system of connections, UBCs of the vestibulocerebellum may act by integrating and prolonging vestibular inputs carried to the granular layer by extrinsic mossy fibers, thus establishing a powerful feedforward excitatory link in the mossy fiber-granule cell-Purkinje cell pathway (Fig. 11). In particular, prolongation of vestibular inputs by the UBC synaptic network could be relevant in the processing of vestibulo-ocular responses.

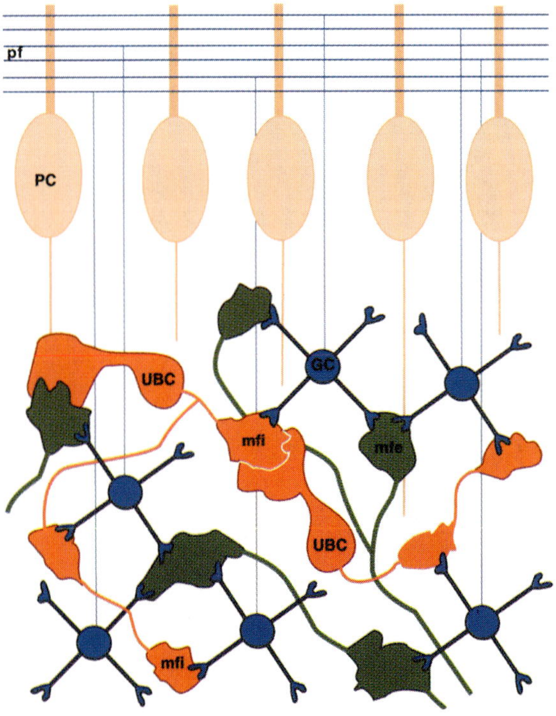

Fig. 11. Schematic diagram of the afferent and efferent connections of UBCs in the vestibulocerebellum.

Acknowledgements

Supported by NIH Grant NS–09904.

References

Anelli, R. and Mugnaini, E. (1998) Characterization of cerebellar unipolar brush cells in vitro. *Soc. Neurosci. Abstr.*, 24: 801.13.

Anelli, R. and Magnaini, E. (1999) Ultrastructural features of unipolar brush cells in vitro. *Soc. Neurosci. Abstr.*, 25: 204.9.

Anelli, R., Kettner, R.E. and Mugnaini, E. (1996). Unipolar Brush Cells are present in cerebellar granule cell in cultures. *Soc. Neurosci. Abstr.*, 22: 640.14.

Anelli, R., Sakane, H. and Mugnaini, E. (1997). Calretinin positive neurons in dissociated cerebellar cultures. *Soc. Neurosci. Abstr.*, 23: 712.12.

Aoki, E., Semba, R. and Kashiwamata, S. (1986) New candidates for GABAergic neurons in the rat cerebellum: an immunocytochemical study with anti-GABA antibody. *Neurosci. Lett.*, 168: 267–271.

Barmack, N.H., Baughman, R.W., Eckenstein, F.P. and Shojaku, H. (1992) Secondary vestibular cholinergic projection to the cerebellum of rabbit and rat as revealed by choline acetyltransferase immunohistochemistry, retrograde and orthograde tracers. *J. Comp. Neurol.*, 317: 250–270.

Berthié, B. and Axelrad, H. (1994) Granular layer collaterals of the unipolar brush cell axon display rosette-like excrescences. A Golgi study in the rat cerebellar cortex. *Neurosci. Lett.*, 167: 161–165.

Braak, E. and Braak, H. (1993) The new monodendritic neuronal type within the adult human cerebellar granule cell layer shows calretinin-immunoreactivity. *Neurosci. Lett.*, 154: 199–202.

Diño, M.R., Sekerková, G., Perachio, A.A. and Mugnaini, E. (1997) Unipolar brush cells are targets of primary vestibular fibers. *Soc. Neurosci. Abstr.*, 23: 712.14.

Diño, M.R., Willard F.H. and Mugnaini, E. (1999) Distribution of unipolar brush cells and other calretinin immunoreactive components in the mammalian cerebellar cortex. *J. Neurocytol.*, 28: 99–126.

Dunn, M.E., Schilling, K. and Mugnaini, E. (1998) Development and fine structure of murine Purkinje cells in dissociated cerebellar culture: dendritic differentiation, synaptic maturation, and formation of cell-class specific features. *Anat. Embryol.*, 197: 31–50.

Floris, A., Diño, M.R., Jacobowitz D.M. and Mugnaini, E. (1994) The unipolar brush cells of the rat cerebellar cortex and cochlear nucleus are calretinin-positive: a study by light and electron microscopic immunocytochemistry. *Anat. Embryol.*, 189: 495–520.

Gerrits, N.M. and Voogd, J. (1982) The climbing fiber projection to the flocculus and adjacent paraflocculus in the cat. *Neuroscience*, 7: 2971–2991.

Glickstein, M. and Voogd, J. (1995) Lodewijk Bolk and the comparative anatomy of the cerebellum. *Trends Neurosci.*, 18: 206–210.

Harris, J., Moreno, S., Shaw, G. and Mugnaini, E. (1993) Unusual neurofilament composition in cerebellar unipolar brush neurons. *J. Neurocytol.*, 22: 1039–1059.

Hess D.T. and Voogd, J. (1986) Chemoarchitectonic zonation of the monkey cerebellum. *Brain Res.*, 369: 383–387.

Jaarsma, D., Wenthold, R.J. and Mugnaini, E. (1995) Glutamate receptor subunits at mossy fiber-unipolar brush cell synapses: light and electron microscopic immunocytochemical study in cerebellar cortex of rat and cat. *J. Comp. Neurol.*, 357: 145–160.

Jaarsma, D., Diño, M.R., Cozzari, C. and Mugnaini, E. (1996) Cerebellar choline acetyltransferase positive mossy fibres and their granule and unipolar brush cell targets: a model for central cholinergic nicotinic neurotransmission. *J. Neurocytol.*, 25: 829–842.

Jaarsma, D., Levey, A.I., Frostholm, A., Rotter, A. and Voogd, J. (1995) Light-microscopic distribution and parasagittal organisation of muscarinic receptors in rabbit cerebellar cortex. *J. Chem. Neuroanat.*, 9: 241–259.

Jaarsma, D., Diño, M.R., Ohishi, H., Shigemoto, R. and Mugnaini, E. (1998) Metabotropic glutamate receptors are associated with non-synaptic appendages of unipolar brush cells in rat cerebellar cortex and cochlear nuclear complex. *J. Neurocytol.*, 27: 303–327.

Kinney, G.A., Overstreet, L.S. and Slater, N.T. (1997) Prolonged physiological entrapment of glutamate in the synaptic cleft of cerebellar unipolar brush cells. *J. Neurophysiol.*, 78: 1320–1333.

Monteiro, R.A.F. (1986) Critical analysis on the nature of synapses en marron of the cerebellar cortex. *J. Hirnforsch.*, 26: 567–576.

Mugnaini, E. and Floris, A. (1994) The unipolar brush cell: a neglected neuron of the mammalian cerebellar cortex. *J. Comp. Neurol.*, 339: 174–180.

Mugnaini, E., Floris, A. and Wright-Goss, M. (1994) Extraordinary synapses of the unipolar brush cell: an electron microscopic study in the rat cerebellum. *Synapse*, 16: 284–311.

Mugnaini, E., Diño, M.R. and Jaarsma, D. (1997) The unipolar brush cells of the mammalian cerebellum and cochlear nucleus: cytology and microcircuitry. *Prog. Brain Res.*, 114: 131–150.

Nieuwenhuys, R. (1967) Comparative anatomy of the cerebellum. *Prog. Brain Res.*, 25: 1–93.

Noda, H. (1991) Cerebellar control of saccadic eye movements: its neural mechanisms and pathways. *Jpn. J. Physiol.*, 41: 351–368.

Nunzi, M.G. and Mugnaini, E. (1998) Cortex instrinsic mossy fibers in the mammalian cerebellum. *Soc. Neurosci. Abstr.*, 24: 262.3.

Nunzi, M.G. and Mugnaini, E. (1999) UBCs axons form a sizeable portion of the mossy fibers in the vestibulocerebellum. *Soc. Neurosci. Abstr.*, 25: 1403: 565.9.

Palay, S.L. and Chan-Palay, V. (1974) *Cerebellar Cortex*. Springer-Verlag, New York.

Rogers, J.H. (1989) Immunoreactivity for calretinin and other calcium-binding proteins in cerebellum. *Neuroscience*, 31: 711–721.

Rossi, D.J., Alford, S., Mugnaini, E. and Slater, N.T. (1995) Properties of transmission at a giant glutamatergic synapse in cerebellum: the mossy fiber-unipolar brush cell synapse. *J. Neurophysiol.*, 74: 24–42.

Sekerková, G. and Mugnaini, E. (1997) Prenatal neurogenesis of cerebellar unipolar brush cells studied by bromodeoxyuridine and cell class specific markers. *Soc. Neurosci. Abstr.*, 23: 712.15.

Slater, N.T., Rossi, D.J. and Kinney, G.A. (1997) Physiology of transmission at a giant glutamatergic synapse in cerebellum. *Prog. Brain Res.*, 114: 151–163.

Spatz, W.B. (1999) Unipolar brush cells in the cochlear nuclei of a primate. *Neurosci. Lett.*, 270: 141–144.

Takács.J., Markova.L., Borostyánköi.Z., Görcs. T.J. and Hámori J. (1999) Metabotropic glutamate receptor type 1a expressing unipolar brush cells in the cerebellar cortex of different species: a comparative quantitative study. *J. Neurosci. Res.*, 55: 733–748.

Voogd, J., Gerrits. N.M., Ruigrok. T.J. (1996) Organization of the vestibulocerebellum. *Ann. N.Y. Acad. Sci.*, 781: 553–579.

Yan, X.X., Garey, L.J. (1996) Calretinin immunoreactivity in the monkey and cat cerebellum: cellular localisation and modular distribution. *J. Hirnforsch.*, 37: 409–419.

SECTION III

Modules and circuits

N.M. Gerrits, T.J.H. Ruigrok and C.I. De Zeeuw (Eds.)
Progress in Brain Research, Vol 124

CHAPTER 11

Topography of cerebellar nuclear projections to the brain stem in the rat

T.M. Teune, J. van der Burg, J. van der Moer, J. Voogd and T.J.H. Ruigrok*

Department of Anatomy, Erasmus University Rotterdam, P.O. Box 1738, 3000 DR Rotterdam, The Netherlands

Introduction

The cerebellum has been implicated to be involved in motor control (Bloedel, 1992; Ito, 1984), cognition (Leiner et al., 1986; Schmahmann, 1996; Thach, 1996), affection and visceral functions (Bradley et al., 1987; Haines et al., 1990; Nisimaru et al., 1991; Waldrop and Iwamoto, 1991). To exert these different functions, the cerebellar cortical output is directed, via the cerebellar nuclei (CN), onto a large number of different brain stem nuclei (see Voogd, 1995) as has been studied in different species [rat: (Chan-Palay, 1977; Faull, 1978; Faull and Carman, 1978; Voogd, 1995); cat: (Voogd, 1964); monkey: (Asanuma et al., 1983)].

The output profile of the cerebellum is organized according to a modular pattern (Voogd, 1964; Oscarsson, 1969; Tolbert et al., 1978; Buisseret-Delmas and Angaut, 1993). A module consists of one or more parasagittally arranged strips of Purkinje cells that project onto a specific subdivision of the CN, both of which receive information from a specific part of the contralateral inferior olive. These olivo-cortico-nuclear modules have been suggested to represent functional cerebellar units (Oscarsson, 1979; Ekerot et al., 1995). In this way, different modules are thought to be involved in different tasks (Godschalk et al., 1994; van der Steen et al., 1994). As such, it would be expected that the output of these modules will reflect this functional heterogeneity.

However, to our knowledge, a detailed and comprehensive anterograde tracer study of the nucleobulbar projections in the rat is not available. Reports by Faull (1978) and Faull and Carman (Faull and Carmen, 1978) provide profiles of the distribution of nucleo-fugal projections, based on lesions and subsequent degeneration of the axons of the superior cerebellar peduncle. Other investigations into the cerebellar nuclear efferents only concerned parts of these projections, such as the projections of the CN to the inferior olive (Ruigrok and Voogd, 1990), the red nucleus (Angaut et al., 1987; Daniel et al., 1987), the nucleus reticularis tegmenti pontis (Cicirata et al., 1982; Angaut et al., 1985a) or the thalamus (Haroian et al., 1981; Angaut et al., 1985b). Only for the output of the lateral cerebellar nucleus an overview of all brain stem projections based on radioactive methionine tracing has been provided by Chan-Palay (1977). A survey of some of the main brain stem targets of the interface between the medial and interposed cerebellar nuclei was recently given by Buisseret-Delmas and colleagues (1998).

This chapter aims to provide a comprehensive overview of the nucleo-fugal projections in the rat, resulting from anterograde tracer deposits in different parts of the CN, thereby verifying and complementing results from the previously mentioned studies.

*Corresponding author. Tel.: +31 10 408 7296; Fax: +31 10 408 9459; e-mail: ruigrok@anat.fgg.eur.nl

Materials and methods

In this overview, the results of thirteen anterograde tracer injections, all made in the CN of adult, purpose-bred, male Wistar rats, will be presented and discussed. They were carefully selected out of a batch of 85 cases. The iontophoretic tracer injections were made with a glass micropipette containing either *Phaseolus vulgaris* Leucoagglutinin (PhaL) or biotinylated dextran amine (BDA). Details of surgical and immunohistochemical procedures are detailed elsewhere (Ruigrok and Voogd, 1990; Teune et al., 1998).

Serially collected transverse sections were analysed with a microscope equipped with a Lucivid™ miniature monitor and Neurolucida™ software. Contours, brain stem nuclei and main fiber tracts of equally spaced sections (spacing 320 μm of levels rostral to the decussation of the superior cerebellar peduncle and 640 μm of levels caudal to this point) were plotted at low magnification (objective 2.5 ×). Subsequently, at high magnification (objective 20 ×), the sections were systematically scanned and the positions of all labeled varicosities were digitized by mouse clicks. Varicosities in both PhaL and BDA labeled fibers have been shown to reflect synaptic boutons (Wouterlood and Groenewegen, 1985; Wouterlood and Jorritsma-Byham, 1993) and thus represent the most likely sites where information is transmitted to postsynaptic structures. In this way, for every plotted section, both the positions and number of plotted terminals was determined with reasonable accuracy. Only in areas with a very dense labeling such as occurring in the CN projections to the inferior olive or red nucleus is it likely that the number of plotted terminals represents an underestimate of the actual number of labeled varicosities. The number of plotted varicosities and the analysed surface area of the section were used to design a bulbar projection profile of a particular case. This profile indicates the relative amount of labeling along the rostrocaudal axis of the brain stem. Table 1, representing a survey of the results of all presented cases, was based on the computerized plots as well as on a supplement of the actual sections.

Nomenclature of brain stem structures is mostly based on Paxinos and Watson (1986). For the nomenclature and the subdivision of the red nucleus and the medial region of the mesodiencephalic junction we have followed Ruigrok and Cella (1995). The nomenclature will be further considered in the discussion. Delineation of the rat CN is slightly modified after Korneliussen (1968; see also: Ruigrok and Voogd, 1990; Buisseret-Delmas and Angaut, 1993; Voogd, 1995).

Results

For easy reference and comparison between selected cases, the location of the injection sites was indicated in standardized transverse drawings of the CN (Fig. 1). Both BDA and PhaL injection sites appeared as rounded, well-defined deposits of tracer substance from which labeled fiber bundles emerged. Within the brain stem these fiber bundles collateralized and showed terminal arborizations which carried many varicosities. The location of the varicosities was indicated in the plots. In accordance with a report on the similarity of BDA and PhaL-tracing (Wouterlood and Jorritsma-Byham, 1993), we noted no obvious differences in staining characteristics. However, based on the amount of labeled varicosities, BDA usually resulted in a smaller number of labeled terminal structures which most likely is related to the somewhat smaller appearance of the injection site. The results presented below will be illustrated with serial plots of actual sections and with the projection profiles which indicate the rostrocaudal levels of the brain stem that receive most prominent projections. An overview of the bulbar projections of all analysed cases is shown in Table 1.

Bulbar projections from the medial cerebellar nucleus

Three cases were selected with injections that were confined to various parts of the medial cerebellar nucleus (MCN: Fig. 1). The injection in case T195 (BDA) was restricted to its rostral half whereas case R100 (PhaL) centered on the caudal part of the nucleus. In an additional case (R127: PhaL) the injection was placed in the lateral aspect of the dorsolateral protuberance (DLP). Together, these cases provide an impression of the targets of the entire nucleus.

TABLE 1

Overview of all described cases indicating in which brain stem regions labeled varicosities were found. Large dot denotes dense labeling, small dot indicates only a minor projection, intermediate dot represents a fair projection. Question mark indicated that the area was not available for analysis; i indicates that the labeling was found ipsilateral to the injection site, otherwise the labeling was observed contralaterally

	MCN			PIN			AIN		DLH	LCN			
	T195	R127	R100	T82	T79	T108	R89	R98	R138	T77	T98	R178	T94
Medulla oblongata													
Inferior olive	•	●	●	⬤	⬤	⬤	⬤	⬤	⬤/●i	⬤	●/•i	●/•i	●/•i
Lateral reticular nucl.			•	•			•i	•i					
Nucl. of the solitary tract							•i						
Parasolitary nucl.	●/•i	●	●										
Medullary reticular nucl.	●/●i	•	•	•				•i	●i				
Paramedian nucl.				•									
Parvocell. retic. nucl.		•			•		•i	●i	•/⬤i	●/⬤i			●i
Gigantocell. retic. nucl.	●/●i	⬤	⬤	●						•	•		⬤/•i
Lat. paragigantocell. nucl.	●			•									⬤/•i
Spinal trigem. nucl., oral part		•					•i	•i	●i	●/⬤i			•i
Spinal trigem. nucl., interpolar part		•					•i	●i	●i	●i			•i
Spinal trigem. nucl., caudal part								•i	●i	•i			•i
Superior vestibular nucl.	•/●i	•/●i	•/•i	•i									
Lateral vestibular nucl.	•/●i		•/•i	•i			•i	•i			•i		
Medial vestibular nucl., rostral	•/•i	•i	•/•i	•i							•i		
Medial vestibular nucl., caudal	•/•i		•/•i	•i									
Spinal vestibular nucl.	●/●i	•	•/•i	●i			•i		•i				
Nucl. prepositus hypoglossi			⬤/•i										
Metencephalon													
Basal pontine nuclei			•				•	•	●		●	●	●
Nucl. retic. tegm. pontis		•	●	●	•		⬤	⬤	●	●	●	⬤	⬤
A5 noradrenergic group (?)		•	•	●				•i	●i	●i			•i
Principal sens. trig. nucl.		•							•				
Caud. pont retic. nucl.	●	⬤	⬤	●			•			•		•	⬤/•i
Oral pont. retic. nucl.	•	●	●	•			●			•	•	●/•i	⬤/•i
Central gray pons		•	⬤/•i		•							●	●
Pedunculopontine tegm. nucl.		●	●		•		•			•i	•i	●	•
Parabrachial nucl.		●/•i	•/•i						•i	•i	•i		
Mesencephalon													
Red nucl., parvicellular	•		•	•	•		•	•	•	●	⬤	●	⬤
Red nucl., magnocellular	•			•	●		⬤	⬤/•i	●	•	●		
Pararubral area		•	●									⬤	⬤
Deep mesenceph. nucl.		⬤	⬤	●	•	●	•	●	●	●	•	●	⬤
Superior colliculus, superficial												●	
Superior colliculus, intermediate			•			●		•				●	•
Superior colliculus, deep		●	●			⬤	•	•				●/•i	●
Ventral tegmental area			•										
Dorsal raphe nucleus			•							•i			•
Ventral tegm. relay zone			•		●				●	●	•	•	●
Medial access. oculomot. nucl.		•	⬤								●	⬤/•i	⬤/•i
Interstitial nucl. of Cajal	●	⬤	●	•	•	•					•	•	●/•i
Nucl. of Darkschewitsch			●								•	●	•
Nucl. parafascicularis prerubr.	●	•	●	●	●	●	●	•	•	●	●	⬤	⬤/•i
Periaquaductal gray		●	⬤		•	•	•					●	⬤
Nucl. of post. commissure			●			●		●				⬤	⬤/•i
Anterior pretectal nucl.			•				●	⬤	•	●	⬤	⬤/•i	⬤
Posterior pretectal nucl.							•	•		●	•	⬤/•i	•

TABLE 1—Continued

	MCN			PIN			AIN		DLH	LCN			
	T195	R127	R100	T82	T79	T108	R89	R98	R138	T77	T98	R178	T94
Diencephalon													
Mammillary nuclei												•	
Lateral hypothal. area			•									●/•i	•
Dorsal hypothal. area			•									●/•i	•
Zona incerta			•	•	•	●	●	●		•	●	●/•i	●/•i
Nucl. fields of Forel	•	•	●	●	•	•	•	•	•	●	●	●/•i	●/•i
Ventral lat. geniculate nucl.						•						•	•
Parafascicular thalam. nucl.		●	●	•			●	•			•	●/•i	●
Central medial thalam. nucl.		•						•	•		•		•
Laterodorsal thalam. nucl.						•	•	•				●	•
Posterior thalamic nucl.		•		•		●	•	•	●	●	•		•
Ventroposterior thalam. group		•				•	•						•
Ventromedial thalam. nucl.	•	●	●	•	•	•	•	•		•	•	●	●/•i
Ventrolateral thalam. nucl.	•	?	?	●	?	●	●	●		●?	●	?	●

Case T195

From the plots of this case (Fig. 2) as well as from the projection profile shown in Fig. 8, it is obvious that most labeling was found in the caudal part of the brain stem. The labeling in the inferior olive (IO) was restricted to the caudolateral part of the medial accessory olive (MAO: not shown in Fig. 2) and subnucleus β contained a few labeled terminals. Main brain stem targets were the vestibular nuclei. In particular, the superior (SVN), lateral (LVN) and spinal (SpVN), and the rostral, magnocellular parts of the medial vestibular nucleus (MVN) were heavily labeled. Within the SVN, MVN and LVN, most labeling was found at the ipsilateral side, whereas the labeling in the SpVN was predominantly observed contralateral to the injected side. The latter labeling extended caudalwards into the parasolitary region. Within the reticular formation, both the ipsilateral and contralateral medullary reticular nuclei and the gigantocellular reticular nucleus (Gi) showed some bilateral terminal labeling. Some sparse labeling was noted in the ipsilateral lateral reticular nucleus (LRN). However, within the pontine reticular formation labeling was restricted to the contralateral side. Some labeling was noted within and directly below the locus coeruleus of both sides (Fig. 2: level 10). The mesencephalon contained little labeling; some labeled terminals were found in the lateral horn of the magnocellular red nucleus (level 8). Within the medial mesodiencephalic region varicosities were noted in the interstitial nucleus of Cajal (INC) and in the area surrounding the fasciculus retroflexus (fr), such as the prerubral field, the rostral interstitial nucleus, the subparafascicular nucleus, and which have been collectively termed the area parafascicularis prerubralis by Carlton et al. (1982). Within the diencephalon, some labeling was noted ventrally in the ventromedial (VM) and laterally in the ventrolateral (VL) thalamic nuclei.

Case R127

In case R127, PhaL had been injected into the lateral aspect of the DLP region of the MCN (Fig. 1). Plots, shown in Fig. 3, as well as the projection profile (Fig. 8) indicate that, compared to T195, a considerable part of the projection from this area is directed to more rostral brain stem areas.

In the IO, terminal labeling was found medially in the caudal MAO, but not in the adjacent subnucleus β. Although some labeling was observed in the caudal part of the contralateral reticular formation, most labeling was noted at more rostral levels. Labeling was especially dense over the parvicellular reticular nucleus (PCRt) but was also distributed to the dorsolateral Gi, as well as to the caudal (PnC) and oral (PnO) parts of the pontine reticular nuclei and extending into the pedunculopontine tegmental nucleus. Relatively few terminals were found in the vestibular complex; only the SVN and the SpVN contained some labeled varicosities. The ipsilateral projection to the

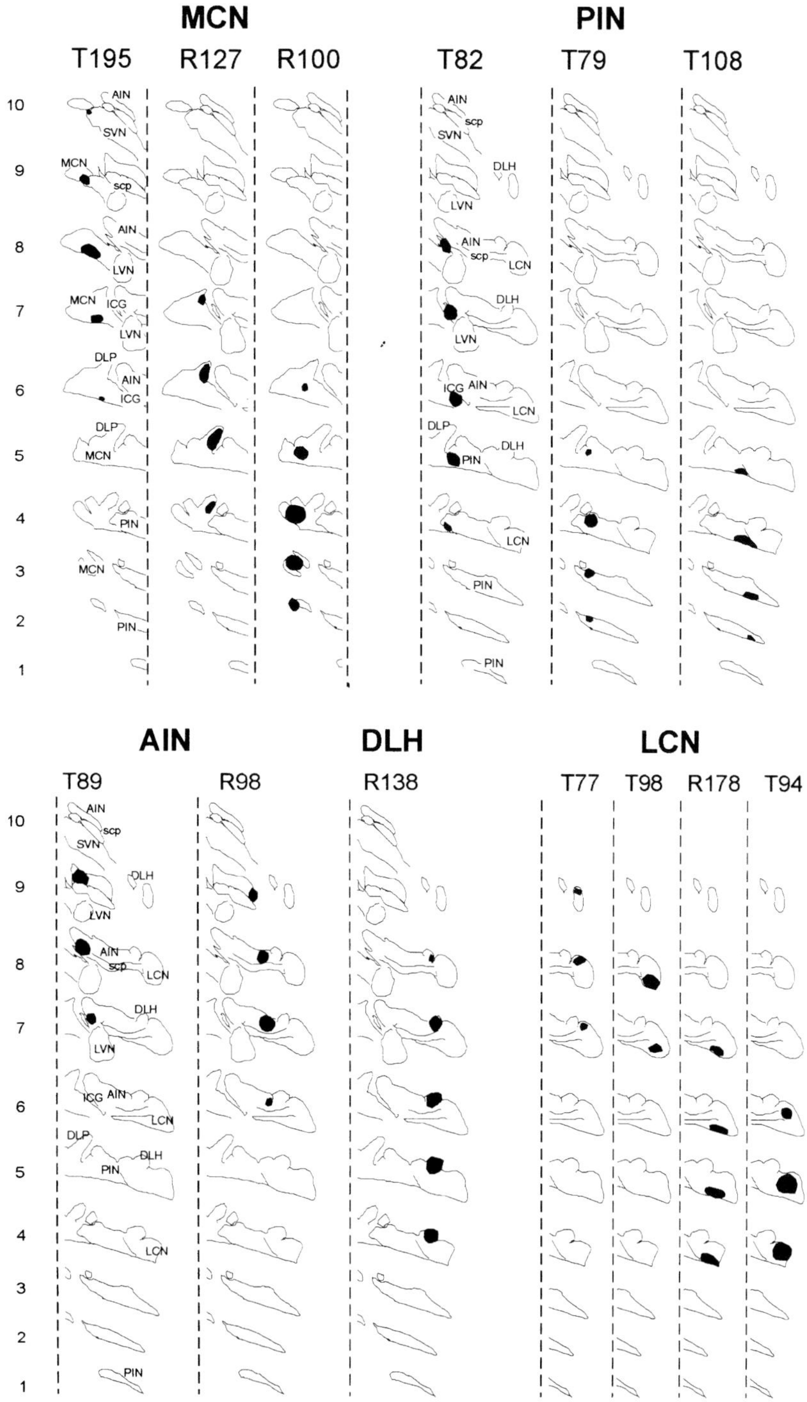

Fig. 1. Diagrams of the injection sites described in this paper. The injection sites are indicated in a standardized, equally spaced (160 μm), series of transverse sections displayed from caudal (1) to rostral (10). For abbreviations, see List.

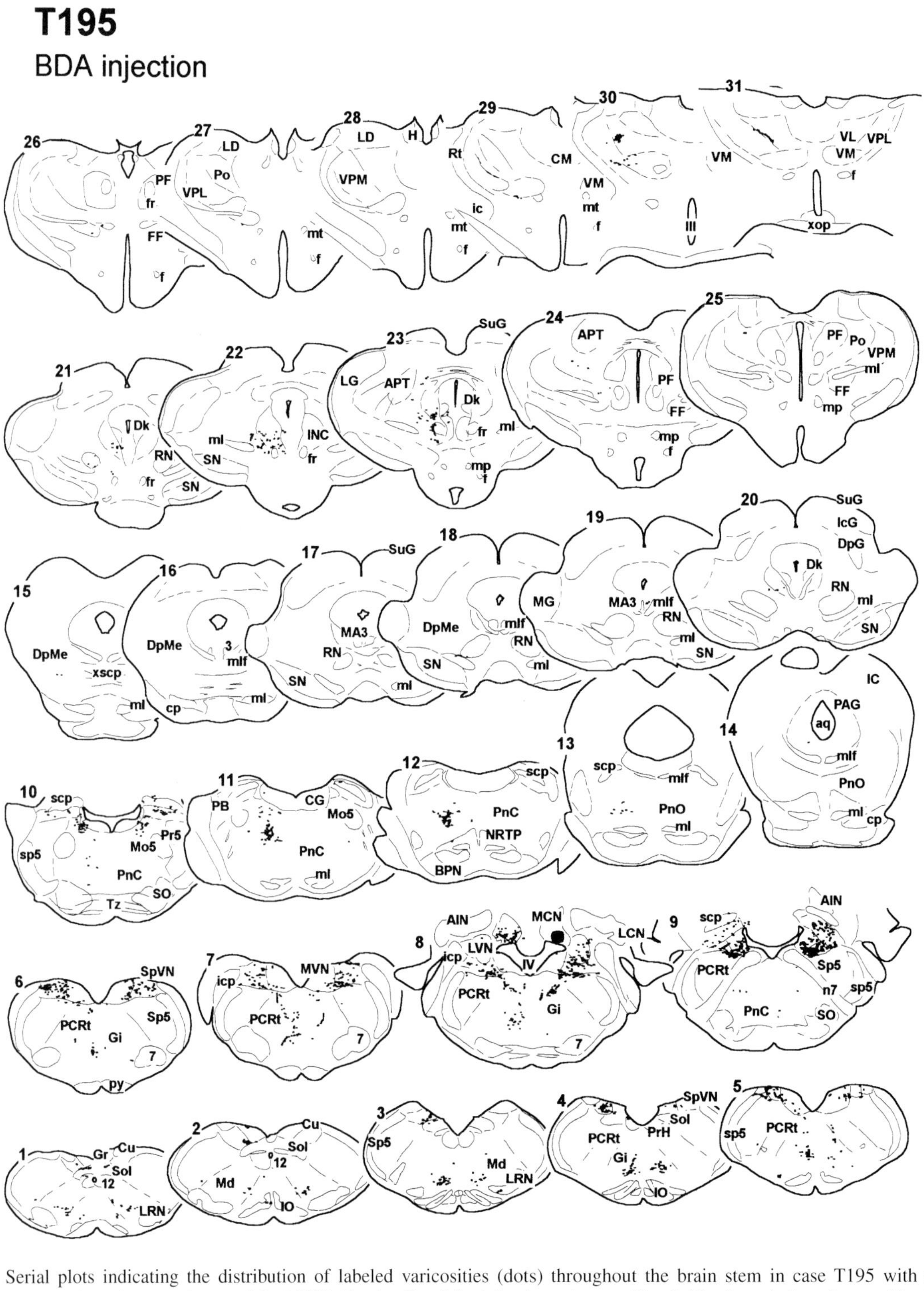

Fig. 2. Serial plots indicating the distribution of labeled varicosities (dots) throughout the brain stem in case T195 with a BDA injection centered on the rostral part of the MCN. For details of the injections site, see Fig. 1. The boundaries of several brain stem nuclei (hatched lines) and fiber tracts (thin lines) are indicated. For abbreviations, see List.

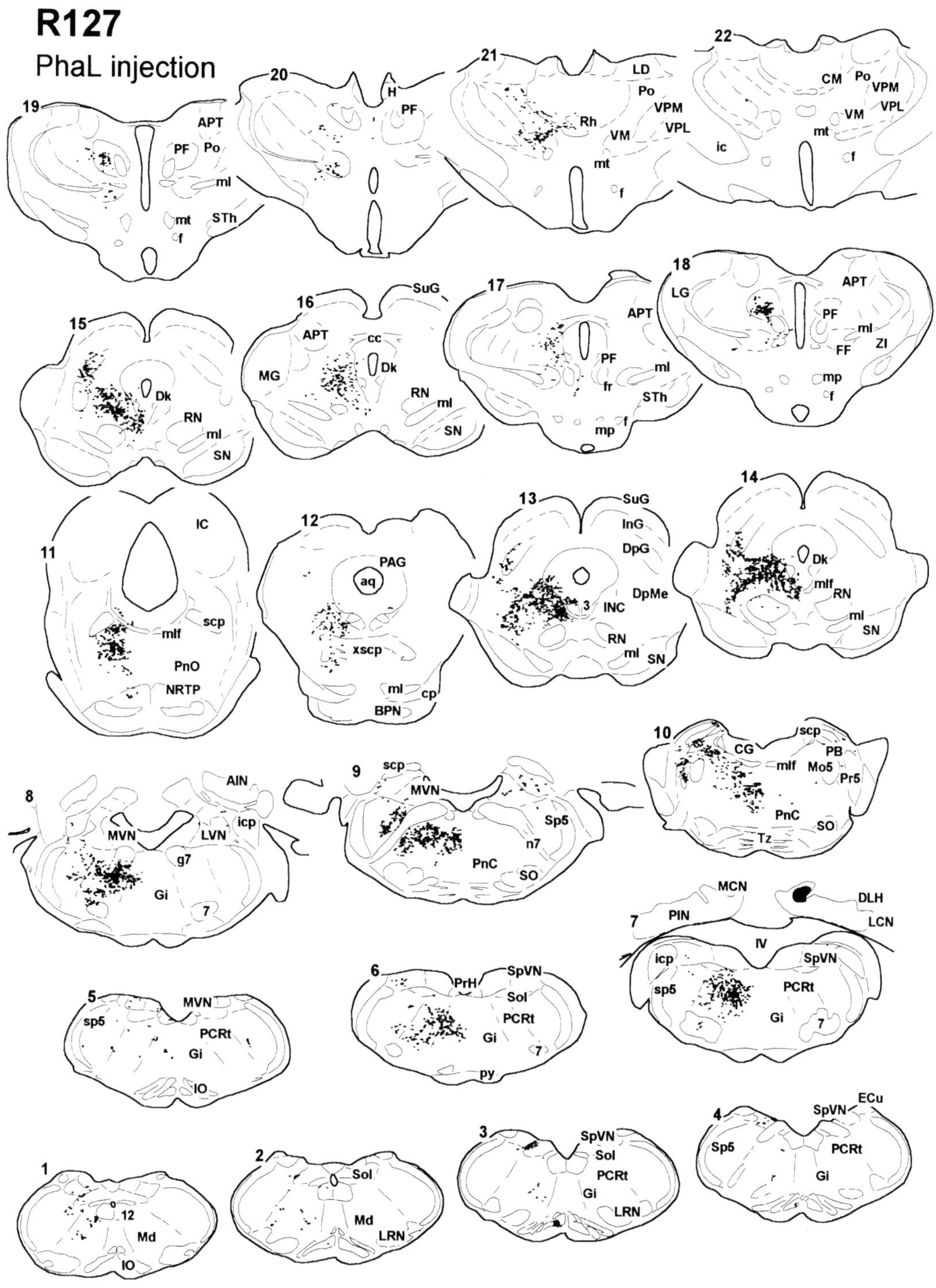

Fig. 3. Serial plots showing the distribution of labeled varicosities in case R127 with a PhaL injection centered on the DLP. Further as in Fig. 2.

SVN predominated over the contralateral projection. The deeper parts of the contralateral spinal trigeminal nucleus (the pars caudalis, the pars interpolaris and, to a lesser extent, the pars oralis) also were provided with input. In addition some labeling was found in the nucleus reticularis tegmenti pontis (NRTP).

A small but significant projection to the parabrachial nuclei (PB) was found, ipsi- as well as contralateral to the injection site, with the contralateral projection predominating. In the midbrain, labeling was only noted contralateral to the injection site. The deep mesencephalic nuclei (DpMe) received a strong projection, however, the red nucleus (RN) remained devoid of labeling. Heavy labeling was observed in the INC. Several additional regions such as the medial accessory oculomotor nucleus (MA3), the nucleus of the posterior commissure, and the prerubral region contained labeled terminals. In addition, varicosities were noted in the deep layers of the superior colliculus (SC). In the diencephalon, terminals filled with PhaL were found in several nuclei such as the parafascicular nucleus (PF), the central medial nucleus (CM) as well as in the VM, VL and posterior thalamic nuclear group (Po).

Case R100

In this case, the injection of PhaL was centered on the caudal aspect of the MCN, but without incorporation of the DLP (Fig. 1). This particular case demonstrated a large yield of labeled varicosities which was distributed over the entire length of the investigated part of the brain stem (Figs 4 and 8).

Within the contralateral IO, labeled terminals were observed in the lateral-most aspect of the caudal MAO as well as in the ventromedial part of subnucleus β and possibly converging on the medial aspect of subnucleus *c* (Akaike, 1992). From the plots of Fig. 4 it is obvious that most prominent labeling was found in the medial part of the contralateral reticular formation at the rostral medullar and pontine levels. In the medullar reticular formation, labeling was mostly restricted to the dorsal part of the Gi where it appeared to be continuous with labeling in the nucleus prepositus hypoglossi (PrH). Labeling within the PnC was continuous with dense projections to the pontine central gray. Some ipsilateral projections to these areas were also noted. Labeling was virtually absent in the PCRt and spinal trigeminal complex. Relatively sparse labeling was observed in the vestibular complex on both sides of the brain. A conspicuous projection to the contralateral parasolitary nucleus was in accordance with other cases of this series. The PB and pedunculopontine tegmental nucleus received a mostly bilateral projection. Labeled fibers carrying varicosities furthermore reached dorsal and medial aspects of the NRTP and basal pontine nuclei (BPN). The mesencephalon also contained numerous labeled terminals, mainly distributed to the contralateral DpMe, MA3 and periaquaductal gray (PAG). Both the magnocellular and the parvicellular RN contained only a few labeled terminals. More dense terminal labeling was observed in the pararubral and prerubral regions, including the nucleus of Darkschewitsch (Dk), the area surrounding the fr, and the nucleus of the posterior commissure. In addition, labeling was noted in the deep and intermediate layers of the SC and extending into the pretectal regions. Ventral to the RN, some terminals reached into the ventral tegmental area. The thalamus contained terminals in the VM, and in the intralaminar nuclei, with most abundant labeling in the PF. Additional labeling was observed in the fields of Forel (FF) which extended into the dorsomedial part of the zona incerta (ZI). In this series of MCN injections, this was the only case that demonstrated terminal labeling that reached the dorsal aspects of the hypothalamus.

Projections from the posterior interposed nucleus and the interstitial cell groups

Three injections were selected to characterize the projections from the posterior interposed nucleus (PIN). Selection of these cases was mainly based on differences in the mediolateral position of the injection site. The most rostromedially placed injection (T82) involved the so-called interstitial cell groups (ICG) which are interspersed between the MCN and both interposed nuclei (Buisseret-Delmas et al., 1993). The injections in cases T79

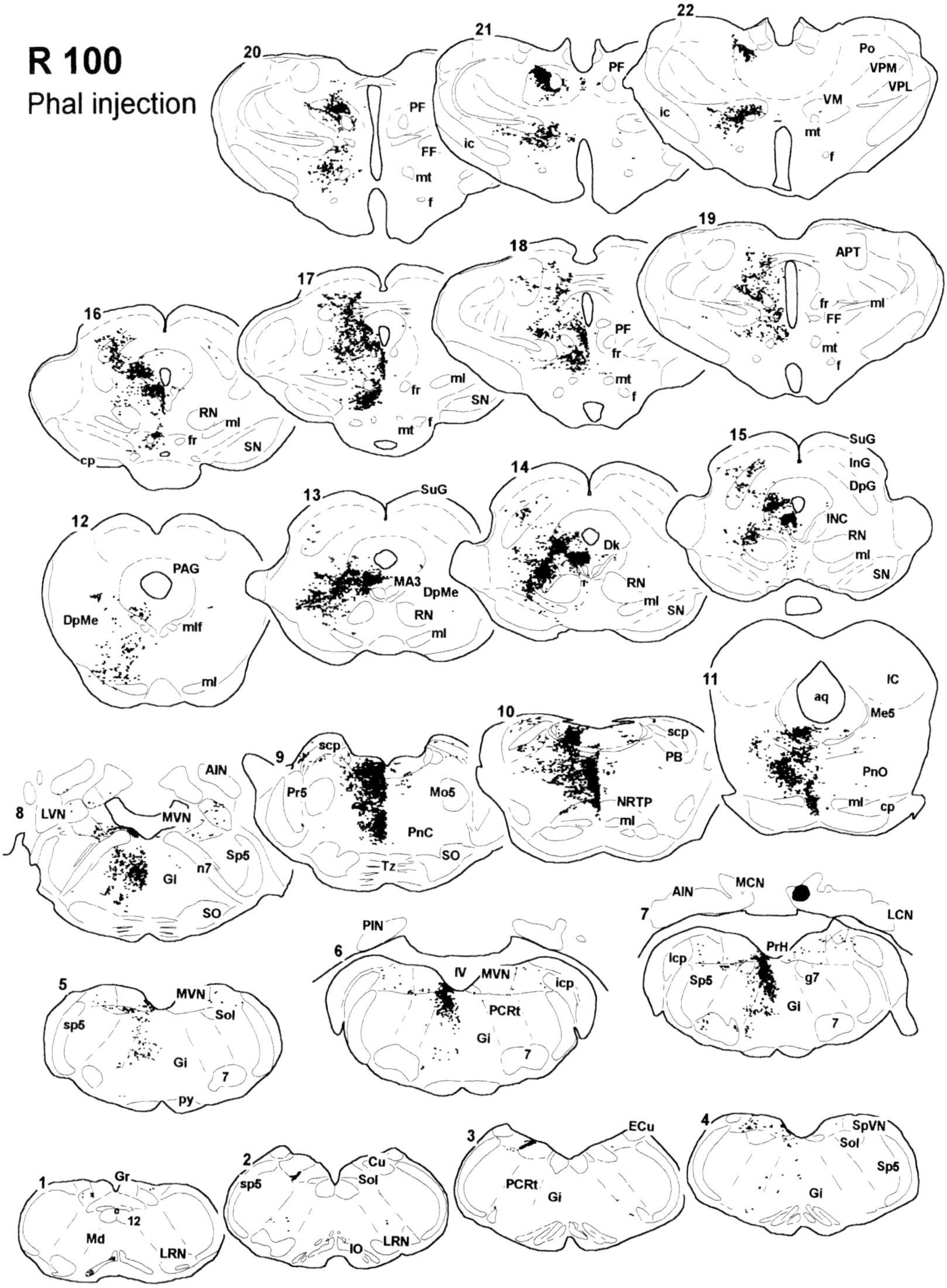

Fig. 4. Serial plots showing the distribution of labeled varicosities in case R100 with a PhaL injection centered on the MCN. Further as in Fig. 2.

and T108 were placed at progressively more lateral levels but remained confined within the boundaries of the PIN (Fig. 1).

Case T82

In this case the BDA injection was centered on the ICG, without any obvious involvement of either the MCN or the anterior interposed nucleus (AIN: Fig. 1). This could be ascertained by the fact that virtually no terminal labeling was observed in the dorsal accessory olive (DAO), indicating that the medial AIN was not involved in the injection site (Ruigrok and Voogd, 1990; Buisseret-Delmas et al., 1998). As shown in the plots of Fig. 5 and in the projection profile of Fig. 8, the projection pattern was more or less equally dispersed over virtually the whole length of brain stem. The only clear concentration of terminals was found in the lateral half of the central MAO. Further medullary labeling was noted in the contralateral Gi and paragigantocellular nucleus and in the ipsilateral vestibular nuclei which did not extend into the parasolitary nuclei. Some labeling was also noted in the NRTP and pontine reticular formation. In the mesencephalon projections were noted to the DpMe, the INC and the prerubral region. In the RN, especially its magnocellular and to a lesser extent its parvicellular part, displayed labeled varicosities that were mostly concentrated along the ventromedial contour of the nucleus. The ventral part of the ZI displayed several labeled terminals as did the PF. More pronounced labeling was noted in the FF, the VM, the Po, and scattered throughout the VL.

Projections originating from the ICG and terminating in the caudoventral RN, the dorsal NRTP, the INC, the vestibular nuclei (especially ipsilaterally) and the IO have recently been described by Buisseret-Delmas et al. (1998). Since the spinal cord was not investigated in this study, the potentially important projections from this area to the cervical cord could not be verified (Horn et al., 1995).

Case T79

BDA had been injected medially in the more rostral parts of the PIN without obvious involvement of either the ICG or AIN. When compared to the former case, the resulting projections may be characterized by the relative lack of labeling in the caudal brain stem (Figs 6 and 8). Olivary labeling was noted caudally in the rostral half of the MAO, at a level somewhat more rostral than in case T82. In the medulla, besides the olivary labeling, only some varicosities were distributed to the contralateral PCRt. A more pronounced labeling was noted in the medial margin of the magnocellular RN. Additional projections were noted to the DpMe, the visual tegmental relay zone (VTRZ) and to part of the nucleus parafascicularis prerubralis located just dorsal to the fr (level 16 of Fig. 6). Unfortunately, rostral thalamic levels were not available for analysis.

Case T108

The injection site (BDA) in T108 was centered on the lateral aspect of the PIN (Fig. 1). In the IO, a dense patch of terminal labeling was found medially in the rostral-most tip of the MAO. Other medullary regions remained devoid of labeling (Figs 7 and 8). A large part of the projections were directed to the SC where especially the deep and intermediate levels contained many labeled varicosities. Projections also reached the DpMe, the INC, the PAG, the nucleus of the posterior commissure, the Dk and, more sparsely, the prerubral area surrounding the fr. The RN was virtually devoid of labeled varicosities. A considerable projection was directed to the ZI, part of the labeling extended into the ventral lateral geniculate nucleus. In the thalamus, further terminal labeling was found in the VM and in the dorsomedial part of the VL.

Projections from the anterior interposed nucleus and dorsolateral hump

In order to describe the bulbar projections from the AIN, a medially (R89) and a laterally (R98) placed injection will be presented together with an injection that was centered on the dorsolateral hump (DLH: R138)). This part of the CN consists of a conspicuous neuronal mass which protrudes somewhat between the lateral parts of the interposed

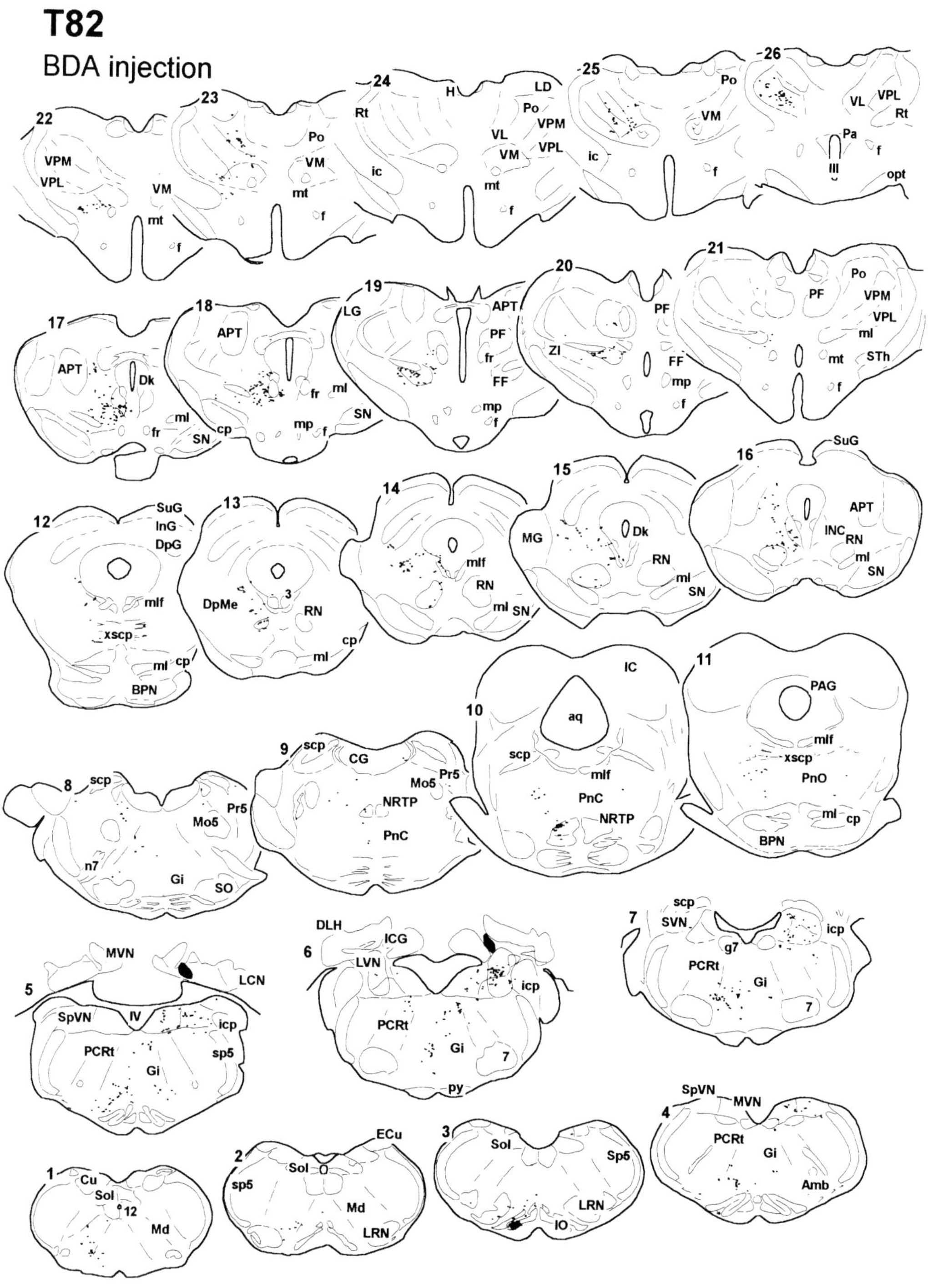

Fig. 5. Serial plots showing the distribution of labeled varicosities in case R82 with a BDA injection centered on the ICG. Further as in Fig. 2.

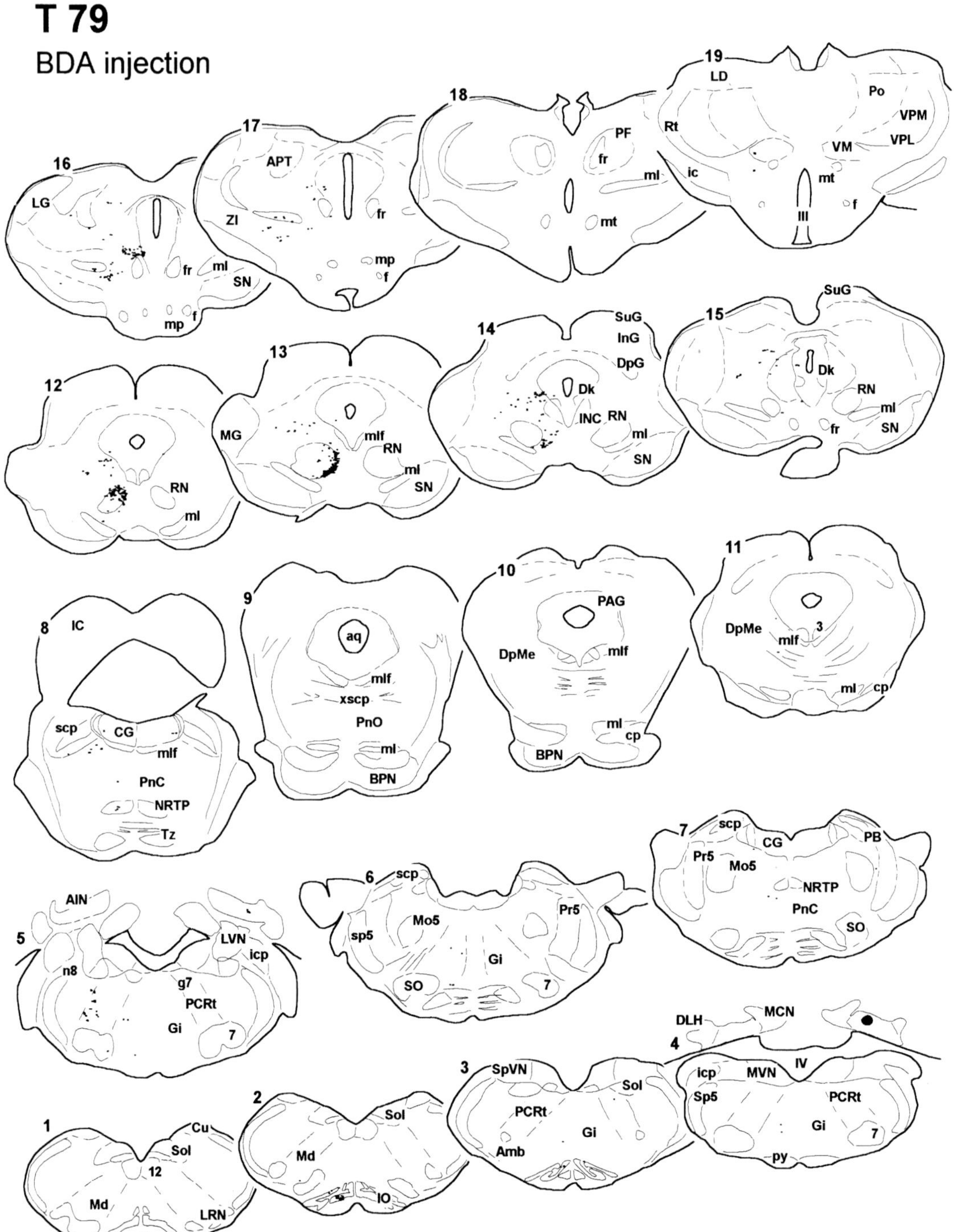

Fig. 6. Serial plots showing the distribution of labeled varicosities in case T79 with a BDA injection centered on the medial aspect of the PIN. Further as in Fig. 2.

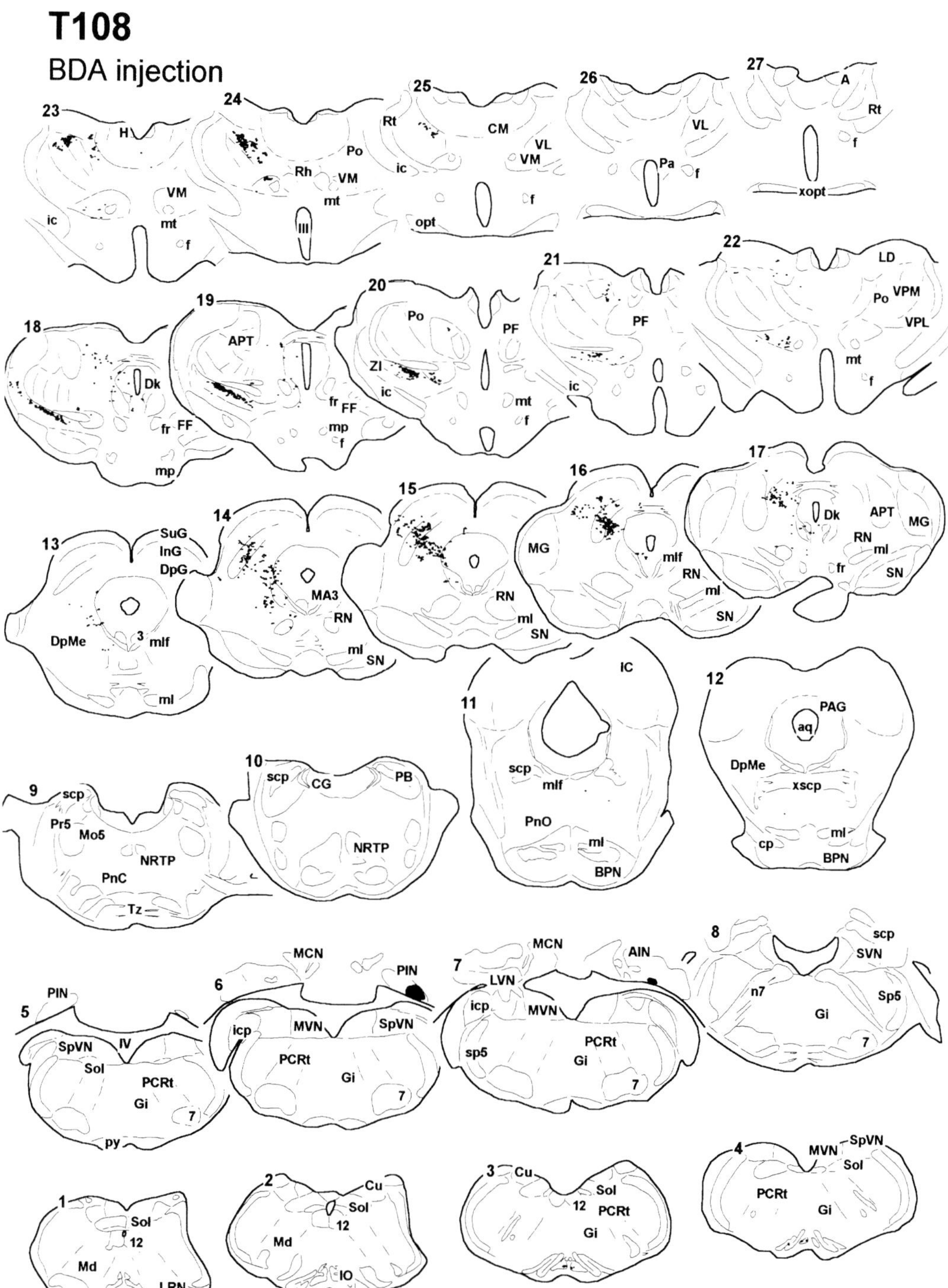

Fig. 7. Serial plots showing the distribution of labeled varicosities in case T108 with a BDA injection centered on the lateral aspect of the PIN. Further as in Fig. 2.

nuclei and the dorsomedial part of the lateral cerebellar nucleus (Korneliussen, 1968).

Case R89

In this case the PhaL injection was confined to the medial part of the rostral half of the AIN (Figs 1 and 9). Although most fibers were directed to the contralateral ascending pathway, some scattered labeled varicosities were found in the ipsilateral vestibular nuclei, particularly within the SpVN, in the PCRt and, sparsely, within the spinal trigeminal complex. An abundant projection, in addition, was noted to the lateral part of the caudal DAO. At pontine levels, the contralateral NRTP and BPN contained labeled terminals. More distributed varicosities were noted in the PnO and pedunculopontine tegmental nucleus. The ventral

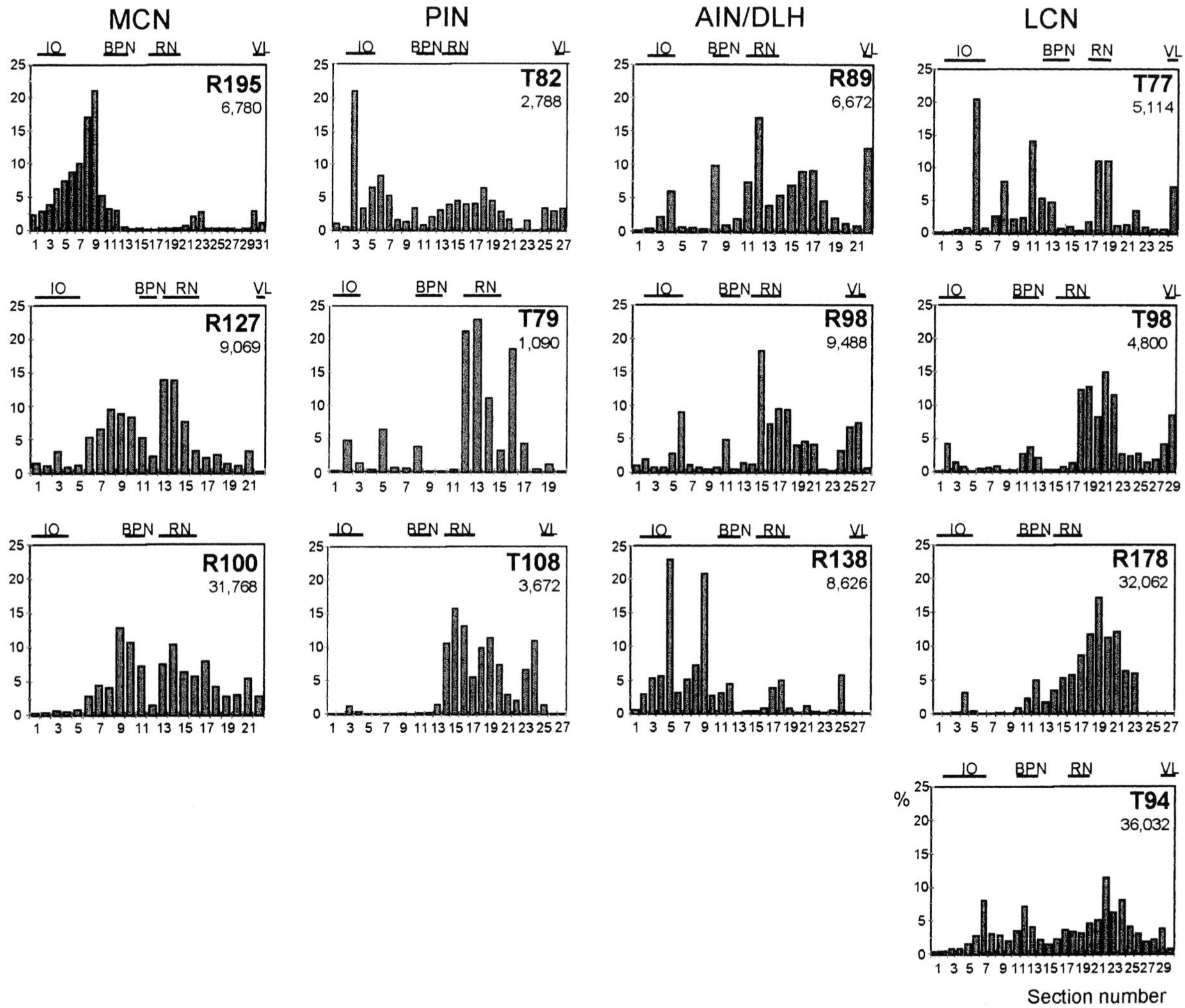

Fig. 8. Bar diagrams indicating rostrocaudal distribution of labeled varicosities of thirteen cases shown in Figs 2–7 and 9–15. The x-axis displays the section number of the corresponding plots (Figs 2.3 to 2.7). The Y-axis displays the percentage of plotted varicosities. The total number of plotted varicosities is indicated below the case number. The approximate rostrocaudal position of four brain stem areas (IO: inferior olive; BPN, basilar pontine nuclei; RN: red nucleus; and VL: ventrolateral thalamic nucleus) is indicated above each graph for easy reference. Note that cases R195 and R199 mostly projects to the caudal brain stem levels whereas cases that involve the DLP and especially the caudal MCN (R100) in addition or rather (R100) supply more rostral brain stem areas.

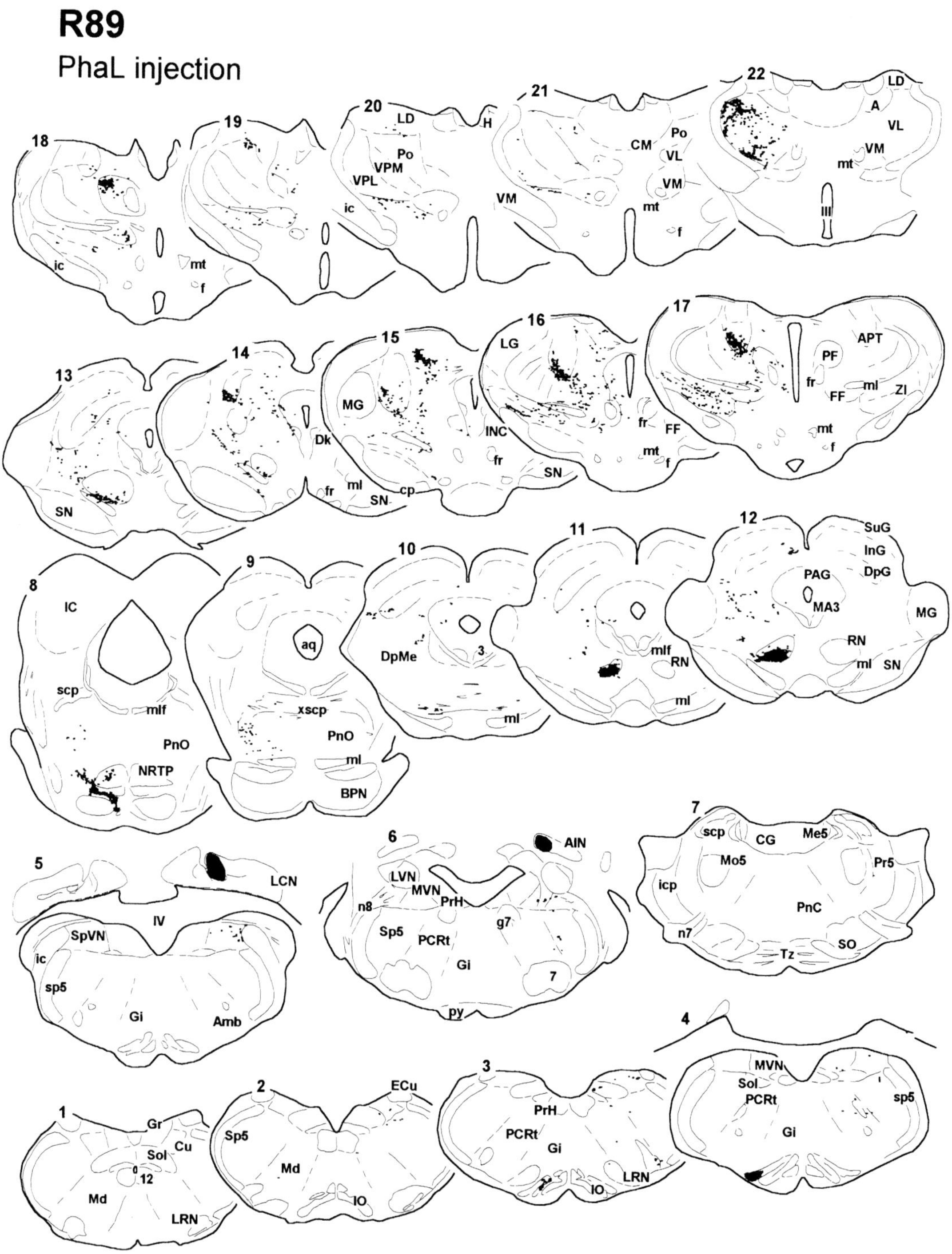

Fig. 9. Serial plots showing the distribution of labeled varicosities in case R89 with a PhaL injection centered on the medial aspect of the AIN. Further as Fig. 2.

part of the contralateral magnocellular RN contained a dense cluster of labeled terminals which extended somewhat into the parvicellular part of the nucleus. Projections were furthermore noted to the DpMe, the SC, the prerubral area, and the PAG; the labeling to the DpMe extended into the ventral part of the anterior pretectal nucleus (APT) where a significant projection was observed; in addition, labeling was noted in the posterior pretectal nucleus. In the diencephalon, both the dorsal and ventral parts of the ZI contained a fair amount of terminals. Furthermore, terminal labeling was observed in the PF, VM, VL, and laterodorsal thalamic cell groups (LD). Within the VL most labeling was found at the periphery of the nucleus.

Case R98

Case R98 was characterized by a PhaL injection that was centered on the lateral part of the AIN, however without involving the DLH (Fig. 1). Plots and projection profile are shown in Figs 10 and 8, respectively. The olivary labeling in this case, contrasting with case T89, was characterized by an impressive number of labeled terminals that were located in the mediorostral part of the DAO. Furthermore, the ipsilateral medulla oblongata, in particular the PCRt and the deep layers of the entire spinal trigeminal nucleus, received a substantial projection from the lateral AIN. Some labeling extended into the subtrigeminal part of the LRN; also, varicosities were noted ventrally in the LVN. At pontine levels, more abundant projections were noted especially to the contralateral NRTP and, to a lesser extent, the BPN. Also, some terminals were observed in the approximate location of the ipsilateral A5 noradrenergic group (Fig. 10, level 10: see also projections that originate from the lateral cerebellar nucleus). More rostrally, the dorsal half of the magnocellular RN contained a dense cluster of labeled terminals, which extended somewhat into the parvicellular part of this nucleus. The DpMe, intermediate layers of the SC, and the prerubral area only received a minor projection; a very prominent projection, however, was noted to the pretectal area, especially to the APT. A mediolateral band of labeled varicosities was furthermore found in the ventral ZI. Most labeling in the thalamus was found in the VL and PF, although some varicosities were also noted in the Po, LD, and VM.

Case R138

The injection site in this experiment particularly involved the rostral part of the DLH, without obvious involvement of either the interposed or lateral cerebellar nuclei (Fig. 1). The IO showed labeling which was found bilaterally in the dorsomedial group (DM), which is usually considered a part of the principal olive (PO: e.g. Buisseret-Delmas and Angaut, 1993, however, see also Ruigrok and Voogd, 1990). From the projection profile (Fig. 8) it is obvious that in this case only a fraction of its connections was directed to levels rostral to the pons. Several structures in the medulla oblongata and metencephalon, ipsilateral to the injection site, contained numerous labeled terminals such as the medullary reticular nuclei (both ventral and dorsal subdivisions), the PCRt, and along the whole length of the spinal trigeminal nucleus. The projection to the PCRt also contained a small contralateral component. At pontine levels, many labeled varicosities were found interspersed between the principal and motor nuclei of the trigeminal complex (Fig. 11, level 10). Ventralwards these fibers appeared to reach the A5 noradrenergic cell group. Furthermore, pontine projections were noted contralaterally to the NRTP and the BPN and ipsilaterally to the PB. In the mesencephalon, both the magnocellular and the parvicellular parts of the RN contained a few labeled terminals, predominantly in their dorsal and dorsomedial aspects. Dorsal to the RN, some terminal labeling was found in the DpMe. In addition, labeling was noted in the VTRZ and in the prerubral area. At thalamic levels only sparse labeling was noted, medially in the Po and extending somewhat into the CM of the intralaminar nuclei. No projection to the VL was observed.

Projections from the lateral cerebellar nucleus

Four cases were selected to analyse the bulbar projections that originate from the lateral cerebellar

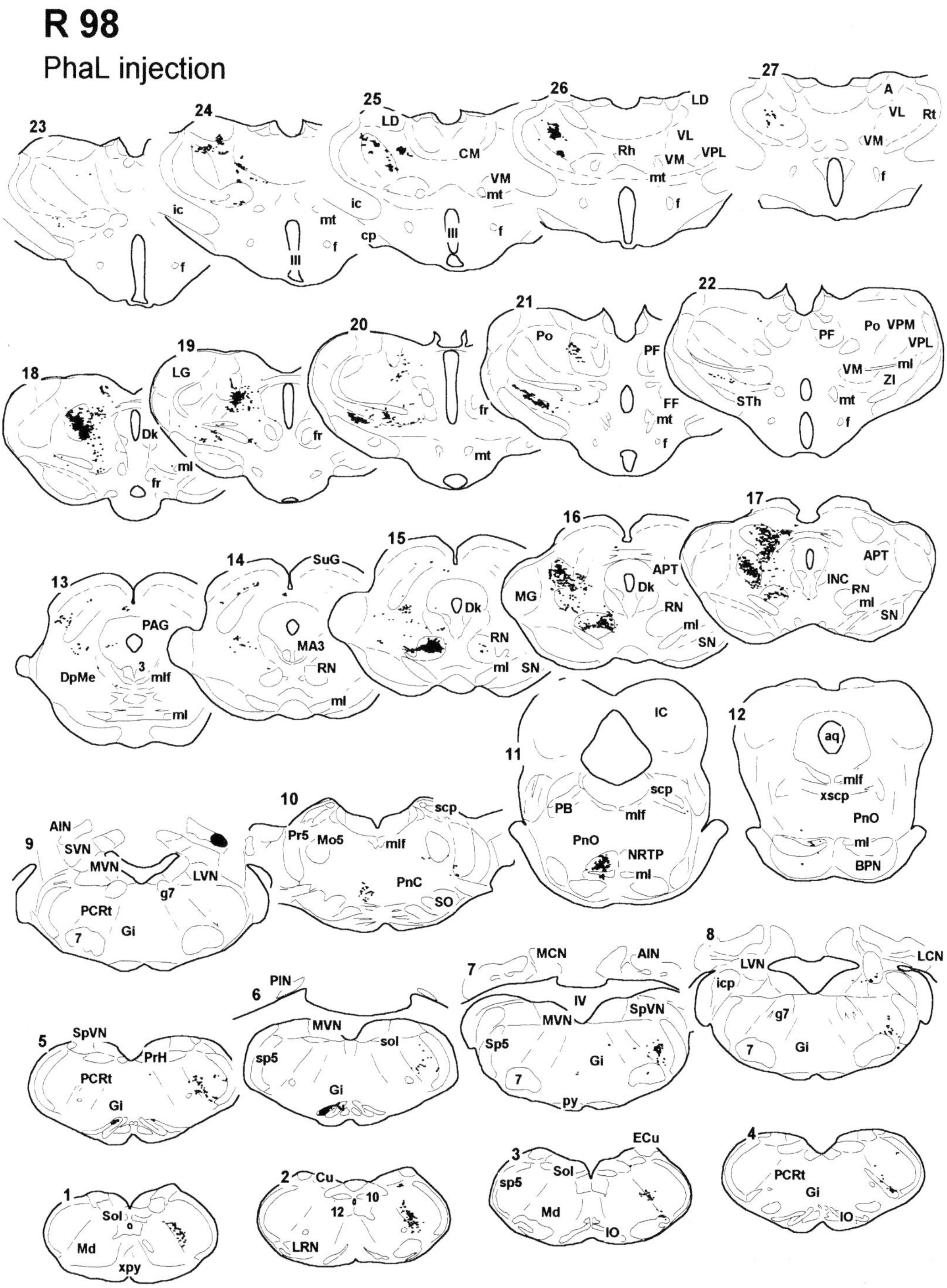

Fig. 10. Serial plots showing the distribution of labeled varicosities in case R98 with a PhaL injection centered on the lateral aspect of the AIN. Further as Fig. 2.

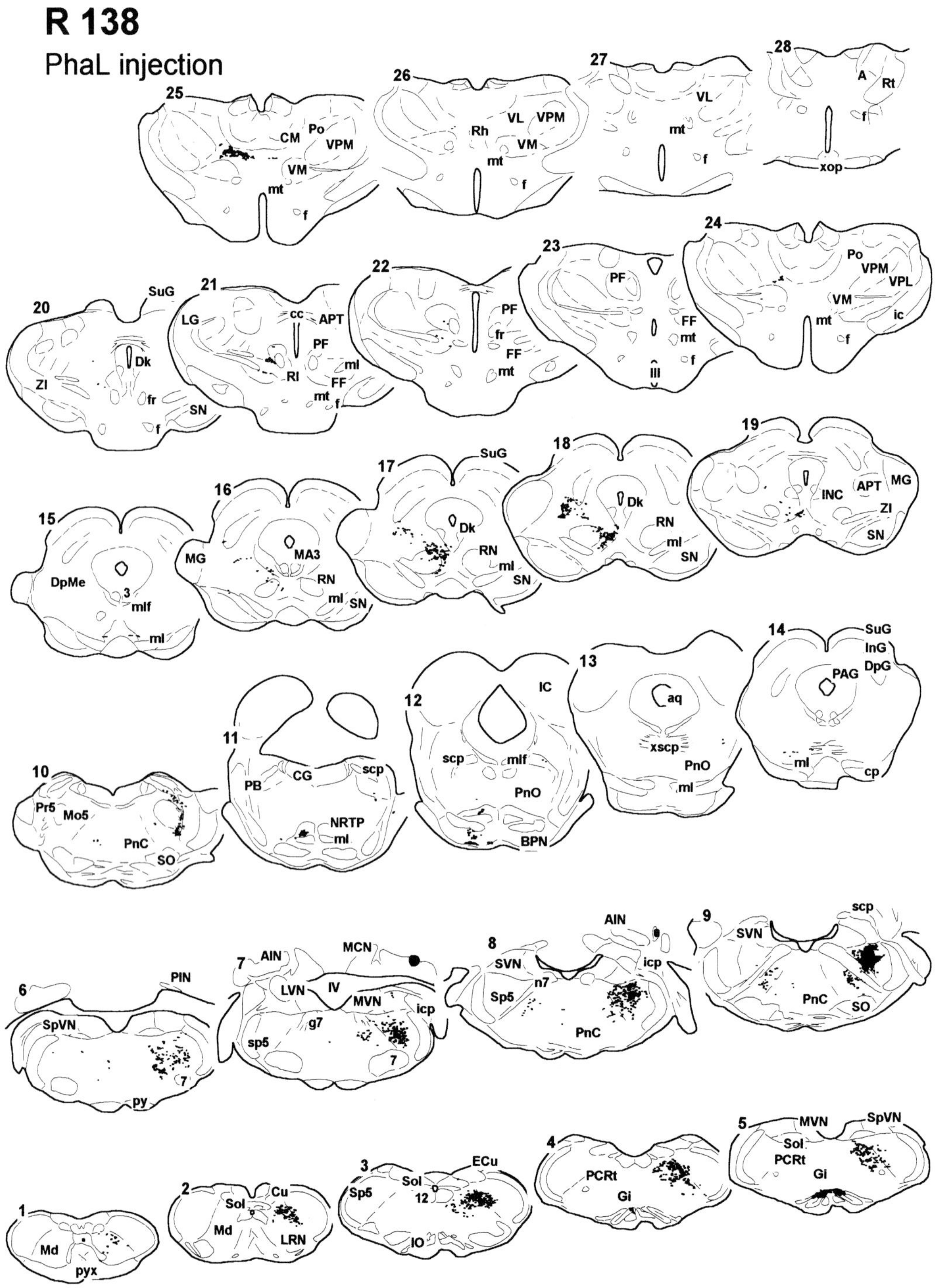

Fig. 11. Serial plots showing the distribution of labeled varicosities in case R138 with a PhaL injection centered on the DLH. Further as Fig. 2.

nucleus (LCN). Two injections were centered on the rostral aspects of the nucleus; one into its dorsal tip (T77) and one covering the ventral half (T98). The other cases were centered on the caudal half of the nucleus; R178 focusing on its ventral, parvicellular, part and T94 covering somewhat more dorsal aspects.

Case T77

In this case, a relatively small injection with PhaL was centered at the dorsal margin of the rostral half of the LCN, however, without involvement of the DLH (Fig. 1). This was corroborated by the observation that the olivary projection was restricted to the dorsal lamellae of the PO, without any labeling of the DM group (Fig. 12, level 5).

In the medulla oblongata, terminal labeling was distributed bilaterally. Apart from the olivary labeling, the most prominent projections were observed in the ipsilateral PCRt, which appeared reminiscent to the projections arising from lateral AIN and DLH, although they did not extend as far caudally. Labeled varicosities were also noted within the confines of the spinal trigeminal nucleus, especially within its oral part. At the pontine level a conspicuous input to the ipsilateral A5 area (Paxinos and Watson, 1986), located just ventral to the motor nucleus of the trigeminal complex (Fig. 12, level 10, 11) was also noted. Dense patches with varicosities were furthermore located within the contralateral NRTP. Some varicosities were noted in the dorsal raphe nucleus (Fig. 12, level 14).

In the mesencephalon, input was specifically noted to the DpMe and extending into the nucleus of the optic tract and caudal part of the APT. In the RN, most terminals were distributed within the dorsomedial aspect of its parvicellular part, extending somewhat into the VTRZ but more prominently into the prerubral region and FF. Within the ZI, only a few scattered varicosities were noted. Thalamic labeling was found in the VM and caudomedial VL. More rostral levels, however, were not available for analysis.

Case T98

Here, the injection (BDA) was centered on the ventral aspect of the rostral LCN (Fig. 1). Within the IO, most prominent labeling was confined to the caudal-most aspect of the PO (Fig. 13, level 2). From the projection profile (Fig. 8) it can be seen that the major part of the projection is directed to the mesodiencephalic transition area (levels 18–22 of Fig. 13). Within the medulla oblongata, only a few labeled terminals were found, mainly ipsilateral to the injection site, in the contralateral Gi and paragigantocellular reticular nucleus. Some scattered varicosities were located in the ipsilateral SVN and MVN and extending rostrally into the PB. A more prominent projection was directed to the contralateral NRTP and BPN.

In the mesodiencephalon, the ventral part of the parvicellular part of the red nucleus was densely labeled. In addition, many terminals were noted in a region covering the rostral parts of the MA3, and extending into the prerubral field and adjacent area surrounding the fr and incorporating FF. Terminal arborizations were furthermore noted in a band aligning the ventrolateral margin of the APT. Additional labeling was noted in the posterior pretectal nucleus. A major projection was observed in the dorsal part of the ZI. Within the thalamus the most prominent labeling was observed in the medial parts of the VL, but, in addition, also within the PF, VM and Po.

Case R178

In this case, a PhaL injection was also centered on the ventral part of the LCN, however at distinctively more caudal levels compared to case T98 (Fig. 1). In fact, the injection appeared to be completely confined to the parvicellular part of the nucleus. In the PO, the ventral part of the lateral bend of the PO was densely labeled in the contralateral IO (Fig. 14, level 4). The ipsilateral IO also displayed a few corresponding labeled terminals. As can be discerned from Fig. 14 and the projection profile displayed in Fig. 8, the medulla oblongata showed no further labeling. The pons contained several structures with labeling contralateral to the injection site. Here, most distinct labeling was found in the caudal and dorsal boundaries of the NRTP, several patches in the lateral and medial parts of the BPN as well as a more scattered projection to the PnO and pedunculopontine tegmental nucleus.

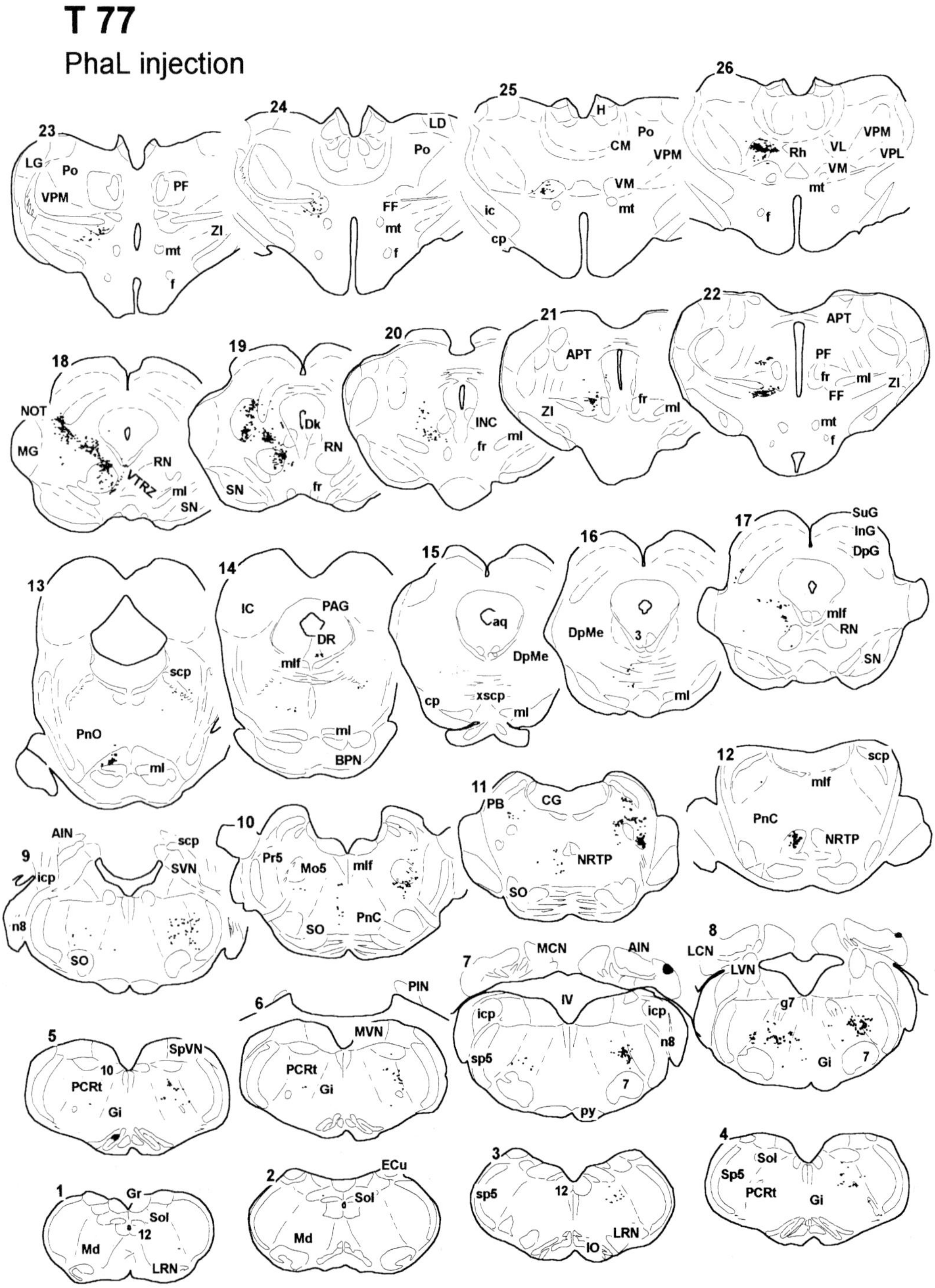

Fig. 12. Serial plots showing the distribution of labeled varicosities in case T77 with a PhaL injection centered on the dorsal tip of the rostral LCN. Further as Fig. 2.

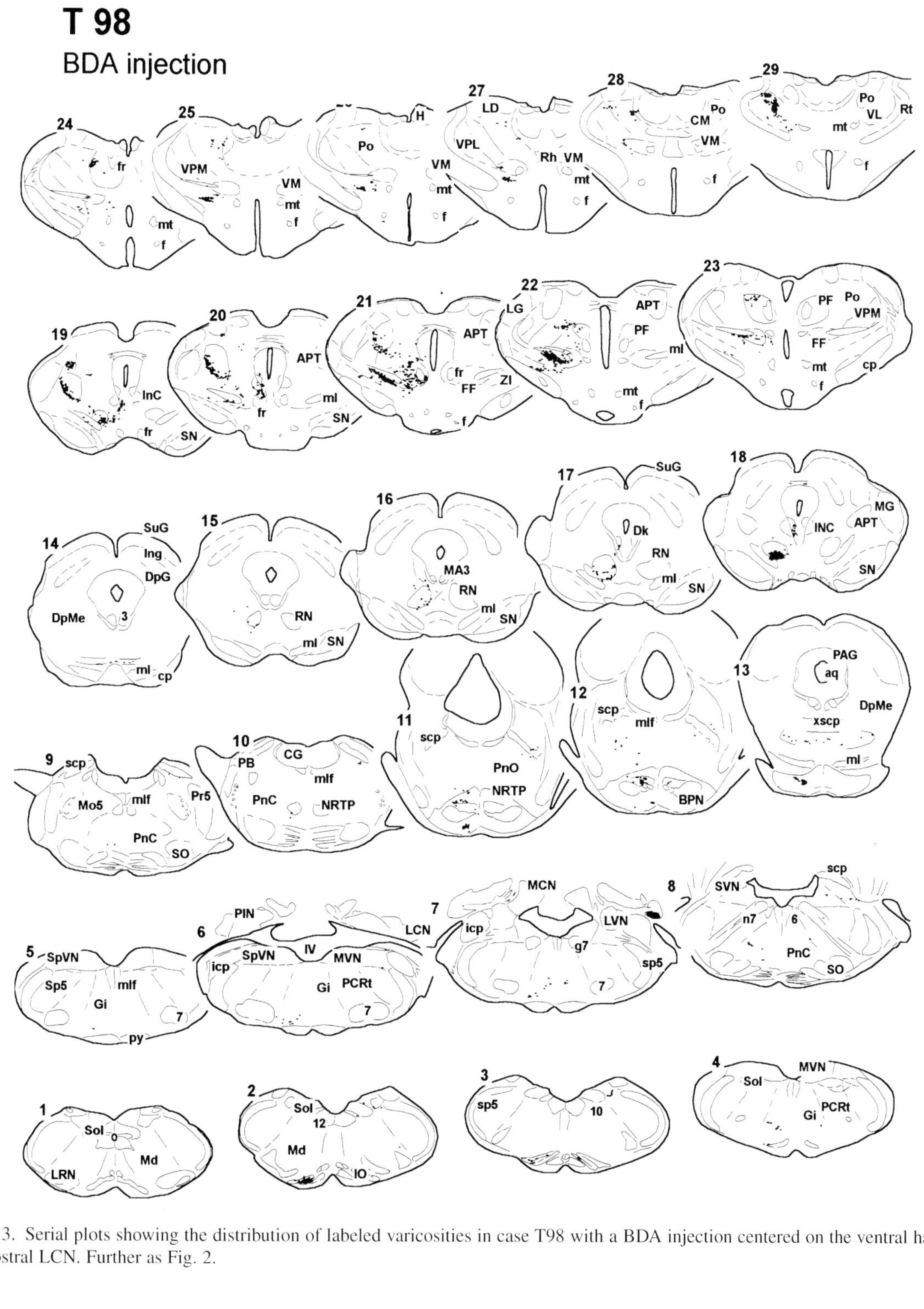

Fig. 13. Serial plots showing the distribution of labeled varicosities in case T98 with a BDA injection centered on the ventral half of the rostral LCN. Further as Fig. 2.

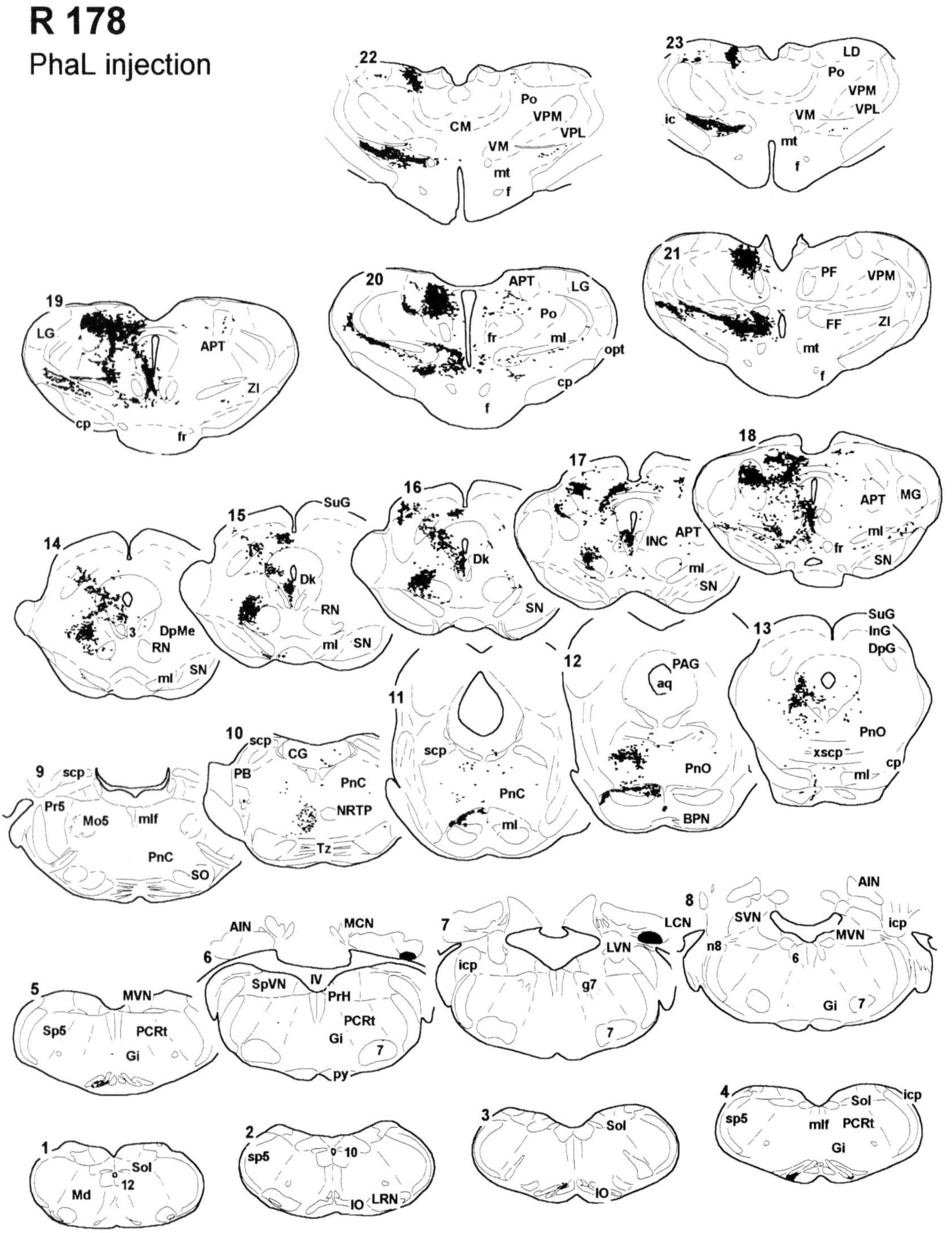

Fig. 14. Serial plots showing the distribution of labeled varicosities in case T178 with a PhaL injection centered on the ventral LCN. Further as Fig. 2.

Most terminals were directed to the mesodiencephalic region. In contrast with former cases, virtually all projections also displayed a small ipsilateral component. Particularly abundant projections were noted to reach all layers of the SC, the DpMe, the dorsal part of the APT, the nucleus of the posterior commissure, the MA3 and the PAG. Within the DpMe, a conspicuous aggregate of labeling was noted directly lateral to the magnocellular part of the RN, which, more rostrally converged with projections to the lateral part of the parvicellular part of the nucleus. For this reason, the area directly lateral to the magnocellular RN has been designated as the pararubral area (Ruigrok and Cella, 1995). Labeling to the parvicellular RN extended into the prerubral region and FF, where a prominent projection was noted (Fig. 14, levels 19–21). Dense terminal fields were also observed in the dorsal part of the ZI, with some patches extending into its ventral part. At the lateral margin of the ZI, labeled fibers bearing varicosities entered the confines of the ventral lateral geniculate nucleus. The ZI projections were especially abundant at the more rostral levels of this area where, medially, the terminal fields included the VM (Fig. 14, level 23). A very dense and salient patch of labeling was noted within the dorsal part of the PF. This terminal field extended into the medial parts of the LD group (levels 22,23). More laterally within this group some additional labeling was noted. Unfortunately, more rostral thalamic levels were not available for analysis. Case R178 also displayed a definite projection to dorsal and lateral hypothalamic regions. A few labeled varicosities were observed in the mammillary nuclei (Fig. 14, level 15).

Case T94

The PhaL injection of case T94 had been made dorsally in the caudal LCN bordering, but not involving, the DLH (Fig. 1). In the IO, the corresponding labeling remained confined to the ventral lamellae of the rostral PO without involvement of the DM group. Plots and the projection profile of this case are shown in Figs. 15 and 8, respectively.

At the medullary levels, this case was characterized by a strong projection to the contralateral Gi and paragigantocellular nucleus, extending ventrally into the region immediately dorsal to the inferior olive (into the ventral gigantocellular nucleus and nucleus gigantocellularis pars alpha of Jones, 1995), and medially to include the raphe nuclei. Some labeling was present in the ipsilateral PCRt and adjacent spinal trigeminal nucleus especially within its ventral margin. Projection patterns to the PnO, pedunculopontine tegmental nucleus, NRTP and BPN were very similar to those observed in case R178, however, projections were also noted in the PnC which appeared as a rostral continuation of the medullar Gi labeling. Patches with labeled varicosities were also seen laterally within the central gray of the pons. Also, some labeling was found in the ipsilateral A5 region (Fig. 15, level 11).

As in the former case, a large contingent of labeled varicosities were observed at the mesodiencephalic transition area. Major projections were noted to the MA3, PAG, pararubral area, DpMe, and APT. Also, the prerubral region and area surrounding the fr were studded with labeled varicosities, extending into the FF, ZI, and nucleus of the posterior commissure. As such the terminal arborizations within these areas appeared to be of a more or less continuous nature. In the thalamus, labeling was found in the PF (both dorsally and medially), within the VM, LD, and, most prominently, within the caudomedial regions of the VL. Finally, as in case R178, the lateral and dorsal hypothalamic area received a projection.

Discussion

The results presented here clearly show that small injections placed into selected parts of the CN invariably result in terminal projection patterns that are distributed over wide areas of the brain stem. It is likely that these output patterns relate to single cerebellar modules or submodules as based on intrinsic and olivocerebellar connections (Buisseret-Delmas and Angaut, 1993; Voogd et al., 1996). Moreover, the material shows that the projections to the IO are usually restricted to a single patch in one of its subdivisions indicating that in these cases the injection relates to a homogeneous olivocerebellar projection zone

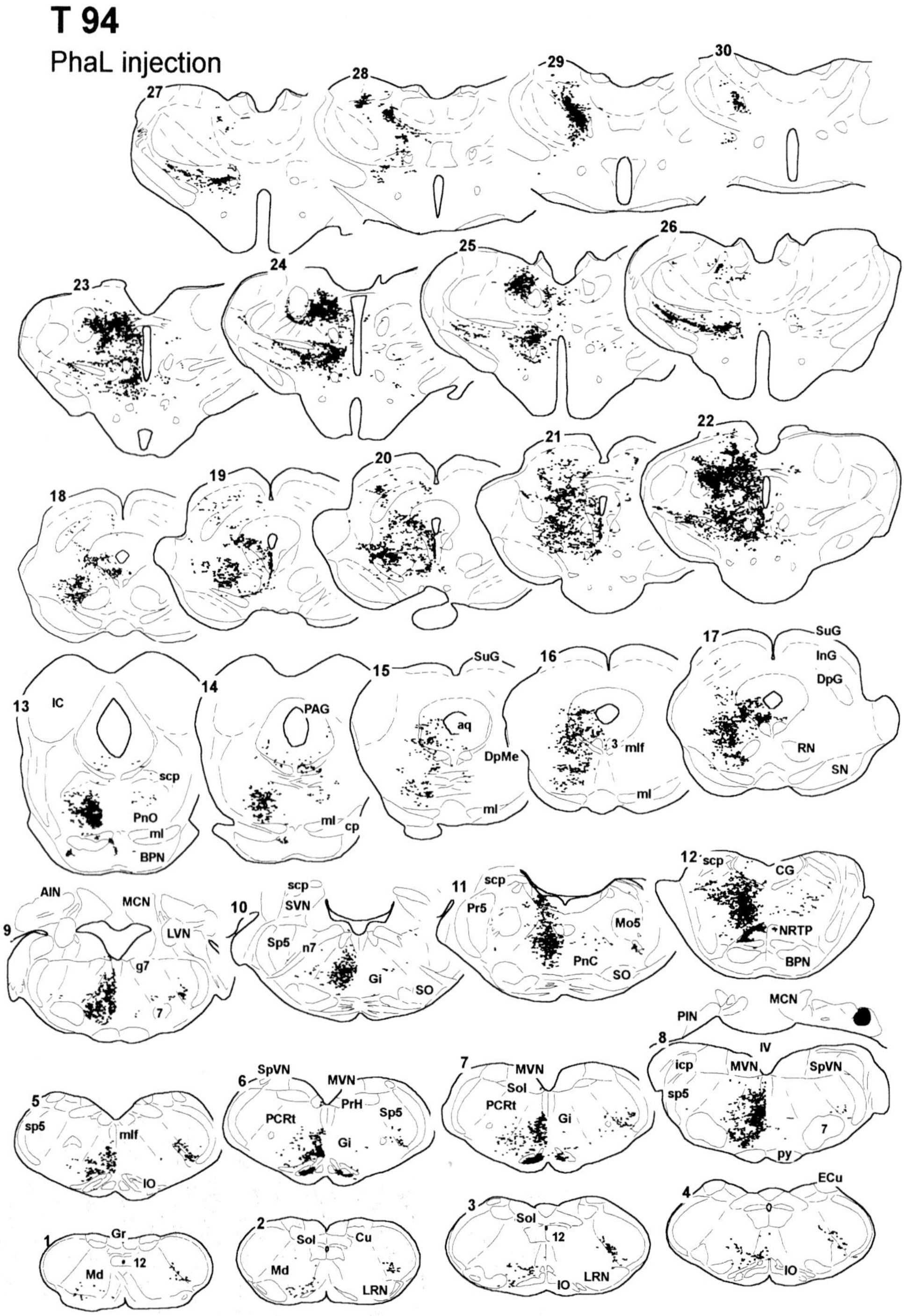

Fig. 15. Serial plots showing the distribution of labeled varicosities in case T94 with a PhaL injection centered on the caudal LCN. Further as Fig. 2.

(Ruigrok, 1997; Voogd, 1995). Hence, we conclude that the modular output is distributed to and processed in parallel in many brain stem areas.

Despite the great variety of brain stem structures that may be affected by the cerebellum, it would appear that most of them may be characterized as belonging to any of five basic groups. (1) The IO obviously functions as the source of cerebellar climbing fibers, with collaterals to all cerebellar target nuclei. The projections to different zones and their target nuclei do not seem to overlap and follow the patterns established previously (Ruigrok and Voogd, 1990; Ruigrok, 1997). (2) More diverse areas, such as the NRTP, BPN, LRN, etc. supply mossy fiber inputs. (3) In addition, as discussed above, several brain stem regions that receive a conspicuous cerebellar input, such as the medial mesodiencephalic region, pretectum, and DpMe, may serve as a source of IO input (Bull and Berkley, 1991; De Zeeuw and Ruigrok, 1994). (4) Several areas may be classified as 'effector' areas, e.g. such as the RN and supraoculomotor nuclei. Some effector regions, such as the vestibular nuclei, the SC and the reticular formation and the trigeminal nuclei, also serve as sources of mossy fiber input and as pre-olivary relay nuclei. (5) Finally, the thalamic projections of the CN should be considered. Some of them can be classified as belonging to cortical effector systems, others are part of complicated loops, involving the basal ganglia and/or precerebellar or effector regions in the brain stem. Since in the present study varicosities were plotted and counted, we noted that, in the rat, only a relatively low percentage ($<10\%$) of terminals were observed in the VL of the thalamus, which, nevertheless, is known as one of the classic termination areas of the CN. In fact, in many cases, most diencephalic terminals were observed in the ZI and/or the parafascicular thalamic nucleus.

From our results, it appears that it is difficult to clearly characterize the output of each cerebellar target nucleus unequivocally. Injections placed at various places in the same nucleus may show large differences in their terminal fields. For example, compare terminal fields and termination profiles of two MCN injections (R100 and T195) and the two PIN injections (T79 and T108). On the other hand, injections in different nuclei (e.g. the LCN in case T94 and the MCN in case R100) may show a basically similar distribution of terminal fields. Nevertheless, especially since the cerebellar modules and/or micromodules have been implicated to act as functional units it will be obvious that, in order to fully understand cerebellar functioning, the distributed nature of its modular output will have to be taken into account.

Cerebellar modules in the cat classically comprise the A, B, C1-C2-C3, D1 and D2 Purkinje cell zones (Voogd, 1964), which are connected to the MCN, lateral vestibular nucleus, AIN, PIN and dorsomedial and ventrolateral LCN, respectively. In the rat, additional zones were distinguished by Buisseret-Delmas and collaborators (Buisseret-Delmas, 1988a, b; Buisseret-Delmas and Angaut, 1989, 1993; Buisseret-Delmas et al., 1993) as will be detailed below. Here we will summarize and shortly review our main results from a modular point of view.

The A-module

The MCN serves as the target nucleus of the medial A zone of the vermis. The dorsolateral protuberance (DLP) of the rat MCN, however, is the target nucleus of a separate zone particular for rodents, the lateral extension of the A zone which is located in the medial hemisphere (Akaike, 1992; Buisseret-Delmas, 1988a). The caudal and rostral MCN and the DLP receive collaterals from different subnuclei of the caudal MAO. Rostral and caudal MCN receive collaterals from lateral subnucleus a, which relays somatosensory information from the cord and the dorsal column nuclei. The caudal MCN, in addition, receives collaterals from the subnucleus β, a vestibular relay nucleus and from subnucleus c (Akaike, 1992). The DLP receives collaterals from a different population of cells in subnucleus *c* of the caudal MAO, which project to the lateral extension of the A zone, but which shares SC afferents with the cells projecting to the vermal visual area located in lobule VII (Akaike, 1992).

The MCN is characterized by bilateral, but mainly crossed projections to the medial medullary and pontine reticular formation. In the case of the rostral MCN these are combined with strong, bilateral projections to the vestibular nuclei. The DLP mostly lacks these vestibular projections. Its strong projections to the reticular formation involve the lateral PCRt and extend rostrally into pons and mesencephalon, where they include the DpMe, the INC, the PAG and the deep layers of the SC. Vestibular projections of the caudal MCN, which receives Purkinje cell axons from the vermal visual area (lobule VII) and lobules IX and X, are mostly restricted to the SVN. Its projections to brain stem includes several visual effector areas such as the nucleus prepositus hypoglossi, the pontine paramedian reticular formation, the supraoculomotor region, the SC, the INC, the MA3 and the rostral interstitial nucleus of the medial longitudinal fascicle, here considered part of the area parafascicularis prerubralis. The caudal MCN and the DLP provide the major thalamic projections of the MCN, with the PF and VM as their main targets. Terminations in precerebellar nuclei providing mossy fibers, such as the NRTP and the LRN, are rather sparse. Some overlap occurs in the projections to the reticular formation of the rostral MCN with the DLP and the caudal MCN. All subdivisions of the MCN share a projection to the parasolitary nucleus, which is a preolivary nucleus projecting to the subnucleus β and the dorsomedial cell column (Barmack et al., 1998). Potential overlap in the projections of the MCN with other CN is present in the medullary reticular formation (for rostral MCN with LCN), in the pontine reticular formation (for DLP and caudal MCN with caudal LCN) and in the DpMe (for DLP and caudal MCN with all CN), and in several of the visual effector areas targeted by the caudal MCN (with lateral PIN and ventral and caudal LCN).

The X-module

The interstitial cell groups (ICG) receive Purkinje cell axons from the X zone, located between the A and B zones of the anterior vermis (Yatim et al., 1995a, b). Its bulbar connections were recently discussed by Buisseret-Delmas and collaborators (Buisseret-Delmas et al., 1998); their conclusion that this nucleus receives climbing fiber collaterals from the DAO, in addition to the MAO, could not be substantiated in our experiments. Its projections to the reticular formation, the vestibular nuclei, the area prerubralis parafascicularis, prerubral field and ventrolateral thalamic nucleus may all overlap with those from other CN. A sparse projection to the caudal RN could be confirmed, and a small focus of termination was observed in the NRTP in our case T82. One of the main, and relatively exclusive, connections of the ICG with the contralateral spinal cord (Bentivoglio and Kuypers, 1982; Horn et al., 1995) was not studied in our material or that of Buisseret-Delmas et al. (1998). According to Bentivoglio and Kuypers (1982) ICG neurons may collateralize to both the spinal cord and the thalamus.

The B-module

The lateral vestibular nucleus of Deiters (LVN), which serves as the target nucleus of the lateral B zone of the anterior vermis, has not been studied in this paper. Its Purkinje cells receive their climbing fiber input from the dorsal fold of the DAO which also provides a collateral projection to the LVN. The LVN gives rise to the lateral vestibulospinal tract (Rubertone et al., 1995).

The C2-module

The PIN is target nucleus of the C2 zone (Voogd and Bigaré, 1980). In the rat, however, the projection of the C1 and C3 zones includes the medial AIN and PIN, and the lateral AIN, respectively. C2 projects to intermediate AIN and lateral PIN (Buisseret-Delmas, 1988b). Climbing fiber afferents result from DAO for C1 and C3 and from rostral MAO for C2. In our material, however, injections placed in both medial (T79) and lateral PIN (T108) project to MAO and not to DAO, whereas injections in medial AIN (R89) failed to result in terminal labeling within the MAO, suggesting that the olivo-cortico-nucleo-olivary circuits may not be closed. The PIN scarcely connects with the medulla and the pons, apart from its reciprocal connections with the rostral MAO. The rostral MAO is one of the main targets of the preolivary neurons in the area parafascicularis prerubralis. The RN receives a projection from the medial PIN, located along the medial border of the nucleus. It is not clear whether these terminals overlap with projections from other CN, such as the AIN or the LCN; a projection to the NRTP seems to be lacking. Several of its projections to mes- and diencephalon, especially those from lateral PIN, overlap with those from the ventral and caudal LCN, and less so, with the caudal MCN (SC, the PAG, the nucleus of the posterior commissure, the area parafascicularis prerubralis and the VL). Its projection to the ZI is shared by most CN, with the exception of the MCN.

The C1- and C3-modules

The AIN is the target nucleus of the C1 and C3 zones. It receives collaterals from the DAO, which is the main somatosensory relay nucleus of the IO. The head is represented in the lateral part of the AIN and the hindlimbs in its medial part (Gellman et al., 1983). Strict somatotopical projections are noted to the magnocellular part of the RN and to the DAO (Daniel et al., 1987); the medial AIN connects to the ventrolateral RN and rostromedial DAO. Other main projections are directed to the NRTP, the APT, the ZI and the thalamus. In the VL nucleus of the thalamus its terminals are typically located laterally, segregated from the other CN; they presumably relay information of a somatosensory nature to the motor cortex. The APT, which is the recipient of a major part of the AIN efferents, gives rise to a projection to the ipsilateral IO, which, in the cat, terminates in the rostral DAO (Bull et al., 1990). Ipsilaterally descending projections to the PCRt and adjacent deep layers of the spinal trigeminal nucleus specifically arise from the lateral AIN (R98).

The D0-module

The dorsolateral hump (DLH) is the target nucleus of the D0 zone (Buisseret-Delmas and Angaut, 1993). There is no clear equivalent of this zone in either cats or monkeys. The DLH is reciprocally connected with the dorsomedial group of the IO (Ruigrok and Voogd, 1990; Ruigrok, 1997). Afferents from the trigeminal nuclei and adjacent reticular formation terminate in this olivary subnucleus (see Ruigrok and Cella, 1995 for a review). The rodent DLH is characterized by a unique ipsilateral projection to the PCRt, the pontine reticular formation and the trigeminal nuclei (Woodson and Angaut, 1984). Neurons with similar projections are present in the adjacent lateral AIN and dorsomedial LCN, some of which also include a contralateral component (Bentivoglio and Kuypers, 1982). Crossed projections to RN, NRTP, DpMe and APT are shared with the adjacent CN; the terminations in the thalamus are mostly restricted to the intralaminar nuclei.

The D1- and D2-modules

The LCN is the target nucleus of the D1 and the D2 zones. In the rat, Purkinje cells of these zones terminate in the rostral and dorsal, and the ventral and caudal parts of the LCN, respectively (Buisseret and Angaut, 1993). The rostral and dorsal parts of the LCN receive collaterals from the dorsal leaf of the PO; the ventral and caudal part of the nucleus from the ventral leaf. The connections of the D1 and D2 zones in rodents and in carnivores appear to be reversed. In the cat, the ventral leaf of the principal olive provides climbing fibers to the medial D1 zone and collaterals to its target nucleus, the caudolateral LCN and the dorsal leaf innervates the lateral D2 zone and its target nucleus, the rostral, dorsomedial LCN (Voogd and Bigaré, 1980).

The rostral and dorsomedial and the caudal and ventral LCN display clear differences in their projection. Bilateral projections of the rostral LCN to the reticular formation and the trigeminal nuclei resemble the adjacent DLH. Strong projections to the ventromedial medullary reticular formation, which overlap with MCN projections, are characteristic of the caudal LCN. All subdivisions of the LCN give rise to projections to the DpMe, the parvicellular part of the RN, the APT, the prerubral field, the parafascicular prerubral area, the ZI and the VM and VL of the thalamus. The terminal field of the rostral LCN in the RN includes its magnocellular part, whereas the caudal LCN provides a strong projection to the pararubral area. The terminal field in the APT extends into lateral and dorsal parts of the nucleus. Projections to visual effector areas mainly arise from the caudal LCN;

these terminations in SC, the MA3, the PAG, and the area parafascicularis prerubralis may partially overlap with the caudal MCN and the lateral PIN. In the paralaminar, medial VL nucleus of the thalamus, the caudal LCN projection overlaps with the PIN (Aumann et al., 1994; Aumann et al., 1996).

Summary and conclusions

The organization of the cerebellum is characterized by a number of parallel and parasagittally ordered olivocorticonuclear modules; as such, the cerebellar nuclei basically function as output system of these modules. The present study provides a comprehensive and detailed description of the organization of the connections from the cerebellar nuclei to the brain stem in the rat. Thirteen small injections with the anterograde tracer *Phaseolus vulgaris* leucoagglutinin or biotinylated dextran amine which were centered on various aspects of the cerebellar nuclear complex are described and are illustrated with serial plots detailing the distribution of labeled varicosities throughout the brain stem. In every case at least 1,000 and up to 36,000 varicosities were plotted.

All injections resulted in some or heavy labeling concentrated within specific regions of the contralateral inferior olivary complex and, usually, in some labeling of the contralateral ventrolateral thalamus. However, apart from these two areas it is shown that the cerebellar projections are generally very widespread and may be found throughout the entire brain stem. Below, only a survey of main projection areas will be given. Terminal arborizations originating from the rostral part of the medial cerebellar nucleus are mostly found in the caudal half of the brain stem with emphasis on the vestibular nuclear complex, whereas its caudal part rather connects to midbrain areas. Terminals that originate from the dorsolateral protuberance of the medial cerebellar nucleus are distributed more evenly throughout the brain stem and are mostly confined to reticular areas. The interstitial cell groups, interspersed between the medial and both interposed cerebellar nuclei, provide major projections to the ipsilateral vestibular nuclear complex and contralateral mesodiencephalic regions. However, reticular areas are also targeted over a large rostrocaudal range. The medial part of the posterior interposed nucleus sends most projections to the caudomedial red nucleus, prerubral regions and parvicellular reticular formation, all contralateral to the injection site. Projections that originate from more laterally placed injections are directed, apart from the inferior olivary complex, to the rostral half of the contralateral brain stem, where most labeled varicosities are found in the superior colliculus and zona incerta. The anterior interposed nucleus specifically targets the inferior olive, the red nucleus, the pontine reticulotegmental nucleus, the prectectum and the ventrolateral thalamic nucleus. More laterally placed injections also project to the ipsilateral parvicellular reticular formation and deep layers of the spinal trigeminal complex. The latter areas are more specifically targeted by the dorsolateral hump. In addition, its projections are found in the red nucleus and pretectum but do not seem to reach the ventrolateral thalamus. Projections from the lateral cerebellar nucleus are all characterized by a widespread distribution of terminals. Especially, the caudal aspect of the nucleus sends, apart from projections to the deep mesencephalic nucleus, red nucleus, periaquaductal gray, pretectum, prerubral area, and several thalamic regions, prominent projections to the caudal brain stem which terminate in the inferior olive and gigantocellular reticular formation. Projections from the ventral, parvicellular part of the nucleus are mostly, but not exclusively, directed to the rostral half of the brain stem and mainly terminate in the pararubral area, accessory oculomotor nuclei, pretectal areas, zona incerta, and in the parafascicular and ventrolateral thalamic nuclei.

We conclude that the impact of the cerebellar nuclei on the brain stem is widespread; projections from different regions of the same cerebellar nucleus may show important differences in distribution of labeled terminals. On the other hand, injections placed in different cerebellar nuclei may result in a similar distribution of at least part of their terminal fields. Hence, complex divergence and convergence between and within the projections of individual cerebellar nuclei appear to be an essential characteristic of the cerebellar nucleobulbar organization.

Acknowledgements

The authors express their gratitude to Mr. E. Dalm and Mrs. E. Sabel-Goedknegt for their excellent technical assistence. T.M. Teune was supported by NWO grant 903–47–005.

List of abbreviations

3	oculomotor nucleus
6	abducens nucleus
7	facial nucleus
10	vagal motor nucleus
12	hypoglossal nucleus
A	anterior thalamic nucleus
A5	A5 noradrenaline cells
AIN	anterior interposed nucleus
Amb	ambiguus nucleus
APT	anterior pretectal nucleus
aq	aquaduct
BDA	biotinylated dextran amine
BPN	basilar pontine nuclei
Cg	central gray
CM	centromedial nucleus
CN	cerebellar nuclei
cp	cerebral peduncle
Cu	cuneate nucleus
DAO	dorsal accessory olive
Dk	nucleus of Darkschewitsch
DLH	dorsolateral hump
DLP	dorsolateral protuberance
DM	dorsomedial group
DpG	deep gray layer SC
DpMe	deep mesencephalic nuclei
Ecu	external cuneate nucleus
f	fornix
FF	field of Forel
fr	fasciculus retroflexus
g7	genu of facial nerve
Gi	gigantocellular reticular nucleus
Gr	gracile nucleus
H	habenula
ic	internal capsule
IC	inferior colliculus
ICG	interstitial cell groups
icp	inferior cerebellar peduncle
III	third ventricle
INC	interstitial nucleus of Cajal
InG	intermediate gray layer SC
IO	inferior olive
IV	fourth ventricle
LCN	lateral cerebellar nucleus
LD	laterodorsal thalamic nucleus
LG	lateral geniculate nucleus
LRN	lateral reticular nucleus
LVN	lateral vestibular nucleus
MA3	medial accessory oculomotor nucleus
MAO	medial accessory olive
MCN	medial cerebellar nucleus
mcp	medial cerebellar peduncle
Md	medullary reticular nucleus
Me5	mesencephalic trigeminal nucleus
MG	medial geniculate nucleus
ml	medial lemiscus
mlf	medial longitudinal fascicle
Mo5	motor trigeminal nucleus
mp	mammillary peduncle
mt	mammillothalamic tract
n7	facial nerve
n8	vestibulocochlear nerve
NOT	nucleus of the optic tract
NRTP	nucleus reticularis tegmenti pontis
opt	optic nerve
Pa	paraventricular hypothalamic nucleus
PAG	periaquaductal gray
PB	parabrachial nuclei
pc	posterior commissure
PCRt	parvicellular reticular formation
PF	parafascicular thalamic nucleus
PhaL	Phaseolus vulgaris-leucoagglutinin
PIN	posterior interposed nucleus
PnC	caudal pontine reticular formation
PnO	oral pontine reticular formation
Po	posterior thalamic nucleus
PO	principal olivary nucleus
Pr5	principal sensory trigeminal nucleus
PrH	prepositus hypoglossal nucleus
py	pyramidal tract
Rh	Rhomboid nucleus
RI	rostral interstitial nucleus
RN	red nucleus
Rt	reticular thalamic nucleus
SC	superior colliculus
scp	superior cerebellar peduncle
SN	substantia nigra

SO	superior olive
Sol	nucleus of solitary tract
Sp5	spinal trigeminal nucleus
sp5	spinal trigeminal tract
SpVN	spinal vestibular nucleus
STh	subthalamic nucleus
SuG	superficial gray layer SC
SVN	superior vestibular nucleus
Tz	trapezoid body
VL	ventrolateral thalamic nucleus
VM	ventromedial thalamic nucleus
VPL	ventral posterolateral thalamic nucleus
VPM	ventral posteromedial thalamic nucleus
VTRZ	visual tegmental relay zone
xopt	optic chiasm
xpy	decussation of py
xscp	decussation of scp
ZI	zona incerta

References

Akaike, T. (1992) The tectorecipient zone in the inferior olivary nucleus in the rat. *J. Comp. Neurol.*, 320: 398–414.

Angaut, P., Cicirata, F. and Pantò, M.-R. (1985a) An autoradiographic study of the cerebellopontine projections from the interposed and lateral cerebellar nuclei in the rat. *J. Hirnforsch.*, 26: 463–470.

Angaut, P., Cicirata, F. and Serapide, F. (1985b) Topographic organization of the cerebellothalamic projections in the rat: an autoradiographic study. *Neuroscience*, 15: 389–401.

Angaut, P., Cicirata, F. and Serapide, M.F. (1987) The dentatorubral projection: an autoradiographic study in rats. *Brain Behav. Evol.*, 30: 272–281.

Asanuma, C., Thach, W.T. and Jones, E.G. (1983) Brainstem and spinal projections of the deep cerebellar nuclei in the monkey, with observations on the brainstem projections of the dorsal column nuclei. *Brain Res. Rev.*, 5: 299–322.

Aumann, T.D., Rawson, J.A., Finkelstein, D.I. and Horne, M.K. (1994) Projections from the lateral and interposed cerebellar nuclei to the thalamus of the rat: a light and electron microscopic study using single and double anterograde labelling. *J. Comp. Neurol.*, 349: 165–181.

Aumann, T.D., Rawson, J.A., Pichitpornchai, C. and Horne, M.K. (1996) Projections from the cerebellar interposed and dorsal column nuclei to the thalamus in the rat: a double anterograde labelling study. *J. Comp. Neurol.*, 368: 608–619.

Barmack, N.H., Fredette, B.J. and Mugnaini, E. (1998) Parasolitary nucleus: a source of GABAergic vestibular information to the inferior olive of rat and rabbit. *J. Comp. Neurol.*, 392: 352–72.

Bentivoglio, M. and Kuypers, H.G.J.M. (1982) Divergent axon colaterals from rat cerebellar nuclei to diencephalon, mesencephalon, medulla oblongata and cervical cord. *Exp. Brain Res.*, 46: 339–356.

Bloedel, J.R. (1992) Functional heterogeneity with structural homogeneity: how does the cerebellum operate? *Behav. Brain Sci.*, 15: 666–678.

Bradley, D.J., Paton, J.F.R. and Spyer, K.M. (1987) Cardiovascular and respiratory responses evoked from the posterior cerebellar cortex and fastigial nucleus in the cat. *J. Physiol. (Lond.)*, 393: 107–121.

Buisseret-Delmas, C. (1988a) Sagittal organization of the olivocerebellonuclear pathway in the rat. I. Connections with the nucleus fastigii and the nucleus vestibularis lateralis. *Neurosci. Res.*, 5: 475–493.

Buisseret-Delmas, C. (1988b) Sagittal organization of the olivocerebellonuclear pathway in the rat. II. Connections with the nucleus interpositus. *Neurosci. Res.*, 5: 494–512.

Buisseret-Delmas, C. and Angaut, P. (1989) Sagittal organization of the olivocerebellonuclear pathway in the rat. III. Connections with the nucleus dentatus. *Neurosci. Res.*, 7: 131–143.

Buisseret-Delmas, C. and Angaut, P. (1993) The cerebellar olivo-cortico-nuclear connections in the rat. *Progr. Neurobiol.*, 40: 63–87.

Buisseret-Delmas, C., Angaut, P., Compoint, C., Diagne, M. and Buisseret, P. (1998) Brainstem efferents from the interface between the nucleus medialis and the nucleus interpositus in the rat. *J. Comp. Neurol.*, 402: 264–275.

Buisseret-Delmas, C., Yatim, N., Buisseret, P. and Angaut, P. (1993) The X zone and CX subzone of the cerebellum in the rat. *Neurosci. Res.*, 16: 195–207.

Bull, M.S. and Berkley, K.J. (1991) Cerebellar projections to the somatic pretectum in the cat. *Somatosens. Motor Res.*, 8: 117–126.

Bull, M.S., Mitchell, S.K. and Berkley, K.J. (1990) Convergent inputs to the inferior olive from the dorsal column nuclei and pretectum in the cat. *Brain Res.*, 525: 1–10.

Carlton, S.M., Leichnetz, G.R. and Mayer, D.J. (1982) Projections from the nucleus parafascicularis prerubralis to medullary raphe nuclei and inferior olive in the rat: a horseradish peroxidase and autoradiography study. *Neurosci. Lett.*, 30: 191–197.

Chan-Palay, V. (1977) *Cerebellar Dentate Nucleus: Organization, Cytology and Transmitters*. Springer-Verlag, Berlin.

Cicirata, F., Panto, M.R. and Angaut, P. (1982) An autoradiographic study of the cerebellopontine projections in the rat. I. Projections from the medial cerebellar nucleus. *Brain Res.*, 253: 303–308.

Daniel, H., Billard, J.M., Angaut, P. and Batini, C. (1987) The interposito-rubrospinal system: anatomical tracing of a motor control pathway in the rat. *Neurosci. Res.*, 5: 87–112.

De Zeeuw, C.I. and Ruigrok, T.J.H. (1994) Olivary projecting neurons in the nucleus of Darkschewitsch in the cat receive

excitatory monosynaptic input from the cerebellar nuclei. *Brain Res.*, 653: 345–350.
Ekerot, C.-F., Jorntell, H. and Garwicz, M. (1995) Functional relation between corticonuclear input and movements evoked on microstimulation in cerebellar nucleus interpositus anterior in the cat. *Exp. Brain Res.*, 106: 365–376.
Faull, R.L.M. (1978) The cerebellofugal projections in the brachium conjunctivum of the rat. II. The ipsilateral and contralateral descending pathways. *J. Comp. Neurol.*, 178: 519–536.
Faull, R.L.M. and Carman, J.B. (1978) The cerebellofugal projections in the brachium conjunctivum of the rat. I. The contralateral ascending pathway. *J. Comp. Neurol.*, 178: 495–518.
Gellman, R., Houk, J.C. and Gibson, A.R. (1983) Somatosensory properties of the inferior olive of the cat. *J. Comp. Neurol.*, 215: 228–243.
Godschalk, M., Van der Burg, J., Van Duin, B. and De Zeeuw, C.I. (1994) Topography of saccadic eye movements evoked by microstimulation in rabbit cerebellar vermis. *J. Physiol.*, 480: 147–153.
Haines, D.E., May, P.J. and Dietrichs, E. (1990) Neuronal connections between the cerebellar nuclei and hypothalamus in *Macaca fascicularis*: cerebello-visceral circuits. *J. Comp. Neurol.*, 299: 106–122.
Haroian, A.J., Massopust, L.C. and Young, P.A. (1981) Cerebellothalamic projections in the rat: an autoradiographic and degeneration study. *J. Comp. Neurol.*, 197: 217–236.
Horn, K.M., Porter, C.M. and Gibson, A.R. (1995) Spinal connections of cat posterior interpositus. *Soc. Neurosci. Abstr.*, 21: 1190.
Ito, M. (1984) *The Cerebellum and Neural Control*, Raven Press, New York.
Jones, B.E. (1995) Reticular formation: cytoarchitecture, transmitters, and projections. In: G. Paxinos (Ed.), *The Rat Central Nervous System*, Academic Press, New York, pp. 155–171.
Korneliussen, H.K. (1968) On the morphology and subdivision of the cerebellar nuclei of the rat. *J. Hirnsforsch.*, 10: 109–122.
Leiner, H.C., Leiner, H.C. and Dow, R.S. (1986) Does the cerebellum contribute to mental skills? *Behav. Neurosci.*, 100: 443–454.
Nisimaru, N., Okahara, K. and Nagao, S. (1991) Olivocerebellar projection to the cardiovascular zone of rabbit cerebellum. *Neurosci. Res.*, 12: 240–250.
Oscarsson, O. (1969) The sagittal organization of the cerebellar anterior lobe as revealed by the projection patterns of the climbing fiber system. In: R. Llinas (Ed.), *Neurobiology of Cerebellar Evolution and Development*, AMA-ERF, Chicago, pp. 525–537.
Oscarsson, O. (1979) Functional units of the cerebellum – sagittal zones and microzones. *TINS*, 2: 143–145.
Paxinos, G. and Watson, C. (1986) *The Rat Brain in Stereotaxic Coordinates*, Academic Press, Sidney.
Rubertone, J.A., Mehler, W.R. and Voogd, J. (1995) The vestibular nuclear complex. In: G. Paxinos (Ed.), *The Rat Central Nervous System*, Academic Press, Sydney, pp. 773–796.
Ruigrok, T.J.H. (1997) Cerebellar nuclei: the olivary connection. In: C.I. De Zeeuw, P. Strata, J. Voogd (Ed.), *The Cerebellum: from Structure to Control*, Elsevier Science B.V., pp. 162–197.
Ruigrok, T.J.H. and Cella, F. (1995) Precerebellar nuclei and red nucleus. In: G. Paxinos (Ed.), *The Rat Central Nervous System*, Academic Press, Sydney, pp. 277–308.
Ruigrok, T.J.H. and Voogd, J. (1990) Cerebellar nucleo-olivary projections in rat. An anterograde tracing study with *Phaseolus vulgaris*-leucoagglutinin (PHA-L). *J. Comp. Neurol.*, 298: 315–333.
Schmahmann, J.D. (1996) From movement to thought: anatomic substrates of the cerebellar contribution to cognitive processing. *Hum. Brain Mapp*, 4: 174–198.
Teune, T.M., van der Burg, J., De Zeeuw, C.I., Voogd, J. and Ruigrok, T.J.H. (1998) Single Purkinje cell can innervate multiple classes of projection neurons in the cerebellar nuclei of the rat: a light microscopic and ultrastructural triple-tracer study in the rat. *J. Comp. Neurol.*, 392: 164–178.
Thach, W.T. (1996) On the specific role of the cerebellum in motor learning and cognition: clues from PET activation and lesion studies in man. *Behav. Brain Sci.*. 19: 411.
Tolbert, D.L., Bantli, H. and Bloedel, J.R. (1978) Organizational features of the cat and monkey cerebellar nucleocortical projection. *J. Comp. Neurol.*, 182: 39–56.
van der Steen, J., Simpson, J.I. and Tan, J. (1994) Functional and anatomic organization of three-dimensional eye movements in rabbit cerebellar flocculus. *J. Neurophysiol.*, 72: 31–46.
Voogd, J. (1964) *The Cerebellum of the Cat: Structure and Fiber Connections*, Van Gorcum, Assen.
Voogd, J. (1995) The cerebellum. In: G. Paxinos (Ed.), *The Rat Central Nervous System*, Academic Press, Sydney, pp. 309–350.
Voogd, J. and Bigaré, F. (1980) Topographical distribution of olivary and cortico nuclear fibers in the cerebellum: a review. In: J. Courville, C. de Montigny, Y. Lamarre (Eds), *The Inferior Olivary Nucleus: Anatomy and Physiology*, Raven Press, New York, pp. 207–234.
Voogd, J., Jaarsma, D. and Marani, E. (1996) The cerebellum: chemoarchitecture and anatomy. In: L.W. Swanson, A. Björklund, T. Hökfelt (Eds), *Handbook of Chemical Neuroanatomy*, Elsevier Sience B.V., pp. 1–369.
Waldrop, T.G. and Iwamoto, G.A. (1991) Cardiovascular responses to chemical stimulation of the inferior olive in the cat. *Brain Res. Bull.*, 26: 667–670.
Woodson, W. and Angaut, P. (1984) The distribution and origin of the ipsilateral descending limb of the brachium conjunctivum. An autoradiographic and horseradish peroxidase study in the rat. *Exp. Brain Res.*, 56: 167–182.
Wouterlood, F.G. and Groenewegen, H.J. (1985) Neuroanatomical tracing by use of Phaseolus vulgaris – leucoagglutinin

(Pha-L): electron microscopy of Pha-L filled neuronal somata, dendrites, axons and axon terminals. *Brain Res.*, 326: 188–181.

Wouterlood, F.G. and Jorritsma-Byham, B. (1993) The anterograde neuroanatomical tracer biotinylated dextran-amine: comparison with the tracer *Phaseolus vulgaris*-leucoagglutinin in preparations for electron microscopy. *J. Neurosci. Meth.*, 48: 75–87.

Yatim, N., Buisseret-Delmas, C., Buisseret, P., Compoint, C. and Angaut, P. (1995a) Nucleus medialis-nucleus interpositus interface: its olivary and cerebellocortical projections in the rat. *J. Comp. Neurol.*, 363: 1–14.

Yatim, N., Compoint, C., Buisseret, P., Angaut, P. and Buisseret-Delmas, C. (1995b) On the caudal extension of the X zone in the cerebellar cortex of the rat. *Neurosci. Res.*, 23: 223–227.

N.M. Gerrits, T.J.H. Ruigrok and C.I. De Zeeuw (Eds.)
Progress in Brain Research, Vol 124

CHAPTER 12

The entire trajectory of single climbing and mossy fibers in the cerebellar nuclei and cortex

Y. Shinoda*, I. Sugihara, H.-S. Wu and Y. Sugiuchi

Department of Systems Neurophysiology, School of Medicine, Tokyo Medical and Dental University, 1-5-45, Yushima, Bunkyo-ku, Tokyo, 113-8519, Japan

Introduction

Afferents of the cerebellum from many different origins, including brain stem precerebellar nuclei except the inferior olive (IO), take the form of mossy fibers in the cerebellar cortex (Cx), while olivocerebellar (OC) axons take the form of climbing fibers (CFs). Since the discovery of CFs in the cerebellar cortex (Ramón y Cajal, 1911), the morphology of these fibers has been extensively studied by Golgi staining, electron-microscopy, and recently anterograde labeling.

Despite abundant anatomical and physiological studies on the OC system, there is little information available on the organization of OC projection at the level of single axons. For example, the exact number of CFs that a single OC axon gives rise to is not known, although the average number of CFs per OC axon (about seven in the rat) was inferred by counting the total numbers of PCs and IO neurons (Schild, 1970). Electrophysiological studies (Armstrong et al., 1973; Ekerot and Larson, 1982) and retrograde labeling studies (Brodal et al., 1980; Rosina and Provini, 1983) have shown that single OC axons project to separate sites in the Cx, but the general extent of the spread of single OC axons in the cerebellar cortex remains unknown. Due to uniformity of the local circuit of the Cx, the projection pattern of afferents to the Cx including the mossy fiber system is significant in determining functional compartmentation and organized action of the cerebellum (Eccles et al., 1967; Palay and Chan-Palay, 1974; Ito, 1984). Recent studies of mapping receptive fields through the olivocerebellar climbing fiber projection have suggested that the Cx is composed of many longitudinal functional compartments (Andersson and Oscarsson, 1978; Ito et al, 1982; Ito, 1984). This electrophysiologically defined organization is consistent with the anatomically defined longitudinal compartments in the Cx (Groenewegen and Voogd, 1977; Voogd et al., 1996). Some mossy fiber projections also seem to show longitudinal organization in the vestibulofloccular system (Sato et al., 1983) and the spinocerebellar system (Oscarsson, 1976). Whether or not there is a common principle in the cortical organization of the projection of mossy fibers conveying somatosensory information is still elusive.

Besides the projection to the Cx, climbing and mossy fiber projections to the deep cerebellar nucleus (DCN) have not been fully investigated. The DCN gives rise to cerebellar output to target structures outside the cerebellum, and efferent neurons in the DCN receive inhibitory input from cortical Purkinje cells (Eccles et al., 1967; Ito, 1984). Therefore, the existence or absence of excitatory inputs to the DCN from the brainstem must be important for our understanding of the

*Corresponding author. Tel.: 81-3-5803-5153; Fax: 81-3-5803-5155; e-mail: yshinoda.phy1@med.tmd.ac.jp

cerebellar function (Shinoda et al., 1993, 1997). It has been generally assumed that excitatory inputs to the DCN are supplied through collaterals of mossy and climbing fiber afferents to the Cx. In spite of a wealth of anatomical reports on the afferent pathways to the Cx, there have been far fewer studies on afferent projections to the DCN until recently. The existence of the projections of mossy fibers to the DCN had been controversial (Chan-Palay, 1977; Dietriches et al., 1983; Ito, 1984), and reliable anatomical data on the projections from the precerebellar nuclei to the DCN had not been available until recently. (see Dietrichs et al. (1983) for the details of the problems). Using modern reliable anatomical staining methods, the projections from the pontine nucleus, nucleus reticularis tegmenti pontis, IO and spinal cord to the DCN have been definitely confirmed (Gerritis and Voogd, 1987; Van der Want et al., 1989; Shinoda et al., 1992; Mihailoff, 1993; Matsushita and Yaginuma, 1995). Intra-axonal staining with horseradish peroxidase of electrophysiologically identified neurons has been successfully used by our group to analyze the trajectories of single axons of various neurons (Futami et al., 1979; Shinoda et al., 1981, 1986, 1992). With this method, we have revealed the detailed morphology of single mossy fibers in the Cx (Krieger et al., 1985) and the existence of collateral projection of mossy fibers of the pontine nucleus and the nucleus reticularis tegmenti pontis to the DCN (Shinoda et al., 1992, 1997). In the pontine nucleus and the nucleus reticularis tegmenti pontis, only a portion of mossy fibers projecting to the Cx have axon collaterals to the DCN, and many mossy fibers project only to the Cx. However, a question still remains as to whether there exist fibers that only project to the DCN without projecting to the Cx. In addition, it is essential for understanding the neural mechanism of the cerebellar function to determine the exact percentage of fibers that have axon collaterals to the DCN in different cerebellar afferent fiber systems.

The purpose of the present study was to analyze the entire trajectories of single OC axons and single axons of the lateral reticular nucleus (LRN) that had been labeled by the injection of biotinylated dextran amine into the IO, and the LRN in the rat, respectively. An anterograde tracer, biotinylated dextran amine dissolved in physiological saline to give a concentration of 10–15% was injected either with pressure (0.001–0.05 μl) or electrophoresis (2 μA positive current pulses of 1 sec duration at 0.5 Hz for 5–20 min to the IO or the LRN). By reconstructing entire axons on serial sections, we examined the detailed morphological characteristics of single OC axons and mossy fibers of the LRN (Wu et al., 1999; Sugihara et al., 1999).

Morphology of single olivocerebellar axons

The trajectory of 16 completely reconstructed OC axons and 46 partially reconstructed axons terminating in the vermis, intermediate area, hemisphere and flocculus were used for the present analysis.

Axons left an injection site in the IO toward the contralateral side, crossed the midline, ran transversely above or through the contralateral IO, and entered the white matter of the inferior cerebellar peduncle (ICP) (Fig. 1). Then they ran longitudinally through the dorsolateral ICP to enter the cerebellum.

After entering the cerebellum, stem axons frequently ramified into many branches in the deep cerebellar white matter; these ramifications often occurred when the stem axons reached the semi-parasagittal plane in which branches terminated as CFs. Stem axons and branches usually ran in the white matter rostral and dorsal to the cerebellar nuclei, but sometimes they ran through or beneath the cerebellar nuclei. Branches entered the folial white matter, ramified and ran along each other in the semi-parasagittal plane to reach the areas for their termination in the cerebellar cortex, in which they further ramified. The branches of OC axons were grouped into thick branches (0.7–1.4 μm in diameter) and thin collaterals (0.2–0.5 μm in diameter). A single OC axon generated 2–17 final thick branches (6.1 ± 3.7, mean ± S.D., $n = 16$), each of which innervated a PC as a CF. We observed thin retrograde collaterals (Scheibel and Scheibel, 1954) originating from a CF terminal arborization and reentering the PC and granular layers, and some swellings of the retrograde collaterals in the PC layer apparently touched a PC which was not the target of the CF terminal arborization.

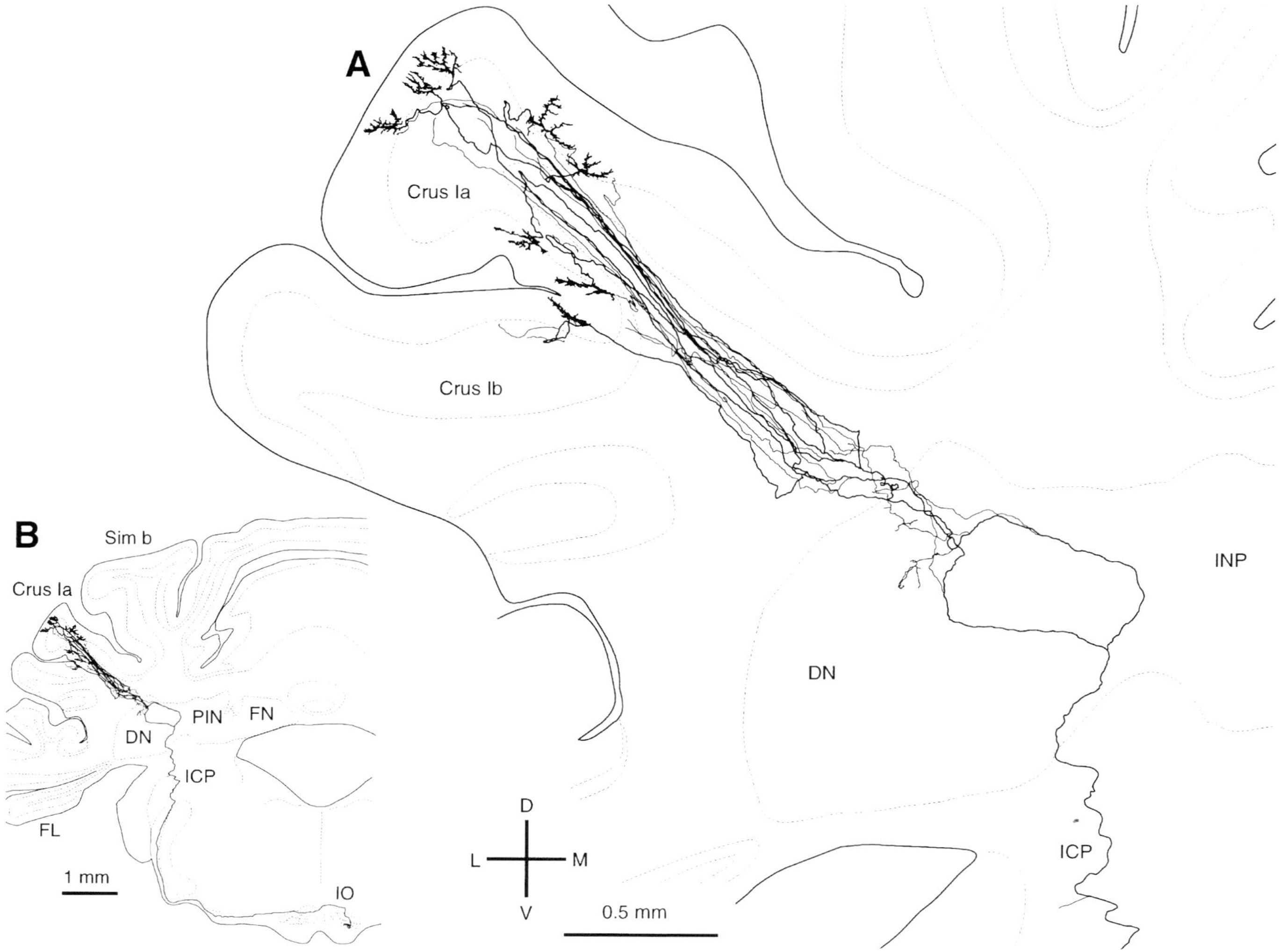

Fig. 1. Frontal view of the trajectory of a single OC axon innervating crus Ia, reconstructed from 32 serial coronal sections. Biotinylated dextran amine was injected into the dorsal lamella of the principal olive. The origin (dorsal lamella of the principal olive) and termination (lateral hemisphere and the dorsal portion of the dentate nucleus) of this axon indicated that it belonged to zone D1 (Buisseret-Delmas and Angaut, 1993; Voogd, 1995). Filled arrowhead indicates a thin collateral in the ICP, an arrow indicates the first bifurcation into two thick branches, and open arrowheads indicate the first two branching points of thin collaterals that terminated in the granular layer. All collaterals could be traced to their terminals (from Sugihara et al., 1999).

(1) Climbing fibers

Most of the reconstructed OC axons terminated on the side contralateral to the IO from which they originated (Fig. 1B). The most distal portion of a thick branch of an OC axon, which was equivalent to a CF, reached a target PC usually at its soma. The number of swellings on the terminal arborization of a single CF was 154–321 (249.8 ± 42.2, $n = 32$ terminal arborizations). The diameter of swellings ranged from 0.5 to 2.0 μm, mostly 1.2–1.6 μm. Assuming 6.1 CFs per OC axon, 249.8 swellings per CF corresponded to 1524 swellings on CF terminal arborizations per OC axon, on average. To examine whether a terminal arborization on a single PC is always formed by a single thick branch of an OC axon, we observed 523 CFs in three experiments. A terminal arborization always originated from a single thick branch of an OC axon.

Spatial distribution of climbing fibers originating from a single olivocerebellar axon had a characteristic feature as expected from the anatomical analysis of the olivocerebellar mass projection. All CFs of a single olivocerebellar axon terminated in a single strip-shaped area in the cerebellar cortex. This area was narrow in the medio-lateral direction (usually about 200 μm) and wide in the rostrocaudal direction, covering single or multiple lobules (Fig. 2). Strip shaped termination areas of single axons originating from olivary neurons located in adjacency in the inferior olive tended to have significant overlap in the cerebellar cortex. These projection patterns of olivocerebellar axons stood in sharp contrast to those of LRN neurons described in the next section.

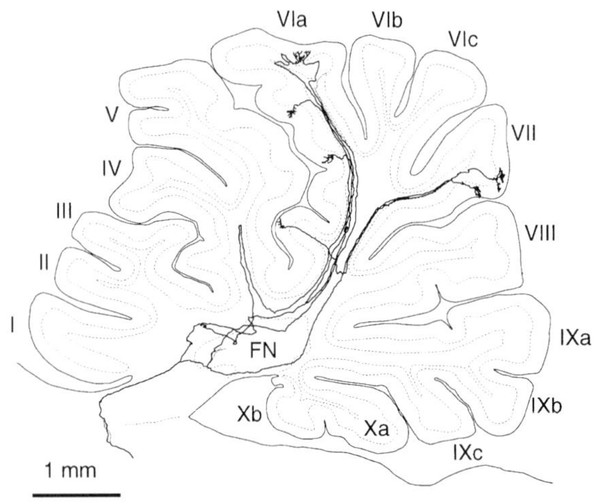

Fig. 2. Lateral view of the entire trajectory of a single OC axon in the cerebellum. The mediolateal extent of this axon in the Cx was less than 300 μm (unpublished data).

(2) Thin collaterals of OC axons

Thin collaterals, which were given off from stem axons and thick branches of OC axons before they formed CF terminal arborizations, were more abundant in terms of the number per OC axon than thick branches. For convenience of description, they were classified into three types according to their main termination sites: (1) white matter in the ICP, (2) the cerebellar nuclei, and (3) the cerebellar granular layer.

When labeled thin axons bearing swellings in the ICP ($n = 23$), cerebellar nuclei ($n = 42$) and the granular layer ($n = 81$) were traced proximally, they were always found to be collaterals of OC axons which gave rise to some CFs. This indicated that all of the axon terminals in the ICP, the cerebellar nuclei and the granular layer were derived from OC axons terminating as CFs in the cerebellar cortex, and none of the OC axons specifically terminated in the ICP, the cerebellar nuclei or the granular layer without sending CFs to the cerebellar cortex. About half of the OC axons reconstructed throughout the ICP had a single or occasionally a few thin collaterals in the ICP.

Thin collaterals to the cerebellar nuclei were observed in 20 out of 22 OC axons that were reconstructed throughout the deep cerebellar white matter. Each of these 20 axons had one ($n = 15$), two ($n = 3$), three ($n = 1$), or six collaterals ($n = 1$) (1.4 ± 1.2 collaterals per OC axon). Each OC axon innervated only a single cerebellar nucleus. Since thin collaterals terminating in the granular layer were also sometimes given off at this portion of OC axons, collaterals could not be identified as terminating in the cerebellar nuclei without tracing them to their terminals. The diameters of nuclear collaterals were 0.2–0.3 μm. Terminal branches of nuclear collaterals bore several en-passant swellings and terminal swellings. The average number of swellings per OC axon was 54.0 ± 66.0 ($n = 22$).

In counterstained sections, some swellings touched somata and proximal dendrites of both large and small neurons. We observed that some OC axons gave rise to thin collaterals terminating in the vestibular nucleus.

Thin collaterals terminating mainly in the granular layer were given off from stem or thick branches of OC axons in the deep white matter and in the folial white matter (Fig. 1A). Some of these thin collaterals terminated in the white matter with a solitary terminal swelling or a few en-passant and a terminal swelling, while the rest of these thin collaterals entered the granular layer. Swellings were most dense in the upper portion of the granular layer, but were seen at all depths in that layer. In counterstained preparations, some swellings in the granular layer seemed to make a contact with a soma of a presumed Golgi cell. Other swellings were located among a dense aggregation of blue-colored granule cells. Thin collaterals usually terminated in the same lobule and in the same thin parasagittal zone as CFs originating from the same OC axons. The number of granular layer thin collaterals ranged from 3 to 16 (average, 8.5) axons per OC axon.

Morphology of single mossy fibers of lateral reticular nucleus neurons

Complete reconstruction of the whole trajectory of single LRN axons could be made from an injection site to terminations of every branch in 11 identified LRN axons, and nearly- complete reconstruction except for distal portions of some branches in the cerebellum was made in 18 identified LRN axons.

A majority of reconstructed axons of LRN neurons passed through the ipsilateral ICP to the cerebellum ($n=25$ out of 29) (Fig. 3), but some axons ($n=4$) passed the midline, ran through or under the contralateral IO, and entered the cerebellum through the contralateral ICP. Generally, an arbor of an LRN axon within the cerebellum could be regarded as consisting of a thick stem axon running transversely, cortical branches of various diameters, and very thin nuclear branches. The stem axons ran medially toward the midline in the deep white matter rostral and dorsal to the DCN. Some stem axons did not cross the midline in the cerebellum and made projections only on the same side as the ICP which they passed through ($n=4$ of 29), but the other stem axons crossed the midline within the cerebellum to make bilateral projections. All completely reconstructed axons had a larger number of terminals in the Cx on the same side as the ICP through which they ran. While running transversely in the cerebellar white matter towards the contralateral side, stem axons gave rise to several primary collaterals to the Cx and DCN. Individual primary collaterals arose at an almost right angle from the stem axon, and ran largely in the parasagittal plane in the Cx. On their way, individual primary collaterals widely ramified mainly in a dorsoventral direction, but did not spread so much in a mediolateral direction, so that each collateral terminated as mossy fiber rosettes in the granular layer in a relatively narrow longitudinal zonal area covering one to four lobules (Fig. 4). The number of cortical primary collaterals given off from a transverse stem axon in the deep cerebellar white matter was 5–9 (7.0 ± 1.0, $n=29$). In each of bilaterally projecting neurons, the number of ipsilateral primary collaterals was 3–5 (3.8 ± 0.8, $n=25$) and the number of contralateral primary collaterals was 2–6 (3.7 ± 1.4, $n=25$). In each of unilateral projecting neurons, the number of primary collaterals was 5–8 (6.0 ± 1.2, $n=4$). These primary collaterals further ramified to innervate 1–4 lobules. Single longitudinal arrangement of terminals in the Cx was originated usually from a single primary collateral, but occasionally from multiple primary collaterals.

(1) Distribution of terminals of single LRN axons in the cerebellar cortex

Each terminal branch of a single LRN axon always terminated as rosette-type terminals. They were borne either at the end (terminal type) or in the middle of the fiber (en-passant type) in the granular layer of the Cx. The number of rosettes per a single LRN axon ranged from 84 to 219 (154.0 ± 37.0, $n=11$). In each of the bilaterally projecting axons, the ratio of axon terminals in the ipsilateral Cx to those in the contralateral Cx ranged from 1.3 to 2.3 (1.7 ± 0.4, $n=7$). In the Cx, the terminations of reconstructed axons were mostly seen in the vermis in the anterior lobe and often in lobule VI, especially VIa.

Cortical primary collaterals sent terminal branches sometimes to a single lobule but often to multiple lobules. Furthermore, terminal branches given off from each cortical primary collateral generally spread rather widely in the semi-parasagittal plane (width, 400–2000 μm, usually more than 1000 μm) within each lobule, and extended to adjacent one to four lobules (Fig. 4, lower drawings). In contrast, the spread of the terminal branches belonging to a single primary collateral was relatively restricted in the transverse direction (mediolateral width, usually 300–650 μm). As a result, the distribution of terminals of a single LRN axon plotted in the coronal plane roughly showed a pattern of multiple longitudinal bands. Some cortical collaterals appeared to give rise to terminal branches which were distributed rather widely up to about 1 mm in the transverse plane. However, the

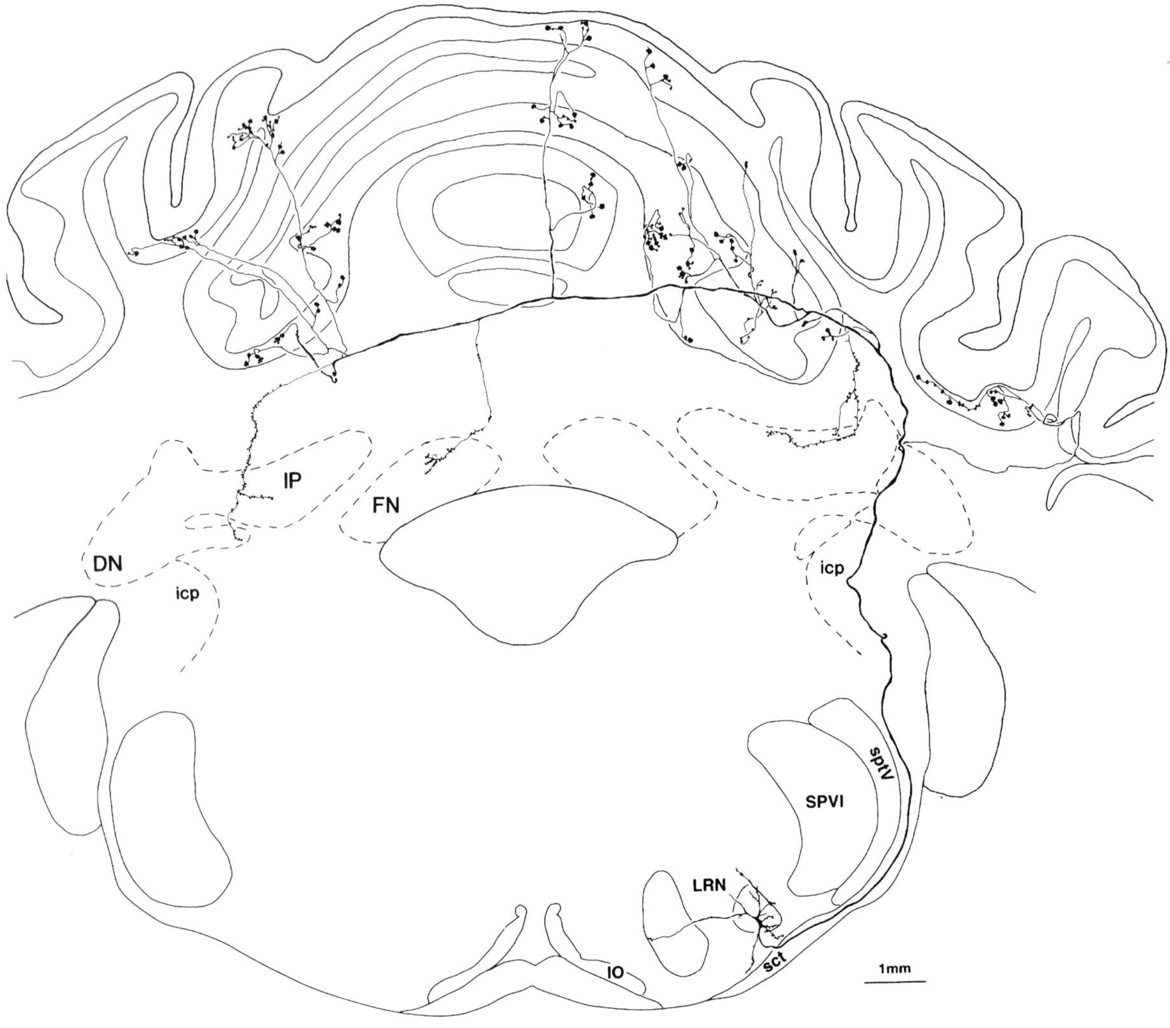

Fig. 3. Reconstruction of the entire trajectory of an axon originating from a lateral reticular nucleus neuron in the cerebellum and brain stem of the rat. The reconstruction of the axon labeled with extracellular iontophoretic application of biotinylated dextran amine was made from 158 serial sections of 50 μm thickness. DN, IP, FN and LRN, dentate, posterior interpositus, fastigial and lateral reticular nucleus; icp, inferior cerebellar peduncle; IO, inferior olive; sct, sptV and SPVI, spinocerebellar tract, spinal trigeminal tract and nucleus interpolaris (Shinoda, 1999).

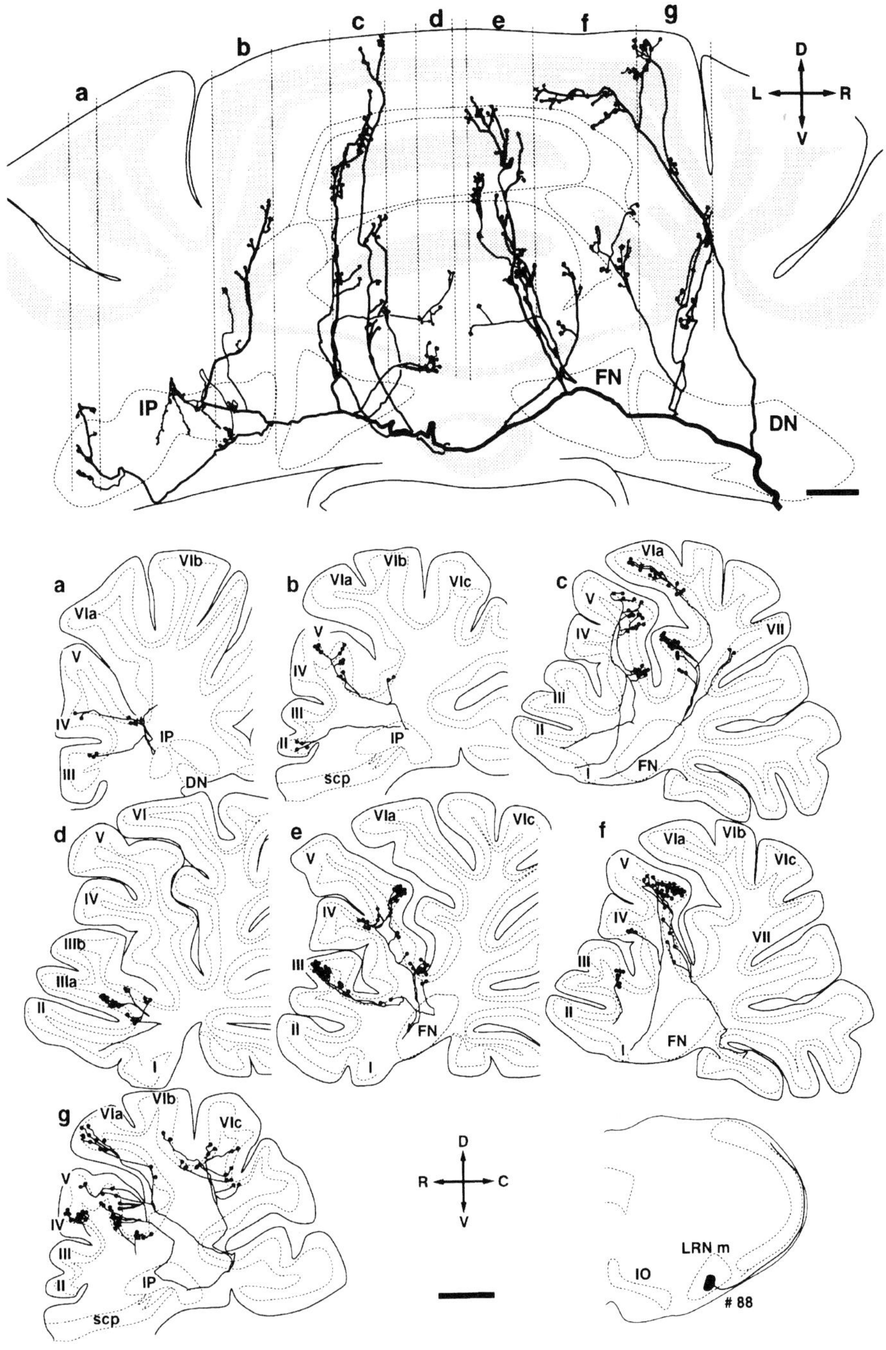

Fig. 4. Frontal (top) and lateral views (bottom) of a completely-reconstructed single LRN axon originating from the caudal part of the right LRN. This reconstruction was made on 148 serial coronal sections. This fiber entered the cerebellum through the ipsilateral icp, projected to the bilateral Cx (lobules II through VII) and DCN, and formed a multiple longitudinal zonal projection pattern by its cortical arborescent collaterals. Bottom drawings show lateral views of individual collaterals as indicated in the top drawing with lower case letters (a–g). Injection site shown in right inset on the bottom. Scale bars, 0.5 mm (top) and 1 mm (bottom) (from Wu et al., 1999).

spread of these terminal branches was not purely in the transverse direction, but also broad in the longitudinal direction which was not shown in the frontal view. Therefore, there seems to be a general tendency for longitudinal spread of terminal branches given off from single cortical primary collaterals. To reveal the nature of the longitudinal distribution of terminals in single LRN axons, terminal distributions of individual primary collaterals were analyzed in more detail, by using the completely reconstructed LRN axons. Figure 5 shows an example of such an analysis on the reconstructed LRN axon in Figure 4. The distances from the midline to individual terminals were measured and plotted on unfolded longitudinal strips of the vermal lobules in the Cx. In Figure 5A, two adjacent primary collaterals terminated in a longitudinal zone, which was labeled as zone c in Figure 4. In this zone, terminals that originated from one primary collateral were distributed in lobule V, and those from the other primary collateral were distributed in lobule VI where two clusters of terminals were slightly separated in lobule VIa and VId. The longitudinal spread of these terminals extended over 6.5 mm, whereas the mediolateral spread of each cluster of terminals was localized between 230 and 400 μm. Even as a whole, these terminals were well aligned in a single longitudinal strip less than 400 μm wide in the transverse plane. Similarly, Figure 5B shows the distribution of terminals of another primary collateral of the same axon (labeled as zone e in Figure 4) on the unfolded cortical parasagittal strip. This collateral bifurcated into two main branches, of which one branch terminated in lobule III, and the other branch terminated in lobules IV and V. Although terminals were distributed widely along a parasagittal longitudinal strip (11.6 mm), the mediolateral spread of the terminals was restricted within 600 μm as a whole. As shown in this example, the general feature of the cortical distribution of terminals of a single LRN axon could be summarized as follows. Terminals belonging to one or sometimes two primary collaterals and spreading in a few lobules made a longitudinal zone in the parasagittal plane that was usually less than 500 μm wide in the transverse plane, and several such longitudinal zones of terminals were arranged in parallel in the mediolateral direction.

(2) Morphology and distribution of collaterals terminating in the DCN and VN

All completely reconstructed LRN axons supplied projections to the DCN by way of their collaterals. Axon collaterals terminating in the DCN were given off from transverse stem axons or proximal parts of cortical primary branches, and ran in the parasagittal direction. A single LRN axon had 2–3 (2.5 ± 0.5, $n = 11$) primary collaterals to the cerebellar (and/or vestibular) nuclei. Each of these primary collaterals was much thinner (diameter, 0.4–1.0 μm) than stem axons (diameter 1.5–2.0 μm) and cortical branches, and had a localized termination area in the DCN. An arbor of a nuclear collateral could be sorted into a primary axon collateral and terminal branches, although the distinction between them was not definite. A primary axon collateral ran toward the target DCN in a relatively straight path without any branching when the branching point was far from the DCN. Within the DCN, a primary collateral bore several en-passant swellings and had only one or two ramifications to produce several (usually one to four) terminal branches. Each terminal branch was usually about 200–500 μm long and bore frequent en-passant swellings and sometimes short branchlets bearing a terminal swelling and occasionally a

Fig. 5. Longitudinal distribution of axon terminals of single primary collaterals of a reconstructed LRN axon in the unfolded Cx. A: Terminals of two primary collaterals plotted on the unfolded Cx (left). Two terminals in lobule VII are not included in the unfolded strip, but they were located in the same parasagittal strip, even though they were far away from the other terminals. B: Terminals borne on a proimary collateral plotted on unfolded lobules III–V (right). The rostrocaludal distance of the Cx was measured at the surface of the molecular layer on a reconstructed parasagittal section, and the locations of axon ternimals in the granular layer were projected to the corresponding sites on the surface of the unfolded Cx. Arabic numbers attached to the portions of the folia correspond to those on the unfolded cortical parasagittal strips. Roman numbers indicate names of lobules. Note the narrow mediolateral and wide longitudinal distributions of terminals belonging to individual primary collaterals. Scale bars = 1 mm (from Wu et al., 1999).

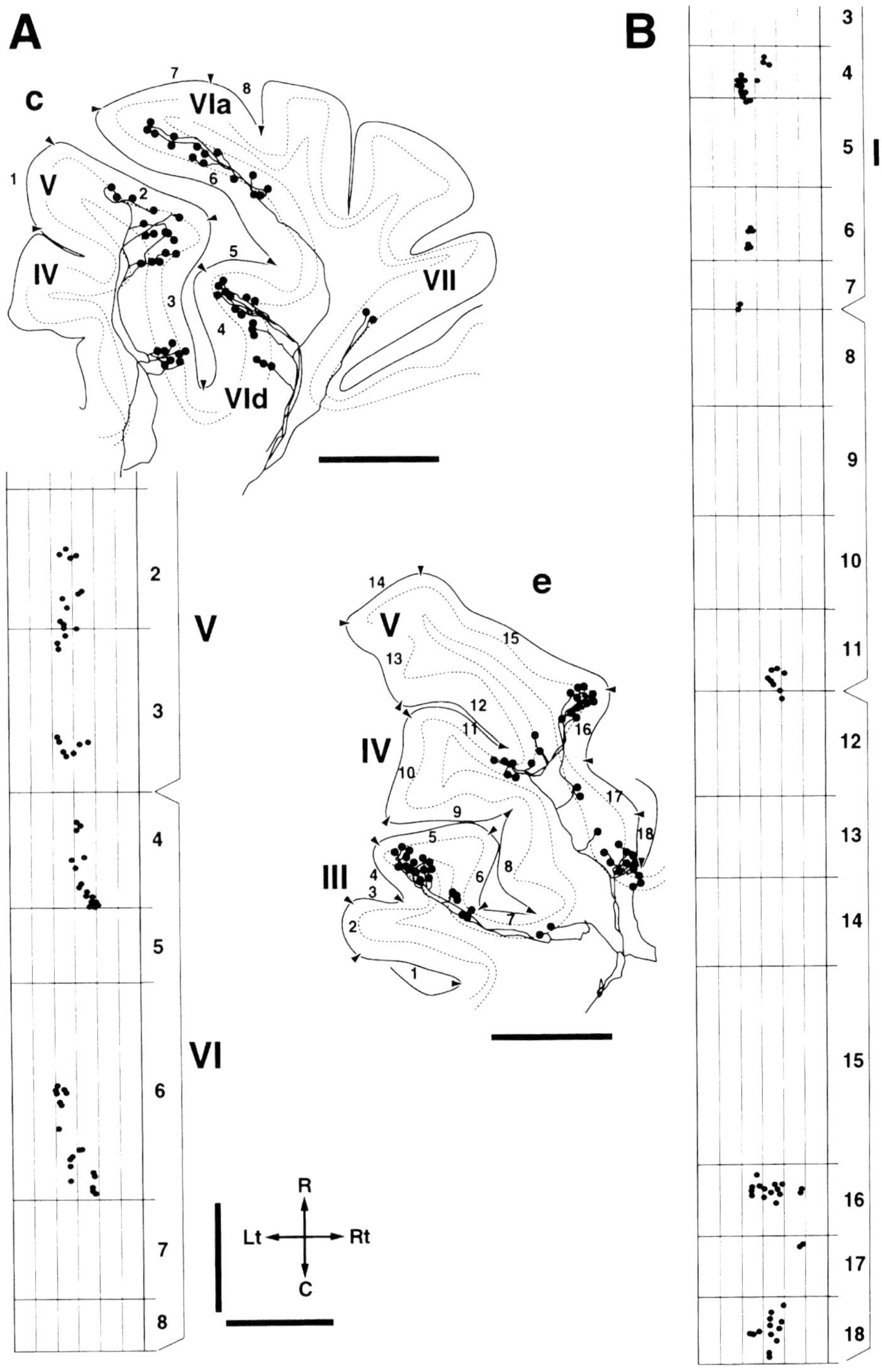
A
B
c
VIa
V
IV
VId
VII
III
VI
e
R
Lt
Rt
C

few en-passant swelling. Usually, a single primary collateral terminated in only one target DCN, but sometimes sent terminal branches into two cerebellar nuclei, or into both the cerebellar and vestibular nuclei.

Projection to the DCN was observed in all reconstructed LRN axons. LRN axons innervated most often the fastigial nucleus (FN) and interpositus nucleus (IP), and occasionally the DN and the IP. There seemed to be a rough correspondence in the cortical and nuclear projections of single LRN axons compatible with the zonal arrangement in corticonuclear projection (Voogd et al., 1996). When an axon had a collateral in the FN, the same axon always had terminations in the vermis on the same side as the FN. When an axon had collateral(s) in the IP and the DN, the same axon often innervated the intermediate area including the most lateral vermis and the hemisphere of the Cx, respectively, although there were occasional exceptions.

Summary and discussion

The present study has revealed that OC axons gave rise to a number of thin collaterals. Due to the abundance of these non-CF thin collaterals, it seems better to make a distinction between the terms CFs and OC axons, as was done in the present paper. The present findings on the innervation of PC dendrites by CFs are basically similar to those in previous reports (Ramón y Cajal, 1911; Palay and Chan-Palay, 1974). The number of swellings on a single CF in the present study ($n=250$) is comparable to a previously measured value in the rat ($n=288$; Rossi et al., 1993) and larger than a value in the frog ($n=$ about 100 beads; Llinás et al., 1969). The average number of CFs per OC axon in this study was close to the number ($n=$ about 7) inferred in the rat by counting the total number of IO neurons and PCs (Schild, 1970).

Contact of interneurons by some swellings of CFs in the molecular layer was emphasized by Scheibel and Scheibel (1954) in their study with Golgi staining. Despite the contact of CF terminals on interneurons, the formation of a synaptic structure between them has been excluded in an electron–microscopic study (Hámori and Szentàothai, 1980). On the other hand, electrophysiological studies have demonstrated a weak excitatory effect of CFs on some interneurons (Eccles et al., 1966). Terminals in the granular layer were originated either from thin collaterals of OC axons or from retrograde collaterals of CF terminal arborizations. The former was the main source of swellings in the granular layer. The morphology of the thin collaterals in the present study was consistent with "globose varicosities connected by a fine thread" as described in Golgi preparations and electron micrograms (Chan-Palay and Palay, 1971). Swellings of thin collaterals (about 1.7 % of the total number of swellings per OC axon) were most abundant in the upper portion of the granular layer just underneath the PC layer, in which Golgi cells are usually located. Furthermore, some of these swellings were observed to touch presumed Golgi cells in the present study, which is consistent with electron–microscopic findings on the innervation of somata of Golgi cells by thin collaterals (Hámori and Szentàothai, 1980; Chan-Palay and Palay, 1971). Inferior olive stimulation has been shown electrophysiologically to have a weak direct excitatory effect on Golgi cells (Eccles et al., 1966).

Ninety-one percent of the OC axons examined had nuclear collaterals; since the possibility of insufficient staining could not be excluded, this percentage may be an underestimation. The ratio of swellings in the cerebellar nuclei versus those of CF terminal arborizations was about 0.036 in individual OC axons in the present study. However, since the volume of the cerebellar nuclei is much smaller than that of the cerebellar cortex, and significant convergence of input from OC axons to cerebellar nucleus neurons is present (Sugihara et al., 1996), cerebellar nucleus projection of OC fibers can still be functionally important. Some swellings seemed to make contact with the soma and the proximal portions of dendrites of large neurons in the present study, which is consistent with the steep rising phase of postsynaptic excitatory potentials in cerebellar nucleus neurons following IO stimulation (Kitai et al., 1977; Shinoda et al., 1987). Although intracellular potentials were presumably recorded only from large output neurons in the cerebellar nuclei, the present study suggested that small neurons were also innervated by OC axons.

The present study revealed that virtually all reconstructed LRN axons projected not only to the Cx as mossy fibers, but also to the DCN including the VN by their axon collaterals. None of the LRN neurons specifically projected to the DCN without projecting to the Cx, namely all axon terminals of LRN neurons in the DCN and VN belonged to axon collaterals of mossy fibers projecting to the Cx. In general, axon collaterals in the DCN and VN had a simple branching pattern and a narrow projection area, whereas axon collaterals in the Cx had a complex branching pattern, and a wide mediolateral spread, usually bilaterally, with multiple longitudinal zones in multiple lobules. Cerebellar cortical projection of single LRN axons is characterized by broadness of the termination area. Single LRN axons simultaneously projected to areas different in mediolateral locations in several lobules, often bilaterally. A significant overlap of termination area of single axons originating from within an injection site was suggested. Similar broadness of cortical innervation has been suggested in axons of other mossy fiber systems such as cerebellar projection from the pontine nucleus (Shinoda et al., 1992). This broadness of cerebellar cortical innervation in single LRN axons makes a good contrast to a single zonal projection of olivocerebellar axons.

Virtually all LRN axons reconstructed in the present study had collaterals terminating in the DCN and also in the VN in some cases. This is the first mossy fiber system in which all cerebellar cortical projection neurons have axon collaterals to the DCN. However, high incidence of nuclear projection via axon collaterals can not be generalized to all mossy fiber systems, since many mossy fibers originating from the pontine nucleus lack nuclear collaterals (Shinoda et al., 1992). We confirmed that virtually all cerebellar nuclear projection from the LRN was via thin collaterals given off from stem axons or primary cortical collaterals. Previous anterograde or retrograde studies have not identified whether nuclear projections of LRN neurons are via collaterals of mossy fibers or there exist specific nuclear projecting neurons. It is very clear that the topographical relationship in the LRN-nuclear projection was much weaker and vaguer than in the olivonuclear projection (Van der Want et al., 1989; Sugihara et al., 1996), in which neurons in a subdivision of the IO project to a relatively localized area in one of the cerebellar nuclei. This is partly because single LRN axons had multiple nuclear collaterals projecting to more than one cerebellar nuclei, and multiple cortical collaterals spreading rather widely in the transverse plane, while single IO neurons had multiple collaterals spreading in a very narrow longitudinal zone in the Cx (Sugihara et al., 1997).

Large injections of BDA into the LRN visualized longitudinal multiple zones of mossy fiber terminals in the Cx in the present study. This finding is consistent with the results obtained by Künzle (1975) and Chan-Palay et al. (1977) using an autoradiographic method in the rat, and by Ruigrok and Cella (1995) using PHA-L. One of the most important findings in this study is the existence of multiple zonal projection of single LRN axons in the Cx. This projection made a sharp contrast with the projection of the IO, which also shows multiple longitudinal zonal pattern, when labeled by a mass injection of anterograde tracer (Chan-Palay et al., 1977; Van der Want et al., 1989). However, at the level of single axons, the cortical projection is generally localized within a single longitudinal zone (Sugihara et al., 1997; Sugihara et al., 1999). The multiple zonal projection may not necessarily contradict with the patchy representation of receptive fields of the facial area (Shambes et al., 1978), since terminals of single axons seemed to be arranged into clusters within each zone. However, analysis of single mossy fiber projection from the trigeminal system would be necessary to clarify the projection pattern.

One of the important remaining questions is how longitudinal arrangements of the LRN projection are related to those zonal components named with A, B, C1, C2, C3, D0, D1, D2, which have been identified with olivocerebellar and corticonuclear projections (Groenewegen and Voogd, 1977; Buissert-Delmas and Angaut, 1993; Voogd, 1995). We attempted in vain to correlate these zones with the present zones of LRN terminals by combining acetylcholine esterase staining and the present staining, since the zones visualized by acetylcholine esterase staining appeared in the posterior vermis. The relationship between LRN projecting multiple zones and zebrin-defined zones which has been addressed by Ruigrok and Cella (1995) and those zones defined by the different methods seem to correspond well to each other. Thus, functional significance of cerebellar compartmentation, which has been so far studied from the viewpoints of labeling with molecular markers, olivocerebellar and corticonuclear projections, will be also considered from the viewpoint of mossy fiber projections.

Excitatory mossy fiber inputs to the Cx is relayed by granule cells and parallel fibers to reach Purkinje cells. Although a parallel fiber spreads in the transverse direction, the most effective input from the parallel fiber is given to Purkinje cells just above the given granule cell through synapses formed by the parallel fiber (Eccles et al., 1967). Therefore, the zonal projection of a single LRN axon may be functionally significant in conveying an excitatory input to Purkinje cells in that specific zone. Individual LRN axons projected to multiple lobules in both the hemisphere and the vermis, and even often bilaterally. Therefore, cerebellar cortical areas innervated by a single mossy fiber from the LRN may not necessarily be related to each other in terms of their functional roles. Then, the question is what is the role of LRN axons sending presumably specific information to different cortical areas. Concerning the nuclear projection, it is obvious that the cortical and nuclear projections of any single reconstructed LRN axon did not exactly follow the so-called corticonuclear topographical relationship, since the cortical projection was generally more widely spread in the transverse direction than the nuclear projection. Then, this situation does not exactly fit the classical micro-complex scheme of Ito (1984), in which the nuclear inputs by a mossy fiber collateral and by a Purkinje cell that receives the same mossy fiber input through granule cells converge onto the same target nuclear output neuron. The nuclear collaterals of mossy fibers are important in activating nuclear neurons as a source for excitation of their target neurons (Shinoda et al., 1992, 1997). To understand the functional interactions at nuclear efferent neurons between Purkinje cell inputs in widely-distributed cortical areas innervated by a single mossy fiber and nuclear inputs by a collateral of the same mossy fiber, we need more detailed information as to how single Purkinje cells spread in the DCN, how Purkinje cells in different cortical areas send their informations to the DCN, and how those informations are integrated as an output in the DCN. Furthermore, it will be also useful for understanding the functional mechanism of whole mossy fiber systems to specify the similarity and the difference in mossy fibers in other systems such as spinocerebellar and cuneocerebellar systems.

Acknowledgements

This research was supported by a research grant from CREST (Core Research for Evolutional Science and Technology) of the Japan Science and Technology Corporation to Y.S., and by Grants-in-Aid for Scientific Research from the Ministry of Education, Science and Culture of Japan to I.S.

References

Andersson, G. and Oscarsson, O. (1978) Climbing fiber microzones in cerebellar vermis and their projection to different groups of cells in the lateral vestibular nucleus. *Exp. Brain Res.*, 32: 564–579.

Armstrong, D.M., Harvey, R.J. and Schild, R.F. (1973) The spatial organization of climbing fibre branching in the cat cerebellum. *Exp. Brain Res.*, 18: 40–58.

Brodal, A., Walberg, F., Berkley, K.J. and Pelt, A. (1980) Anatomical demonstration of branching olivocerebellar fibres by means of a double retrograde labelling technique. *Neuroscience*, 5: 2193–2202.

Buisseret-Delmas, C. and Angaut, P. (1993) The cerebellar olivo-corticonuclear connections in the rat. *Prog. Neurobiol.*, 40: 63–87.

Chan-Palay, V. and Palay, S.L. (1971) Tendril and glomerular collaterals of climbing fibers in the granular layer of the rat's cerebellar cortex. *Z. Anat. Entwickl-Gesch.*, 133: 247–273.

Chan-Palay, V., Palay, D.L., Brown, J.T. and Van Itallie, C. (1977) Sagittal organization of olivocerebellar and reticulocerebellar projections: autoradiographic studies with 35S-methionine. *Exp. Brain Res.*, 30: 561–576.

Dietrichs, E., Bjaalie, J.G. and Brodal, P. (1983) Do pontocerebellar fibers send collaterals to the cerebellar nuclei? *Brain Res.*, 259: 127–131.

Eccles, J.C., Ito, M. and Szentágothai, J. (1967) *The Cerebellum as a Neuronal Machine.* Springer-Verlag, Berlin-Heidelberg-New York, pp. 227–261.

Eccles, J.C., Llinás, R. and Sasaki, K. (1966) The inhibitory interneurones within the cerebellar cortex. *Exp. Brain Res.*, 1: 1–16.

Ekerot, C.-F. and Larson, B. (1982) Branching of olivary axons to innervate pairs of sagittal zones in the cerebellar anterior lobe of the cat. *Exp. Brain Res.*, 48: 185–198.

Futami, T., Shinoda, Y. and Yokota, J. (1979) Spinal axon collaterals of corticospinal neurons identified by intracellular injection of horseradish peroxidase. *Brain Res.*, 164: 279–284.

Gerrits, N.M. and Voogd, J. (1987) The projection of the nucleus reticularis tegmenti pontis and adjacent regions of the pontine nuclei to the central cerebellar nuclei in the cat. *J. Comp. Neurol.*, 258: 52–62.

Groenewegen, H.J. and Voogd, J. (1977) The parasagittal zonation within the olivocerebellar projection. I. Climbing fiber distribution in the vermis of the cat cerebellum. *J. Comp. Neurol.*, 174: 417–488.

Hámori, J. and Szentáothai, J. (1980) Lack of evidence of synaptic contacts by climbing fibre collaterals to basket and stellate cells in developing rat cerebellar cortex. *Brain Res.*, 186: 454–457.

Ito, M., Orlov, I. and Yamamoto, M. (1982) Topographical representation of vestibulo-ocular reflex in rabbit cerebellar flocculus. *Neuroscience*, 7: 1657–1664.

Ito, M. (1984) *The Cerebellum and Neural Control*. New York: Raven Press.

Kitai, S.T., McCrea, R.A., Preston, R.J. and Bishop, G.A. (1977) Electrophysiological and horseradish peroxidase studies of precerebellar afferents to the nucleus interpositus anterior. I. Climbing fiber system. *Brain Res.*, 122: 197–214.

Krieger, C., Shinoda, Y. and Smith, A.M. (1985) Labelling of cerebellar mossy fiber afferents with intra-axonal horseradish peroxidase. *Exp. Brain Res.*, 59: 414–417.

Künzle, H. (1975) Autoradiographic tracing of the cerebellar projections from the lateral reticular nucleus in the cat. *Exp. Brain Res.*, 22: 255–266.

Llinàs, R., Bloedel, J.R. and Hillman, D.E. (1969) Functional characterization of neuronal circuitry of frog cerebellar cortex. *J. Neurophysiol.*, 32: 847–870.

Matsushita, M. and Yaginuma, H. (1995) Projections from the central cervical nucleus to the cerebellar nuclei in the rat, studied by anterograde axonal tracing. *J. Comp. Neurol.*, 353: 234–246.

Mihailoff, G.A. (1993) Cerebellar nuclear projections from the basilar pontine nuclei and nucleus reticularis tegmenti pontis as demonstrated with PHA-L tracing in the rat. *J. Comp. Neurol.*, 330: 130–146.

Oscarsson, O. (1976) Spatial distribution of climbing and mossy fibre inputs into the cerebellar cortex. In: O. Creutzfeldt (Ed.), *Experimental Brain Research Suppl. 1: Afferent and Intrinsic Organization of Laminated Structures in the Brain*, Berlin: Springer Verlag, pp. 36–42.

Palay, S.L. and Chan-Palay, V. (1974) *Cerebellar Cortex. Cytology and Organization*. New York: Springer-Verlag.

Ramón y Cajal S. (1911) *Histologie du Systeme Nerveux de l'Homme et des Vertbré*, Vol. II. Paris: Maloine.

Rosina, A. and Provini, L. (1983) Somatotopy of climbing fiber branching to the cerebellar cortex in cat. *Brain Res.*, 289: 45–63.

Rossi, F., Borsello, T., Vaudano, E. and Strata, P. (1993) Regressive modifications of climbing fibres following Purkinje cell degeneration in the cerebellar cortex of the adult rat. *Neuroscience*, 53: 759–778.

Ruigrok, T.J.H. and Cella, F. (1995) Precerebellar nuclei and red nucleus. In: G. Paxinos (Ed.), *The Rat Nervous System, Vol. III. Brain Stem and Cerebellum*, Sydney: Academic Press, pp. 281–286.

Sato, Y., Kawasaki, T. and Ikarashi, K. (1983) Afferent projections from the brainstem to the floccular three zones in cats. II. Mossy fiber projections. *Brain Res.*, 272: 37–48.

Scheibel, M.E. and Scheibel, A.B. (1954) Observations on the intracortical relations of the climbing fibers of the cerebellum. A Golgi study. *J. Comp. Neurol.*, 101: 733–763.

Schild, R.F. (1970) On the inferior olive of the albino rat. *J. Comp. Neurol.*, 140: 255–260.

Shambes, G.M., Beermann, D. H. and Welker, W. (1978) Multiple tactile areas in cerebellar cortex: another patchy cutaneous projection to granule cell columns in rats. *Brain Res.*, 157: 123–128.

Shinoda, Y. (1999) Visualization of the entire trajectory of long axons of single mammalian CNS neurons. *Brain Res. Bull.*, 50: 387–388.

Shinoda, Y., Sugiuchi, Y. and Futami, T. (1993) Organization of excitatory inputs from the cerebral cortex to the cerebellar dentate nucleus. *Can. J. Neurol. Sci.*, 20 (Suppl. 3): S19–S28.

Shinoda, Y., Izawa, Y., Sugiuchi Y. and Futami, T. (1997) Functional significance of excitatory projections from the precerebellar nuclei to interpositus and dentate nucleus neurons for mediating motor, premotor and parietal cortical inputs. In: C.I. deZeeuw, P. Strata, and J. Voogd (Eds), *The Cerebellum: from Structure to Control. Progress in Brain Research, Vol. 114*, Amsterdam: Elsevier, pp. 193–207.

Shinoda, Y., Ohgaki, T. and Futami, T. (1986) The morphology of single lateral vestibulospinal tract axons in the lower cervical spinal cord of the cat. *J. Comp. Neurol.*, 249: 226–241.

Shinoda, Y., Yokota, J. and Futami, T. (1981) Divergent projection of individual corticospinal axons to motoneurons of multiple muscles in the monkey. *Neurosci. Lett.*, 23: 7–12.

Shinoda, Y., Sugiuchi, Y. and Futami, T. (1987) Excitatory inputs to cerebellar dentate nucleus neurons from the cerebral cortex in the cat. *Exp. Brain Res.*, 67: 299–315.

Shinoda, Y., Sugiuchi, Y., Futami, T. and Izawa, R. (1992) Axon collaterals of mossy fibers from the pontine nucleus in the cerebellar dentate nucleus. *J. Neurophysiol.*, 67: 547–560.

Sugihara, I., Wu, H. and Shinoda, Y. (1996) Morphology of axon collaterals of single climbing fibers in the deep cerebellar nuclei of the rat. *Neurosci. Lett.*, 217: 33–36.

Sugihara, I., Wu, H.-S. and Shinoda, Y. (1997) Projection of climbing fibers originating from single olivocerebellar neurons in the rat cerebellum. *Soc. Neurosci. Abstr.*, 23: 1830.

Sugihara, I., Wu, H. and Shinoda, Y. (1999) Morphology of single olivocerebellar axons labeled with biotinylated dextran amine in the rat. *J. Comp. Neurol.*, 414: 131–148.

Van der Want, J.J.L. and Voogd, J. (1987) Ultrastructural identification and localization of climbing fiber terminals in the fastigial nucleus of the cat. *J. Comp. Neurol.*, 258: 81–90.

Van der Want, J.J. L., Wiklund, L., Guegan, M., Ruigrok, T. and J. Voogd (1989) Anterograde tracing of the rat olivocerebellar system with Phaseolus vulgaris leucoagglutinin (PHA-L). Demonstration of climbing fiber collateral innervation of the cerebellar nuclei. *J. Comp. Neurol.*, 288: 1–18.

Voogd, J. (1995) Cerebellum. In: G. Paxinos (Ed.), *The Rat Nervous System*, 2nd edn. Sydney: Academic Press, pp. 309–350.

Voogd, J., Jaarsma, D. and Marani, E. (1996) The cerebellum, chemoarchitecture and anatomy. In: L.W. Swanson, A. Bjsrklund, and T. Hskfelt (Eds), *Integrated Systems of the CNS, Part III. Cerebellum, Basal Ganglia, Olfactory System. Handbook of Chemical Neuroanatomy Vol. 12*, Amsterdam: Elsevier, pp. 1–369.

Wu, H., Sugihara, I. and Shinoda, Y. (1999) Projection patterns of single mossy fibers originating from the lateral reticular nucleus in the rat cerebellar cortex and nuclei. *J. Comp. Neurol.*, 411: 97–118.

N.M. Gerrits, T.J.H. Ruigrok and C.I. De Zeeuw (Eds.)
Progress in Brain Research, Vol 124

CHAPTER 13

Micro-organisation of cerebellar modules controlling forelimb movements

Martin Garwicz*

Department of Physiological Sciences, Lund University, Sölvegatan 19, S-223 62 Lund, Sweden

Introduction

More than thirty years ago, Voogd and Oscarsson presented the first evidence that the cerebellar cortex is divided into a number of sagittally oriented zones, each defined by its specific afferent and efferent connections (reviewed by Oscarsson, 1980; Voogd and Bigaré, 1980). According to recent accounts, the anterior lobe of the cat is divisible into at least nine zones of Purkinje cells on each side of the midline (Fig. 1A). The significance of this specificity is apparent in light of the uniform design of the basic neuronal circuit across the cerebellar cortical sheet. Such an arrangement implies that while the neural processing carried out by different cortical regions is likely to be similar in principle, the function subserved by any particular region will be to a large extent determined by the characteristics of its input, its output and the relationship between the two.

That regional function is determined by regional connectivity indicates that function can be understood only by taking connectivity into account. This underscores the importance of identifying single cells with respect to functional localisation when systematising data and interpreting neuronal discharge patterns in studies of behaving animals. It also poses a challenge to incorporate specific input-output characteristics of the cerebellar operational units into any computational model of cerebellar function that aims at approaching the 'realistic'.

The above points are further emphasised by the fact that beyond the zones there is an even finer-grain functional compartmentalisation of the cerebellum (e.g. Andersson and Oscarsson, 1978; Ekerot and Larson, 1979; Ekerot et al., 1991a; Garwicz et al., 1992). This chapter focuses on the 'micro-organisation' of a control system that includes the C1, C3 and Y zones in the cerebellar pars intermedia. As illustrated in Fig. 1B, the forelimb areas of these zones receive a partly shared climbing fibre input from common neurones located in the rostral dorsal accessory olive (rDAO; e.g. Groenewegen et al., 1979; Trott and Apps, 1991) and all in turn project to nucleus interpositus anterior (NIA; e.g. Voogd and Bigaré, 1980; Voogd 1982; Trott and Armstrong, 1987; see also Garwicz and Ekerot, 1994). Via its nucleofugal projections to the rubrospinal and the corticospinal tracts this cerebellar system contributes to the control of forelimb movements (e.g. Armstrong and Edgley, 1984a, b; Edgley and Lidierth, 1988; Thach et al., 1992; Kitazawa et al., 1993; van Kan et al., 1993a, b).

Modular organisation of olivo-cortico-nuclear circuitry

The climbing fibre receptive field as a marker

When dissecting the connectivity of a circuitry spanning three neuronal elements in series and with

*Corresponding author. Tel.: +4646 222 77 55; Fax: +4646 222 45 46; e-mail: martin.garwicz@mphy.lu.se

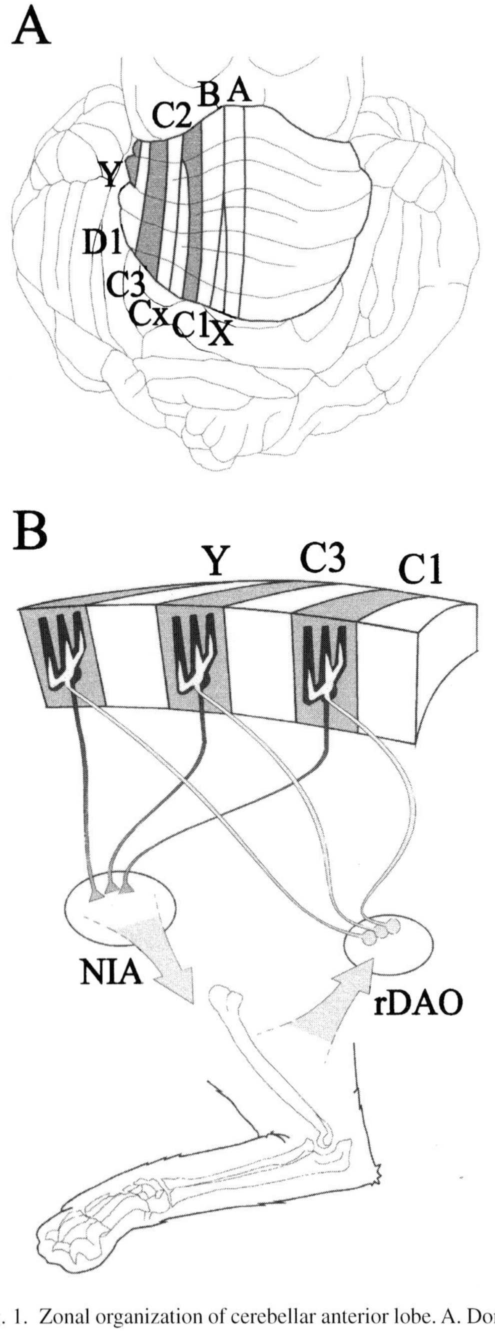

Fig. 1. Zonal organization of cerebellar anterior lobe. A. Dorsal view of the cat cerebellum. Thick lines demarcate the extent of the anterior lobe and the borders of schematically drawn sagittal zones A through Y on its left-hand side. Zones C1, C3 and Y indicated by shading. B. Schematic representation of the olivo-cortico-nuclear circuitry controlling forelimb movements via the rubrospinal and corticospinal tracts. Cx zone not shown. rDAO: rostral dorsal accessory olive; NIA: nucleus interpositus anterior.

a number of different types of input in parallel, it is necessary to have a frame of reference that allows consistent systematisation of data. Given the variability at the 'sub-zonal' level of resolution between individual animals, neither anatomical landmarks nor stereotaxic co-ordinates are sufficiently reliable. Instead, a physiological marker is required for functional localisation within the system. In the studies reviewed here, receptive fields of climbing fibres served this purpose (Ekerot et al., 1991a; Garwicz and Ekerot, 1994; Ekerot et al., 1995; Jörntell et al., 1996; Garwicz et al., 1998; see also Garwicz et al., 1996; Apps and Garwicz, 2000).

The quality of receptive field mapping was in turn dependent to a large extent on the use of nociceptive input to climbing fibres. By contrast to the phasic responses to tactile stimulation, noxious pinch of the skin evokes repetitive climbing fibre responses at relatively high frequencies, with a graded input from different parts of the skin (Ekerot et al., 1987; 1991a). Combined with on-line auditory monitoring of evoked responses, nociceptive input is therefore ideal for detailed quantitative mapping of climbing fibre receptive fields.

Receptive field classes, microzones and zonal topography

The characteristics of receptive fields and the zonal distribution of climbing fibre input inform both the analysis of connectivity and the analysis of what signals are mediated by the climbing fibre system. The latter issue will be dealt with in later sections. Here we highlight findings pertaining to the climbing fibre receptive fields as a population and aspects of the topographical organisation of climbing fibre input that will form the basis for the delineation of the olivo-cortico-nuclear modules.

First, an accumulated sample of more than 500 climbing fibres demonstrates that the different types of receptive fields are limited in number and display stepwise changes in extent and focus (Ekerot et al., 1991a; Jörntell et al., 1996; Garwicz et al., 1998). In the initial study, close to one hundred climbing fibre units were characterised using a standardised pinch to allow quantitative comparisons and subsequent correlation analysis with regard to similarity in spatial organisation of receptive fields supported the existence of a number of distinct clusters. In the most recent count, ten classes with a total of more than thirty-five subclasses of receptive fields were identified. A subclass representative of each of the ten classes is illustrated in Fig. 2.

Second, climbing fibres with 'same subclass' receptive fields terminate in a clustered fashion in the cortex, forming sagittally oriented microzones. Similar receptive fields occupy corresponding mediolateral positions in different folia, suggesting that microzones may be continuous throughout the interfolial fissures. In the mediolateral dimension, microzonal boundaries are usually clear-cut and in some cases receptive fields of neighbouring microzones differ substantially (Ekerot et al., 1991a, b). Together with the discontinuities between receptive fields in the pooled population, this indicates that the C3 zone is composed of distinct sub-units (see also Ekerot and Larson, 1979). Despite some variability in the representation of different microzones the overall topographical map of the zone, shown schematically in Fig. 3 (top panel; Ekerot et al., 1991a), is recognisable in most individual animals.

The olivo-cerebellar and cortico-nuclear projections

Physiological characteristics of climbing fibre input have previously been used successfully to guide neuronal tracer injections into specific zones or even parts of zones (e.g. Trott and Apps, 1991). Therefore, the presence of a microzonal map in theory makes possible a similarly detailed anatomical assessment of the topography of the olivo-cerebellar and cortico-nuclear projections. In practice, however, tracer injections restricted to single, properly identified microzones are not feasible. In the medial part of the C3 zone only four subdivisions can be reliably distinguished, each division corresponding to a handful of different microzones (Garwicz et al., 1996).

However, even given these constraints, combined electrophysiological and fluorescent tracer technique reveals a clear topographical organisation of the olivo-cerebellar projection to the medial C3

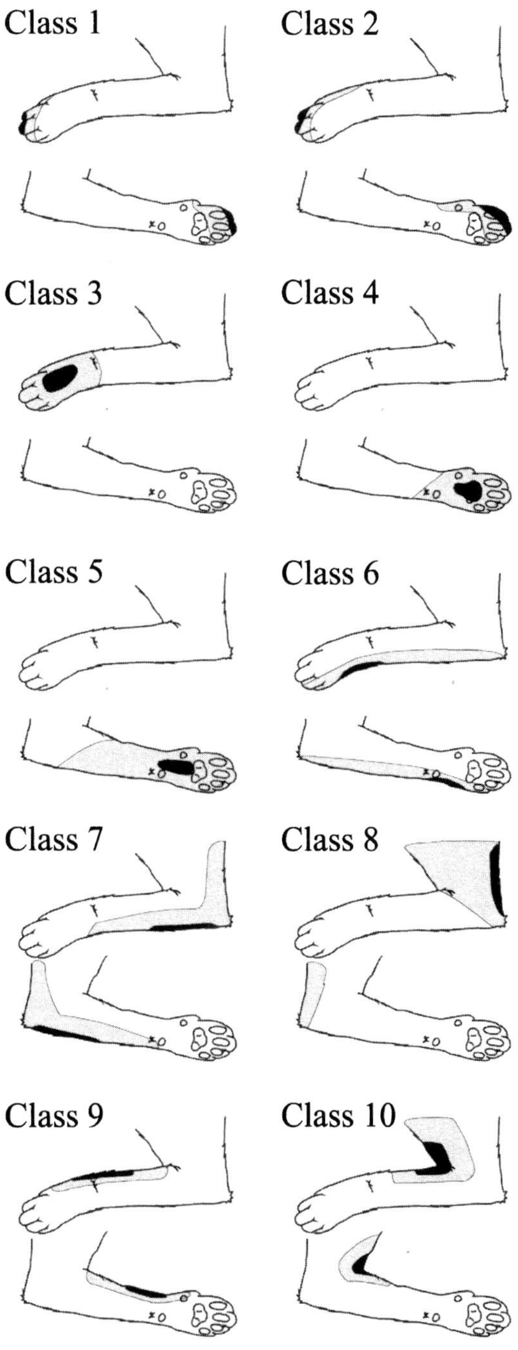

Fig. 2. Examples of climbing fibre receptive fields classified according to their spatial characteristics. A representative subclass was chosen for each of the ten classes. Each pair of figurines shows dorsal (top) and ventral (bottom) view of the left forelimb of the cat. Shading and blackening indicate extent of receptive field and area of maximal sensitivity, respectively.

zone. Medial-to-lateral in the olive is mapped lateral-to-medial in the zone whereas rostral-to-caudal in the olive appears to be mapped caudal-to-rostral in the zone (Fig. 3, lower panel; Garwicz et al., 1996). If extrapolated, these findings suggest that the aforementioned compartmentalisation of the cerebellar cortex is

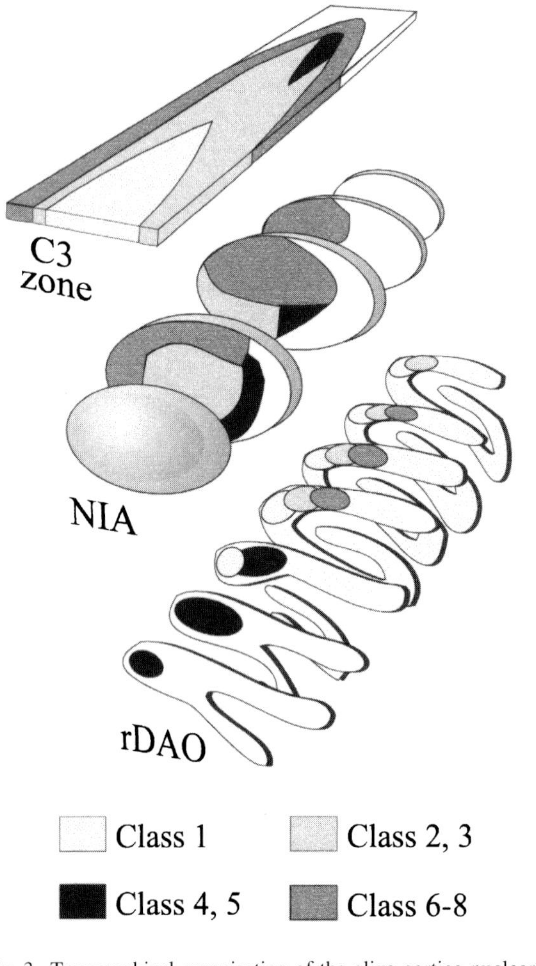

Fig. 3. Topographical organisation of the olivo-cortico-nuclear circuitry depicted schematically. The three panels show from top to bottom, left-hand C3 zone and nucleus interpositus anterior (NIA) and right-hand rostral dorsal accessory olive (rDAO). The perspective into the depth of the figure is from caudal to rostral. Different shadings indicate different classes of receptive fields, as outlined in the key (cf. Fig. 2). Classes 9 and 10 (not shown) occupy mainly caudal parts of the lateral C3 zone (Garwicz et al., 1998). The olivary topography was derived from olivo-cerebellar connections to the medial C3 zone only.

already founded in the organisation of the inferior olive and the olivo-cerebellar projection.

Insights into the cortico-nuclear topography gained from fluorescent tracer studies are less conclusive and the level of specificity of this projection is better appreciated using instead an electroanatomical approach (Garwicz and Ekerot, 1994). The technique is based on the observation that tactile or nociceptive stimulation of the skin evokes positive potentials in NIA. These potentials reflect extracellularly recorded currents at the inhibitory synapses between cerebellar cortical Purkinje cells and nuclear neurones, generated as a result of synchronous Purkinje cell activation by climbing fibres. Consequently, to map 'receptive fields' of positive potentials at any particular nuclear site means in effect to map the climbing fibre receptive fields of the Purkinje cells projecting to that site. Thus, the technique can be used for a detailed assessment of the nuclear termination areas of different cortical microzones and, with previous maps of zonal topography as reference, of the topographical organisation of the cortico-nuclear projection (Fig. 3, middle panel; Garwicz and Ekerot, 1994).

The majority of receptive fields previously identified for single climbing fibres in the C3 zone are distinguishable also in NIA. This indicates a rather modest degree of convergence in the cortico-nuclear projection between microzones with different receptive fields. On the other hand, the fact that each type of receptive field appears to be represented in a single continuous region of the nucleus suggests that microzones with similar, 'same subclass' receptive fields converge in the cortico-nuclear projection, regardless of their location in the cortex. These findings demonstrate that the compartmentalisation of the olivo-cerebellar and cerebellar cortical circuitry is preserved in the projection to NIA, as outlined in Fig. 3.

Interim conclusion and comments

There is now rather compelling evidence that the NIA system is composed of at least twenty (Garwicz and Ekerot, 1994), but probably more than thirty-five functionally distinct sub-units, or modules. As defined here, a module comprises a cluster of olivary neurones with very similar cutaneous receptive fields, the cortical microzones innervated by these olivary neurones and a group of deep cerebellar nuclear cells upon which those microzones converge. Since the axons of some olivary cells branch to innervate the medial C3 and C1 zones or the lateral C3 and Y zones (Ekerot and Larson, 1982) and, in addition, other non-branching cells may share common peripheral input, microzones with identical receptive fields are not necessarily confined to one zone only. According to the working hypothesis, such microzones, regardless of zonal identity, converge onto the same group of nuclear neurones (Garwicz, 1992; Garwicz and Ekerot, 1994). A recent investigation specifically addressing this issue provides support for the hypothesis (Apps and Garwicz, 2000).

Although the modular organisation provides a structural framework for understanding how information is processed by the cerebellar circuitry, issues expected to constitute the core of any viable theory of cerebellar function remain outstanding. For example, what kinds of signals are mediated by the climbing fibre system and what specific role do these signals play in processing of information in the cerebellar cortex (cf. Simpson et al., 1996)? What are the typical patterns of convergence between climbing fibre and mossy fibre input (Ekerot and Larson, 1980) and how do signals mediated via the two systems interact (e.g. Ito, 1989)? Last but not least, what is the relationship between input (van Kan et al., 1993b) and output (van Kan et al., 1993a) in the cerebellar modules? Some of these questions are addressed in the following sections.

Information processing in cerebellar modules

Signals mediated by the olivo-cerebellar climbing fibre system

Although the high sensitivity to cutaneous stimulation would enable the climbing fibre system to faithfully mediate information about specific peripheral events, the multi-modal convergence between cutaneous nociceptive, tactile (Ekerot et al., 1987; Ekerot et al., 1991a) and muscle afferents (Jörntell et al., 1996) onto single olivary neurones appears to be at odds with such a role. We have proposed, therefore, that the climbing fibre signals are not primarily 'sensory' in the classical sense of the term, but reflect instead activity in (or afferent input to) spinal reflex circuits (Ekerot et al., 1991a; see also Oscarsson, 1973).

The most direct support for this notion stems from a comparison between cutaneous receptive fields of climbing fibres and receptive fields of nociceptive withdrawal reflexes of single forelimb muscles (Garwicz et al., 1999, 2000). While the analysis is still preliminary, the similarities with regard to spatial organisation of the two types of receptive fields are rather striking, as illustrated in Fig. 4A. Combined with the strong nociceptive input to climbing fibres, the characteristics of their receptive fields thus suggest a functional link with nociceptive withdrawal reflexes (Schouenborg et al., 1994). More importantly, the findings reveal what may be a key principle in the organisation of olivary input to the NIA system, regardless of whether or not it is related to reflexes (see Garwicz et al., 2000). Each cutaneous climbing fibre receptive field seems to correspond to the area of skin withdrawn from an external stimulus upon contraction of a specific muscle when the limb is in the standing position (Fig. 4B; see also Levinsson et al., 1999). Such an organisation would be compatible also with other features of the climbing fibre input.

First, the discreteness of receptive field subclasses and the modular design of the olivo-cortico-nuclear circuitry is likely to reflect something similarly discrete/modular in nature. Skin areas moved by the contraction of single muscles have this property (Schouenborg and Weng, 1994). Second, proximal borders of individual receptive fields are usually located close to borders between limb segments, i.e. at the metacarpo-phalangeal joints, at the wrist, the elbow or the shoulder and the area of maximal sensitivity is often located eccentrically within the receptive field (e.g. Ekerot et al., 1991a). Again, skin areas moved by the contraction of single muscles have these properties. Taken together, these characteristics suggest that the system operates within a frame of reference based on elemental movements.

A final suggestive piece of evidence is provided by a re-evaluation of the functional relationship

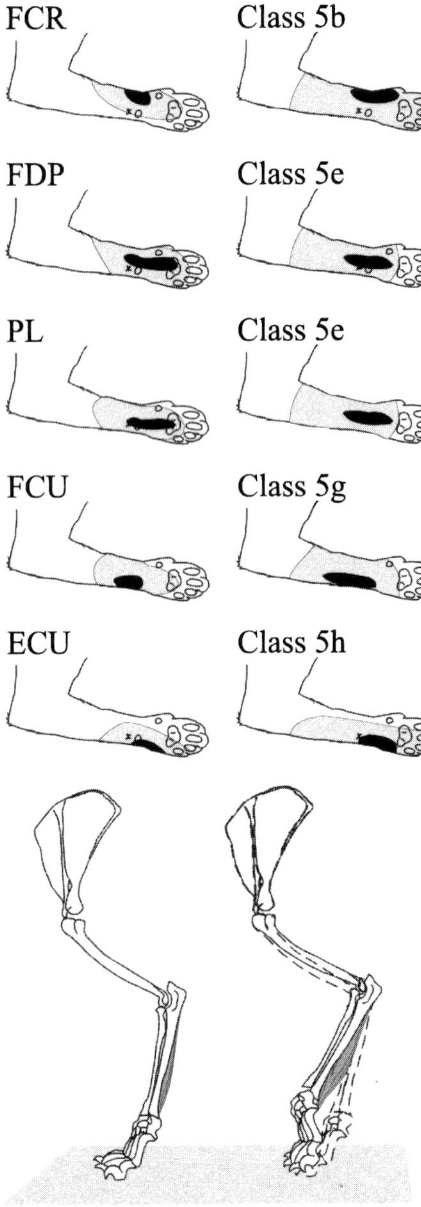

Fig. 4. Upper panel: Comparison between receptive fields of withdrawal reflexes and climbing fibres. Conventions as in Fig. 2. Left: schematically depicted nociceptive receptive fields of forelimb muscles acting at the wrist and in one case on the digits (redrawn from Garwicz et al., 2000). FCR: m. flexor carpi radialis; FDP: m. flexor digitorum profundus; PL: m. palmaris longus; FCU: m. flexor carpi ulnaris; ECU: m. extensor carpi ulnaris. Right: examples of class 5 receptive field subclasses. Lower panel: action of wrist palmar flexors when the limb is in the standing position. Left: muscle relaxed. Right: muscle contracted (see Levinsson et al., 1999).

between the cutaneous receptive field and the muscle afferent input converging onto the same olivary neurone (Fig. 5). Several of the cutaneous receptive fields are associated with input from muscles, the action of which was originally interpreted as moving the receptive field towards a stimulus applied to the skin. This was based on the assumption that the limb is in a free position, i.e. off the ground (Jörntell et al., 1996). If, however, inspired by the organisational principles of withdrawal reflexes one assumes instead that the limb is in the standing position, i.e. on the ground, it appears that the action of at least some of these muscle(s) would be to withdraw the cutaneous receptive field from an external stimulus (Fig. 4B; Levinsson et al., 1999). This re-interpretation also provides a better fit with the actual data (cf. Jörntell et al., 1996). Hence, it seems that a given module via the climbing fibres receives cutaneous input from a skin area corresponding to the receptive field of a particular withdrawal reflex pathway and muscle afferent input from the muscle constituting the output element of the very same reflex (Garwicz et al., 2000).

Receptive fields and topography of mossy fibre input

The role of mossy fibre signals with respect to their effects in the cerebellar cortex is less controversial than that of climbing fibres. Whereas a variety of modulatory roles have been assigned to climbing fibre signals with regard to their effects on the cortical circuitry (Simpson et al., 1996), it is generally agreed that activity in the mossy fibre–granule cell route provides the main 'throughput' of the cortex (see Ito, 1984). In the following account, the receptive field characteristics and topographical organisation of mossy fibres conveying cutaneous input to the C3 zone, presumably mainly via the cuneocerebellar tract (e.g. Cooke et al., 1971), will be compared to the climbing fibre input discussed above.

The majority of cutaneous mossy fibre units are readily activated by tactile stimuli and units with the highest sensitivity respond to displacement of single hairs or even to weak airflow, whereas nociceptive input appears to be lacking almost

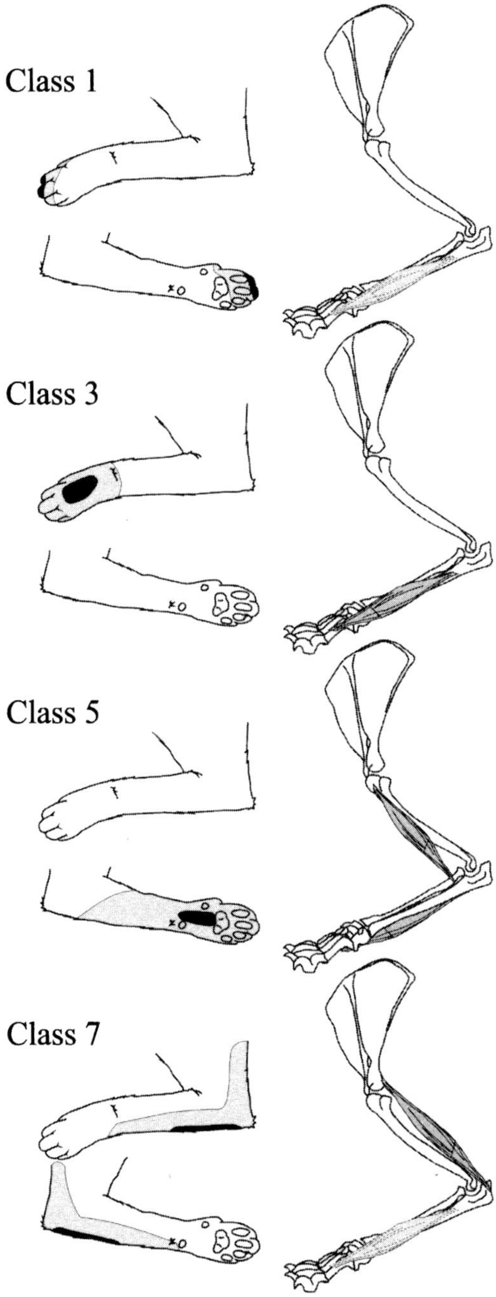

Fig. 5. Patterns of convergence between cutaneous and muscle afferent input. Left panel: Some example classes represented by their cutaneous receptive fields. Conventions as in Fig. 2. Right panel: The group(s) of muscles providing the associated muscle (group II) afferent input is shown schematically (wrist dorsiflexors/palmar flexors; elbow flexors/extensors). Dashed outline and lighter grey (top and bottom figurines) indicates weak input.

completely. The spatial organization of individual mossy fibre receptive fields resembles that of climbing fibres and although there is an overall bias for receptive fields on the radial side of the forearm most mossy fibre receptive fields fit the climbing fibre classification scheme (Garwicz et al., 1998).

Mossy fibres with similar receptive fields terminate in a relatively clustered fashion in the granule cell layer and if material is pooled between animals the volume defined by a particular set of receptive field appears rather clearly delineated in the mediolateral, rostrocaudal and in some cases also dorsoventral dimensions. There is also an approximate correspondence between mossy fibre and climbing fibre receptive fields in the same parts of the zone, as shown in Fig. 6(A and B).

However, by contrast to the discrete organisation of climbing fibre input, different mossy fibre terminal fields overlap to a substantial extent. As a result of this overlap, mossy fibres found underneath any given microzone display a variety of different receptive fields, only some of which are identical to the climbing fibre receptive field defining the microzone. In spite of this, the area of skin most prominently represented by the *population* of mossy fibres often strongly resembles the climbing fibre receptive field above (Fig. 6a′ and b′).

Nucleofugal output to forelimb muscles

The efferent limb of olivo-cortico-nuclear modules can be investigated using micro-stimulation in NIA in a decerebrate, unanaesthetised preparation. Each module is first identified by mapping of receptive fields of positive field potentials in the nucleus, as described above. The set of forelimb muscles controlled by the module via the rubrospinal tract is then assessed by inspection of movements evoked on microstimulation of the same nuclear site (Ekerot et al., 1995). Both the type of preparation and the stimulation technique used call for caution when interpreting what the modules actually control. Nevertheless, some key points are outlined below.

The movements usually involve two or three segments of the limb and are not reduced to single joint movements even at threshold stimulation,

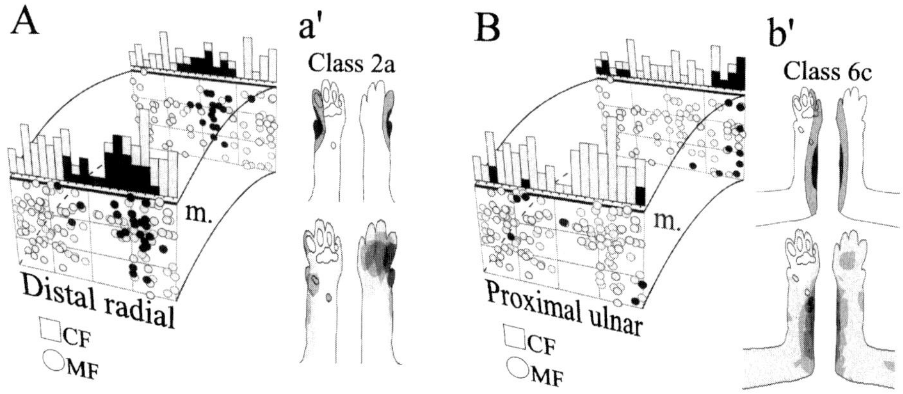

Fig. 6. Topographical organisation of mossy fibre input – some examples (modified from Garwicz et al., 1998). A: Two cross-sections of the granular layer in caudal and middle (in depth of figure) folia of the forelimb area in the C3 zone. Distribution of mossy fibre units with receptive fields on distal radial paw (black) relative to total material (grey) shown for reference. m.: medial, CF: climbing fibre units, MF mossy fibre units. a′: Schematic receptive field of climbing fibre subclass (microzone) 2a (top) and superimposed receptive fields of mossy fibre units terminating underneath this microzone (bottom). B and b′: Same as A and a′, but for proximal ulnar and subclass 6c, respectively.

suggesting that the control exerted from these modules is organized in simple muscle synergies. On stimulation in the nucleus, movements evoked at proximal joints are usually flexion movements, i.e. retraction of the limb at the shoulder and flexion at the elbow. Movements at the wrist display similar proportions of flexion and extension whereas distal movements consist mainly of dorsiflexion of central or ulnar digits (Ekerot et al., 1995).

Although there is not a simple one-to-one relationship for individual modules, the pooled material reveals a relative specificity between the efferent ‘muscle field’ and the cutaneous climbing fibre receptive field of the different groups of modules, as shown in Fig. 7. For example, stimulation of sites with distal receptive fields evokes the highest proportion of digit movements, whereas sites with proximal receptive fields have the lowest proportion of such movements. More importantly, movements often display a ‘directional’ specificity in relation to the receptive field. Stimulation of sites with climbing fibre receptive fields on the dorsum of the paw frequently evokes palmar flexion at the wrist, whereas stimulation of sites with receptive fields on the ventral side of the forearm and paw evokes dorsiflexion at the wrist (Ekerot et al., 1995). In other words, the efferent group of muscles typically includes an antagonist of the muscle(s) providing input to the olivary neurones of the module (see Fig. 5).

Main conclusions, uncertainties and some perspectives

The data on the cerebellar rDAO-C1-C3-Y-NIA system reviewed here can be summarised by the following three points, which will be discussed in part with reference to an example module, illustrated in Fig. 8. The module represents the least incompletely characterised group of modules, namely those with ventral receptive fields, and should in this respect be regarded as ‘model’ rather than ‘typical’.

- *The forelimb part of the rDAO-C1-C3-Y-NIA system consists of approximately thirty-five olivo-cortico-nuclear modules. Each module is defined by the spatial organisation of the cutaneous tactile and nociceptive receptive field of its olivary neurones and is characterised by olivo-cortical divergence and cortico-nuclear convergence.*

Approximately thirty-five different receptive field subclasses/microzones have been identified in the cerebellar cortex whereas, strictly speaking, only some twenty different receptive fields were identified in NIA. This could be due to limitations of the methods used, but may also suggest that some convergence exists between microzones with non-identical albeit similar receptive fields.

The main structural differences between the cerebellar micro-organisation described here and the uni-zonal olivo-cortico-nuclear 'micro-complexes' discussed by Ito (1984; cf. Oscarsson 1979)

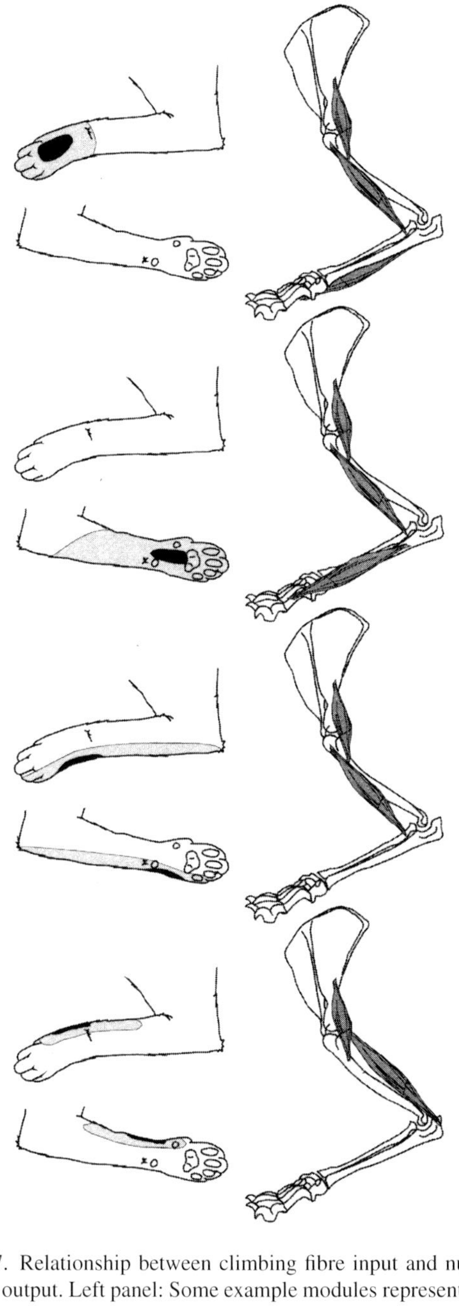

Fig. 7. Relationship between climbing fibre input and nucleofugal output. Left panel: Some example modules represented by their cutaneous receptive fields. Conventions as in Fig. 2. Right panel: The group(s) of muscles activated upon stimulation of the associated sites in NIA shown schematically (shoulder flexors; elbow flexors/extensors; wrist dorsiflexors/palmar flexors).

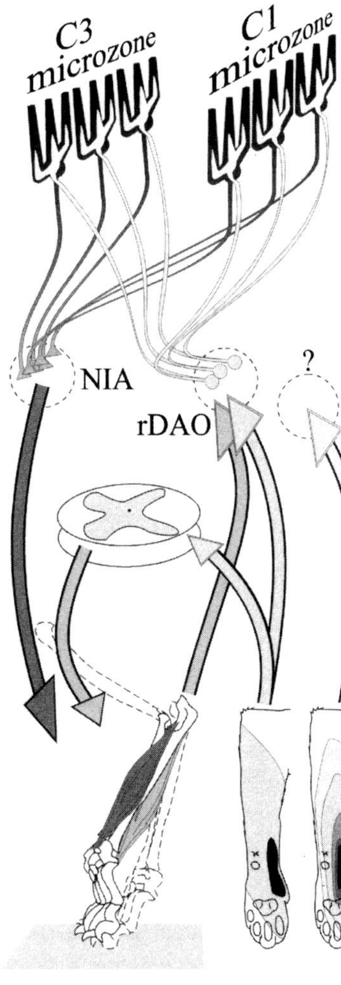

Fig. 8. Functional organisation of cerebellar modules. Upper part: Schematically, patterns of divergence and convergence within the olivo-cortico-nuclear connections. Lower part: input–output characteristics of a spinal nociceptive withdrawal reflex (short arrows, light grey muscle; cf. Fig. 4) and its relationship to the cerebellar module (long arrows to rDAO). Mossy fibre input far right (relay not identified, '?'), nucleofugal output far left (dark grey muscle). Proximal muscles not shown for simplicity (cf. Figs. 5 and 7). rDAO: rostral dorsal accessory olive; NIA: nucleus interpositus anterior.

is the distributed nature of the modules. Due to the branching of olivary axons (Ekerot and Larson, 1982), identical microzones must exist in different zones and recent data appear to confirm that these microzones in turn converge onto the same group of nuclear neurones (Apps and Garwicz, 2000). One consequence of such a distributed organisation is an increase in the number of neighbouring microzones of each module. This structural feature could contribute to multiplying 'lateral' interactions between modules along the longitudinal axes of the folia, presumably mediated by the parallel fibres.

The receptive field of our example module, subclass '5b' (Ekerot et al., 1991a) extends on the ventral forearm and paw, with a centre overlying the wrist close to the radial edge of the limb (Fig. 8). The '5b' microzones are located predominantly in rostral parts of the forelimb area of the C3 zone (lobule Va; Fig. 3, upper) and, as can be inferred from branching patterns of olivary axons, in lobule Vb/c of the C1 zone. These microzones receive common climbing fibre input from olivary neurones located in relatively caudal parts of the contra-lateral rDAO (Fig. 3, lower) and project in turn to the medio-ventral portion of the forelimb part of NIA (Fig. 3, middle; see Garwicz et al., 1996; Apps and Garwicz, 2000).

- *The system operates in a movement based frame of reference, such that the climbing fibre receptive field defining a module corresponds to the skin area withdrawn (from an external reference point) upon contraction of a specific muscle when the limb is in the standing position. The same muscle is the most prominent, albeit not the unique source of group II afferent input converging with the cutaneous input onto the olivary cells of the module.*

The concept that each cutaneous receptive field subclass is related to the action of a single muscle accommodates most key features of the climbing fibre input and is compatible also with the approximate number of muscles in the cat forelimb. The variability within receptive field subclasses (e.g. Fig. 4 in Ekerot et al., 1991a) is most likely due to sampling from different animals and/or fluctuations of excitability in afferent pathways. As previously outlined in Fig. 4, the configuration of the subclass 5b receptive field (Fig. 8) appears to correspond approximately to the skin area withdrawn by the contraction of m. flexor carpi radialis when the limb is in the standing position.

The straightforward relationship between receptive fields and muscle action in the standing position should not detract attention from the importance of the rDAO-C1-C3-Y-NIA system in controlling 'off the ground' limb movements, such as reach-to-grasp (e.g. Thach et al., 1992; Kitazawa et al., 1993; van Kan et al., 1993a). The particularities of receptive field architecture probably reflect the phylogenetic origin of this control system and do not appear to be limiting with respect to function.

The most prominent group II afferent input converging with the cutaneous input onto the module '5b' olivary neurones stems from m. flexor carpi radialis (Fig. 8; see Garwicz et al., 2000). In addition, there is muscle afferent input from other palmar flexor muscles and from flexors at the elbow (Fig. 5; not shown in Fig. 8). The significance of convergence between cutaneous and muscle afferent input to climbing fibres remains obscure. We have previously proposed that the muscle afferent input may constitute one of the signals contributing to the gating of cutaneous input during movement (Jörntell et al., 1996; Apps et al., 1997; see also Horn et al., 1996). The re-interpretation of the functional relationship between cutaneous and muscle afferent input with respect to limb position 'on the ground' and 'off the ground' further specifies the conditions for interaction between the two types of input.

- *Each module receives input from mossy fibres with a variety of individual receptive fields, but the area of skin most prominently represented by this input resembles the climbing fibre receptive field defining the module. On the efferent side, the module controls a small group of muscles via the rubrospinal tract. As a group, these muscles usually act on more than one limb segment and typically include an antagonist of the muscle(s) providing input to the olivary neurones of the module.*

Since most receptive fields of individual mossy fibres fit the climbing fibre classification scheme, it is likely that also the mossy fibre receptive fields reflect in some sense single muscle actions. However, by contrast to the climbing fibres, the mossy fibre input does not appear to have 'microzonal' patterns of termination within the C3 zone. It is conceivable that as a result of the overlapping terminal fields of any individual mossy fibre receptive field the mossy fibre input as a whole, although topographically aligned with the modular organisation of climbing fibre input, provides a rather continuous representation of the skin surface.

Interestingly, the more extensive mossy fibre representation of the skin surface in any given module appears at least in principle to match more closely the multi-joint nature of the nucleo-fugal output from the modules.

Few behavioural studies have been concerned with topographical aspects of the single unit output from NIA and it is therefore difficult to make comparisons with regard to modular organisation as suggested by the present data. However, the striking observation that in the context of reaching movements NIA neurones discharge only when the reach is coupled to a grasp (reviewed by Gibson et al., 1996) appears to be compatible with the multi-joint nature of the nucleo-fugal output. The '5b' module, for example, controls dorsiflexors at the wrist and in addition flexors at the shoulder and elbow (cf. Fig. 7, not shown in Fig. 8). A divergent output acting on both proximal and distal limb segments could constitute a neuronal substrate contributing to an appropriately timed coupling between the deceleration of the reach and the initiation of the grasp. The predominance of output to flexor muscles at proximal limb segments ties in with this idea.

At present, one of the most important limiting factors in understanding the information processing in the rDAO-C1-C3-Y-NIA modules is the lack of data concerning excitatory input to NIA (see also Dietrichs and Walberg, 1987). Until the receptive field characteristics and topographical organisation of mossy fibre collaterals have been described in some detail, the description of functional connectivity and hence the mode of operation of the modules will remain incomplete.

Acknowledgements

The experiments reviewed here were carried out in collaboration with friends and colleagues in Lund and Bristol, as listed in the References. Projects carried out at the laboratory in Lund were supported by the Swedish Medical Research Council, the Medical Faculty of Lund University, the Swedish Society of Medicine, the Swedish Society for Medical Research, the Royal Physiographic Society in Lund, Magn. Bergvalls Stiftelse, Dr.P. Håkanssons Stiftelse, Greta och Johan Kocks Stiftelser, Segerfalks Stiftelse and Crafoordska Stiftelsen.

References

Andersson, G. and Oscarsson, O. (1978) Climbing fibre microzones in cerebellar vermis and their projection to different groups of cells in the lateral vestibular nucleus. *Exp. Brain Res.*, 32: 565–579.

Apps, R. and Garwicz, M. (2000) Precise matching of olivo-cortical divergence and cortico-nuclear convergence between somatotopically corresponding areas in the medial C1 and medial C3 zones of the paravermal cerebellum. *Eur. J. Neurosci.*, 12: 205–214.

Apps, R., Atkins, M.J. and Garwicz, M. (1997) Gating of cutaneous input to cerebellar climbing fibres during a reaching task in the cat. *J. Physiol.*, 502.1: 203–214.

Armstrong, D.M. and Edgley, S.A. (1984a) Discharges of nucleus interpositus neurones during locomotion in the cat.J. Physiol., 351: 411–432.

Armstrong, D.M. and Edgley, S.A. (1984b) Discharges of Purkinje cells in the paravermal part of the cerebellar anterior lobe during locomotion in the cat. *J. Physiol.*, 352: 403–424.

Cooke, J.D., Larson, B., Oscarsson, O. and Sjölund, B. (1971) Origin and termination of cuneocerebellar tract. *Exp. Brain Res.*, 13: 339–358.

Dietrichs, E. and Walberg, F. (1987) Cerebellar nuclear afferents – where do they originate? A re-evaluation of the projections from some lower brain stem nuclei. *Anat. Embryol.*, 177: 165–172.

Edgley, S.A. and Lidierth, M. (1988) Step-related discharges of Purkinje cells in the paravermal cortex of the cerebellar anterior lobe in the cat. *J. Physiol.*, 401: 399–415.

Ekerot, C.-F. and Larson, B. (1979) The dorsal spino-olivocerebellar system in the cat. II. Somatotopical organization. *Exp. Brain Res.*, 36: 219–232.

Ekerot, C.-F. and Larson, B. (1980) Termination in overlapping sagittal zones in cerebellar anterior lobe of mossy and climbing fibre paths activated from dorsal funiculus. *Exp. Brain Res.*, 38: 163–172.

Ekerot, C.-F. and Larson, B. (1982) Branching of olivary axons to innervate pairs of sagittal zones in the cerebellar anterior lobe of the cat. *Exp. Brain Res.*, 48: 185–198.

Ekerot, C.-F., Oscarsson, O. and Schouenborg, J. (1987) Stimulation of cat cutaneous nociceptive C fibres causing tonic and synchronous activity in climbing fibres. *J. Physiol.*, 386: 539–546.

Ekerot, C.-F., Garwicz, M. and Schouenborg, J. (1991a) Topography and nociceptive receptive fields of climbing fibres projecting to the cerebellar anterior lobe in the cat. *J. Physiol.*, 441: 257–274.

Ekerot, C.-F., Garwicz, M. and Schouenborg, J. (1991b) The postsynaptic dorsal column pathway mediates cutaneous nociceptive information to cerebellar climbing fibres in the cat. *J. Physiol.*, 441: 275–284.

Ekerot, C.-F., Jörntell, H. and Garwicz, M. (1995) Functional relation between corticonuclear input and movements evoked on microstimulation in cerebellar nucleus interpositus anterior in the cat. *Exp. Brain Res.*, 106: 365–376.

Garwicz, M. (1992) Cerebellar control of forelimb movements: modular organization revealed by nociceptive and tactile climbing fibre input. Doctoral Thesis, Lund University.

Garwicz, M. and Ekerot, C.-F. (1994) Topographical organization of cerebellar cortical projection to nucleus interpositus anterior in the cat. *J. Physiol.*, 474: 245–260.

Garwicz, M., Ekerot, C.-F. and Schouenborg, J. (1992) Distribution of cutaneous nociceptive and tactile climbing fibre input to sagittal zones in cat cerebellar anterior lobe. *Eur. J. Neurosci.*, 4: 289–295.

Garwicz, M., Apps, R. and Trott, J.R. (1996) Micro-organization of olivocerebellar and corticonuclear connections of the paravermal cerebellum in the cat. *Eur. J. Neurosci.*, 8: 2726–2738.

Garwicz, M., Jörntell, H. and Ekerot, C.-F. (1998) Cutaneous receptive fields and topography of mossy fibres and climbing fibres projecting to cat cerebellar C3 zone. *J. Physiol.*, 512.1: 277–293.

Garwicz, M., Levinsson, A. and Schouenborg, J. (1999) Spinal nociceptive withdrawal reflexes and the olivo-cerebellar climbing fibre system – common organizational features suggest a functional link. (Abstract) *Acta Physiol. Scand.*, 165: 4:A1..

Garwicz, M., Levinsson, A. and Schouenborg, J. (2000) Kinetotopic organisation of input to cerebellar climbing fibres – A novel principle of sensory encoding (manuscript in preparation)

Gibson, A.R., Horn, K.M., Stein, J.F. and van Kan, P.L. (1996) Activity of interpositus neurons during a visually guided reach. *Can. J. Physiol. Pharmacol.*, 74(4): 499–512.

Groenewegen, H.J., Voogd, J. and Freedman, S.L. (1979) The parasagittal zonation within the olivocerebellar projection. II. Climbing fibre distribution in the intermediate and hemispheric parts of cat cerebellum. *J. Comp. Neurol.*, 183: 551–602.

Horn, K.M., van Kan, P.L.E. and Gibson, A.R. (1996) Reduction of rostral dorsal accessory olive responses during reaching. *J. Neurophysiol.*, 76: 4140–4151.

Ito, M. (1984) *The Cerebellum and Neural Control*, Raven Press, New York.

Ito, M. (1989) Long-term depression. *Ann. Rev. Neurosci.*, 12: 85–102.

Jörntell, H., Garwicz, M. and Ekerot, C.-F. (1996) Relation between cutaneous climbing fibre receptive fields and muscle afferent input to climbing fibres projecting to the cerebellar C3 zone in the cat. *Eur. J. Neurosci.*, 8: 1769–1779.

Kitazawa S., Goto T. and Urushihara Y. (1993) Quantitative evaluation of reaching movements in cats with and without cerebellar lesions using normalized integral of jerk. In: N. Mano, I. Hamada and M.R. DeLong (Eds), *Role of the Cerebellum and Basal Ganglia in Voluntary Movement*, Elsevier, pp. 11–19.

Levinsson, A., Garwicz, M. and Schouenborg, J. (1999) Sensorimotor transformation in cat nociceptive withdrawal reflex system. *Eur. J. Neurosci.*, 11: 4327–4332.

Oscarsson, O. (1973) Functional organization of spinocerebellar paths. In: A. Iggo (Ed.), *Handbook of Sensory Physiology, Vol. II: Somatosensory System*, Springer Verlag, New York, pp. 339–380.

Oscarsson, O. (1979) Functional units of the cerebellum–sagittal zones and microzones. *TINS*, 2/6: 143–145.

Oscarsson, O. (1980) Functional organization of olivary projection to cerebellar anterior lobe. In: J. Courville, C. de Montigny and Y. Lamarre (Eds), *The Inferior Olivary Nucleus: Anatomy and Physiology*, Raven Press, New York, pp. 279–289.

Schouenborg, J. and Weng, H.-R. (1994) Sensorimotor transformation in a spinal motor system. *Exp. Brain Res.*, 100: 170–174.

Schouenborg, J., Weng, H.-R. and Holmberg, H. (1994) Modular organization of spinal nociceptive reflexes. *NIPS*, 9: 261–265.

Simpson, J.I., Wylie, D.R. and De Zeeuw, C.I. (1996) On climbing fibre signals and their consequence(s). *Behav. Brain Sci.*, 19: 384–398.

Thach, W.T., Goodkin, H.P. and Keating, J.G. (1992) The cerebellum and adaptive coordination of movement. *Ann. Rev. Neurosci.*, 15: 403–442.

Trott, J.R. and Apps, R. (1991) Lateral and medial subdivisions within the olivocerebellar zones of the paravermal cortex in lobule Vb/c of the cat anterior lobe. *Exp. Brain Res.*, 87: 126–140.

Trott, J.R. and Armstrong, D.M. (1987) The cerebellar corticonuclear projection from lobule Vb/c of the cat anterior lobe: a combined electrophysiological and autoradiographic study.I. Projections from the intermediate region. *Exp. Brain Res.*, 66: 318–338.

van Kan, P.L.E., Houk, J.C. and Gibson, A.R. (1993a) Output organization of intermediate cerebellum of the monkey. *J. Neurophysiol.*, 69/1: 57–73.

van Kan, P.L.E., Gibson, A.R. and Houk, J.C. (1993b) Movement-related inputs to intermediate cerebellum of the monkey. *J. Neurophysiol.*, 69: 74–94.

Voogd, J. (1982) The olivocerebellar projection in the cat. In: S.L. Palay and V. Chan-Palay (Eds), *The Cerebellum, New Vistas (Exp. Brain Res. Suppl. 6)*, Springer, Berlin, Heidelberg, New York, pp. 134–161.

Voogd, J. and Bigaré, F. (1980) Topographical distribution of olivary and cortico nuclear fibers in the cerebellum: a review. In: J. Courville, C. de Montigny and Y. Lamarre (Eds), *The Inferior Olivary Nucleus: Anatomy and Physiology*, Raven Press, New York, pp. 207–234.

N.M. Gerrits, T.J.H. Ruigrok and C.I. De Zeeuw (Eds.)
Progress in Brain Research, Vol 124

CHAPTER 14

Gating of climbing fibre input to cerebellar cortical zones

Richard Apps*

Department of Physiology, School of Medical Sciences, University of Bristol, Bristol BS8 1TD UK

Introduction

Cerebellar zones

Since the pioneering anatomical investigations of Professor Jan Voogd (see for example Voogd, 1964, 1969) there has been a wealth of studies providing evidence in support of the existence of cerebellar cortical zones, including many that document clear differences in their anatomical connections (for reviews see Brodal and Kawamura, 1980; Armstrong, 1990; Buisseret-Delmas and Angaut, 1993; Voogd and Glickstein, 1998). Indeed, the number of chapters in this volume that refer to zones bears witness to the far reaching impact of his original findings.

It is now generally accepted that at least eight 'Voogd' zones are present on either side of the cerebellar midline, each about 1–2 mm in width but extending across the transverse lobules for many millimeters in the rostrocaudal axis. They are denoted from medial to lateral as the A, X and B zones in the vermis, the C1, C2 and C3 zones in the paravermis and the D1 and D2 zones in the hemisphere. Each is defined by its climbing fibre input from a specific part of the inferior olive and by the projection of its Purkinje cells to a particular region within the deep cerebellar or vestibular nuclei, thereby forming a series of olivo-cortico-nuclear compartments or 'modules'.

Such a scheme is likely, however, to be an oversimplification because physiological studies have revealed further divisions of individual cortical zones. In particular, various ascending spinal pathways have been described that relay in the inferior olive before terminating as climbing fibres in the cortical zones (for reviews see Oscarsson, 1980; Ito, 1984; Garwicz et al. 1998). These spino-olivocerebellar paths (SOCPs) convey information mainly from limb cutaneous, joint and muscle afferents and detailed electrophysiological mapping experiments have shown that subpaths within the SOCPs relay climbing fibre input with similar receptive field characteristics to subzones or microzones located within the broader Voogd zones (Andersson and Oscarsson, 1978; Ekerot and Larson, 1979b, 1982; Ekerot et al., 1991). The paravermal C3 zone, for example, contains 30 or more such microzones (Ekerot et al., 1991) and detailed anatomical tracing studies have revealed a corresponding micro-organization within its climbing fibre input from the inferior olive (Garwicz et al., 1996; see also Apps, 1990; Atkins and Apps, 1997; Apps and Garwicz, 2000). The cerebellum is thought therefore to be composed of a large number of olivo-corticonuclear 'microcomplexes', each (or in some cases groups of anatomically linked complexes) serving a distinct function in motor control (cf. Oscarsson, 1979; Ito, 1984).

Studies in the awake animal

The cortical zones (and microzones) are considered to be a fundamental organizational feature of the cerebellum, yet much less is known of any functional differences that may exist between them, particularly in the awake behaving animal. Fur-

*Corresponding author. Tel.: +44(0)117 928 7803; Fax: +44(0)117 928 8923; e-mail: r.apps@bristol.ac.uk

thermore, the functional significance of the climbing fibres remains speculative (see Simpson et al., 1996 for a review of various theories). Nevertheless, one general finding is that when an animal is sitting quietly at rest, the complex spike discharges (generated as a result of activity in the climbing fibre afferents) of individual Purkinje cells located in a number of different cortical zones are usually readily evoked by light taps or brushes of their peripheral receptive fields.

For the majority of Purkinje cells located within the paravermal zones, receptive fields in the awake animal have been found to include the distal part of the limbs (e.g. Armstrong et al., 1988). However, during active movements such as locomotion it has been found that the same cells (irrespective of their zonal location) discharge complex spikes essentially randomly with respect to events of the step cycle, despite the fact that the same limb afferents are likely to be activated during stepping (Armstrong et al., 1982, 1988; Andersson and Armstrong, 1987; see also Smith, 1998). Similar negative findings have been obtained for other voluntary movements, including studies of olive cell activity during performance of a reaching task (Horn et al., 1996; see also Gellman et al., 1985).

Central gating of climbing fibre inputs to the cerebellum

The single unit recordings in awake animals suggest that tactile inputs relayed via the SOCPs are very effective at generating complex spike discharges in a number of cortical zones in the passive animal, yet very similar inputs are rather ineffective at generating complex spikes in the active animal. A central gating mechanism has therefore been proposed which is thought to operate in a context-dependent fashion to regulate access to the cerebellum of climbing fibre inputs (cf. Gellman et al., 1985; for a review Apps, 1999).

Such a regulatory mechanism is unlikely, however, to always gate-out climbing fibre signals during active movements. In the same studies outlined above it was also found that, if the voluntary movement was subjected to an unexpected external perturbation, the discharges of olive cells or the complex spikes they generate in the cortical Purkinje cells could increase their probability of occurrence (Gellman et al., 1985; Andersson and Armstrong, 1987; see also Lou and Bloedel, 1992). The postulated gating mechanism may serve therefore at least two functions. On the one hand, to gate-out during active movements 'predictable' self-generated climbing fibre inputs while on the other, to gate-in sensory 'errors' generated as a result of unexpected external perturbations (cf. the error detector hypothesis originally proposed by Miller and Oscarsson, 1970).

In relation to the functional significance of the cortical zones it is also of interest to note that complex spike discharges of Purkinje cells located in the vermal B zone have been found, during locomotion in the awake cat, to increase their probability of occurrence when perturbations were imposed during the *stance* phase of the step cycle (Andersson and Armstrong, 1987). By comparison, similar studies in the decerebrate ferret have found that complex spike activity in the paravermis (most probably in the C1 zone) could occur with increased probability in response to perturbations during the *swing* phase of the step cycle (Lou and Bloedel, 1992). Thus different zones may process climbing fibre sensory 'error' signals arising at different times during an active movement, raising the possibility that one or more gating mechanism(s) act differentially on the SOCPs that target individual zones so that each is 'permitted' climbing fibre input only at specific times during a movement.

The aim of the present chapter is to describe a series of studies in which this gating phenomenon has been studied in detail and the possibility explored that climbing fibre pathways that target different cerebellar cortical zones (the C1, C2 and C3 zones in the paravermis) are gated-in at different times during voluntary movements. Clearly, any such regulation will have important implications as to the function of the climbing fibres since changes in transmission will set limits to when information is forwarded to the different cortical zones.

Experimental arrangements

The series of studies under consideration have all employed a similar methodology, whereby low intensity (non-noxious) electrical stimulation of a forelimb cutaneous nerve (the superficial radial, SR) has been used as a probe to monitor gating of transmission in climbing fibre pathways in the awake behaving animal. Since the techniques have been fully described in previous papers (see for example, Apps, 1990; Apps et al.,1997) they are outlined here only in brief.

Two model movements in intact cats have been used, namely walking on a moving belt and performance of a voluntary reaching task. Each animal was trained to perform one or other behavioural task and then, at an aseptic operation under pentobarbitone general anaesthesia, received a series of implants. These included an array of about 20 platinum-iridium recording microwires inserted to a depth of 1–2 mm within the different paravermal cortical zones; two pairs of bipolar cuff electrodes around the ipsilateral SR nerve for stimulation and nerve volley recording respectively, and in locomotor studies, a pair of EMG recording leads inserted into the ipsilateral triceps brachii muscle (see Apps et al., 1990, 1995 for further details). In reaching experiments, a unipolar 'contact' lead was also implanted into the ipsilateral forelimb to monitor paw lift-off (see Apps et al., 1997 for further details).

Delivery of electrical stimuli to the peripheral nerve

Following recovery from the operation, single pulse electrical stimuli were applied to the ipsilateral SR nerve both during task performance and during periods of rest. The stimulation set up volleys in the ascending SOCPs which were recorded as evoked climbing fibre field potentials at the different microwire recording sites in the cerebellar cortex (see Fig. 1). The size of these fields was then used as a measure of changes in transmission in the different climbing fibre pathways during rest as compared to different phases of the movement. Since the stimulation also evoked activity in ascending mossy fibre pathways a number of criteria were used to identify responses attributable to activity in the climbing fibres, including onset latency, waveform, and pattern of response to a 'paired pulse' test (cf, Armstrong and Harvey, 1966; Eccles et al., 1966).

During each recording session the timing of the peripheral nerve stimulus was 'pseudo-randomized' so that climbing fibre responses were evoked as a result of stimuli distributed evenly throughout the step cycle or the reaching task. To aid analysis further the two movements were also divided into a number of epochs. For locomotion, the step cycle was arbitrarily divided into tenths relative to onset of EMG activity in the triceps brachii muscle in the ipsilateral forelimb (which corresponds approximately to the onset of the stance phase of the step cycle) while for reaching, the task was divided into three epochs: 'reaction', when the animal responded to a cue to perform the task; 'reach', when the animal lifted its limb from the ground towards the target, and 'grasp', when the animal's

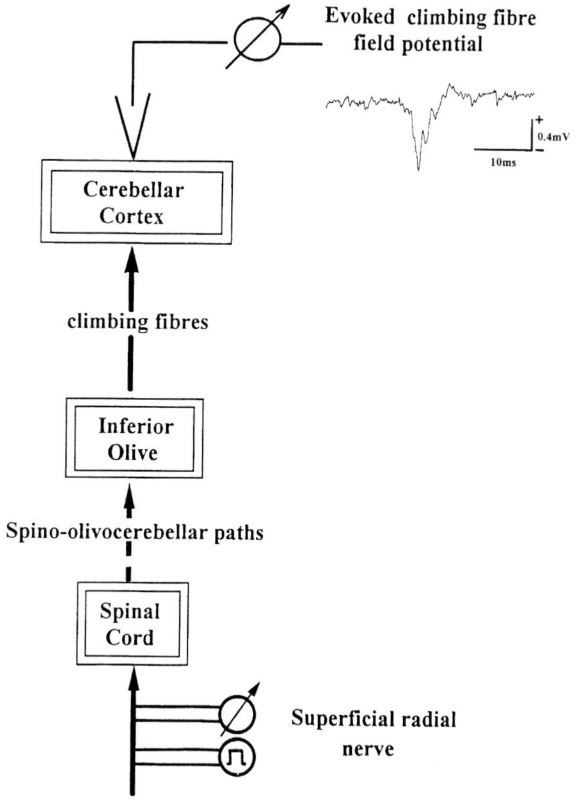

Fig. 1. Schematic diagram to show the experimental arrangements used in the awake cat. Top right, example cerebellar climbing fibre response recorded in the C2 zone evoked by stimulation of the ipsilateral superficial radial nerve and generated as a result of activity in the spino-olivocerebellar paths.

paw reached the target to retrieve a morsel of food. Climbing fibre responses evoked by stimuli delivered in each epoch were separately pooled and averaged to generate histograms showing fluctuations in mean size of response during the course of the movement (e.g. Figs. 3 and 4).

Gating of transmission in SOCPs during reaching

Figure 2 shows examples of averaged climbing fibre responses evoked at a C2 zone recording site during different phases of the reaching movement. Clearly, the size of the response was reduced during the task as compared to rest, and most notably during the grasp phase. In fact, single climbing fibre responses during this phase of the movement

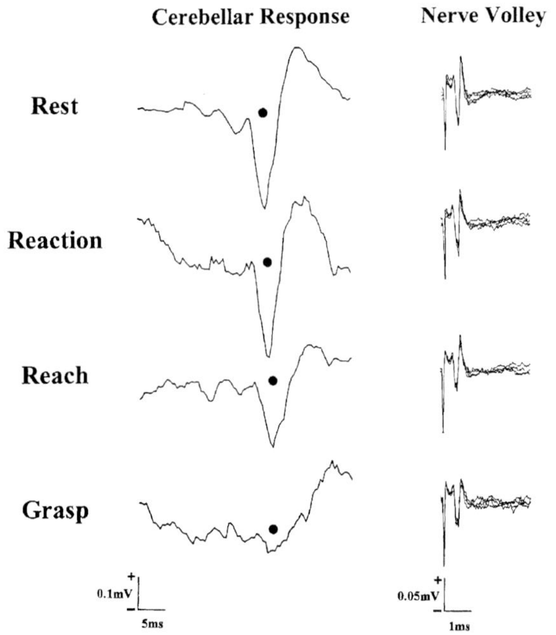

Fig. 2. Example responses from a C2 zone recording site (anterior lobe, lobule V) during performance of a reach to grasp movement (site RW304). From top to bottom, climbing fibre responses (indicated by a filled circle) recorded during rest as compared to the three phases of the task (each trace average of $n=49$, 7, 14 and 6 sweeps, respectively). To the right is shown superimposed sweeps of the compound action potential in the ipsilateral superficial radial nerve ($n=4$ example traces in each case) (modified from Apps et al., 1997).

were often reduced below the level of detection, despite the relative constancy of the nerve volley as shown to the right. Figure 3A displays data in histogram format from a different C2 zone site recorded in another animal to emphasize that the typical finding was a progressive reduction in mean response size during the course of the movement.

Similar findings were also obtained for neighbouring C1 and C3 zone sites (see for example, Fig. 3B) and the results from all three zones have been pooled in the frequency histogram of Figure 3C. This displays for each recording site the mean size of response in the grasp phase expressed as a percentage of rest and shows that by far the most common finding was a reduction in response size, implying that transmission in the relevant SOCPs was reduced, or even in some cases, entirely blocked at this time.

Because the peripheral nerve was monitored in each case and any variation in its size was found not to be correlated to the variations in size of the cerebellar responses (see Apps et al., 1997 for further details), it can be concluded that the latter was attributable to modulatory mechanisms operating centrally rather than any variation in the afferent input. Such data are therefore consistent with the presence of a gating mechanism which operates to reduce pathway transmission, especially during the grasp phase of the movement. Note however that the gating-out of inputs was not without exceptions and that at some sites, response size during the grasp phase was greater than rest, implying a degree of heterogeneity between recording loci (see below).

Gating of transmission in SOCPs during locomotion (anterior lobe)

While the pattern of gating in SOCPs targeting the paravermal zones was rather similar during reaching, by comparison, studies of gating during locomotion revealed striking differences between the C1–C3 zones and the intervening C2 zone in the anterior lobe (lobule V). At most paravermal recording sites pronounced step-related fluctuations in response size were present with the phasing being similar at C1 and C3 zone sites. In such cases the largest and smallest responses (i.e. best and worst pathway transmission) consistently occurred respectively during the swing and stance phases of the step cycle in the ipsilateral forelimb (Lidierth and Apps, 1990; Apps et al., 1995, see Fig. 4A for an example step histogram and the summary frequency distributions of Fig. 5A and B). By

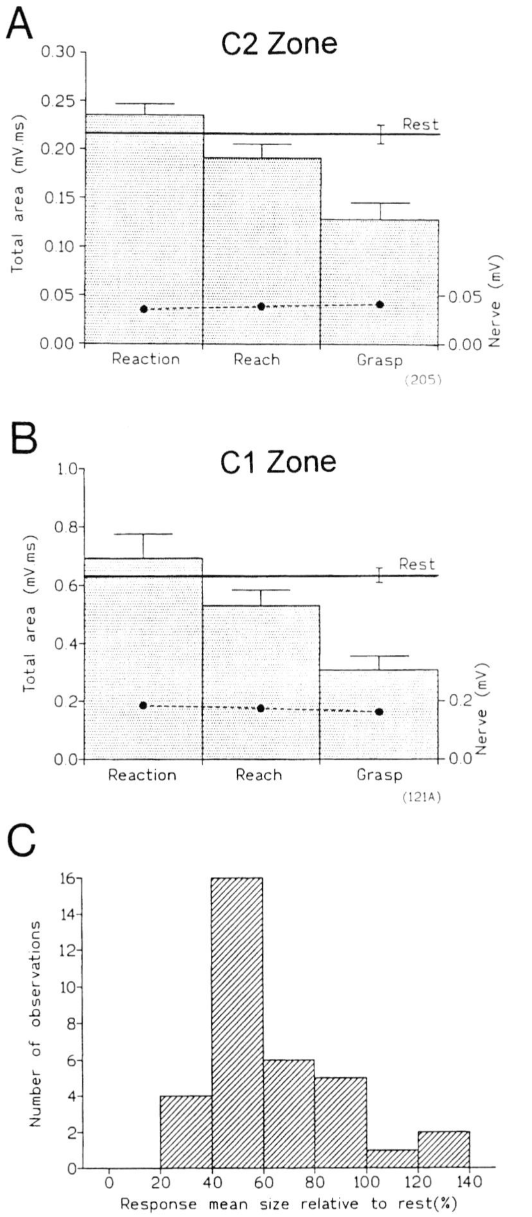

Fig. 3. (A) Example histogram from a C2 zone recording site (RW205) to show climbing fibre response mean size for the three phases of the reaching task (+S.E.M). Horizontal line shows mean size at rest. Dashed line shows mean size of nerve volley for each phase of the task. (B) Same as A but for an example C1 zone recording site. (C) Frequency histogram to show for all anterior lobe paravermal recording sites, response mean size during the grasp phase of the task expressed as a percentage of rest ($n=34$ recording sessions obtained from a total of 16 C1, C2 and C3 zone sites) (modified from Apps et al., 1997).

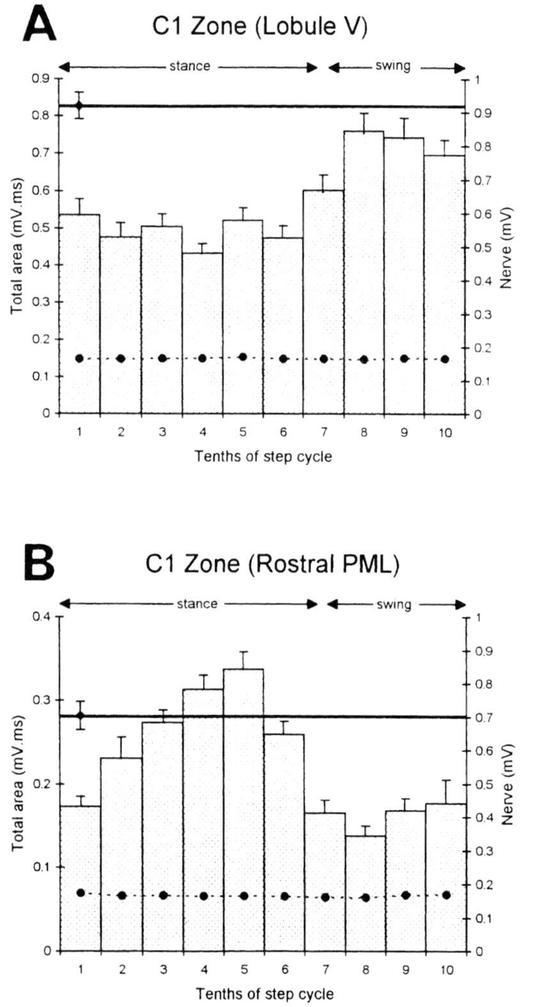

Fig. 4. Step analysis for two C1 zone recording sites. In each histogram the step cycle has been divided into tenths relative to onset of locomotor EMG in the ipsilateral triceps brachii muscle and plots the mean size of climbing fibre response for each step tenth (+S.E.M.). Horizontal line, mean size of response at rest, dashed line, mean size of the nerve volley. (A) Step histogram for a C1 zone site in the anterior lobe (lobule V). (B) Step histogram for a C1 zone site in rostral paramedian lobule recorded simultaneously with the site shown in A.

marked contrast, at C2 zone sites timing of best transmission, although more variable between sites, nevertheless occurred most frequently during the stance phase of the step cycle (Fig. 5C), with time of worst transmission occurring either during early stance or swing (Fig. 5D).

As in the reaching task the peripheral nerve volley was monitored during locomotion and any variation in its size was found not to be correlated to the step-phase dependent variations in size of the cerebellar responses (for further details see Apps et al., 1995). The SOCPs targeting the C2 zone as compared to the neighbouring C1/C3 zones are subject therefore to very distinct temporal patterns of central gating during locomotion, implying a marked functional difference between zones in the awake behaving animal.

Functional implications of the step-related gating

For C2 zone recording sites it is difficult to gauge the behavioural significance of the step-related gating because the changes in SOCP pathway transmission are likely to relate to a complex integration of inputs from all four limbs (cf. Larson et al., 1969). By comparison, the C1 and C3 zones receive climbing fibre input exclusively from the ipsilateral limbs (see for example Ekerot and Larson, 1979a) and are presumably concerned with its control alone. For these sites pathway transmis-

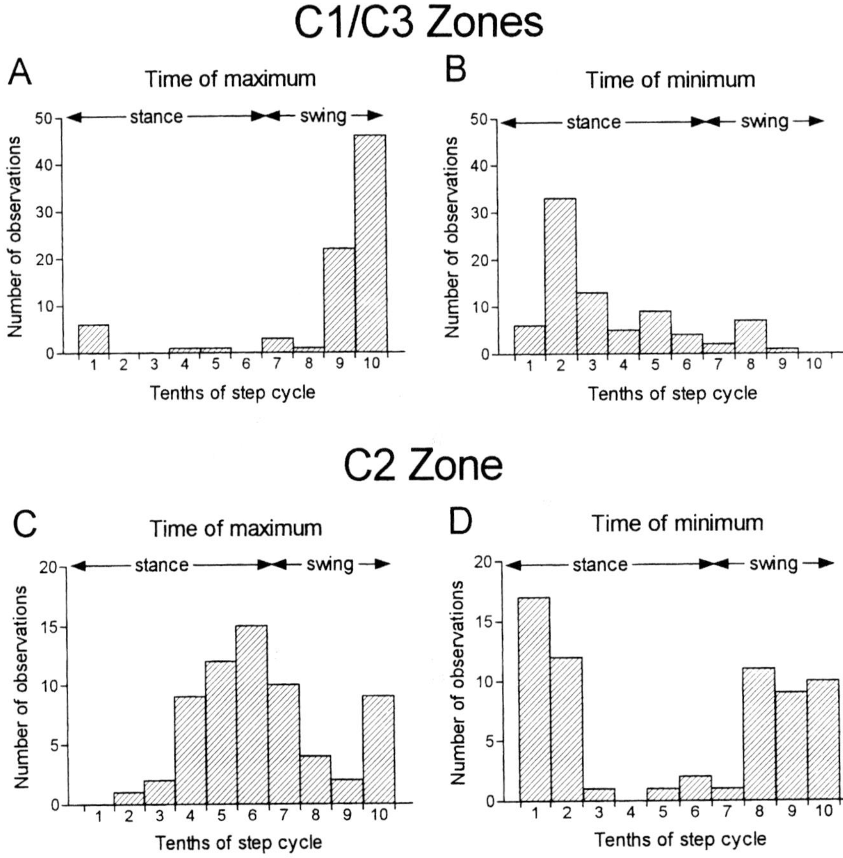

Fig. 5. Frequency distributions for times during the step cycle in the ipsilateral forelimb when the largest (A and C) and smallest (B and D) climbing fibre responses at paravermal recording sites (anterior lobe, lobule V) were evoked by ipsilateral superficial radial nerve stimulation. (A, B) Timings for $n = 80$ recording sessions obtained from a total of $n = 28$ different C1 and C3 zone recording sites. (C, D) Timings for $n = 64$ recording sessions obtained from a total of $n = 25$ different C2 zone sites (modified from Apps, 1999).

sion was usually gated-out during the stance phase of the step cycle (Fig. 5B) which is a time when the greatest self-generated input from the paw might be expected to occur. The results are therefore consistent with the proposal that one possible function of the gating is to prevent 'predictable' inputs from being forwarded to the cerebellum.

Gating-in of pathway transmission during locomotion

The gating mechanism(s) did not, however, always reduce transmission of cutaneous inputs during locomotion. At most sites it was possible to compare the size of the responses evoked during walking with those evoked by similar stimuli delivered while the animal was sitting quietly at rest and such a comparison revealed a further difference between zones. At many C1 sites the largest responses during locomotion were often considerably larger than rest and similar findings were obtained for the more limited data available for the C3 zone (Fig. 6A). By contrast, at C2 zone sites the largest responses during locomotion were usually similar in size to rest (Fig. 6B). Thus climbing fibre input to the C1 and C3 zones can be facilitated during the swing phase of the step cycle while at C2 zones this is only rarely the case. The results therefore provide evidence that the gating also operates to enhance transmission of climbing fibre inputs to selected zones during certain types of active behaviour (see Apps et al., 1995 for further details).

Such findings are compatible with (although of course do not prove) the 'error detector' hypothesis inasmuch as they suggest that sensory traffic in the ipsilateral SR nerve is most likely to be forwarded via the SOCPs to the C1 and C3 zones in the anterior lobe (lobule V) during the late swing phase of the step cycle, i.e. when the limb is flexed forward just prior to touch down. This is the time during the step cycle under natural conditions when the part of the limb supplied by the SR nerve (the paw dorsum and dorsal parts of the distal forelimb, see Kitchell et al., 1982) is most likely to make unexpected contact to obstacles to progression. The SOCPs targeting the C1 and C3 zones are therefore gated-in at the most appropriate time to signal external sensory 'errors' during locomotion.

Intrazonal differences in pattern of gating

A number of observations in the present experiments suggest a degree of functional heterogeneity between different parts of the same zones. For example, during reaching individual sites within a given zone varied considerably in the extent to which their responses were reduced relative to rest (Fig. 3C) while during locomotion, individual C2 zone recording sites were found to display some-

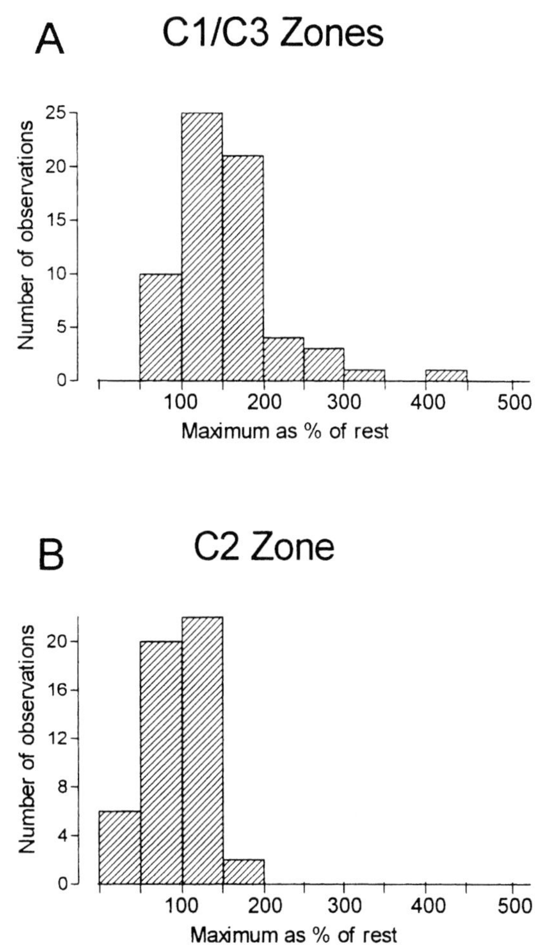

Fig. 6. Frequency distributions of response size during locomotion relative to rest for all anterior lobe paravermal sites. For each recording session the step cycle tenth with the largest mean size of response is shown expressed as a percentage of mean size in the resting animal. (A) Data for C1 and C3 zone sites ($n = 72$ sessions obtained from a total of $n = 22$ different recording sites). (B) Data for C2 zone sites ($n = 51$ sessions obtained from a total of $n = 14$ different recording sites) (modified from Apps, 1999).

times very different patterns of step-related gating (Fig. 5C and D). Such differences suggest that different parts of individual paravermal zones are in receipt of climbing fibre input relayed by separate subpaths within the respective SOCPs (cf. Ekerot and Larson, 1979a, b; Ekerot et al., 1991), raising the possibility that sites at different rostrocaudal loci within a given zone may be subject to different patterns of gating during active movements.

Gating of SOCPs during locomotion (posterior lobe)

The findings so far described have been confined to recording sites located within the forelimb receiving area in the anterior lobe paravermis (lobule V). Such studies have now been extended (see Apps and Lee, 1999) to include investigation of step-related changes in size of climbing fibre responses evoked at 'corresponding' C1/C3 zone sites located in the posterior lobe paravermis (in the rostral paramedian lobule which is also a 'forelimb' receiving area of the cerebellar cortex).

Figure 4B displays an example step histogram obtained from a C1 zone site in rostral paramedian lobule (PML). Typical of the material as a whole, the largest mean size of response (i.e. best transmission) occurred during the stance phase in the ipsilateral forelimb while the smallest mean size of response occurred during swing. The step cycle phasing of the modulation was therefore strikingly different from the pattern for responses recorded under similar experimental conditions but obtained from C1/C3 zone sites in the anterior lobe (Figs 4A and 5A, B).

In fact, the step histogram in Fig. 4A is an example of a C1 zone site located in the anterior lobe (lobule V, see above) which was recorded simultaneously with the rostral PML C1 zone site illustrated in Fig. 4B. In such cases the responses at each site were evoked by the same peripheral nerve stimulation, thereby controlling for any differences in intensity of the stimulation used in different experiments (a factor which has been shown to have a significant influence on the pattern of modulation at individual sites cf. Apps et al., 1990, 1995). Overall, such a comparison between anterior and posterior lobe C1/C3 zone sites was possible in a total of 13 pairs of cases, and in each pair similar differences in gating pattern were found (for further details see Apps and Lee, 1999).

The rostal PML findings are summarized in the frequency histogram of Figure 7 which shows that the time of best (Fig. 7A) and worst (Fig. 7B) transmission at C1/C3 zone sites occurred, respectively, during the stance and swing phases of the step cycle in the ipsilateral forelimb (cf. Fig. 5A and B). Thus different parts within the same zones (possibly equating to different microzones) can display very different patterns of step-related gating, raising the possibility that they monitor

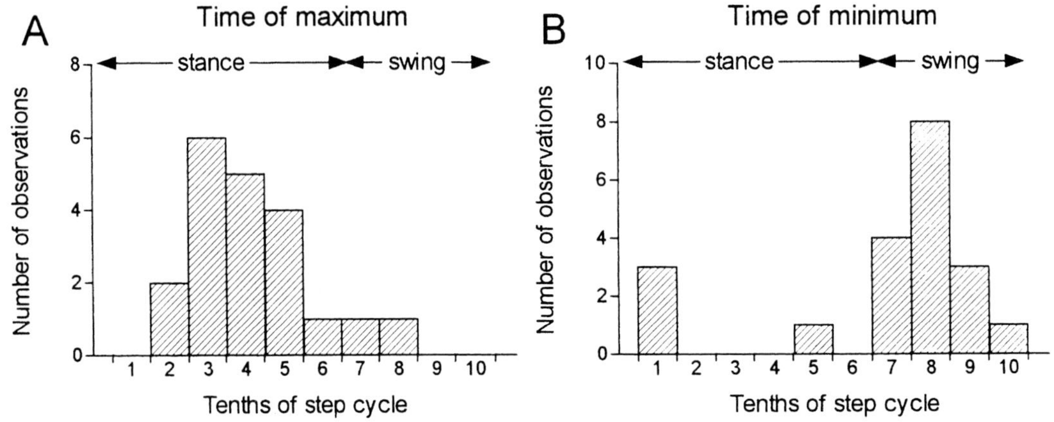

Fig. 7. Frequency distributions to show for all available C1 and C3 zone sites in the rostral paramedian lobule the timing during the step cycle when mean size of response to ipsilateral superficial radial stimulation was (A) largest and (B) smallest ($n = 20$ sessions obtained from a total of $n = 8$ different recording sites) (modified from Apps and Lee, 1999).

climbing fibre signals at different times during the step cycle.

Anatomical studies of intrazonal differences

The striking difference in temporal pattern of step-related gating for different rostrocaudal sites within the C1/C3 zones implies that anterior and posterior regions of the same zones receive at least partly independent climbing fibre inputs. Such a possibility receives support from anatomical and physiological studies which indicate that the C1 and C3 zones in the anterior lobe (lobule V) should in fact be considered as composed of medial and lateral subzones (Ekerot and Larson, 1982; Campbell and Armstrong, 1985; Trott and Apps, 1991).

In particular, combined electrophysiological mapping and neuroanatomical tracing experiments (Trott and Apps, 1991) have shown that injections of WGA–HRP into the medial half of the C1 zone in lobule V produce retrograde labelling in the rostral dorsal accessory olive (DAO), while injections of tracer into the lateral half of the zone result in retrograde labelling in mid/rostral parts of the medial accessory olive (MAO, Fig. 8A). By contrast, similar injections of WGA–HRP into the C1 zone in rostral PML only produce labelling in rostral DAO, suggesting that a zone equivalent to the lateral C1 zone does not exist within rostral PML (Fig. 8B, Trott and Apps, 1993).

Furthermore, although inspection of Figures 8A and B shows that the territories within rostral DAO that target both cortical regions overlap heavily, nevertheless, at more rostral levels of the olive, cells projecting to the C1 zone in lobule V are located progressively more lateral within DAO than the cells projecting to the corresponding zone in rostral PML. The olivocerebellar projection to the two 'corresponding' regions of the cerebellar cortex cannot therefore be considered as anatomically uniform.

The partial overlap in the two olivary territories in DAO implies that different parts of the same zone receive some common information (notably via olive cells whose axons branch to innervate both regions, cf. Armstrong et al., 1973a, b, c), while other information is targeted to only one of the two sites. Such a possibility is supported by the findings from double fluorescent retrograde tracing studies which have shown that about one third of the cells in rostral DAO have axons that branch to supply climbing fibres to paravermal lobule V and rostral PML (Rosina and Provini, 1983, 1987; Apps and Trott, 1991; Apps, 2000). Thus branching within the olivocerebellar projection (at least to apical folia of the cerebellar cortex) is likely to involve only a subgroup of the total population of cells that target the C1 and C3 zones in the anterior lobe (lobule V) and rostral PML, providing ample scope for independent lines of communication.

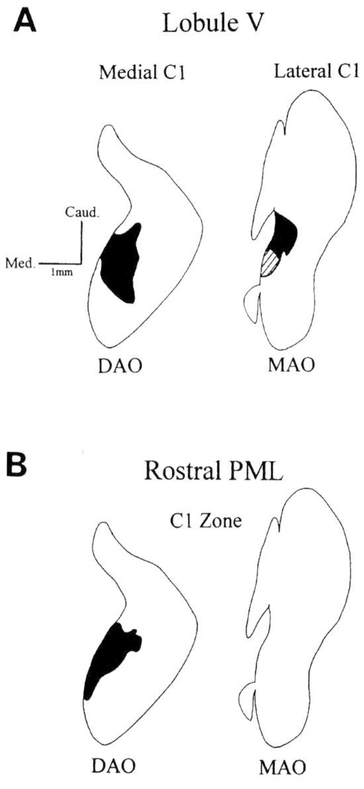

Fig. 8. Diagrams of the unfolded dorsal accessory olive (DAO) and medial accessory olive (DAO) to show the location of pooled retrograde labelling arising from injections of WGA-HRP into (A) the medial and lateral C1 zones in the anterior lobe, lobule V ($n = 6$ cases) and (B) the C1 zone in the rostral paramedian lobule ($n = 6$ cases). In each diagram the filled area represents the 'core' territory which contained labelled cells in at least 3 cases. The hatched area in A shows the core area within MAO which it was not possible to identify specifically as projecting to the lateral C1 zone or the neighbouring C2 zone (modified from Trott and Apps, 1993), see also Trott and Apps (1991) (Med., medial; Caud., caudal).

Concluding comments

The findings outlined in this chapter demonstrate that powerful, task-dependent gating mechanisms operate during active movements to regulate transmission in climbing fibre pathways targeting the paravermal cerebellar cortical zones. The pattern of this gating depends on the particular zone under study, providing the first clear evidence in the awake behaving animal in support of the zonal hypothesis of cerebellar organization. A conclusion which seems a fitting tribute to Professor Voogd on the occasion of his retirement.

Acknowledgements

The financial support of the MRC (Senior Research Fellowship to RA) and the Wellcome Trust is gratefully acknowledged. Many thanks are also due to Miss Rachel Bissett and Miss Clare Everard for their help in the preparation of the illustrations.

References

Andersson, G. and Armstrong, D.M. (1987) Complex spikes in Purkinje cells in the lateral vermis (b zone) of the cat cerebellum during locomotion. *J. Physiol. (Lond.)*, 385: 107–134.

Andersson, G. and Oscarsson, O. (1978) Climbing fiber microzones in cerebellar vermis and their projection to different groups of cells in the lateral vestibular nucleus. *Exp. Brain Res.*, 32: 565–579.

Apps, R. (1990) Columnar organisation of the inferior olive projection to the posterior lobe of the rat cerebellum. *J. Comp. Neurol.*, 302: 236–254.

Apps, R. (1999) Movement-related gating of climbing fibre input to cerebellar cortical zones. *Prog. Neurobiol.*, 57: 537–562.

Apps, R. (2000) Rostrocaudal branching within the climbing fibre projection to forelimb-receiving areas of the cerebellar cortical C1 zone. *J. Comp. Neurol.*, 419: 193–204.

Apps, R., Atkins, M.J. and Garwicz, M. (1997) Gating of cutaneous input to cerebellar climbing fibres during a reaching task in the cat. *J. Physiol. (Lond.)*, 502: 203–214.

Apps, R. and Garwicz, M. (2000) Precise matching of olivo-cortical divergence and cortico-nuclear convergence between somatotopically corresponding areas in the medial C1 and medial C3 zones of the paravermal cerebellum. *Eur. J. Neurosci.*, 12: 205–214.

Apps, R., Hartell, N.A. and Armstrong, D.M. (1995) Step phase-related excitability changes in spino-olivocerebellar paths to the C1 and C3 zones in cat cerebellum. *J. Physiol. (Lond.)*, 483: 687–702.

Apps, R. and Lee, S. (1999) Gating of transmission in climbing fibre paths to cerebellar cortical C1 and C3 zones in the rostral paramedian lobule during locomotion in cat. *J. Physiol. (Lond.)*, 516.3: 875–883.

Apps, R., Lidierth, M. and Armstrong, D.M. (1990) Locomotion-related variations in excitability of spino-olivocerebellar paths to cat cerebellar cortical C2 zone. *J. Physiol. (Lond.)*, 424: 487–512.

Apps, R. and Trott, J.R. (1991) Branching in the rostrocaudal axis within the projection from the inferior olive to the C1 zone in the cat cerebellum. *Soc. Neurosci. Abs.*, 17: 627.8.

Armstrong, D.M. (1990) Topographical localisation in the projections from the inferior olive to the paravermal cortex of the anterior lobe and paramedian lobule in the cerebellum of the cat. *Arch. Ital. Biol.*, 128: 183–207.

Armstrong, D.M., Campbell, N.C., Edgley, S.A., Schild, R.F. and Trott, J.R. (1982) Investigations of the olivo-cerebellar and spino-olivary pathways. In: V. Chan-Palay and S.L. Palay (Eds), *The Cerebellum: New Vistas*, Springer Verlag, New York, pp. 195–232.

Armstrong, D.M., Edgley, S.A. and Lidierth, M. (1988) Complex spikes in Purkinje cells of the paravermal part of the anterior lobe of the cat cerebellum during locomotion. *J. Physiol. (Lond.)*, 400: 405–414.

Armstrong, D.M. and Harvey, R.J. (1966) Responses in the inferior olive to stimulation of the cerebellar and cerebral cortices in the cat. *J. Physiol. (Lond.)*, 187: 553–574.

Armstrong, D.M., Harvey, R.J. and Schild, R.F. (1973a) Spino-olivocerebellar pathways to the posterior lobe of the cat cerebellum. *Exp. Brain. Res.*, 18: 1–18.

Armstrong, D.M., Harvey, R.J. and Schild, R.F. (1973b) Cerebello–cerebellar responses mediated via climbing fibres. *Exp. Brain Res.*, 18: 19–39.

Armstrong, D.M., Harvey, R.J. and Schild, R.F. (1973c) Branching of inferior olivary axons to terminate in different folia, lobules or lobes of the cerebellum. *Brain Res.*, 54: 365–371.

Atkins, M.J. and Apps, R. (1997) Somatotopical organisation within the climbing fibre projection to the paramedian lobule and copula pyramidis of the rat cerebellum. *J. Comp. Neurol.*, 389: 249–263.

Brodal, A. and Kawamura, K. (1980) Olivocerebellar projection: a review. In: A. Brodal, W. Hild and J. van Limborgh. (Eds), *Advances in Anatomy, Embryology and Cell Biology*, Springer-Verlag, Berlin, Heidelberg, New York, pp. 1–140.

Buisseret-Delmas, C. and Angaut, P. (1993) The cerebellar olivo-corticonuclear connections in the rat. *Prog. Neurobiol.*, 40: 63–87.

Campbell, N.C. and Armstrong, D.M. (1985) Origin in the medial accessory olive of climbing fibres to the x and lateral C1 zones of the cat cerebellum: a combined electrophysiological/WGA-HRP investigation. *Exp. Brain. Res.*, 58: 520–531.

Eccles, J.C., Llinas, R. and Sasaki, K (1966) The excitatory synaptic action of climbing fibres on the Purkinje cells of the cerebellum. *J. Physiol. (Lond.)*, 182: 268–296.

Ekerot, C.F. and Larson, B. (1979a) The dorsal spino-olivocerebellar system in the cat. I. Functional organization and termination in the anterior lobe. *Exp. Brain Res.*, 36: 201–217.

Ekerot, C.F. and Larson, B. (1979b) The dorsal spino-olivocerebellar system in the cat. II. Somatotopical organization. *Exp. Brain Res.*, 36: 219–232.

Ekerot, C.F. and Larson, B. (1982) Branching of olivary axons to innervate pairs of sagittal zones in the cerebellar anterior lobe of the cat. *Exp. Brain Res.*, 48: 185–198.

Ekerot, C.F., Garwicz, M. and Schouenborg, J. (1991) Topography and nociceptive receptive fields of climbing fibres projecting to the cerebellar anterior lobe in the cat. *J. Physiol. (Lond.)*, 441: 257–274.

Garwicz, M., Apps, R. and Trott, J.R. (1996) Micro-organization of olivocerebellar and corticonuclear connections of the paravermal cerebellum in the cat. *Eur. J. Neurosci.*, 8: 2726–2738.

Garwicz, M., Ekerot, C.F. and Jörntell (1998) Organizational principles of cerebellar neuronal circuitry. *News Physiol. Sci.*, 13: 26–32.

Gellman, R.S., Gibson, A.R. and Houk, J.C. (1985) Inferior olivary neurons in the awake cat: detection of contact and passive body displacement. *J. Neurophysiol.*, 54: 40–60.

Horn, K.M., Van Kan, P.L.E. and Gibson, A.R. (1996) Reduction of rostral dorsal accessory olive responses during reaching. *J. Neurophysiol.*, 76: 4140–4151.

Ito, M. (1984) *The Cerebellum And Neural Control*, Raven Press. New York.

Kitchell, R.L., Canton, D.D., Johnson, R.D. and Maxwell, S.A. (1982) Electrophysiologic studies of the cutaneous nerves of the forelimb of the cat. *J. Comp. Neurol.*, 210: 400–410.

Larson, B. Miller, S. and Oscarsson, O. (1969) A spinocerebellar climbing path activated by the flexor reflex afferents from all four limbs. *J. Physiol. (Lond.)*, 203: 641–649.

Lidierth, M. and Apps, R. (1990) Gating in the spino-olivocerebellar pathways to the C1 zone of the cerebellar cortex during locomotion in the cat. *J. Physiol. (Lond.)*, 430: 453–469.

Lou, J.-S. and Bloedel, J.R. (1992) Responses of sagittally aligned Purkinje cells during perturbed locomotion: synchronous activation of climbing fibre inputs. *J. Neurophysiol.*, 68: 570–580.

Miller, S. and Oscarsson, O. (1970) Termination and functional organization of spino-olivocerebellar paths. In: W.S. Fields and W.D. Willis (Eds), *The Cerebellum in Health and Disease*, Warren H Green, St Louis, pp. 171–200.

Oscarsson, O. (1979) Functional units of the cerebellum-sagittal zones and microzones. *Trends Neurosci.*, 2: 143–145.

Oscarsson, O. (1980) Functional organization of olivary projection to the cerebellar anterior lobe. In: J. Courville, C. De Montigny and Y. Lamarre (Eds), *The Inferior Olivary Nucleus: Anatomy and Physiology*, Raven Press: New York, pp. 279–289.

Rosina, A. and Provini, L. (1983) Somatotopy of climbing fiber branching to the cerebellar cortex in the cat. *J. Comp. Neurol.*, 289: 45–63.

Rosina, A. and Provini, L. (1987) Spatial distribution of axon collaterals of single inferior olive neurones. *J. Comp. Neurol.*, 256: 317–328.

Simpson, J.I., Wylie, D.R. and De Zeeuw, C.I. (1996) On climbing fibre signals and their consequences. *Behav. Brain Sci.*, 19: 384–398.

Smith, S.S., (1998) Step cycle-related oscillatory properties of inferior olivary neurons recorded in ensembles. *Neuroscience*, 82: 69–81.

Trott, J.R. and Apps, R. (1991) Lateral and medial sub-divisions within the olivo-cerebellar zones of the paravermal cortex in lobule Vb/c of the cat anterior lobe. *Exp. Brain Res.*, 87: 126–140.

Trott, J.R. and Apps, R. (1993) Zonal organization within the projection from the inferior olive to the rostral paramedian lobule of the cat cerebellum. *Eur. J. Neurosci.*, 5: 162–173.

Voogd, J. (1964) *The Cerebellum of the Cat: Structure and Fibre Connexions*, Van Gorcum, Assen.

Voogd, J. (1969) The importance of fiber connections in the comparative anatomy of the mammalian cerebellum. In: R. Llinas (Ed.), *Neurobiology of Cerebellar Evolution and Development*, American Medical Association, Chicago, pp. 493–541.

Voogd, J. and Glickstein, M. (1998) The anatomy of the cerebellum. *Trends Neurosci.*, 21: 370–375.

N.M. Gerrits, T.J.H. Ruigrok and C.I. De Zeeuw (Eds.)
Progress in Brain Research, Vol 124

CHAPTER 15

GABAergic modulation of olivary oscillations

Anna Devor* and Yosef Yarom

Department of Neurobiology, Life Sciences Institute, Hebrew University, Jerusalem, 91904, ISRAEL

Introduction

The uniqueness of the olivo-cerebellar system lies in its precise morphological organization (Voogd and Glickstein, 1998), reliable, high safety factor synaptic transmission (Eccles et al., 1966), low firing rate (Armstrong, 1974) and accurate sub- and supra-threshold rhythmic activity (Llinás and Yarom, 1981a, b). The rhythmic activity has been attributed to intrinsic properties of the inferior olivary (IO) nucleus. Olivary neurons are endowed with a variety of ionic channels. Their properties and spatial distribution over the somatic and dendritic membrane furnish these neurons with the natural tendency to oscillate (Llinás and Yarom, 1986). Subthreshold sinusoidal-like membrane potential oscillations result from this intrinsic resonance as well as from electrotonic coupling between olivary neurons (Lampl and Yarom, 1997; Manor, et al., 1997). Suprathreshold activity in IO neurons reflects the interaction between high threshold dendritic Ca^{2+} spikes, somatic low threshold Ca^{2+} spikes and Ca^{2+} dependent potassium conductances (Llinás and Yarom, 1981a, b).

There are number of hypotheses concerning the role of climbing fibers (the output of olivary neurons) in cerebellar physiology. Despite their differences, a common denominator among these hypotheses is the importance of precise timing of climbing fiber input to the cerebellum.

*Corresponding author. Tel.: 972-2-6585172; Fax: 972-2-6586296; e-mail: annakn@lobster.ls.huji.ac.il

Marr, Albus and Ito in the early 70s proposed ‘motor learning hypothesis’ of cerebellar function (Marr, 1969; Albus, 1971; Ito, 1998). According to this hypothesis climbing fibers act by modifying the response of Purkinje cells to mossy fiber input as an essential step in the acquisition of new motor skills. Thompson adapted the motor learning hypothesis to explain associative learning during eyeblink conditioning (Thompson et al. 1997). The results of numerous experiments carried out over the years in Thompson's laboratory show that a signal associated with unconditioned stimuli is conducted to the cerebellum via the IO nucleus (Thompson, 1989). Although controversial, the learning theory stresses the importance of precise timing of olivary output.

The observation that during active movement the firing of olivary cells becomes less reliable gave rise to a third hypothesis whereby the IO nucleus plays the role of a detector of ‘errors’ or an ‘unexpected events’ (Gellman et al., 1985; Andersson and Armstrong, 1987). While it is still unclear what exactly the olive detects, we would like to point out that a particular stimulus can serve as an ‘error’ or not, depending on the time of its arrival.

Llinás and colleagues proposed a different idea, where the IO nucleus plays the role of a ‘timing device’ for movement execution (Welsh and Llinás, 1997). According to this hypothesis climbing fiber activity serves as a pacemaker producing ‘a highly significant and rhythmic modulation of the motor systems as movement proceeds’ (Llinás and Welsh, 1993). Experimental evidence in support of the ‘timing hypothesis’ includes synchronous Purkinje cell firing (Llinás and Sasaki, 1989), rhythmic inhibitory potentials at the level of the deep cerebellar nuclei (Llinás and Muhlethaler, 1988) and temporal correlation between rhythmic olivo-cerebellar activity and rhythmic movement (Welsh et al., 1995). The importance of timing of olivary output is inherent in the name of the hypothesis itself.

There are two theories that stand apart from others, but have two postulates in common: (1) the cerebellum is involved in sensory acquisition and (2) climbing fibers act to reset the computational process of Purkinje cells. One of the hypotheses is the 'state estimator hypothesis' of Paulin, whereby the cerebellum 'is involved in constructing neural representation of moving systems' (Paulin, 1993). The second is Bower's hypothesis of the cerebellum as a sensory acquisition device (Bower, 1997). According to both hypotheses, climbing fibers act to bring the Purkinje cells back to their default level. This reset signal obviously has to come at the right time to prevent computational errors; hence, here too, timing is of the essence.

Irrespective of whether the olive is an 'error detector', a 'teacher', a 'timing device' or a 'reset operator', all the theories above assume that olivary output must be delivered to the cerebellum at precisely designed times. We have shown that interactions between synaptic input and subthreshold oscillations in the IO are so powerful that they can alter the apparent synaptic delay over a range of tens of milliseconds (Lampl and Yarom, 1993). As a result, olivary output is more likely to occur during a narrow time window at the peak of the depolarizing wave of the oscillations. Thus, the IO nucleus has all the timing mechanisms necessary to enable its neurons to discriminate between two identical stimuli on the basis of their time of arrival. In this chapter we further discuss the subthreshold oscillations of olivary neurons, their network behavior and functional clustering of neurons in the nucleus.

Electrotonic coupling

Electrotonic coupling has a central role in the generation of subthreshold oscillations in the IO (Llinás and Yarom, 1986; Yarom, 1991; Bleasel and Pettigrew, 1992; Manor et al., 1997). Understanding the physiology of the nucleus requires exploration of the extent of synchronicity among IO neurons, spatial organization of the electrical connections among them and their relative strength. To address these issues we performed simultaneous double whole-cell patch recordings from cell bodies of two IO neurons. The coupling was bi-directional and non-symmetrical (Fig. 1). The lack of symmetry is either due to differences in cell size or asymmetric location of the coupling junction. No evidence for nonlinear coupling was found. In fact, all apparent nonlinearity can be attributed to the voltage dependence of the input impedance of the post-junctional cell. In cell pairs that showed coupling, the coupling coefficient ranged between 0.2 and 0.01. It should be noted that the lower level of this range is an overestimation of the lowest value, as it is limited by the noise level of the recording system. This wide range of coupling strength can be attributed either to non-homogeneity of the electrotonic connections, or, more likely, to indirect coupling through chains of IO neurons. Indeed, the strength of the coupling depended on distance between the recorded neu-

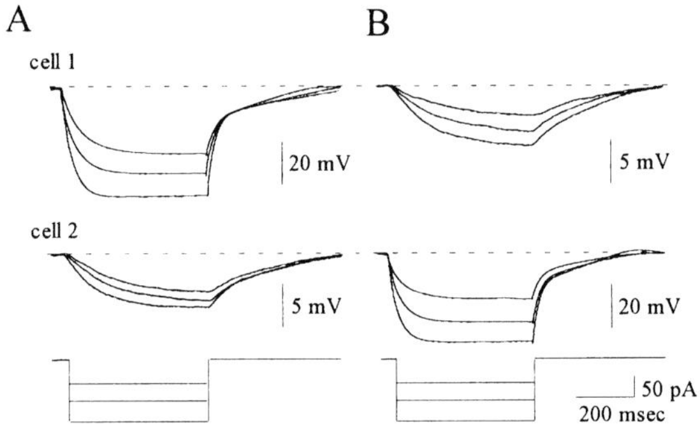

Fig. 1. Bi-directional electrical coupling between a pair of olivary neurons. Two cells (cell 1 and cell 2) were recorded simultaneously in the whole cell patch configuration from rat brain-stem slice. Slices were prepared from young (3–4 week old) rats. Experiments were performed at constant temperature of 35°C. A. Three current steps of various intensities were injected into cell 1 (bottom traces) and the voltage responses were recorded in cell 1 (top trace) and in cell 2 (middle trace). B. As in A, for current injection into cell 2. Note that voltage changes were observed in both cells, demonstrating direct, bi-directional current flow. Coupling coefficients of 0.177 and 0.166 were calculated for A and B respectively. Note the different voltage calibration for the pre- and post- junctional cells.

rons. The further the distance between the cells, the more cell pairs showed no electrotonic coupling. Coupling was sufficient to support synchronous subthreshold oscillations, but was not strong enough to evoke an action potential in the post-junctional cell.

In agreement with electrophysiological observations neurobiotin injections into one IO neuron often resulted in indirect labeling of 1–6 neighboring neurons by transfer of the dye through gap junctions (Fig. 2). All the directly labeled neurons showed fully stained dendritic trees up to the level of single spines. Indirectly stained neurons, on the other hand, showed much lighter staining. In agreement with previous reports, IO neurons can be divided into two groups according to their dendritic morphology (Scheibel and Scheibel, 1955; De Zeeuw et al., 1998). Some show curly dendrites, which tend to extend back toward the cell body; others have straight, sparsely bifurcating dendrites. Recording from neurons of both types, we found no difference in the basic electrophysiological properties: resting potential, input impedance, oscillatory activity and spiking behavior. Although it was difficult to identify the dendritic morphology of the indirectly labeled neurons, our data suggest that coupling exists only between neurons of the same type.

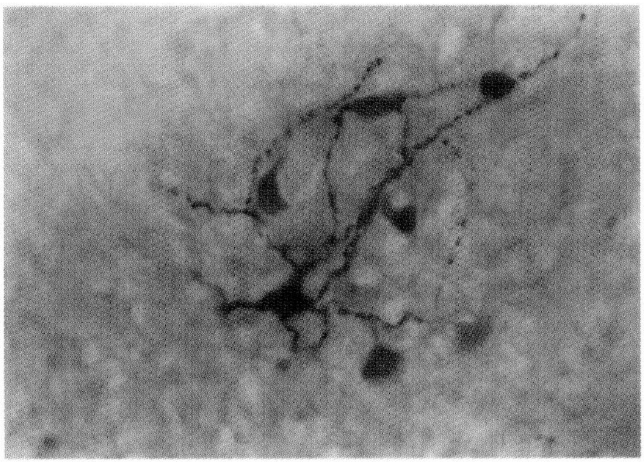

Fig. 2. Dye coupling between olivary neurons. A single cell (lower right) was injected with neurobiotin. The electrode was filled with 0.5% neurobiotin in intracellular solution. Preparation was developed according to standard procedures using biotinylated horseradish peroxidase conjugated to avidin. Six other cells were stained indirectly by diffusion of neurobiotin through the gap junctions.

Subthreshold oscillations

We divided IO neurons into three categories according to their oscillatory behavior: continuously oscillating cells (Fig. 3A), intermittently oscillating cells (Fig. 3B), and silent (non-oscillating) cells. In the continuously oscillating cells, the oscillation amplitude ranged between 1 and 25 mV, and the frequency between 1 and 9 Hz. Fast Fourier Transforms of consecutive time periods in these cells revealed random, spontaneous shifts in the main frequency component (Fig. 3C). These ran-

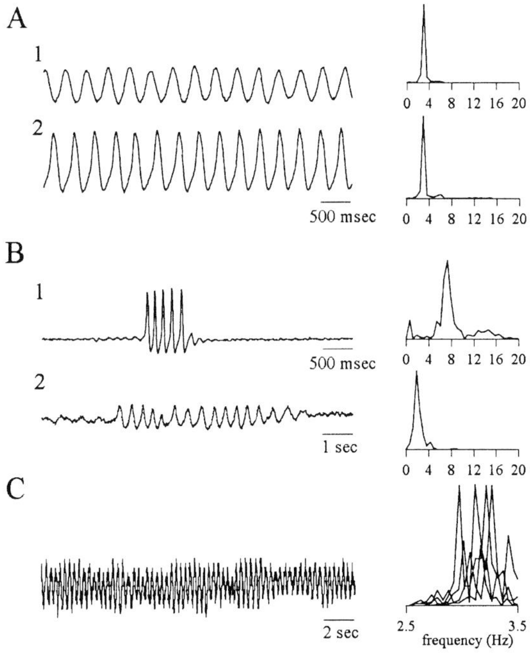

Fig. 3. Types and stability of subthreshold oscillations. Neurons in the IO display two types of oscillatory behavior: continuous and intermittent. A. Two examples of continuously oscillating cells. The results of the Fourier analyses show a sharp peak at 3 Hz, indicating that within the time period sampled an accurate stable oscillations occurred. B. Two examples of intermittently oscillating cells, where epochs of oscillations were alternated with silent periods. In these examples a peak frequencies of 8 and 2.5 Hz was revealed by the Fourier analyses for cell 1 and 2 respectively. C. Instability of the frequency of the oscillations was observed when prolonged time periods were analyzed. Fourier analyses of six consecutive time periods of approximately 20 seconds revealed random shifts of up to 0.5H z in the main frequency component. One of these time periods is shown on the left. Each of the examples shown was recorded in different brain slice.

dom shifts in frequency suggest that either the frequency of the oscillation encodes information, or that frequency shifts reflect dynamic control of the oscillations by synaptic input. Intermittent cells had an 'on–off' rhythm composed of epochs of oscillations separated by silent periods; epochs of oscillation lasted in different cells from tens of seconds to several minutes.

The coherence of subthreshold activity in the IO is best illustrated in simultaneous recordings from pairs of IO neurons. The examples shown in Fig. 4 demonstrate the variability that we found in the coherent activity of different cell pairs. Closely adjacent continuously oscillating neurons usually displayed subthreshold oscillations with the same frequency and identical phase (Fig. 4A). Differences in either phase or frequency of subthreshold oscillations were more frequently observed when the distance between the recorded neurons was relatively long, 70 μm and more (Fig. 4B). Coherence of the subthreshold oscillations was also a frequent feature in closely adjacent intermittently oscillating cells. The identical distribution of the numerous peaks in the Fourier analysis of coupled neurons is shown in Fig. 4C. Despite constant variations in the oscillatory pattern, coupled cells showed remarkable similarity in their subthreshold behavior. This finding provides additional evidence that subthreshold activity is a manifestation of the behavior of a network of electrotonicaly coupled cells.

It is interesting to note that cluster organization of the IO, a morphological feature well documented in higher mammals such as the cat (Armstrong, 1974), is not readily apparent in the rat where the size of a cluster is more or less a whole subnucleus. However, our data on synchronized behavior of neighboring IO neurons suggest that in the rat, too, the IO might be organized in functionally coupled clusters.

GABAergic control of the subthreshold oscillations

The morphological correlates of electrotonic coupling, gap junctions, are frequently found in clusters of spines in olivary glomeruli (Sotelo et al., 1974; De Zeeuw et al., 1990b) surrounded by both

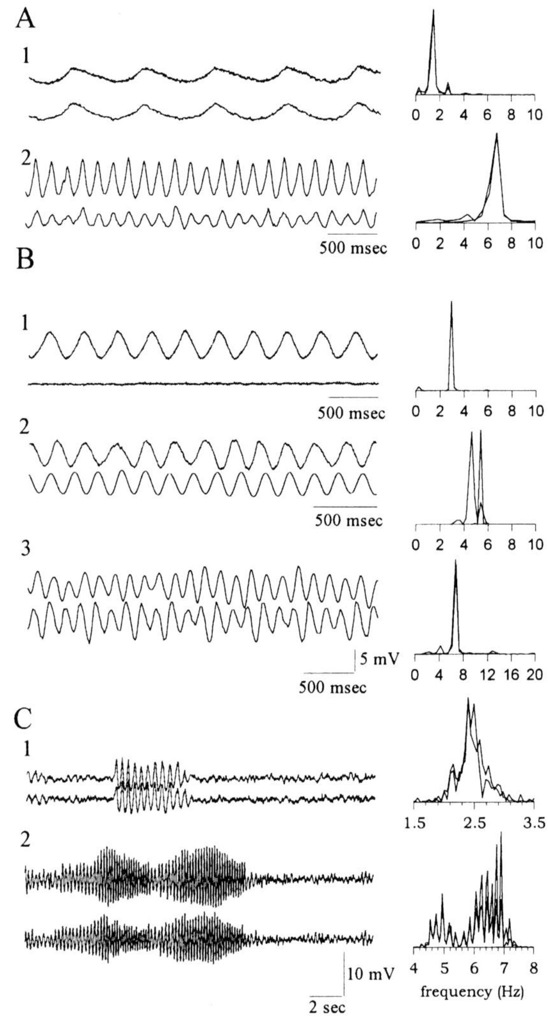

Fig. 4. The coherence of the oscillatory activity. Several examples of simultaneously recorded cell pairs showing different oscillatory patterns. A. Two examples of cell pairs displaying synchronous oscillations. The distance between the recorded neurons was 15 μm in A1 and 20 μm in A2. The Fourier analysis, shown on the right, reveals almost identical frequencies. B. Three examples of pairs where the neurons showed different oscillatory behavior: (1) only one neuron oscillated, (2) the neurons oscillated at different frequencies, (3) the neurons had a constant phase shift. Fourier analyses of each case are shown on the right. The distance between the recorded neurons was 92 μm in (1), 76 μm in (2) and 70 μm in (3). C. Two cell pairs with a complex pattern of intermittent oscillations were recorded, and the frequency distribution was analyzed. Note that despite constant changes in the pattern of subthreshold oscillations, in each case (1 and 2) both coupled neurons had the same distribution of frequency components.

excitatory and inhibitory synaptic terminals of extrinsic origin (De Zeeuw et al., 1990a). Based on this unique ultrastructural feature of olivary glomeruli, Llinás proposed in the mid-70s (Llinás, 1974) that the inhibitory feedback from the cerebellar deep nuclei produces a 'synaptic modulation of the electrical coupling between olivary neurons by shunting current away from gap junctions' (Llinás and Welsh, 1993). Therefore, GABAergic input from the cerebellum would reorganize the IO by 'fragmenting an entirely coupled nucleus into discrete clusters of coupled neurons'. Consistent with this hypothesis physiological experiments in vivo have demonstrated the modulatory effect of GABA on the degree of synchronization of complex spike activity in Purkinje cells. According to these experiments, both an administration of GABA-receptor antagonist to the IO, and a lesion of deep cerebellar nuclei produces global synchrony of climbing fiber discharge (Llinás and Sasaki, 1989; Lang et al., 1996).

We investigated the effect of local application of GABA on olivary subthreshold oscillations. Local pressure application of GABA (1 mμ, durations 20–500 msec) near the cell body or in the dendritic area revealed two distinct types of GABA responses which differed in their level of inactivation. Both responses were reversibly blocked by 100 μM bicuculline, indicating the $GABA_A$ nature of the receptors. Whereas somatic application was characterized by a prolonged response, the response to dendritic application was brief and almost independent of the duration of the stimulus (Fig. 5A). The different location of the receptors was confirmed by measuring the reversal potential for the two responses; the fast desensitizing dendritic response had, as expected, a more negative reversal potential.

The effect of GABA application on oscillating cells varied as a function of the type of oscillation, the location of the application and the amount of GABA applied. In continuously oscillating cells, prolonged (at least 200 msec) GABA application often completely blocked the oscillations (Fig. 5B1). With shorter duration pulses, a significant reduction in the amplitude of the subthreshold oscillations was observed; these effects were accompanied by membrane hyperpolarization. In

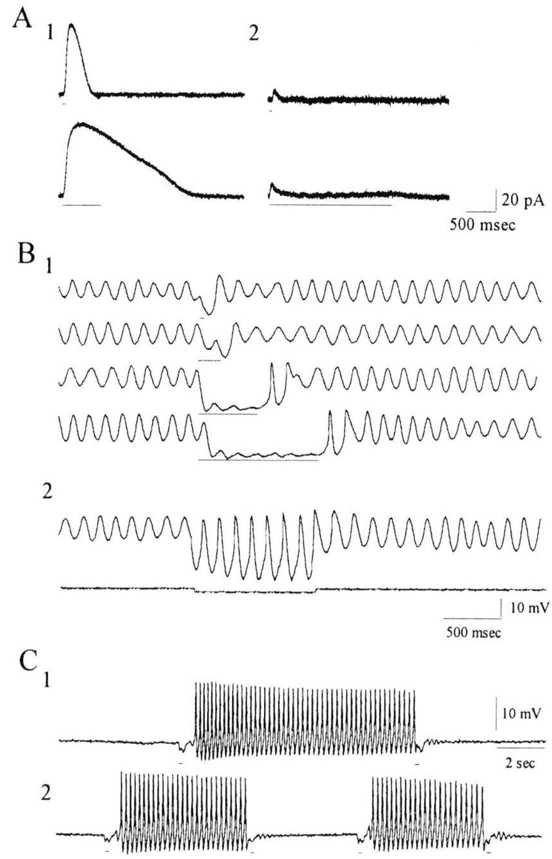

Fig. 5. GABAergic modulation of subthreshold oscillations. A. $GABA_A$ receptors on distal dendrites differ from those on the cell body in their rate of inactivation. GABA-induced currents were recorded from IO neurons in voltage clamp mode. GABA was applied by pressure via a pipette filled with 1 mM GABA in physiological solution. When GABA was applied near the cell body (1) an increase in GABA pulse duration resulted in an increase in the duration of the response. When GABA was applied in the area of distal dendrites (2) fast inactivation diminished the response. Bars mark the time and duration of GABA application. B. A continuously oscillating IO neuron was recorded in current clamp mode. GABA application during subthreshold oscillations decreased the amplitude of the oscillations and hyperpolarized the membrane potential (1). The decrease in amplitude and the amount of hyperpolarization depended on the duration of GABA application (indicated by bars). An injection of hyperpolarizing current to mimic GABA-induced hyperpolarization failed to block the oscillations, but rather increased their amplitude (2). C. Intermittently oscillating IO neuron recorded in current clamp mode. Consecutive pulses of GABA (marked by bars) turned the oscillations on and off. The effect was preserved under a range of inter-pulse intervals (1, 2).

continuously oscillating cells, the oscillations as well as the membrane potential resumed their pre-pulse behavior almost immediately at the end of the pulse. Current injection into the cell body, sufficient to mimic the hyperpolarization produced by a pulse of GABA, failed to reduce the amplitude of the oscillations (Fig. 5B2). On the contrary, it yielded an increase in amplitude due to hyperpolarization and consequent enhancement of low threshold Ca^{2+} conductance.

In intermittently oscillating cells the effect of GABA depended on the phase of the 'on-off' rhythm. A pulse of GABA given during silent periods usually triggered an epoch of the oscillations following the recovery from the GABA-induced hyperpolarization (Fig. 5C). Two types of responses were observed when GABA was applied during the oscillating phase. One type of response resembled that seen in continuously oscillating cells; i.e. GABA application reduced oscillations amplitude (Fig. 6A2) or turned them off entirely for the duration of the pulse (Fig. 6B2). In the second type of response GABA also blocked oscillations, but the blockade outlasted the GABA

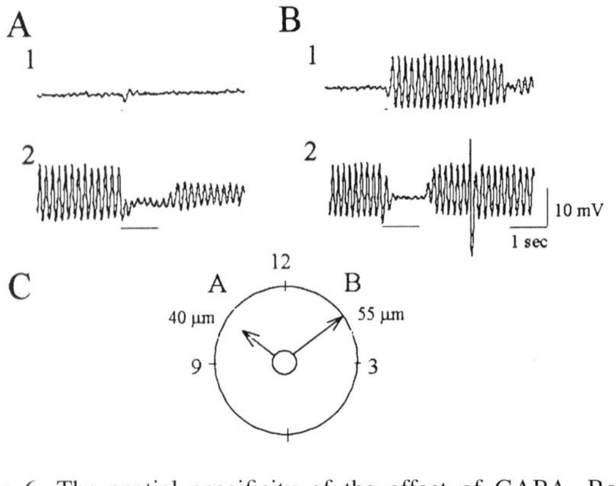

Fig. 6. The spatial specificity of the effect of GABA. Both suppression of oscillations, and induction of oscillations during quiescent periods, were especially robust when the injection was made in particular areas in the dendritic field (not necessarily close to the cell body). The cell shown in the figure was recorded in current clamp mode. GABA injections were made at two locations (A and B). At location B both effects were more robust than at location A. The locations are shown in C where a small circle represents the cell body at the center of a clock. Note that in B, where GABA was applied close to the cell body, its effects on the oscillations were weaker than when it was applied at A.

pulse (Fig. 5C). Thus, in this type of intermittently oscillating neurons, an on-off response to GABA application was observed (Fig. 5C). During the silent period GABA induced oscillations while during the oscillating period GABA turned them off.

The suppression and induction of subthreshold oscillations by GABA in continuously and intermittently oscillating cells usually were spatially specific. In the location marked A in Fig. 6 only a small hyperpolarizing response was observed during the silent period (Fig. 6A1), and a decrease in the amplitude of the oscillations was observed during the oscillatory period (Fig. 6A2). In location B, which was further from the cell body, GABA application during the silent period evoked an epoch of the oscillations (Fig. 6B1). In contrast, when GABA was applied during an oscillatory periods, a complete blockade of the oscillations was recorded (Fig. 6B2).

In conclusion, our results show two main effects of GABA on subthreshold oscillations: ongoing oscillations are suppressed during GABA pulse and alternatively, oscillations are induced in quiescent networks at the end of a GABA pulse. Both of these effects can be explained by mixed activation of GABA receptors located in various cellular compartments, spines on distal dendrites, proximal dendrites and the cell body.

The suppression of the oscillations can be attributed either to uncoupling of olivary neurons, or to an increase in membrane conductance that shunts the oscillatory current. The largest decrease in input impedance was observed when GABA was applied directly to the cell body. On the other hand, significant suppression of oscillations could also be observed during GABA application at remote dendritic locations. Therefore, the effect of GABA could be at least partially mediated by the activation of GABA receptors in IO glomeruli with consequent uncoupling of IO neurons.

The mechanism that initiates subthreshold oscillations at the end of a GABA pulse is probably related to the synchronous hyperpolarization of several coupled IO neurons. This hyperpolarization removes the inactivation of the low threshold Ca^{2+} conductance, enabling simultaneous activation of low threshold Ca^{2+} spikes in several cells, and initiating an epoch of the oscillations. This possibility is consistent with the morphological organization of the IO. The cells are closely packed and their dendrites intermingled (Ruigrok et al., 1990). Therefore, pressure injection of GABA is bound to affect several nearby neurons simultneously.

Closing remarks

The IO receives an inhibitory input from a number of sources: cerebellar deep nuclei, dorsal column nuclei and reticular formation (Scheibel and Scheibel, 1955; King, 1980; De Zeeuw et al., 1988; Nelson and Mugnani, 1989). Non-selective activation of GABAergic receptors may be responsible for the complexity of the effects that was observed in our experiments. In vivo, on the other hand, these inputs have different origins and are differentially regulated. We propose that both the suppression of subthreshold oscillations during GABA application and the induction of subthreshold oscillations in quiescent networks, occur dynamically during olivary functioning in vivo.

GABAergic input from deep cerebellar nuclei terminate mostly in glomeruli located on distal dendrites (Sotelo et al., 1986; Fredette and Mugnnaini, 1991). The resultant localized hyperpolarization should not activate low threshold Ca^{2+} conductance located at the soma and proximal dendrites. Therefore, the main predicted effect of GABAergic input from deep cerebellar nuclei is uncoupling of IO neurons. The relatively fast desensitization of the dendritic receptors observed in our experiments suggests that if uncoupling occurs, it will be limited to a short time period. It is yet to be demonstrated that this short term uncoupling is sufficient to block the oscillatory activity and functionally reorganizes the IO nucleus.

GABAergic input from sources other than deep cerebellar nuclei is not focused on olivary glomeruli, but synapse also on proximal dendrites and the cell soma (De Zeeuw et al., 1989). Activation of these inputs will lead to synchronization of IO cells, and hence to oscillatory behavior through GABA-induced hyperpolarization. Taken together, GABAergic inputs to the IO would control both functional clustering of IO neurons and the oscillatory state of the IO network.

Acknowledgement

This study was supported by the European Commission.

References

Albus, J.S. (1971) A theory of cerebellar function. *Math. Biosci.*, 10: 25–61.

Andersson, G. and Armstrong, D.M. (1987) Complex spikes in Purkinje cells in the lateral vermis of the cat cerebellum during locomotion. *J. Physiol. (Lond.)*, 385: 107–134.

Armstrong, D.M. (1974) Functional significance of connections of the inferior olive. *Physiol. Rev.*, 54: 358–417.

Bleasel, A.F. and Pettigrew, A.G. (1992) Development and properties of spontaneous oscillations of the membrane potential in inferior olivary neurons in the rat. *Devel. Brain Res.*, 65: 43–50.

Bower, J.M. (1997) Control of sensory acquisition. In: J.D. Schmahmann (Ed.), *The Cerebellum and Cognition*, Academic Press, San Diego, pp. 489–513.

De Zeeuw, C.I., Holstege, J.C., Calkoen, F., Ruigrok, T.J.H. and Voogd, J. (1988) A new combination of WGA–HRP anterograde tracing and GABA-immunocytochemistry applied to afferents of the cat inferior olive at the ultrastructural level. *Brain Res.*, 447: 369–375.

De Zeeuw, C.I., Holstage, J.C., Ruigrok, T.J.H. and Voogd, J. (1989) Ultrastractural study of the GABAergic, cerebellar, and mesodiencephalic innervation of the cat medial accessory olive: anterograde tracing combined with immunocytochemistry. *J. Comp. Neurol.*, 284: 12–35.

De Zeeuw, C.I., Holstege, J.C., Ruigrok, T.J.H. and Voogd, J. (1990a) Mesodiencephalic and cerebellar terminals terminate upon the same dendritic spines in the glomeruli of the cat and rat inferior olive: an ultrastructural study using a combination of [^{3}H]leucine and wheat germ agglutinin coupled horseradish peroxidase anterograde tracing. *Neuroscience.*, 34: 645–655.

De Zeeuw, C.I., Ruigrok, T.J.H., Holstege, J.C., Jansen, H.G. and Voogd, J. (1990b) Intracellular labeling of neurons in the medial accessory olive of the cat: 2. Ultrastructure of the dendritic spines and their GABAergic innervation. *J. Comp. Neurol.*, 300: 478–494.

De Zeeuw, C.I., Simpson, J.I., Hoogenraad, C.C., Galjart, N., Koekkoek, S.K.E. and Ruigrok, T.J.H. (1998) Microcircuitry and function of the inferior olive. *Trends Neurosci.*, 21: 391–400.

Eccles, J.C., Llinás, R. and Sasaki, K. (1966) The excitatory synaptic action of climbing fibers on the Purkinje cells of the cerebellum. *J. Physiol. (Lond.)*, 182: 268–296.

Fredette, B.J. and Mugnnaini, E. (1991) The GABAergic cerebello–olivary projections in the rat. *Anat. Embriol.*, 184: 225–243.

Gellman, R.S., Gibson, A.R. and Houk, J.C. (1985) Inferior olivary neurons in the awake cat: detection of contact and passive body displacement. *J. Neurophysiol.*, 54: 40–60.

Ito, M. (1998) Cerebellar learning in the vestibulo-ocular reflex. *Trends Cogn. Sci.*, 2: 313–321.

King, J.S. (1980) Synaptic organization of the inferior olivary complex. In: J. Courville (Ed.), *The Inferior Olivary Nucleus: Anatomy and Physiology*, Raven Press, New York, pp. 1–32.

Lampl, I. and Yarom, Y. (1993) Subthreshold oscillations of the membrane potential: a functional synchronizing and timing device. *J. Neurophysiol.*, 70: 2181–2186.

Lampl, I. and Yarom, Y. (1997) Subthreshold oscillations and resonant behavior: two manifestations of the same mechanism. *Neuroscience.*, 78: 325–341.

Lang, E.J., Sugihara, I. and Llinás, R. (1996) GABAergic modulation of complex spike activity by the cerebellar nucleoolivary pathway in rat. *J. Neurophysiol.*, 76: 255–275.

Llinás, R. (1974) Eighteenth Bowdwich lecture. Motor aspects of cerebellar control. *Physiologist*, 17: 19–46.

Llinás, R. and Muhlethaler, M. (1988) Electrophysiology of guinea-pig cerebellar nuclear cells in the in vitro brainstem–cerebellar preparation. *J. Physiol. (Lond.)*, 404: 241–258.

Llinás, R. and Sasaki, K. (1989) The functional organization of the olivocerebellar system as examined by multiple Purkinje cell recordings. *Europ. J. Neurosci.*, 1: 587–602.

Llinás, R. and Welsh, J.P. (1993) On the cerebellum and motor learning. *Curr. Opin. Neurobiol.*, 3: 958–965.

Llinás, R. and Yarom, Y. (1981a) Electrophysiology of mammalian inferior olivary neurons in vitro, different types of voltage-dependent ionic conductances. *J. Physiol. (Lond.)*, 315: 549–567.

Llinás, R. and Yarom, Y. (1981b) Properties and distribution of ionic conductances generating electro-responsiveness of mammalian inferior olivary neurons in vitro. *J. Physiol. (Lond.)*, 315: 569–584.

Llinás, R. and Yarom, Y. (1986) Oscillatory properties of guinea pig inferior olivary neurons and their pharmacologcal modulaton: an in vitro study. *J. Physiol. (Lond.)*, 376: 163–182.

Manor, Y., Rinzel, J., Segev, I. and Yarom, Y. (1997) Low-amplitude oscillations in the inferior olive: a model based on electrical coupling with heterogeneous channel densities. *J. Neuphysiol.*, 77: 2736–2752.

Marr, D. (1969) A theory of cerebellar cortex. *J. Physiol. (Lond.)*, 202: 437–470.

Nelson, B.J. and Mugnani, E. (1989) Origins of GABAergic inputs to the inferior olive. In: P. Strata (Ed.), *The Olivo-cerebellar System in Motor Control*, Springer-Verlag, New York, pp. 265–282.

Paulin, M.G. (1993) The role of the cerebellum in motor control and perception. *Brain Behav. Evolut.*, 41: 39–50.

Ruigrok, T.J., De Zeeuw, C.I., Burg, J.V.D. and Voogd, J. (1990) Intracellular labeling of neurons in the medial accessory olive of the cat: 1. Physiology and light microscopy. *J. Comp. Neurol.*, 300: 462–477.

Scheibel, M.E. and Scheibel, A.B. (1955) The inferior olive – a Golgi study. *J. Comp. Neurol.*, 102: 77–132.

Sotelo, C., Gotow, T. and Wassef, M. (1986) Localization of glutamic-acid-decarboxylase-immunoreactive axon terminals in the inferior olive of the rat, with special emphasis on anatomical relations between GABAergic synapses and dendrodendritic gap junctions. *J. Comp. Neurol.*, 252: 32–50.

Sotelo, C., Llinás, R. and Baker, R. (1974) Structural study of inferior olivary nucleus of the cat: morphological correlates of electrotonic coupling. *J. Neurophysiol.*, 37: 541–559.

Thompson, R.F. (1989) Role of inferior olive in classical conditioning. In: P. Strata (Ed.), *The Olivo-cerebellar System in Motor Control*, Springer-Verlag, New York.

Thompson, R.F., Bao, S., Chen, L., Cipriano, B.J., Grethe, J.S., Kim, J.J., Thompson, J.K., Tracy, J.A., Weninger, M.S. and Kupra, D.J. (1997) Associative learning. In: J.D. Schmahmann (Ed.), *The Cerebellum and Cognition*, Academic Press, San Diego, pp. 152–189.

Voogd, J. and Glickstein, M. (1998) The anatomy of the cerebellum. *Trends Neurosci.*, 21: 370–375.

Welsh, J.P., Lang, E.J., Sugihara, I. and Llinás, R. (1995) Dynamic organization of motor control within the olivo-cerebellar system. *Nature*, 374: 453–457.

Welsh, J.P. and Llinás, R. (1997) Some organizing principles for the control of movement based on olivo-cerebellar physiology. *Prog. Brain Res.*, 114: 449–461.

Yarom, Y. (1991) Rhythmogenesis in a hybrid system – interconnecting an olivary neuron to an analog network of coupled oscillators. *Neuroscience.*, 44: 263–275.

N.M. Gerrits, T.J.H. Ruigrok and C.I. De Zeeuw (Eds.)
Progress in Brain Research, Vol 124

CHAPTER 16

Analysis of the discharge pattern of floccular Purkinje cells in relation to vertical head and eye movement in the squirrel monkey

Y. Hirata and S.M. Highstein*

Department of Otolaryngology, Box 8115, Washington University School of Medicine, 4566 Scott Ave., St. Louis MO 63110, USA

Introduction

The vestibulo-ocular reflex (VOR) is compensatory in both the spatial and temporal domains. Eye movements resultant from this reflex help to keep the viewed object on the fovea of the retina during head movement to ensure clear vision. The gain of this reflex, defined as eye velocity/head velocity is plastic or adaptable and changes in the reflex gain are a kind of learning called motor learning.

When visual–vestibular mismatch conditions are employed within the laboratory the squirrel monkey (*Saimiri sciurius*) VOR gain can be adapted from normal values of about 1, to values between 0.2 and 1.6. There is extensive documentation of the neuronal substrate for the VOR, particularly for the vertical (V) VOR in this species. This substrate includes the extra ocular motor nuclei, the vestibular nuclei, several brainstem nuclei, and the flocculus and ventral paraflocculus of the cerebellum that project the axons of their Purkinje cells (P-cells) to brainstem target neurons (FTNs). The individual nerve cells in these interconnecting neuronal circuits control VOR behavior usually characterized by gain and phase. It is important to examine the discharge properties of single cells within these circuits to attempt to correlate animal behavior (VOR gain state) with the discharge modulation of these cells. Previously this strategy has been extremely effective when the central nervous system (CNS) neurons chosen for study have been close to the output of the brain. For example, the discharge properties of single extra ocular motor neurons can be reconstructed to accurately compile, predict, or document the eye movements caused by their discharges, even on a trial to trial basis without averaging. A simple linear equation relating extra ocular motor neuronal discharge with time can account for more than 85% of the behavior of the eyes (Fuchs et al., 1988) and a fourth order differential equation can capture most eye movement behavior (Fuchs et al., 1988). However, when similar studies are extended to neurons a few synapses further from the periphery, predicting behavior from neuronal discharge becomes problematic. This is particularly the case for cerebellar Purkinje cells (P-cells) that project into vertical VOR pathways because of the irregular nature of the firing modulation of these neurons.

Previously the behavior of neurons within the Y-group of the vestibular nuclei has been documented during and following gain changes of the VVOR (Partsalis et al., 1995a, b). Y neurons are fairly regularly discharging cells whose discharge patterns grossly follow the sinusoidal waveform of most of our stimuli (see Fig. 3). Y neurons change their gain and phase with changing VVOR gain and have been hypothesized to be causal in these gain

*Corresponding author. Tel.: (314) 362-1012; Fax: (314) 747-3444; e-mail: highstein@medicine.wustl.edu?

changes, at least in part (Partsalis et al., 1995b). With normal VVOR gain Y neurons hardly modulate during VVOR in darkness but when the gain is high, (ca. 1.6) they modulate 180 degrees out of phase with head movement. The converse is true for low gain adaptation. These changes in Y neurons are acquired over a period of hours to days during visual–vestibular mismatch training that leads to VOR gain change. Because floccular P-cells of the V zones project to Y-group neurons, and because the flocculus has been implicated in VOR plasticity, it was natural to study P-cells during VOR gain change to clarify whether the observed changes in the firing patterns of Y neurons are caused by changes in P-cell firing patterns or are intrinsic to Y neurons.

A striking feature of P-cell discharge is its irregularity and the lack of conformity to the underlying sinusoidal pattern of the input stimuli, i.e. head movement and/or optokinetic stimulus movement. Instantaneous firing rate of neurons that project to the flocculus (FPNs), Y neurons and superior vestibular nuclear (SVN) neurons that are the targets of floccular P-cells (FTNs), are regular and predictable, and the sine wave shape of the head and visual surround motion are readily visible in their discharge modulation. By contrast P-cells are highly irregular showing apparent variability in their modulation within and between cycles. Thus the usual analysis paradigms such as using a least squares routine or fast Fourier transform (FFT) to fit neuronal discharges with a sine wave while quite precise for FPNs and FTNs may be inadequate to describe the modulation of P-cells.

The output of floccular P-cells has been thought to encode the dynamic aspects of eye movements, i.e. eye velocity and acceleration (Shidara et al., 1993; Gomi et al., 1998). P-cells were hypothesized to construct these signals from visual and vestibular sensory input and oculomotor efference copy input via mossy fibers and by visual sensory input via climbing fibers (Maekawa and Simpson, 1973). Knowing how this input–output transformation is modified during the process of VOR adaptation is important in understanding the mechanism of VOR motor learning.

Herein we embark on a description of our experimental data studying floccular V-zone P-cells during VVOR gain adaptation. We have employed a variant of the multiple linear regression technique utilizing the concept of system identification theory to quantify P-cell firing and to understand the signal processing executed in the flocculus during VOR adaptation. This technique enables us to evaluate the irregularity of P-cell firing and to study the relationship of P-cell discharge to retinal slip, head motion, and oculomotor efference copy signals. Although not dealt with in this communication, changes in the sensitivity of P-cell firing to retinal slip or other input signals mentioned above might be responsible for providing the stimuli resulting in VOR gain change during visual-vestibular mismatch. In addition, the amount of retinal slip at various VOR gains was evaluated to see if slip changes in parallel with VOR gain and if it might provide an adequate index as an error signal required for VOR adaptation. Results concerning the visual following paradigm are reported in this communication.

Methods

Figure 1 is a diagram of the known synaptic linkage within the VVOR pathway. Motion of the head is sensed by the semicircular canals and distributed by axons of the VIIIth nerve to several cell types within the vestibular nuclei. These vestibular cells relay this information to central nervous system sites including the extra ocular motor nuclei, the cerebellar flocculus, and to the Y-group of the vestibular nuclei. In turn axons of floccular P-cells project back to the Y-group and to some of the same vestibular nuclear neurons that receive vestibular input. The targets of floccular axons are called FTNs while neurons projecting to flocculus are called FPNs. Visual input arises from the retina and is relayed via the pontine nuclei (DLPN) and the lateral terminal nucleus (LTN) of the accessory optic system. The output of this interconnecting neuronal system is an eye movement and is called a VOR.

Input to the floccular P-cells arises from motion of the head, motion of the visual surround, or both. Head velocity input is conveyed to P-cells by FPNs as mossy fiber terminals within the cerebellar cortex (Zhang et al., 1995a, b), and visual signals or retinal slip input is by DLPN mossy fiber terminals. Additional input to the flocculus is an efference copy of oculomotor commands relayed by terminals of dorsal raphe nuclear (DR) neurons. The efference copy of eye movements is conveyed to the raphe nuclei by axon collaterals of ascending vestibular nuclear neurons (McCrea et al., 1987). Eye movement is also fed back via an external physical feedback loop and is combined with the optokinetic stimulus (OKS) movement and head movement as an input to the retina. The Y group receives head motion information via SVN interneurons (Blazquez et al., in press) and Y-group (Partsalis et al., 1995a, b) and SVN-FTNs receive the terminals of flocculus P-cells (Langer et al., 1985). The output of the Y group and the FTNs projects to the extra ocular motor neurons that send the final command to the oculomotor muscle plant to move the eye.

Adult male squirrel monkeys were utilized and were placed in a primate chair daily for several weeks before any surgery was performed to acclimatize them to the experimental setup. All surgery was performed in a sterile operating suite using induction by Ketamine and inhalation anesthesia utilizing Isofluorane. For head fixation, a stainless steel bolt was secured to the occiput using small, stainless steel screws and dental cement.

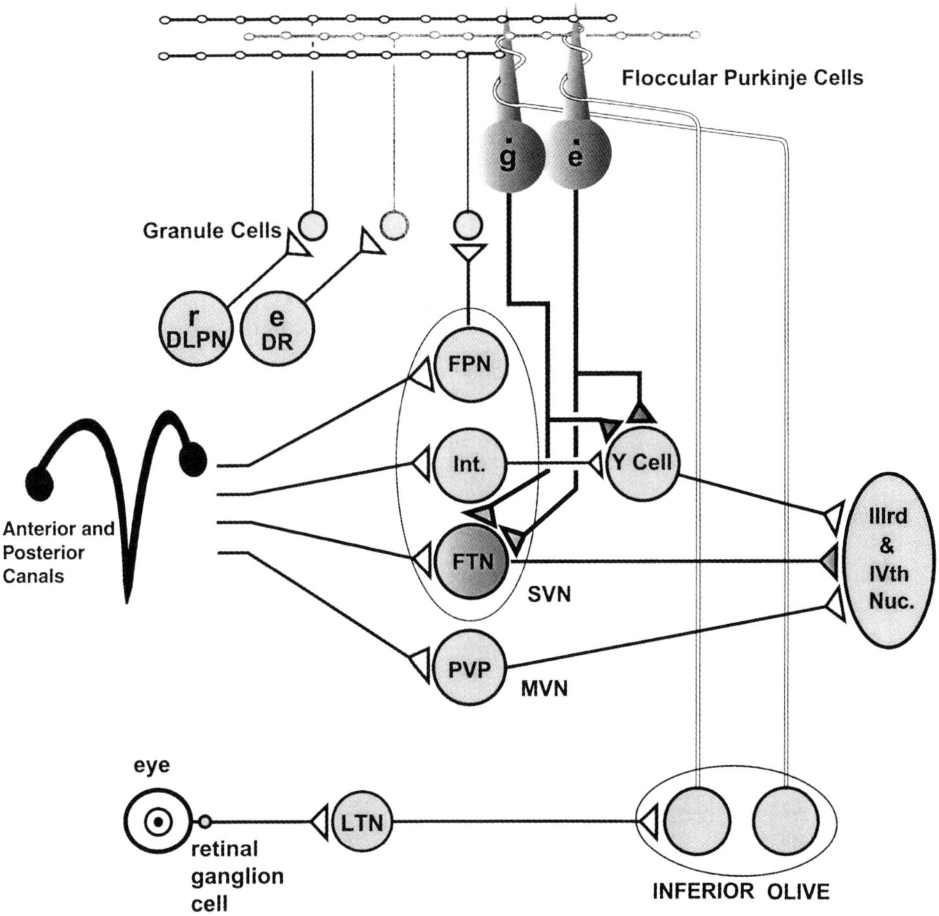

Fig. 1. Schematic of the vertical VOR circuitry. LTN, lateral terminal nucleus, SVN, MVN, superior and medial vestibular nuclei, respectively, r retinal slip via the DLPN, dorsal lateral pontine nuclei, e, efference copy of oculomotor command via the DR, dorsal raphe nuclei, FPN, FTN, flocculus projecting and receiving (target) neurons, respectively, PVP, position vestibular pause neuron, Int. interneuron, e, g, within the Purkinje neuron indicates eye dominated and gaze dominated cell, respectively.

This bolt fit into a receptacle on the monkey chair to fix the head in the center of a magnetic field generated by two sets of field coils driven in quadrature. For eye movement recording, a prefabricated eye coil (2.1 cm dia.) constructed of Teflon-insulated stainless steel wire (Cooner) was implanted under the conjunctiva at the limbus of either eye, sutured to the sclera and the twisted ends of the coil wire led to an occipital plug. A stainless steel recording chamber aimed at the cerebellar flocculus was implanted and secured with dental cement over a fenestra in the skull. All wounds were treated daily with antibiotics; if animal distress was apparent 1 mg of morphine sulfate was injected to alleviate pain. All procedures and experiments were approved by the Animal Welfare and Use Committee of Washington University.

Figure 2 illustrates the experimental setup. Animals were seated in a primate chair with their heads fixed. The chair was placed in the center of a white cylinder on which black random dots were projected. The cylinder was 1 m in diameter, extending 36 cm above and 50 cm below the

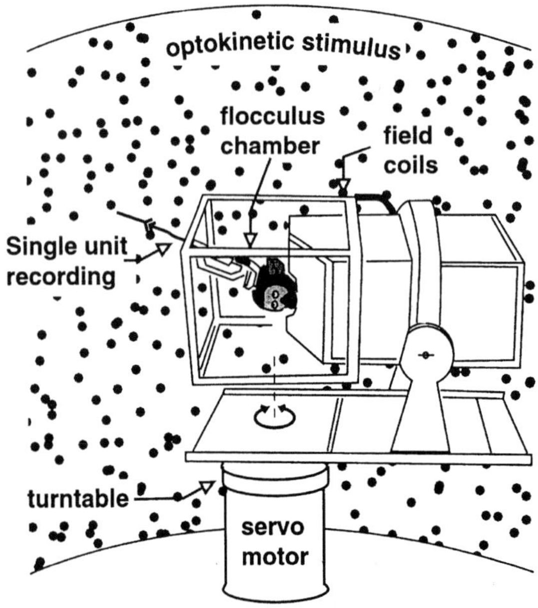

Fig. 2. Experimental setup. The squirrel monkey was placed with its right or left side tilted down 90 degrees on a turntable in the center of a white cylindrical screen. A black, random dot pattern was projected onto the screen to surround the monkey by 360 degrees.

animal's inter aural line. The right or left side of the animal was placed down so that rotation of the OKS about an earth vertical axis produced upward or downward nystagmus. The eye coil output was led to a phase-locked detector whose output gave signals proportional to horizontal and vertical eye position. Vertical and horizontal eye position and velocity were calibrated. Horizontal and vertical eye positions, OKS velocity, and chair velocity were continuously digitized at sampling frequency 200 Hz, 100 Hz, and 100 Hz, respectively with the use of a CED 1401 plus interface (Cambridge Electronic Design) for display and storage using the Spike–2 program. The output of a threshold detector, signaling the occurrence of action potentials, was sampled with a resolution of 0.01 msec and stored in the same way. Raw data were also stored on VCR tape (Neurodata PCM).

Experimental paradigms

Stimuli used for all paradigms in the present experiment were an 0.5 Hz sinusoid to compare the present results with previous related experiments (Partsalis et al., 1995a, b). To change the animals' VOR gain, suppression of the VOR (VORs) or reversal of the VOR (VORr) and enhancement of the VOR (VORe) paradigms were utilized for low gain and high gain training, respectively. In VORs paradigm, the chair moves in phase with the OKS (velocity peak amplitude 40, 80 deg/s). In VORr, the chair moves in phase with the OKS but with a velocity amplitude 2-fold that of chair (chair: 40 deg/s, OKS: 80deg/s). In VORe, the chair moves 180 degrees out of phase with OKS at the same or half speed (chair: 40 deg/s, OKS: 40, 80 deg/s). During the training, P-cells were isolated at various stages of VOR gain adaptation. Visual following (VF), VOR in darkness (VORd), VORr, VORs, VORe, and no stimulus in light (OKS and chair velocities are 0, OKS light is on) and no stimulus in dark (chair velocity is 0, OKS light is off) paradigms were performed for about 60 seconds each. For many cells recorded during high gain training, VORr and VORs were not performed because these paradigms interfere with the ongoing VOR gain adaptation training. For the same reason, VORe paradigm was omitted for some cells recorded during low gain training. Some cells could be continuously recorded and were evaluated at several VOR gains.

Unit recording, identification of P-cells

Floccular V zone P-cells were identified by the occurrence of complex spikes and by their characteristic discharge patterns during vertical VORr, VORs or VORe. Complex spikes could not always be recorded simultaneously during simple spike (SS) recording, but were usually seen in close physical proximity to the sites of SS recording. This report will only be concerned with SS responses.

Data Handling

Data analysis was performed on Matlab (Mathworks) running on a PC. Vertical eye velocity, eye acceleration, head acceleration, and OKS acceleration were calculated from vertical eye position, eye velocity, head velocity and OKS velocity, respectively. Retinal slip velocity was calculated as eye velocity – OKS velocity – chair velocity. P-cell instantaneous firing rate (IFR) was calculated as the reciprocal of each inter-spike interval.

Saccades and postsaccadic drifts if any were eliminated from eye movement traces by using an automated de-saccading algorithm and manually checked on the computer screen. Data periods corresponding to saccades and the post-saccadic slide were eliminated from both head and OKS movements as well as from P-cell IFR.

Gain of the VOR was measured as eye velocity/head velocity. More precisely, the following regression was executed and coefficient c was referred to as the gain of the VOR:

$$vel_x(t) = c\, vel_h(t - \tau) + d + \xi(t)$$

where $vel_x(t)$ and $vel_h(t)$ denote desaccaded vertical eye velocity and head velocity traces during VORd, respectively, d denotes the dc term which was almost 0 in most cases, and $\xi(t)$ is the error term. τ[s] is a delay between eye velocity and head velocity from which phase of the VOR at 0.5 Hz was calculated as 180τ [deg].

P-cell IFR was reconstructed by the following multiple linear regression model to evaluate contribution of each input signal to P-cell IFR:

$$\begin{aligned} f(t) = {} & \alpha_h acc_h(t - \tau_h) + \beta_h vel_h(t - \tau_h) \\ & + \alpha_h acc_r(t - \tau_r) + \beta_h vel_r(t - \tau_r) \\ & + \beta_e ecopy(t - \tau_e) + \delta + \varepsilon(t) \end{aligned}$$

where α_h[spikes/s per deg/s^2], β_h[spikes/s per deg/s], α_r[spikes/s per deg/s^2, β_h[spikes/s per deg/s], β_e[spikes/s per deg/s], and δ[spikes/s] are regression coefficients and denote sensitivities of P-cell IFR to desaccaded head acceleration $acc_h(t)$, head velocity $vel_h(t)$, retinal slip acceleration $acc_r(t)$, retinal slip velocity $vel_r(t)$, efference copy signal $ecopy(t)$, and the dc component of P-cell IFR, respectively. τ_h[s], τ_r[s], and τ_e[s] denote delays between head movement and P-cell activity, retinal slip and P-cell activity, and efference copy signal and P-cell activity, respectively. $\varepsilon(t)$ is the error term whose mean is 0. For the efference copy signal, eye velocity traces were used in this study. To confirm validity of this model, statistical properties of residual ($\varepsilon(t)$) following the regression were evaluated by comparing them with those of P-cell IFR during no stimulus conditions.

In addition, the amounts of retinal slip in relation to VOR gain during VF, VORe and VORr paradigms were evaluated to see whether they change in parallel with VOR gain adaptation or not. Only results of the VF paradigm are presented here.

Results

Figure 3 is selected traces of a P-cell IFR in comparison with the IFR of an VIIIth nerve anterior semicircular canal afferent fiber, and a Y-group neuron during an 0.5 Hz VORs paradigm. Note the trend towards increasing irregularity of discharge from the nerve fiber to the P-cells. The VIIIth nerve fiber and the Y neuron generally follow the waveform of the stimulus applied and the envelope of their individual firing patterns can be fairly well approximated or fit with a sine wave (either least squares or FFT derived). However the P-cell appears to deviate from this condition because of its irregularity.

Figure 4 illustrates the behavioral paradigms presently employed and the accompanying P-cell responses. In the top panel VF, the top trace is eye velocity. We are concerned with the slow envelope of the response. The sharp deflections are fast phases of nystagmus and will not be considered further here. The second trace is the IFR of a V-zone P-cell in spikes/s and the third traces are the velocities of the OKS (gray) and head movement (black) at 0.5 Hz. Note that the head is stationary in this paradigm. The other three sets are in the same format and are the VORd, VORl, and VORs paradigms.

Note the very irregular firing pattern of the P-cell and that it does not describe a sine wave in the VF

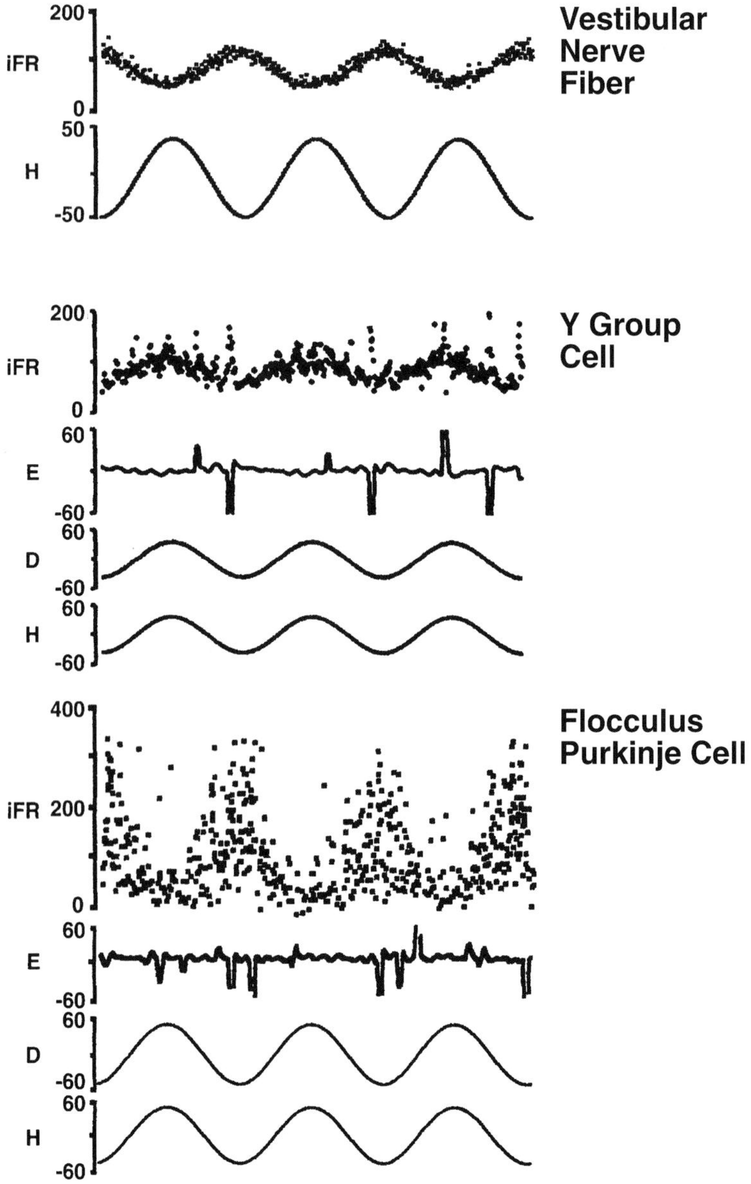

Fig. 3. VIIIth nerve, Y-group, and floccular Purkinje cell discharge. The top traces are of a vestibular semicircular canal afferent, the middle traces from a Y-group neuron, and the bottom traces from a floccular Purkinje cell. All sinusoidal stimuli are at 0.5 Hz. H is head velocity, E eye velocity, and D optokinetic stimulus velocity. iFR is instantaneous firing rate. iFR in the Y-group and Purkinje cell experiments is during VOR suppression.

or VORs paradigms. For example, in the case of VF, the IFR is nicely modulated during the first cycle, the second cycle is skewed to the right, in the third cycle the P-cell ceases firing at the peak eye velocity, and the fifth cycle is flat-topped and of low amplitude. In the VORd paradigm the IFR varies from 0 to 200 spikes/s and the firing rate fluctuates greatly, even though there is no overt response. The VORs IFR is also fairly irregular. The spontaneous IFR of P-cells in experiments where no overt stimuli are presented either in the light or dark is also highly variable (not illustrated).

We employed multiple linear regressions against the input parameters of retinal slip, head motion, and an efference copy of the oculomotor command to extract any stimulus related modulation from the P-cell discharge. To reduce the number of variables we employed this strategy during VORd where there is no retinal slip and VF where there is no head movement. Following this strategy we could demonstrate that the spectral content and amplitude distribution of the remaining signal following the extraction of any stimulus related modulation were almost identical to that of the P-cell IFR in experiments where no stimuli were presented (Hirata et al., in preparation). Sample results of this analysis are illustrated below.

Figure 5 illustrates the contributions of retinal slip, head motion and efference copy signals to the discharge of P-cells as determined by the multiple

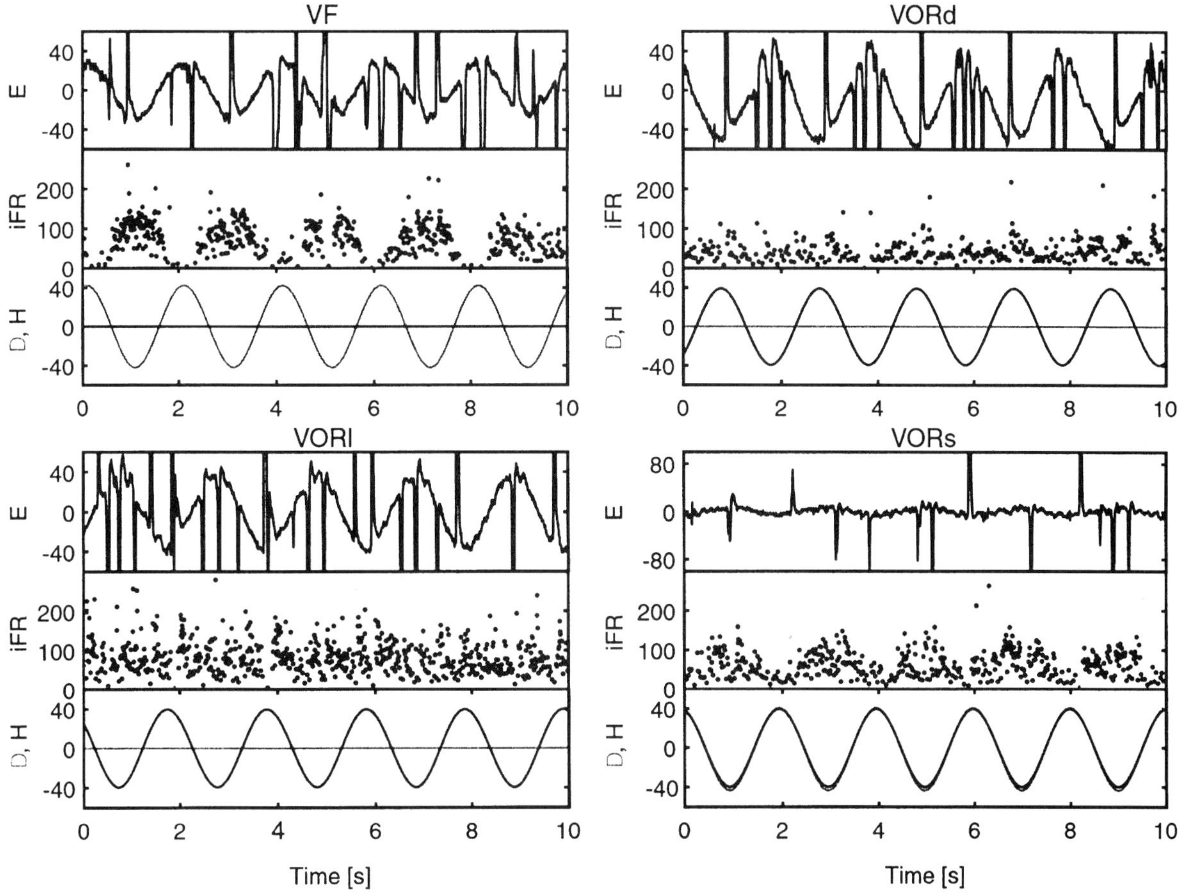

Fig. 4. Eye movements and corresponding Purkinje cell simple spike activity during the VF, VORl, VORd, and VORs experimental paradigms. E is vertical eye velocity in deg/s, iFR is Purkinje cell instantaneous firing rate in spikes/s, D and H are optokinetic stimulus (OKS) and chair velocity in deg/s, respectively. The OKS and chair velocities are illustrated in the same panels in gray and black lines, respectively. Ten seconds of about 60 seconds of each experiment are illustrated.

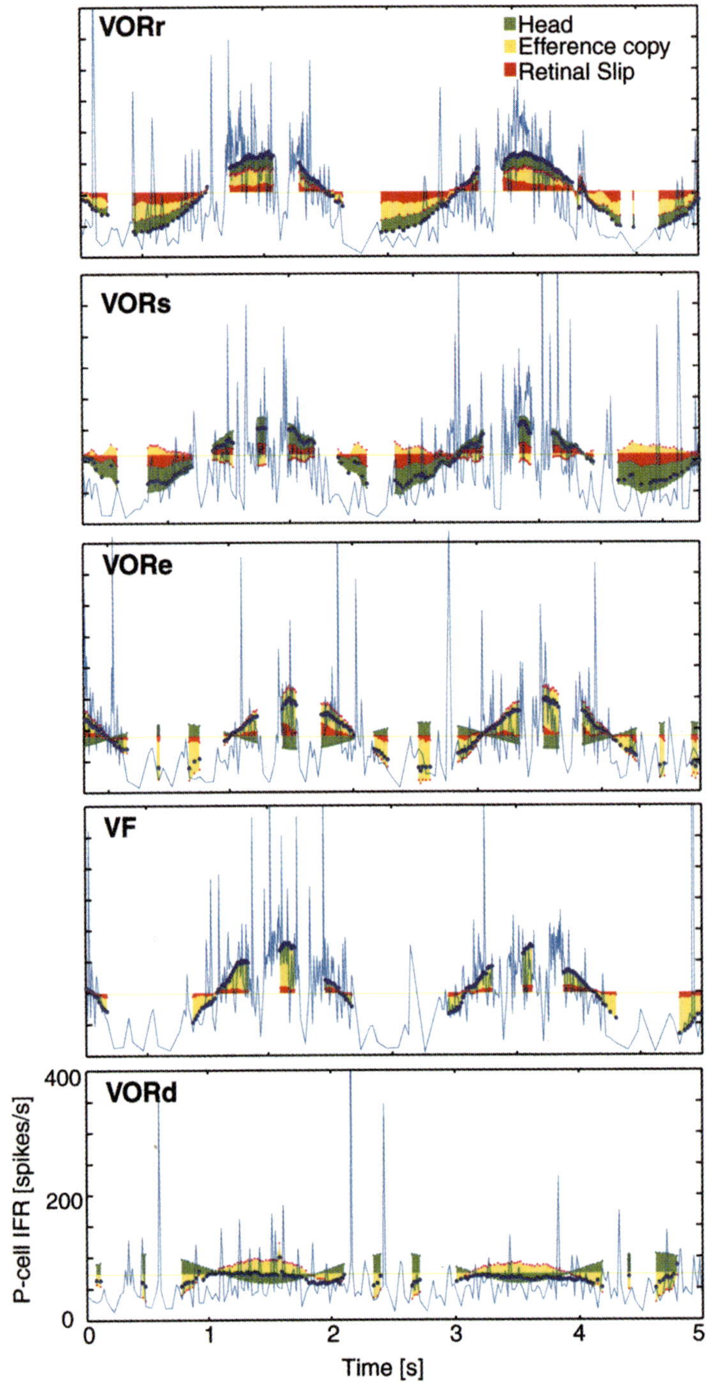

Fig. 5. Prediction and reconstruction of Purkinje cell IFR and decomposition of the predicted and reconstructed IFR during VORr, VORs, VORe, VF, and VORd paradigms. In each panel the spikey cyan trace is the original Purkinje cell IFR and dark blue dots are the predicted IFR (VORr, VORs, VORe) or the reconstructed IFR for VF, VORd by the multiple linear regression model. The green, yellow and red areas indicate the contributions of head, efference copy, and retinal slip to the predicted or reconstructed Purkinje cell IFR.

linear regression method. The top panel is VORr (chair peak velocity amplitude: 40 deg/s, OKS: 80 deg/s), the second VORs (chair and OKS: 80 deg/s), third VORe (chair and OKS: 40 deg/s), forth VF (OKS: 80 deg/s) and the lowest is VORd (chair: 40 deg/s). The head movement component is indicated in green, efference copy of the oculomotor command in yellow, and retinal slip in red. The light blue traces are the original P-cell IFR and the dark blue dots are the reconstructed IFR generated by the multiple linear regression model. Time periods in which only the original P-cell IFR is shown are desaccaded periods. Note that model parameters were determined by using only VF and VORd paradigms, thus the top three panels demonstrate the prediction by the model of P-cell IFR during unknown visual–vestibular mismatch paradigms. During the VORr paradigm the contributions from all of the three inputs sum to produce the reconstruction of P-cell IFR. During VORs (panel 2) the major contributor to P-cell firing originated from the head movement as there was little eye movement or retinal slip during this paradigm. During VORs the efference copy component contributes negatively to P-cell modulation while the retinal slip contribution is positive. During VORe the efference copy and retinal slip components are additive and contribute positively to the P-cell modulation while the head movement contribution is negative. In the VF paradigm the efference copy and slip components are additive. During VORd the head movement and efference copy components subtract from each other resulting in the minimal modulation seen. Note that the reconstruction fits the P-cell IFR very well in each panel except for the noise superimposed on the baseline modulation. As mentioned above, the amplitude distribution and spectral content of this noise are almost identical to those of the P-cell discharge when no external inputs are present.

Figure 6 shows some properties of retinal slip during the VF paradigm in relation to VOR gain. Figure 6A illustrates OKS velocity versus retinal slip velocity at a low VOR gain state, showing that retinal slip velocity is fairly linearly related to OKS velocity within the stimulus range employed (±80 deg/s). The slope of the regression line fit to the individual data points is the gain of retinal slip (retinal slip velocity/OKS velocity). Figure 6B illustrates that the gain of retinal slip is constant at VORd gains between 0.6 and 1.4 as the slope of the line fit to the individual data points is 0 ($F_0 = 0.0623$, $p > 0.804$). On the other hand, P-cell discharge versus retinal slip velocity is plotted as gray dots in Fig. 6C. The black line is a sigmoidal curve fit to the gray dots. Downward slip increases the P-cell firing while upward slip decreases it. Between ±10 deg/s the relationship is roughly linear, however, there is a negative and a positive saturation below and above this range. Change in the shape of the sigmoidal curve in relation to VOR gain will be reported elsewhere (Hirata et al, in preparation).

Discussion

P-cell's irregular firing pattern

The highly irregular discharge of P-cells is a feature that has received little attention by vestibular and oculomotor systems scientists. Apparently most assume that the termination by multiple P-cells upon single target neurons will result in the averaging of this irregular activity by the membrane of the individual FTNs (Partsalis et al., 1995a) thus removing this noisy feature of P-cell discharge from the computational process. Indeed inactivation of the flocculus causes FTNs to become slightly more irregular than in the control condition (Partsalis et al., 1995b) indicating that the activity of the multiple P-cells that converge upon a single Y neuron has been summed and smoothed. However true this may be it does little to aid in understanding the information content of single P-cell discharge. Any raster or trial by trial display of single P-cell activity clearly indicates the variability in discharge both within and between trials of a single neuron. When a population of these cells is considered the variability is multiplied. However, from the P-cell perspective it is interesting to consider the origins of this noise and any possible effects of the noise upon the P-cell computational process.

In the VIIIth nerve the most sensitive fibers with the highest gains have the largest coefficients of variation (CV) of their interspike intervals (Goldberg and Fernandez, 1971). Generally noisy processes are concomitant with increased sensitivity. Thus noisy P-cell discharge may be related to the sensitivity and/or threshold of these neurons. The noise on individual P-cells most likely increases the possible richness of computation by these neurons, but this remains to be experimentally tested.

We have developed a method to extract stimulus and motor output-related modulation from the noisy floccular P-cell firing patterns. The method employs multiple linear regression in concert with experimental paradigms that reduce the number of linearly related variables during a given experiment. For example the stimulus-related modulation that we sought was related to retinal slip, head movement, and oculomotor efference copy (eye velocity). In the VF paradigm the head movement-related variables were removed, during VORd retinal slip is not a factor thus reducing the number of variables that show colinearity for analysis during the performance of each of these paradigms. Combining the results of this analysis of the firing of P-cells during these different paradigms we were able to extract the stimulus related P-cell modulation, and were able to decompose the firing pattern into components caused by the input signals. This allows us to evaluate changes in each component of P-cell firing during the process of VOR adaptation. We demonstrated that the spectrum and amplitude distribution of the remaining signal following the extraction of stimulus-related modulation was almost identical to that of the spontaneous firing of P-cells when no stimulus was applied, thus demonstrating that we had extracted the relevant signal encoded in the irregular discharge of P-cells (Hirata et al., in preparation). This also suggests that the apparent irregularity is intrinsic noise which may be unrelated to visual and/or vestibular stimuli. However, this does not mean that the conventional

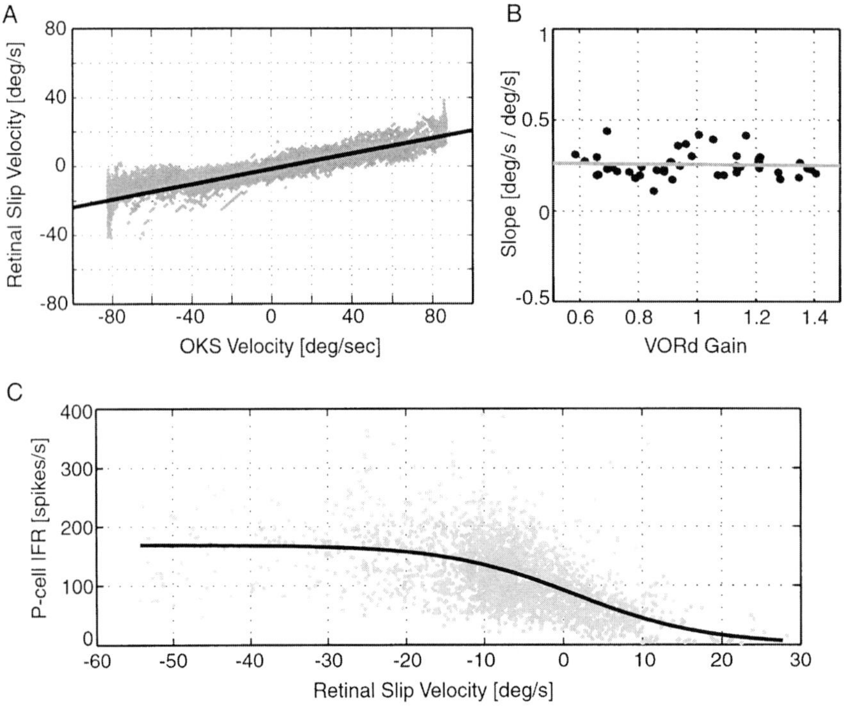

Fig. 6. Retinal slip during the VF paradigm. A. is retinal slip velocity plotted as a function of OKS velocity. The gray dots are individual traces and the black line is a fit to the data. B. illustrates that the gain of retinal slip velocity during visual following (the slope of the line in A) is independent of the VOR gain. The null hypothesis that slope of the regression line (gray) is 0 was not rejected ($F_0 = 0.0623$, $p > 0.804$). C. illustrates the IFR of a Purkinje cell during retinal slip velocity from -60 to $+30$ deg/s. Each dot is the IFR of a Purkinje cell and the solid line is a sigmoidal curve fit to the data.

'averaging over cycles' method is valid for analysis of the P-cell firing pattern, as the noise is not Gaussian and may cause some bias in the resultant averaged traces.

P-cells are reported to receive over 100,000 inputs on the spines of their dendritic trees. However individual dendrites have been shown to fire all or no impulses or dendritic spikes and these spikes can initiate somatic and initial segment action potentials. Apparently P-cells are irregularly discharging because of their modes of somatic impulse initiation by these dendritic spikes and because the somatic and initial segment membranes lack the potassium conductances that are present in regularly firing neurons and regularize discharge (Hille, 1984). Because individual dendrites can cause the cell to fire, the number of synaptic inputs that influences the characteristic of the P-cell spike train is not the sum of all dendritic 100,000 inputs but is reduced relative to those on a neuron that integrates inputs over its entire dendritic tree and then fires when an initial segment threshold is reached. Each dendritic compartment must form its own computational unit and these units interact at the initial segment to generate the spike train. It can be speculated that evolution might have decreased P-cell CV if the intrinsic noise conferred no computational advantage. However, such arguments must be tempered by the realization that decreasing the P-cell CV would ultimately result in remodeling individual P-cell cytoarchitecture and electrical architecture, thus one would no longer have P-cells recognizable in their present form. The value of the high P-cell CV is then still an open question.

Retinal slip gain during visual following

We have shown that retinal slip is linearly related to OKS velocity during the VF paradigm over the stimulus range employed (±80 deg/s) and that retinal slip gain during the VF paradigm does not change when the VVOR gain is adapted. The retinal slip gain defined as retinal slip velocity/OKS velocity can be rewritten as 1-VF gain where VF gain is defined as eye velocity/OKS velocity. Therefore, the current result shows that VF gain does not change along with the VVOR adaptation. This is consistent with the results of Lisberger et al. (1981) who showed that during VOR adaptation the indirect component of OKN was adapted or changed along with the horizontal VORd gain while the direct component did not change. Since we are using 0.5 Hz stimuli that invoke OKN mechanisms involving the direct pathway via the flocculus, we can confirm their study of the horizontal system. The fact that the retinal slip gain or VF gain does not change after the VOR adaptation does not necessarily refute the hypothesis that the flocculus is responsible for VOR adaptation. To study this issue, it is important to evaluate the P-cell's behavior in response to each input signal during VOR adaptation. The proposed method enables this evaluation by decomposing the firing pattern into retinal slip, head movement and efference copy components.

Summary

Purkinje (P) neurons are highly irregular in their discharge patterns having a high coefficient of variation. We have developed a method to extract the signal from these noisy neurons employing multiple linear regression against the input signals that are causal in P-cell firing, namely retinal slip, head motion parameters, and the efference copy of the oculomotor commands. This method was shown to extract all of the experimental stimulus-related signal as the residual following extraction had the same amplitude distribution and spectral content as those of the spontaneous discharge of P-cells when no external stimuli were applied. The method enables us to decompose P-cell firing modulation into components mediated by mossy fiber inputs to flocculus, that is, retinal slip, vestibular and oculomotor efference copy signals. By evaluating these components in relation to VOR adaptation, one can argue that one role of the flocculus is in converting these sensory and efference copy input signals into eye movement-related signals in conjunction with VOR motor learning.

Acknowledgements

Supported by NIH National Eye Institute grant EY–05433. We thank Dr. Gavin Perry for computer and electronics support and Ms. Pat Keller for histological support.

References

Blazquez, P, Partsalis, A. and Highstein, S.M. Synaptic linkage of anterior and posterior vestibular semicircular canal interneurons carrying head velocity information to the dorsal Y-group of the vestibular nuclei. *J. Neurophysiol.* In press.

Fuchs, A., Scudder, C.A. and Kaneko, C.R.S. (1988) Discharge patterns and recruitment order of identified motoneurons and internuclear neurons in the monkey abducens nucleus. *J. Neurophysiol.*, 60: 1874–1895.

Goldberg, J.M. and Fernandez, C. (1971) Physiology of peripheral neurons innervating semicircular canals of the squirrel monkey. I. Resting discharge and response to constant linear accelerations. *J. Neurophysiol.*, 34: 635–660.

Gomi, H., Shidara, M., Takemura, A., Inoue, Y., Kawano, K. and Kawato, M. (1998) Temporal firing patterns of Purkinje cells in the cerebellar ventral paraflocculus during ocular following responses in monkeys. I. Simple spikes. *J. Neurophysiol.*, 80: 818–831.

Hille, B. (1984) *Ionic Channels of Excitable Membranes.* Sinauer Associates Inc. Sunderland, MA 1–425.

Hirata, Y., Arikan, R. and Highstein, S.M. Analysis of the discharge pattern of cerebellar Purkinje cells before and during adaptation of the vestibulo ocular reflex in the squirrel monkey. In preparation.

Langer, T., Fuchs, A.F., Chubb, M.C., Scudder, C.A. and Lisberger, S.G. (1985) Floccular efferents in the rhesus macaque as revealed by autoradiography and horseradish peroxidase. *J. Comp. Neurol.*, 235: 26–37.

Lisberger, S.G., Miles, F.A., Optican, L.M. and Eighmy, B.B. (1981) Optokinetic response in monkey: underlying mechanisms and their sensitivity to long term adaptive changes in VOR. *J. Neurophysiol.*, 45: 869–890.

Maekawa, K. and Simpson, J.I. (1973) Climbing fiber responses evoked in vestibulo-cerebellum of rabbit from visual system. *J. Neurophysiol.*, 36: 649–667.

McCrea, R.A., Strassman, A. and Highstein, S.M. (1987) Anatomical and physiological characteristics of vestibular neurons mediating the vertical vestibulo ocular reflex in the squirrel monkey. *J. Comp. Neurol.*, 264: 571–594.

Partsalis, A.M., Zhang, Y. and Highstein, S.M. (1995a) Dorsal Y-Group in the squirrel monkey. I. Neuronal responses during instantaneous and long-term modifications of the vertical VOR. *J. Neurophysiol.*, 73: 615–631.

Partsalis, A.M., Zhang, Y. and Highstein, S.M. (1995b) Dorsal Y-Group in the squirrel monkey. II. Contribution of the cerebellar flocculus to neuronal responses in normal and adapted animals. *J. Neurophysiol.*, 73: 632–650.

Shidara, M., Kawano, K., Gomi, H. and Kawato, M. (1993) Inverse-dynamics encoding of eye movements of Purkinje cells in the cerebellum. *Nature*, 365: 50–52.

Zhang, Y., Partsalis, A.M. and Highstein, S.M. (1995a) Properties of superior vestibular nucleus flocculus target neurons in the squirrel monkey. I. General properties in comparison to flocculus projecting neurons. *J. Neurophysiol.*, 73: 2261–2278.

Zhang, Y., Partsalis, A.M. and Highstein, S.M. (1995b) Properties of superior vestibular nucleus flocculus target neurons in the squirrel monkey. II. Signal components revealed by reversible flocculus inactivation. *J. Neurophysiol.*, 73: 2279–2292.

SECTION IV

Models and learning

N.M. Gerrits, T.J.H. Ruigrok and C.I. De Zeeuw (Eds.)
Progress in Brain Research, Vol 124

CHAPTER 17

Hypotheses about the neural trigger for plasticity in the circuit for the vestibulo-ocular reflex

Jennifer L. Raymond[1,*] and Stephen G. Lisberger[2]

[1] *Department of Neurobiology, Stanford University School of Medicine, Fairchild D-251, Stanford, CA 94305, USA*
[2] *Howard Hughes Medical Institute, Department of Physiology, W.M. Keck Foundation Center for Integrative Neuroscience, and Sloan Center for Theoretical Neurobiology, University of California, San Francisco, CA 94143, USA*

Introduction

The cerebellum has long been implicated in motor learning. We performed a series of experiments to examine the neural events associated with the induction of one particular form of cerebellum-dependent learning, motor learning in the vestibulo–ocular reflex (VOR). The VOR stabilizes images on the retina by using vestibular signals to generate compensatory smooth eye movements in the opposite direction from head motion. Motor learning calibrates the reflex to maintain proper function (e.g. Gonshor and Melvill Jones, 1973). When image motion is present during head turns, motor learning adjusts the VOR to reduce image motion during subsequent head turns. This form of learning depends on a region of the cerebellum called the floccular complex, comprising the flocculus and ventral paraflocculus (Gerrits and Voogd, 1989). We looked for patterns of neural activity that were consistently present in the floccular complex during stimuli that induce motor learning in the VOR and which might therefore serve as the neural trigger for induction of plasticity in vivo.

Motor learning can alter a number of features of the VOR, including its amplitude, dynamics and direction (e.g. Gonshor and Melvill Jones, 1973; Robinson, 1976; Schultheis and Robinson, 1981; Lisberger et al., 1984; Powell et al., 1996; Raymond and Lisberger, 1996; Angelaki and Hess, 1998). We focus on learned changes in the amplitude, or gain, of the reflex. The *gain of the VOR* is defined as the ratio of eye movement amplitude to head movement amplitude during head turns in total darkness. The VOR is measured in the dark to isolate eye movements driven by vestibular stimulation (the VOR) from visually-driven, tracking eye movements that can occur in the light. In rhesus monkeys, the eye movements driven by the VOR are normally opposite in direction and approximately equal in amplitude to head movements, for a VOR gain close to 1.0.

Learned changes in the gain of the VOR can be induced by pairing movements of the head with movements of a visual stimulus. In general, when the head and visual stimulus move in the same direction, a decrease in gain is induced; whereas when the head and visual stimulus move in opposite directions, an increase in gain is induced. We performed experiments to look at the neural signals that trigger these learned changes in gain.

In the laboratory, the head movements used to induce motor learning in the VOR can be delivered with a servo-controlled turntable, and head movements with a sinusoidal velocity profile are often used. Motor learning is induced by pairing these head movements with the movement of a projected visual stimulus or optokinetic drum (for detailed

*Corresponding author. Tel.: (650)–725–9201; Fax: (650)–725–3958; e-mail: jenr@stanford.edu

methods, see Raymond and Lisberger, 1996). We consider two specific stimulus configurations (Fig. 1). In the $\times 0$ *configuration*, the visual stimulus moves in the same direction and at the same speed as the head. This configuration is called $\times 0$ since learning must decrease the gain of the VOR toward a value of zero to improve image stability. In the $\times 2$ *configuration*, the visual stimulus moves at the same speed as the head but in the opposite direction, and learning must increase the gain of the VOR toward a value of 2.0 to improve image stability.

Our experiments were aimed at determining the neural mechanisms by which sensory events such as $\times 0$ and $\times 2$ visual-vestibular stimuli trigger changes in the circuit for the VOR that alter the gain of the reflex. Like all sensory stimuli, the $\times 0$ and $\times 2$ visual-vestibular stimuli are encoded in patterns of neural activity in the brain. Some feature of the patterns present during the $\times 0$ stimuli must be able to trigger changes in the circuit for the VOR that result in a decrease in gain. Likewise, some feature of the patterns present during the $\times 2$ stimuli must be able to trigger changes that increase the gain.

Previous hypotheses proposed two different patterns of neural activity as the trigger for plasticity during motor learning in the VOR. The first hypothesis, known as the flocculus hypothesis, suggested that plasticity in the circuit for the VOR was triggered by coincident activity in climbing fibers carrying visual signals and parallel fibers carrying vestibular signals (Ito, 1972, 1982). A second hypothesis suggested that plasticity was triggered by coincident activity in Purkinje cells and brainstem vestibular pathways (Miles and Lisberger, 1981). Both hypotheses were based on observations of the patterns of neural activity present in the cerebellum during $\times 0$ and $\times 2$ stimuli with low-frequency (≤ 0.5 Hz) sinusoidal velocity profiles. We recorded the patterns of neural activity present during a broader range of visual-vestibular stimuli to discriminate among hypotheses about the neural mechanisms of learning in the VOR.

In particular, we recorded from the cerebellum during $\times 0$ and $\times 2$ stimuli at a range of higher sinusoidal frequencies. Behavioral studies revealed that $\times 0$ and $\times 2$ stimuli at frequencies up to 10 Hz are effective at inducing motor learning in the VOR (Raymond and Lisberger, 1996). Our recordings from the cerebellum during the high-frequency (≥ 5 Hz) stimuli place constraints on the neural mechanisms of learning that could not be derived

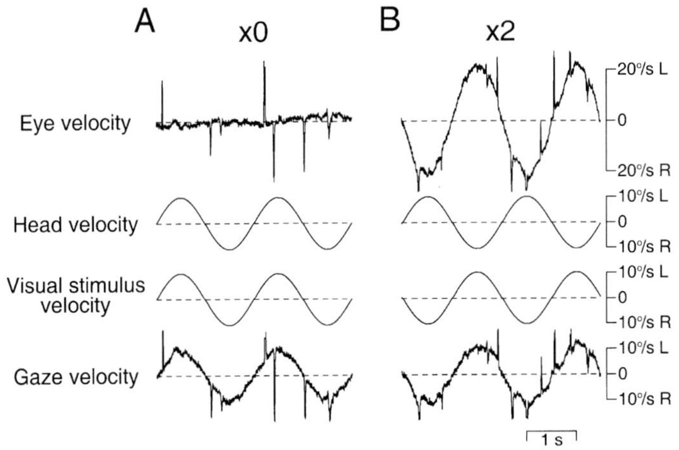

Fig. 1. Stimuli that induce learned decreases ($\times 0$, A) and increases ($\times 2$, B) in the gain of the VOR. From top to bottom, the traces are eye velocity with respect to the orbit, angular head velocity in space, visual stimulus velocity in space, and gaze velocity in space. Gaze velocity was computed as the sum of head velocity in space plus eye velocity in the orbit. In all traces, upward deflections represent leftward position or velocity (L); downward deflections represent rightward position or velocity (R). The brief deflections in the eye and gaze velocity traces are caused by saccadic eye movements; their amplitudes have been cropped. The frequency of the stimuli is 0.5 Hz (from Raymond and Lisberger, 1998).

from observations during low-frequency stimuli (Raymond and Lisberger, 1997, 1998). Neither of the previous hypotheses about motor learning in the VOR can account for the induction of learning by the high-frequency stimuli. We propose two alternative hypotheses.

Motor learning induced by visual-vestibular stimuli of different frequencies

Our first step in analyzing the events associated with induction of motor learning in the VOR was to examine the ability of a range of visual-vestibular stimuli to induce learning. In particular, we examined the induction of learning by sinusoidal $\times 0$ and $\times 2$ stimuli at a broad range of stimulus frequencies (Raymond and Lisberger, 1996).

Figure 2 illustrates the changes in gain induced by $\times 0$ and $\times 2$ stimuli at frequencies from 0.5 to 10 Hz. In each experiment, learning was induced by a 3-hour presentation of a $\times 0$ or $\times 2$ visual-vestibular stimulus at a single sinusoidal frequency, but in different experiments that frequency varied from 0.5 to 10 Hz. Before and after the 3-hr training period the gain of the VOR was measured using sinusoidal head movements in the dark at frequencies from 0.2 to 10 Hz. Learning was computed as the ratio of VOR gain after training to the gain before training (post/pre) and plotted as a function of the frequency of the test stimulus. In these plots, a value greater than 1.0 represents a learned increase in the gain of the VOR, and a value less than one represents a learned decrease in the gain of the VOR.

The results reveal that $\times 0$ (open symbols) and $\times 2$ stimuli (filled symbols) at frequencies up to 5 Hz are very effective at inducing changes in VOR gain in the adaptive direction (i.e. decrease in gain for $\times 0$ stimuli, increase in gain for $\times 2$ stimuli). $\times 0$ and $\times 2$ stimuli at frequencies of 8 or 10 Hz induced changes in the gain of the VOR that were less consistent but generally in the adaptive direction. In all experiments, changes in the phase of the VOR (phase of eye movement relative to head movement) were less than 10 degrees (data not shown).

Patterns of neural activity in the cerebellum during motor learning in the VOR

The behavioral analysis provides a framework for evaluating the neural events that trigger the

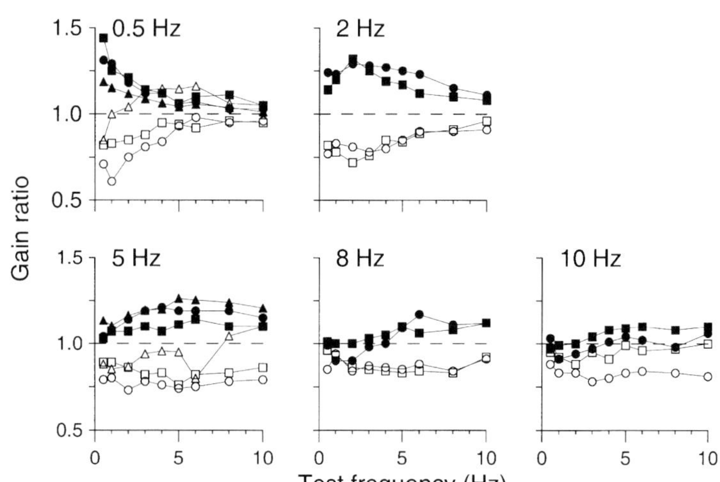

Fig. 2. Learned changes in the gain of the VOR induced by sinusoidal visual-vestibular stimuli. Each plot shows the results for a single training frequency (0.5, 2, 5, 8, or 10 Hz, indicated at the top of the plot). Changes in VOR gain (post-training/pretraining gain ratio) are plotted as a function of the frequency of the sinusoidal vestibular test stimuli used to measure the VOR. Open symbols, $\times 0$ stimulus conditions; filled symbols, $\times 2$ stimulus conditions. Different symbol shapes show results from different monkeys (from Raymond and Lisberger, 1996).

induction of plasticity in the circuit for VOR. The behavioral results indicate that some feature of the patterns of neural activity present during ×0 stimuli at frequencies up to 10 Hz must be capable of triggering changes in the circuit for the VOR that decrease the gain of the VOR. Likewise, the patterns present during ×2 stimuli at frequencies up to 10 Hz must be capable of inducing circuit changes that increase the gain of the VOR.

We recorded from the floccular complex of the cerebellum during ×0 and ×2 stimuli at frequencies from 0.5 to 10 Hz (Raymond and Lisberger, 1997, 1998). In these recordings, the idea was to compare the neural signals present in individual neurons during several different visual-vestibular stimuli, all of which would induce motor learning in the VOR if presented for a long enough period. We then evaluated which patterns of Purkinje cell, climbing-fiber, and vestibular signals might be able to guide learning across the range of stimulus frequencies. These neural signals were identified previously as candidates for the signals that trigger plasticity in the circuit for the VOR.

We used two criteria to evaluate which patterns of neural activity might trigger plasticity in the circuit for the VOR during learning: First, signals that guide learning must discriminate ×0 stimuli, which decrease the gain of the VOR, from ×2 stimuli, which increase the gain of the VOR. Patterns of neural activity that are the same during both stimulus configurations would contain no information about whether the gain should decrease or increase and therefore could not be responsible for the opposite learned changes in gain induced by the ×0 and ×2 stimuli. Second, if there is a single mechanism that mediates learning in the VOR, then there should be similar patterns of neural activity present during all stimuli that induce similar learned changes in gain. For example, if an increase in the gain of the VOR is induced by coincident activity in two neural pathways, then activity in those pathways should be coincident during ×2 stimuli at all frequencies from 0.5 to 10 Hz. Neural signals that meet our two criteria would provide unambiguous information about whether to increase or decrease the gain of the VOR and thus could serve as the neural trigger for plasticity in the circuit for the VOR.

Recordings from horizontal gaze velocity Purkinje cells

We focus on an extensively studied group of Purkinje cells in the floccular complex called the horizontal gaze velocity Purkinje cells, or HGVPs (Lisberger and Fuchs, 1978; Miles et al., 1980). A Purkinje cell receives two kinds of excitatory inputs. It receives numerous parallel fiber inputs, which drive simple spikes in the Purkinje cell, and it receives a powerful input from a single climbing fiber, whose spikes drive complex spikes in the Purkinje cell in a one-to-one manner. In HGVPs, the simple-spike activity carries a combination of vestibular, visual, and eye movement signals (for review, see du Lac et al., 1995). Climbing fiber inputs to the HGVPs carry signals related to visual image movement (for review, see Stein and Glickstein, 1992, Krauzlis, 1994). There is considerable evidence that HGVPs contribute to the VOR (for review, see du Lac et al., 1995.

Simple spikes

The histograms in Fig. 3 show the simple-spike activity in a typical HGVP during ×0 and ×2 stimuli that, if prolonged, would have induced learned changes in the gain of the VOR. We compare the patterns of simple spike activity present during ×0 and ×2 stimuli at 0.5 and 5 Hz. All four stimuli caused modulation of the simple-spike activity around a fairly high average firing rate of approximately 75 spikes/sec. Because of the high spontaneous firing rate of the HGVPs, we analyzed the modulation of simple-spike activity rather than absolute levels of activity. At 0.5 Hz, the timing of elevated simple-spike activity relative to the vestibular stimulus discriminated the ×0 stimulus configuration from the ×2 stimulus configuration. When a 0.5 Hz vestibular stimulus was paired with ×0 visual stimulus motion, peak simple-spike activity in the HGVP coincided with ipsiversive head motion (Fig. 3A, vertical dashed line, Ipsi). When the same vestibular stimulus was paired with ×2 visual stimulus motion, peak simple-spike activity coincided with contraversive head motion (Fig 3C, vertical dashed line, Contra). Similar results have been obtained by a number of

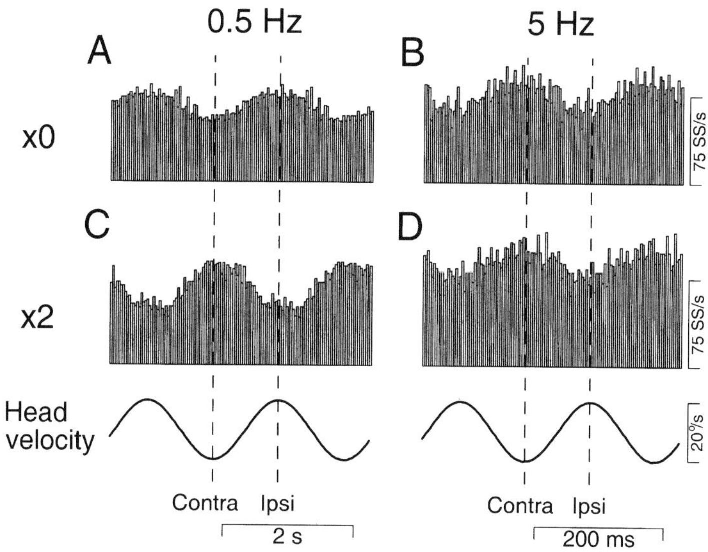

Fig. 3. Histograms showing the simple-spike activity in a representative Horizontal Gaze Velocity Purkinje cell during visual-vestibular stimuli that induce learned decreases (×0, A, B) or increases (×2, C, D) in the gain of the VOR. Head velocity, Angular head velocity in the horizontal plane. Vertical dashed lines mark peak contraversive and ipsiversive head velocity. Note the different time scales in the left (A, C) and right (B, D) panels, which show data for sinusoidal stimuli at 0.5 Hz and 5 Hz, respectively. Twenty to 500 stimulus cycles were averaged to obtain each histogram (from Raymond and Lisberger, 1998).

laboratories (e.g. Lisberger and Fuchs, 1978; Miles et al., 1980; Watanabe, 1984; Highstein et al., 1997). Based on such results, Miles and Lisberger (1981) suggested that when simple-spike activity increased during ipsiversive head movement, that activity would trigger synaptic changes that decrease the gain of the VOR; but when simple-spike activity increased during contraversive head movement, that activity would trigger synaptic changes that increase the gain of the VOR. A mechanism of plasticity in the circuit for the VOR would discriminate these two conditions by comparing simple-spike activity in Purkinje cells with information about the direction of head movement carried in the activity of vestibular pathways.

This hypothesis is consistent with observations at low stimulus frequencies, but it cannot account for the induction of learning by high-frequency visual-vestibular stimuli. Fig. 3B, D show the simple-spike responses of the same HGVP during ×0 and ×2 stimuli at 5 Hz. The simple-spike activity exhibits some modulation during the 5 Hz stimuli, however its timing relative to the head movement fails to discriminate the ×0 stimulus configuration (Fig. 3B) from the ×2 stimulus configuration (Fig. 3D). The simple-spike response was the same regardless of whether the stimulus induces a decrease in the gain of the VOR (×0) or an increase in the gain of the VOR (×2). Therefore, a plasticity mechanism that compared simple-spike activity in Purkinje cells with the activity in vestibular pathways could not produce opposite changes in response to the ×0 and ×2 stimuli.

The simple-spike responses shown in Fig. 3 were typical of those recorded in the population of HGVPs (Raymond and Lisberger, 1998). Figure 4 summarizes the results from 54 HGVPs. The simple spike responses are plotted in polar coordinates, at a distance from the origin corresponding to the amplitude of response modulation and an angle corresponding to the phase of peak simple-spike activity relative to the vestibular stimulus. Simple-spike responses in phase with peak ipsiversive head velocity are plotted to the right of the origin, responses in phase with peak contraversive head velocity are plotted to the left of the origin, and clockwise rotation around the graph represents increased phase lead. Thus, in these polar plots, the phase reflects the relative timing of the simple-spike and vestibular signals. The key question is whether the responses to ×0 and ×2 stimuli are different and plot in separate parts of each graph or whether the responses during ×0 and ×2 stimuli are similar and plot together. In other words, do the patterns of simple spike and vestibular signals meet our first criterion that signals that guide learning should discriminate stimuli that increase the gain of the VOR from those that decrease the gain of the VOR?

At 0.5 Hz (Fig. 4A) and 2 Hz (Fig. 4B), the patterns of simple-spike and vestibular signals present during the ×0 stimuli (Fig. 4, + and × symbols) were clearly different from the signals present during the ×2 stimuli (Fig. 4, filled symbols). Responses to the ×0 and ×2 stimuli form distinguishable, almost non-overlapping populations that plot on opposite sides of each graph. Thus, the timing of the simple spikes relative to the vestibular stimulus can discriminate ×0 from ×2 stimuli at these low frequencies. In contrast, at 5 Hz (Fig. 4C) and 10 Hz (Fig. 4D), the patterns of simple-spike and vestibular signals present during the ×0 stimuli were quite similar to those present during the ×2 stimuli. The polar plots in Fig. 4 show extensive overlap in the response populations, even for the 5 Hz stimuli, which produced simple-spike responses of amplitudes similar to those produced by lower frequency stimuli. Thus, the relative timing of simple-spike and vestibular signals cannot distinguish the ×0 stimuli from the ×2 stimuli at frequencies of 5 Hz or above.

We must conclude that the patterns of simple-spike and vestibular signals cannot be responsible for triggering the opposite changes in the gain of the VOR induced by the ×0 and ×2 stimuli at 5 Hz. If the correlation of simple-spike and vestibular activity contributes to motor learning in the VOR, it must do so only during low-frequency and not high-frequency visual-vestibular stimuli.

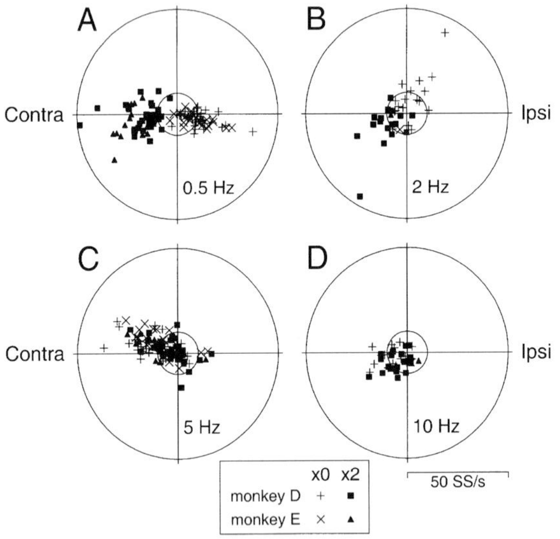

Fig. 4. Summary of simple-spike responses in Horizontal Gaze Velocity Purkinje cells, plotted relative to the vestibular stimulus. Each plot compares responses during ×0 stimuli and ×2 stimuli at the single frequency indicated in the bottom right quadrant. Each point represents the simple-spike activity recorded in a single Purkinje cell, plotted in polar coordinates with distance from the origin corresponding to the amplitude of response modulation and an angle corresponding to the phase shift between peak simple-spike activity and head velocity. Responses in phase with peak ipsiversive head velocity are plotted to the right of the origin, responses in phase with peak contraversive head velocity are plotted to the left of the origin, and clockwise rotation around the graph represents increased phase lead. An individual Purkinje cell contributed two symbols for each stimulus frequency: a + (monkey D) or × (monkey E) symbol for the ×0 stimulus, and a filled square (monkey D) or filled triangle (monkey E) for the ×2 stimulus. The plots in A–D are on the same scale, with inner and outer circles representing 10 simple spikes/sec (SS/s) and 50 simple spikes/sec, respectively (from Raymond and Lisberger, 1998).

Climbing fiber responses

The correlation of climbing fiber and vestibular activity is another candidate for the neural signals that guide motor learning in the VOR (Ito, 1972, 1982). The histograms in Fig. 5 show the responses of a typical climbing fiber to low- and high-frequency ×0 and ×2 stimuli. We recorded the climbing fiber responses as complex spikes in an HGVP. During the 0.5 Hz, ×0 stimulus, which induces a decrease in VOR gain, peak climbing fiber activity coincided with contraversive head velocity (Fig. 5A); but during the 0.5 Hz, ×2 stimulus, which induces an increase in VOR gain, peak climbing fiber activity coincided with ipsiversive head velocity (Fig. 5C). Similar results have been obtained by a number of laboratories (e.g. Simpson and Alley, 1974; Ghelarducci et al., 1975; Watanabe, 1984). Based on results like these, and the theoretical work of Marr (1969) and Albus (1971), Ito (1972, 1982) proposed that when

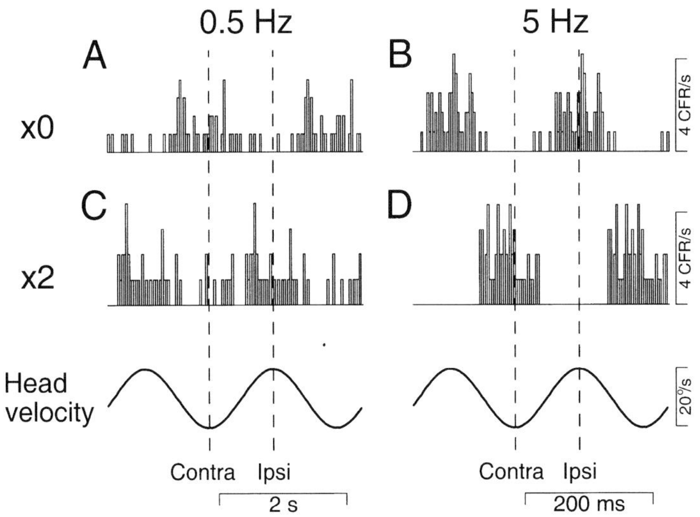

Fig. 5. Histograms showing activity in the climbing-fiber input to an HGVP during ×0 and ×2 stimuli at 0.5 and 5 Hz. Histograms showing climbing-fiber activity during stimuli that induce learned decreases (×0, A, B) and increases (×2, C, D) in the gain of the VOR. Climbing-fiber activity was recorded as complex spikes in the Purkinje cell whose simple-spike responses are shown in Fig. 3. Climbing-fiber responses from 20–500 stimulus cycles were averaged to obtain each histogram. Note the different time scales in the left (A, C) and right (B, D) panels, which show data for sinusoidal stimuli at 0.5 and 5 Hz, respectively. Vertical dashed lines mark peak contraversive and ipsiversive head velocity (from Raymond and Lisberger, 1998).

climbing fiber activity coincided with ipsiversive head movement (as reported by elevated activity in the vestibular parallel fibers), synaptic changes should be induced that increase the gain of the VOR; but when climbing fiber activity coincided with contraversive head movement (as reported by reduced activity in the vestibular parallel fibers), synaptic changes should be induced that decrease the gain of the VOR. He further proposed that the form of synaptic plasticity that mediated such changes was the long-term depression of synapses from vestibular parallel fibers onto Purkinje cells, driven by coincident activity in the parallel fibers and climbing fibers (cerebellar LTD; Ito, 1989). This hypothesis is known as the flocculus hypothesis.

Whereas the flocculus hypothesis can account for the observations at low frequencies, it cannot account for the induction of learning by the high-frequency stimuli. Fig. 5B, D show the responses of the climbing fiber to ×0 and ×2 stimuli at 5 Hz. The activity of the climbing fiber is well modulated during both of the 5 Hz stimuli, and the timing of the climbing fiber activity is different during the ×0 and ×2 stimuli. Nevertheless, the flocculus hypothesis cannot account for the induction of appropriate learned changes in gain by these stimuli. According to the flocculus hypothesis, the coincidence of climbing fiber activity with ipsiversive head movement (and hence with elevated activity in vestibular parallel fibers) should trigger synaptic changes that increase the gain of the VOR. Therefore the flocculus hypothesis would predict that the 5 Hz, ×0 stimulus should increase the gain of the VOR, which is opposite to the change that is actually induced by this stimulus. Furthermore, climbing fiber activity is not coincident with ipsiversive head movement during the ×2 stimulus at 5 Hz, which actually does trigger an increase in VOR gain

The responses shown in this example were typical of those recorded in climbing fiber inputs to HGVPs (Raymond and Lisberger, 1998). Figure 6 uses polar plots to summarize the correlation of the vestibular stimulus with activity in the climbing fiber inputs to 26 HGVPs. For each stimulus frequency, the climbing fiber responses to the ×0 and ×2 stimuli plot on opposite sides of the graph, forming distinguishable, almost non-overlapping populations. However, in the different graphs, the phase of the climbing fiber responses relative to the vestibular stimulus rotated as a function of stimulus frequency, indicating that the relative phase of climbing-fiber and vestibular signals was not similar for all stimuli that produced similar learned changes in the gain of the VOR. During the $\times 2$ stimuli (Fig. 6, filled symbols), for example, peak

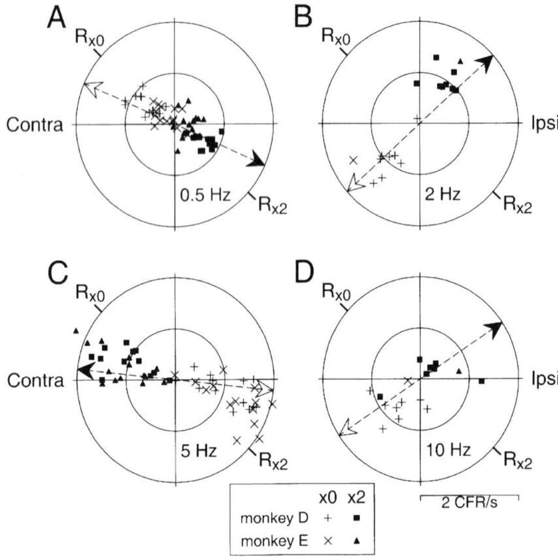

Fig. 6. Summary of climbing-fiber responses, plotted relative to the vestibular stimulus. A–D, Climbing-fiber responses to stimulus frequencies of 0.5, 2, 5, and 10 Hz. Responses are plotted in polar coordinates, with distance from the origin corresponding to the amplitude of response modulation and an angle corresponding to the phase shift between peak climbing-fiber activity and head velocity. Responses in phase with peak ipsiversive head velocity are plotted to the right of the origin, responses in phase with peak contraversive head velocity are plotted to the left of the origin, and clockwise rotation around the graph represents increased phase lead. An individual climbing fiber contributed two symbols for each stimulus frequency: a + (monkey D) or × (monkey E) symbol for the $\times 0$ stimulus, and a filled square (monkey D) or a filled triangle (monkey E) for the $\times 2$ stimulus. $R_{\times 0}$ marks the point in the vestibular stimulus leading peak contraversive head velocity by 46°, and $R_{\times 2}$ marks the point in the vestibular stimulus leading peak ipsiversive head velocity by 46°, Open arrows show the predicted phase at each frequency for a fixed delay in the climbing-fiber response of 122 msec from $R_{\times 0}$. Filled arrows show the predicted phase at each frequency for a fixed delay in the climbing-fiber response of 122 msec from $R_{\times 2}$. In all panels, inner and outer circles represent 1 climbing-fiber spike per second (CFR/s) and 2 climbing-fiber spikes/sec, respectively. Note the difference in scale from Fig. 4. (from Raymond and Lisberger, 1998).

climbing fiber activity led ipsiversive head velocity slightly at 0.5 Hz (Fig. 6A), it lagged ipsiversive head velocity at 2 and 10 Hz (Fig. 6B, D), and it was almost exactly out of phase with ipsiversive head velocity at 5 Hz (Fig. 6C). Similarly, climbing fiber responses to $\times 0$ stimuli at different frequencies peaked during different phases of the vestibular stimulus (Fig. 6, + and × symbols). These data demonstrate that coincident activity in vestibular and climbing-fiber pathways could not serve as a reliable neural trigger for motor learning in the VOR across the range of effective stimulus frequencies. Computational analysis of the data further indicates that this conclusion is valid regardless of the particular form of the vestibular signals at the site of plasticity (Raymond and Lisberger, 1998).

Recordings from other groups of Purkinje cells

The recordings from HGVPs suggest that neither of the previous hypotheses about the patterns of simple-spike, climbing-fiber, and vestibular activity that trigger plasticity can account for the induction of motor learning in the VOR by the high-frequency visual-vestibular stimuli. However, it was possible that another group of Purkinje cells in the floccular complex could exhibit responses consistent with these hypotheses. Therefore, we recorded the responses of a heterogeneous group of 45 Purkinje cells in the floccular complex during $\times 0$ and $\times 2$ stimuli at different frequencies (Raymond and Lisberger, 1997). These Purkinje cells responded to horizontal head or eye movements, but failed to meet the criteria for classification as HGVPs.

The results from these non-HGV Purkinje cells, like the results from the HGVPs, failed to support either of the previous hypotheses about the neural mechanisms of motor learning in the VOR. During high-frequency stimuli, the phase of the simple-spike responses relative to head movement failed to discriminate $\times 0$ from $\times 2$ stimuli, which is difficult to reconcile with the Miles-Lisberger hypothesis (1981). Furthermore, the phase of the climbing fiber response relative to head movement depended not just on stimulus configuration (x0 vs. $\times 2$), but also varied with stimulus frequency, which is difficult to reconcile with the flocculus hypothesis (Ito, 1972, 1982).

New hypotheses about the neural mechanisms of motor learning in the VOR

Our recordings of the patterns of activity present in Purkinje cells and their climbing fiber inputs during high-frequency visual-vestibular stimuli place constraints on the mechanisms of motor learning in the VOR that could not be derived from observations during low-frequency stimuli. These constraints suggest that both of the previous hypotheses about motor learning in the VOR require revision. We propose two alternative hypotheses.

> *The first hypothesis suggests that motor learning in the VOR is mediated by a single form of plasticity in which climbing fiber activity induces a change in the strength of vestibular synapses active about 100 ms earlier.*

Examination of the recording data reveals two patterns of neural activity that potentially could serve as reliable neural triggers for increases and decreases in VOR gain across stimulus frequency. The filled arrows in Fig. 6 show a fixed time delay of 122 ms from the point in the head movement labeled $R_{\times 2}$*. The fixed time delay translates into a progressively larger phase shift at higher frequencies, and provides a good fit to the phase of the actual climbing fiber responses to $\times 2$ stimuli. $R_{\times 2}$ roughly coincides with the time of peak activity in the ipsilateral vestibular nerve (Fernandez and Goldberg, 1971; Lisberger and Pavelko, 1986). Therefore, during $\times 2$ stimuli at all frequencies, climbing fiber activity peaks about 122 ms after peak activity in the ipsilateral vestibular nerve. This represents a consistent pattern of neural signals that potentially could be used to reliably trigger synaptic changes that increase the gain of the VOR. Similarly, the climbing fiber responses during $\times 0$ stimuli are well fit by a fixed delay of 122 ms from the point in the head movement labeled $R_{\times 0}$ (open arrows). $R_{\times 0}$ coincides with a period of reduced activity in the ipsilateral vestibular nerve. Therefore learned decreases in the gain of the VOR could be reliably triggered by a plasticity mechanism that is activated whenever climbing fiber activity peaks 122 ms after a period of inhibition in the ipsilateral vestibular nerve. These two patterns of neural activity meet both of our criteria for the neural signals that guide learning: they discriminate $\times 0$ from $\times 2$ stimuli, and similar patterns of neural signals are present during stimuli that induce similar changes in the behavior. Thus, a new hypothesis is that motor learning in the VOR is mediated by a single form of plasticity in which climbing fiber activity induces a change in the strength of vestibular synapses active approximately 100 ms earlier, rather than a plasticity mechanism driven by coincident activity in these pathways. Computational analysis (Raymond and Lisberger, 1998) suggests that this 100 ms gap between vestibular and climbing-fiber activity must be bridged by the biochemical properties of the plasticity mechanism itself, and not by a delay in the vestibular pathways to the site of plasticity.

The suggestion of a plasticity mechanism tuned to a 100 msec delay between activity in vestibular pathways and activity in climbing fibers is difficult to reconcile with the hypothesis that cerebellar LTD, as it is presently understood, can account for all motor learning in the VOR. Studies of cerebellar LTD generally have found that plasticity is most effectively induced when climbing-fiber stimulation is simultaneous with or precedes parallel-fiber stimulation (Ito et al., 1982; Ekerot and Kano, 1985, 1989; Sakurai, 1987; Schreurs and Alkon, 1993; Lev-Ram et al., 1995, 1997). In contrast, our results suggest that a plasticity mechanism in the circuit for the VOR would need to be most effectively driven by climbing fiber activity that followed activity in vestibular parallel fibers by approximately 100 ms. Even reports that depression can be induced in the cerebellar cortex when climbing-fiber stimulation follows parallel-fiber stimulation by 100–250 msec (Chen and Thompson, 1995; Shreurs and Alkon, 1996) cannot account for motor learning in the VOR. It is not sufficient for a 100 msec delay between vestibular and climbing-fiber activity to be effective at inducing plasticity. It must be the *most* effective

* For details on how the values for $R_{\times 2}$, $R_{\times 0}$ and the fixed delay of 122 ms were obtained, see Raymond and Lisberger, 1998.

input configuration for inducing plasticity; that is, it must induce the largest synaptic changes. A plasticity mechanism that produced the same synaptic changes in vestibular parallel fibers active 100 msec before climbing-fiber activity as it did in vestibular parallel fibers whose activity coincided with activity in the climbing fiber would fail to produce changes in the gain of the VOR during $\times 0$ and $\times 2$ stimuli at 5 Hz. Thus, if a plasticity mechanism driven by climbing fiber and vestibular parallel fiber activity is to account for all motor learning in the VOR, then the timing contingencies of this plasticity mechanism must be different from those that have been described for cerebellar LTD. Alternatively, there could be a form of plasticity with appropriate timing contingencies that is driven by activity in climbing fibers and vestibular pathways other than the parallel fibers (e.g. inhibitory neurons in the cerebellar cortex).

A second hypothesis suggests that there are multiple mechanisms of motor learning in the VOR

It has generally been assumed that the simple form of learning exhibited by the VOR has a single neural mechanism. However, it is possible that visual-vestibular stimuli of different frequencies utilize different plasticity mechanisms in inducing motor learning in the VOR. This hypothesis is supported by the behavioral observation of different patterns of generalization when learning is induced with high- vs. low-frequency visual-vestibular stimuli (Fig. 2). When learning was induced with 0.5 Hz stimuli, the learned changes in gain were frequency-selective. The largest increases or decreases in gain were measured at test frequencies near 0.5 Hz, and progressively smaller changes were observed at higher test frequencies. In contrast, when motor learning was induced with high-frequency stimuli (e.g. 5 Hz), the changes induced in VOR gain were more similar across test frequency. Thus, the changes induced in the circuit for the VOR by the high-frequency stimuli must be somewhat different than the changes induced by the low-frequency stimuli. Therefore, different neural mechanisms may be involved when learning is induced by stimuli of different frequencies.

A plasticity mechanism that depended on the timing of simple-spike activity relative to the vestibular stimulus potentially could guide motor learning in the VOR at low stimulus frequencies but not at high stimulus frequencies. Additional analysis of the recordings reveals that the correlation of simple-spike and climbing-fiber activity contains information that could be used to guide learning at high stimulus frequencies but not low frequencies (Raymond and Lisberger, 1998). It is possible that different combinations of neural signals each contribute to the induction of motor learning in the VOR, but under a limited range of stimulus conditions. It may be that multiple mechanisms are required to ensure effective motor learning in the VOR across a wide range of stimulus frequencies.

Conclusion

We used motor learning in the VOR as a model system for analyzing the neural signals in the cerebellum that trigger plasticity. By recording from the cerebellum during a broad range of stimuli that induce learning, we obtained results that challenge both of the previous hypotheses about the neural mechanisms of motor learning in the VOR (Ito, 1972, 1982; Miles and Lisberger, 1981), and suggest two new hypotheses that can be tested in future studies.

Motor learning in the VOR is a simple form of learning, but it has many parallels at the behavioral and circuit level with other forms of cerebellum-dependent learning (e.g. Raymond et al., 1996). Given these parallels, and the highly uniform architecture of the cerebellum, it is possible that by studying a simple and tractable system like the VOR, we may uncover general principles that apply to the cerebellum as a whole, and that such principles may eventually be applied to understanding the role of the cerebellum in more complex behaviors and perhaps even non-motor functions.

Acknowledgements

We thank I-han Chou and Holger Rambold for helpful comments on an earlier draft of the manuscript. This work was supported by National Institutes of Health Grants EY10198, EY03878 (S.G.L.), and DC03342 (J.L.R.), and a National

Aeronautics and Space Administration Research Associate Fellowship (J.L.R.). S.G. Lisberger is an investigator of the Howard Hughes Medical Institute.

References

Albus, J.S. (1971) A theory of cerebellar function. *Math. Biosci.*, 10: 25–61.

Angelaki, D.E. and Hess, B.J. (1998) Visually induced adaptation in three-dimensional organization of primate vestibulo–ocular reflex. *J. Neurophysiol.*, 79: 791–807.

Chen, C. and Thompson, R.F. (1995) Temporal specificity of long-term depression in parallel fiber Purkinje synapses in rat cerebellar slice. *Learn. Mem.*, 2: 185–198.

duLac, S., Raymond, J.L., Sejnowski, T.J. and Lisberger, S.G. (1995) Learning and memory in the vestibulo–ocular reflex. *Annu. Rev. Neurosci.*, 18: 409–441.

Ekerot, C.-F. and Kano, M. (1985) Long-term depression of parallel fibre synapses following stimulation of climbing fibres. *Brain Res* 342:357–360.

Ekerot, C.-F. and Kano, M. (1989) Stimulation parameters influencing climbing fibre induced long-term depression of parallel fibre synapses. *Neurosci. Res.*, 6: 264–268.

Fernandez, C. and Goldberg, J.M. (1971) Physiology of peripheral neurons innervating semicircular canals of the squirrel monkey. II. Response to sinusoidal stimulation and dynamics of peripheral vestibular system. *J. Neurophysiol.*, 34: 661–675

Gerrits, N.M. and Voogd, J. (1989) The topographical organization of climbing and mossy fiber afferents in the flocculus and ventral paraflocculus in the rabbit, cat, and monkey. *Exp. Brain Res. Suppl.*, 17: 26–29.

Ghelarducci, B., Ito, M. and Yagi, N. (1975) Impulse discharges from flocculus Purkinje cells of alert rabbits during visual stimulation combined with horizontal head rotation. *Brain Res.*, 87: 66–72.

Gonshor, A. and Melvill Jones, G. (1973) Changes of human vestibulo-ocular response induced by vision-reversal during head rotation. *J. Physiol.*, 234: 102–103.

Highstein, S.M., Partsalis, A. and Arikan, R. (1997) Role of the Y-group of the vestibular nuclei and flocculus of the cerebellum in motor learning of the vestibulo-ocular reflex. *Prog. Brain Res.*, 114: 383–397.

Ito, M. (1972) Neural design of the cerebellar motor control system. *Brain Res.*, 40: 81–84.

Ito, M. (1982) Cerebellar control of the vestibulo–ocular reflex – around the flocculus hypothesis. *Annu. Rev. Neurosci.*, 5: 275–298.

Ito, M. (1989) Long-term depression. *Annu. Rev. Neurosci.*, 12: 85–102.

Ito, M., Sakurai, M. and Tongroach, P. (1982) Climbing fibre induced depression of mossy fibre responsiveness and glutamate sensitivity of cerebellar Purkinje cells. *J. Physiol. (Lond.)*, 324: 113–124.

Krauzlis, R.J. (1994) The visual drive for smooth eye movements. In: A.T. Smith and R.J. Snowden (Eds), *Visual Detection of Motion*, London: Academic Press.

Lev-Ram, V., Makings, L.R., Keitz, P.F., Kao, J.P.Y. and Tsien, R.Y. (1995) Long-term depression in cerebellar Purkinje neurons results from coincidence of nitric oxide and depolarization-induced Ca^{2+} transients. *Neuron*, 15: 407–415.

Lev-Ram, V., Jiang, T., Wood, J., Lawrence, D.S. and Tsien, R.Y. (1997) Synergies and coincidence requirements between NO, cGMP, and Ca^{2+} in the induction of cerebellar long-term depression. *Neuron*, 18: 1025–1038

Lisberger, S.G. and Fuchs, A.F. (1978) Role of primate flocculus during rapid behavioral modification of the vestibulo–ocular reflex I. Purkinje cell activity during visually guided horizontal smooth pursuit eye movements and passive head rotation. *J. Neurophysiol.*, 41: 733–763.

Lisberger, S.G. and Pavelko, T.A. (1986) Vestibular signals carried by pathways subserving plasticity of the vestibulo-ocular reflex in monkeys. *J. Neurosci.*, 6: 346–354.

Lisberger, S.G., Miles, F.A. and Zee, D.S. (1984) Signals used to compute errors in monkey vestibulo–ocular reflex: possible role of the flocculus. *J. Neurophysiol.*, 52: 1140–1153.

Marr, D. (1969) A theory of cerebellar cortex. *J. Physiol.*, 202: 437–470.

Miles, F.A. and Lisberger, S.G. (1981) Plasticity in the vestibulo-ocular reflex: a new hypothesis. *Ann. Rev. Neurosci.*, 4: 273–299.

Miles, F.A., Fuller, J.H., Braitman, D.J. and Dow, B.M. (1980) Long-term adaptive changes in primate vestibulo–ocular reflex. III. Electrophysiological observations in flocculus of normal monkeys. *J. Neurophysiol.*, 43: 1437–1476.

Powell, K.D., Peterson, B.W. and Baker, J.F. (1996). Phase-shifted direction adaptation of the vestibulo–ocular reflex in cat. *J. Vestib. Res.*, 6: 277–293.

Raymond, J.L. and Lisberger, S.G. (1996) Behavioral analysis of signals that guide learned changes in the amplitude and dynamics of the vestibulo–ocular reflex. *J. Neurosci.*, 16: 7791–7802.

Raymond, J.L. and Lisberger, S.G. (1997) Multiple subclasses of Purkinje cells in the primate floccular complex provide similar signals to guide learning in the vestibulo–ocular reflex. *Learn. Mem.*, 3: 503–518.

Raymond, J.L. and Lisberger, S.G. (1998) Neural learning rules for the vestibulo–ocular reflex. *J. Neurosci.*, 18: 9112–9129.

Raymond, J.L., Lisberger, S.G. and Mauk, M.D. (1996). The cerebellum: a neuronal learning machine? *Science*, 272: 1126–1131.

Robinson, D.A. (1976) Adaptive gain control of vestibulo–ocular reflex by the cerebellum. *J. Neurophysiol.*, 39: 954–969.

Sakurai, M. (1987) Synaptic modification of parallel fibre-Purkinje cell transmission in in vitro guinea-pig cerebellar slices. *J. Physiol. (Lond.)*, 394: 463–480.

Schreurs B.G. and Alkon D.L. (1993) Rabbit cerebellar slice analysis of long-term depression and its role in classical

conditioning. *Brain Res.*, 631: 235–240.
Schreurs, B.G. and Alkon, D.L. (1996) Pairing-specific long-term depression of Purkinje cell excitatory postsynaptic potentials results from a classical conditioning procedure in the rabbit cerebellar slice. *J. Neurophysiol.*, 75: 1051–1060.
Schultheis, L.W. and Robinson, D.A. (1981). Directional plasticity of the vestibulo–ocular reflex in the cat. *Ann. NY Acad. Sci.*, 374: 504–512.
Simpson, J.I. and Alley, K.E. (1974) Visual climbing fiber input to rabbit vestibulo-cerebellum: a source of direction-specific information. *Brain Res.*, 82: 302–308.
Stein, J.F. and Glickstein, M. (1992) Role of the cerebellum in visual guidance of movement. *Physiol. Rev.*, 72: 967–1017.
Watanabe, E. (1984) Neuronal events correlated with long-term adaptation of the horizontal vestibulo–ocular reflex in the primate flocculus. *Brain Res.*, 297: 169–174.

N.M. Gerrits, T.J.H. Ruigrok and C.I. De Zeeuw (Eds.)
Progress in Brain Research, Vol 124

CHAPTER 18

On the nature of gain changes of the optokinetic reflex

M.A. Frens*, A.L. Mathoera and J. Van der Steen

Department of Physiology, Erasmus University Rotterdam, P.O. Box 1738, 3000 DR Rotterdam, The Netherlands

Introduction

Compensatory eye movements serve to minimize retinal slip. By rotating the eyes in a direction opposite to movements of the head, the orientation of the eyes in space remains relatively stable, which prevents blurring of the retinal image by movements of the visual scene relative to the head. In head-restrained afoveate species, such as the rabbit, these reflexive eye movements (most importantly the vestibulo–ocular reflex (VOR) and the optokinetic reflex (OKR)) form the sole oculomotor output. Therefore the rabbit is an ideal animal to study compensatory eye movements in isolation, without interference of pursuit or saccades.

Both the VOR and the OKR are highly plastic, and changing conditions can modify the performance of the reflexes within short periods of time. Nonetheless, somewhat surprisingly the gain (the ratio of eye movement amplitude and stimulus amplitude) of the reflexes of the rabbit is usually less than one. In other words, the response is not perfect but rather shows a consistent undershoot (Collewijn, 1969). Only during combined vestibular and visual stimulation the gain reaches a value that is close to unity.

Plasticity of compensatory eye movements has been most extensively studied in the VOR (e.g. Collewijn and Grootendorst, 1979; Demer et al., 1989; De Zeeuw et al., 1998). For instance, a *gain-increase* paradigm in which the animal is sinusoidally rotated, while the visual environment is rotated in the opposite direction, leads to an increased amount of retinal slip with respect to a rotation in a stable environment. The VOR adapts to this new condition within hours by increasing the amplitude of the oculomotor response (Collewijn and Grootendorst, 1979). This increase persists even when the animal is subsequently rotated in the dark. Likewise, a *gain-decrease* paradigm in which the visual environment moves in phase with the vestibular stimulation leads to a decrease in the amplitude of the eye movements. In the rabbit VOR-adaptation is somewhat specific for the frequency of the adapting stimulus. At other stimulus frequencies the response changes less (Collewijn and Grootendorst, 1979).

Also the OKR has plastic properties. This has been shown when the animal is subjected to prolonged optokinetic stimulation. The gain of the OKR increased from 0.65 to 0.8 after four hours of stimulation at a frequency of 1/6 Hz (Collewijn and Kleinschmidt, 1975). This type of plasticity is, similar to the VOR, frequency specific. Furthermore it not only increases the performance of the OKR but also of the VOR.

Although there is controversy about the exact neural substrate of VOR-plasticity, it is generally accepted that the flocculus of the cerebellum plays an important role. The flocculus receives two major inputs. One source is the *parallel fiber* system that consists of afferents from the vestibular nuclei. This input consists of mixed visual and vestibular information, as well as eye velocity signals. Furthermore, the flocculus receives input from the

*Corresponding author. Fax: (31) 10 408 9457;
e-mail: frens@fys.fgg.eur.nl

inferior olive through *climbing fibers* that carry an – almost purely visual – retinal slip signal (e.g. Simpson and Alley, 1974; Graf et al., 1988). Since retinal slip is the parameter that should be minimized by compensatory eye movements, the climbing fiber input has been considered an 'error signal' (Ito, 1970). Both the parallel and climbing fibers connect, directly and indirectly through floccular interneurons, with the Purkinje-cells. These neurons form the sole output of the flocculus. The inputs result in different activity in the Purkinje-cells. The parallel fiber input induces 'conventional' Na^{+}-spikes in the Purkinje-cells ('simple spikes'), whereas climbing fiber activity results in large Ca^{++}-spikes with a long duration ('complex spikes').

In vitro, a form of plasticity has been demonstrated at the level of the Purkinje-cells (Ito et al, 1982; Linden, 1994). When climbing fiber input arrives at the Purkinje cell simultaneously with parallel fiber input, the strength of the synapse between that parallel fiber and the Purkinje cell is reduced. Thus the error signal changes the efficacy of synapses, and therefore the strength of specific inputs. This phenomenon is known as long term depression (LTD).

Cerebellar LTD could very well be the mechanism that is responsible for inducing the plasticity in the VOR in paradigms such as described above. Here, the vestibular input is accompanied by an abnormal amount of retinal slip. As a consequence, there is a different amount of climbing fiber input, which causes the parallel fiber activity to change accordingly. Recent data support this view. Knockout mice that do not express cerebellar LTD in vitro are also not capable of VOR adaptation within hours of stimulation (De Zeeuw et al., 1998).

It is much less understood if and how LTD could be responsible for OKR-plasticity. During the prolonged presentation of an optokinetic stimulus no mismatch is created between the stimulus, the oculomotor response, and the resulting retinal slip. The climbing fiber input at the beginning of an adaptation paradigm is identical to the activity that is encountered by the animal during the normal movements it makes. Rather it seems that during the adaptation phase the stimulus becomes more salient, and the *setpoint* for the proper response to be made to a certain stimulus changes. In other words, the errors that are created by the natural undershoot of the response are apparently tolerated in normal life, but are not tolerated anymore by the system after prolonged optokinetic stimulation.

Not only prolonged optokinetic stimulation can change the response of the optokinetic system. Microinjections of cholinergic agonists and antagonists in the flocculus affect the gain of the OKR (Tan and Collewijn, 1991). For instance the non-specific cholinergic agonist Carbachol has been shown to increase the gain within minutes after injection. This increase, which never causes the gain to rise above unity, persists for at least several hours. Similar to the stimulus induced gain changes, the effect of cholinergic substances is to a lesser extent, present in the VOR.

There is anatomic evidence for cholinergic mossy fibers that stem from the vestibular complex and the nucleus prepositus hypoglossi (Barmack et al, 1986; Barmack et al 1992b). Furthermore there is evidence for a thin-beaded network of afferents, terminating in granular and molecular layers (Ojima et al, 1989; Illing, 1990; Barmack et al, 1992a). Nicotinic receptors have been demonstrated on several types of floccular interneurons (Jaarsma et al, 1997). Electrophysiological studies have shown that iontophoretic application of Carbachol increases the amplitude of the simple spike modulation of Purkinje cells (Van der Steen and Tan, 1997). However, at present it is not clear what the exact working mechanism of these cholinergic agents in the flocculus is.

In this paper we investigate the functional effect of floccular micro-injections of Carbachol. Therefore we injected Carbachol into the flocculus, and applied subsequent optokinetic stimulation. If Carbachol and prolonged optokinetic stimulation act on the same mechanism, the setpoint for the response should be changed. Therefore optokinetic stimulation that is presented *after* the Carbachol-injection should either not change the oculomotor response or increase the gain even more.

Alternatively, if the setpoint is unaffected by Carbachol, a mismatch is created between subsequent stimulation and the oculomotor response. Thus, a situation arises that is similar to the VOR-adaptation paradigm, and it is predicted that a plastic mechanism (e.g. LTD) changes the gain back to its original value. Note that in the original experiments of Tan and Collewijn (1991), the rabbits were kept in a stable visual environment after the injection. Therefore the optokinetic system in these experiments had no means to evaluate its gain.

Materials and methods

Animal preparation

Six female Dutch belted rabbits were used in this study. All experimental and surgical procedures were in accordance with the *Principles of Laboratory Animal Care* (NIH publication no. 86–23, revised 1985), and were approved by the ethical committee of the Erasmus University.

The rabbits were equipped with a headholder that provided stable head fixation in the experimental setup, as well as induction coils in both eyes for recording eye position (for surgical details see Mathoera et al, 1999). In addition the skull and dura mater were removed above both paramedial lobules of the cerebellum. Stainless steel recording chambers were placed above both openings.

A few days after surgery we started localizing the VA-zone of both flocculi. The VA-zone is the region in the flocculus where units respond optimally to stimulation about the vertical axis (Graf et al, 1988). To that means, we recorded single unit activity with glass micropipettes filled with 4 M NaCl. The VA-zone was identified on the basis of activity of Purkinje-cells that modulated their activity in response to vertical axis optokinetic stimulation.

When the VA-zone was thus localized, the position of the tip of the electrode was reconstructed. A guide tube was subsequently cemented in the recording chamber, in such a way that, during the experiments, it allowed an injection canula of the proper length to enter the VA-zone (for procedural details see Tan and Collewijn, 1991).

Experimental Protocol

When a rabbit was provided with two guide tubes we applied the following protocol.

Pre-injection test

We started by testing the oculomotor response of the rabbit to optokinetic and vestibular stimulation. The animal was first vestibularly stimulated in the dark with a sinusoidal stimulus at a frequency of 0.1 Hz (amplitude 5° p.p.). Then optokinetic stimulation was given at frequencies of 0.05, 0.1 and 0.2 Hz (amplitude 10°, 5°, and 2.5° p.p. respectively). The amplitudes of stimulation were chosen in such a way that the maximum stimulus velocity was always 1.7°/s. All stimuli were about the vertical axis, and presented over 11 cycles. A detailed description of the stimulus apparatus is given in Mathoera et al. (1999).

Injections

Subsequently we injected 1.0 μl of Carbachol (1 g/l; Carbamylcholine Chloride, Sigma Chemical Co., St. Louis, USA) bilaterally in the VA-zones of the flocculi. To that means a canula was filled with the Carbachol solution and attached by a microtube to a water-filled 1.0 μl syringe (Hamilton, Reno, USA). After the canula was placed in the guide tube the Carbachol was applied by pressure injection. The volume was injected in roughly 10 to 15 sec. in order to avoid pressure artifacts on the neural tissue. Both injections were made within 3 min.

Post-injection test

In order to assess the effect of these floccular injections of Carbachol on the compensatory eye movements, we performed the post-injection test immediately after the injections. The stimuli that were applied were identical to those in the pre-injection test.

Prolonged optokinetic stimulation

Subsequently, the animal was subjected to constant sinusoidal optokinetic stimulation about the vertical axis. The stimulus that was given had a frequency of 0.1 Hz at an amplitude of 5° p.p. Stimulus duration was 120 min.

Post-stimulation test

Immediately after stimulation, the same stimuli as in the pre- and post-injection tests were given once more. This was done in order to investigate the frequency-specificity of the effects of optokinetic stimulation, as well as a possible transfer to the VOR. In order to rule out that (part of) the effects were due to fatigue or other such factors, the animal was then kept stationary in the light in a stable visual environment for ten min, and a second post-stimulation test was given. However, since no significant differences were observed between both post-stimulation tests, we only present the data of the first post-stimulation test in this chapter.

Controls

Two types of control experiments were performed.

1. In all rabbits we performed an experiment that was identical to the protocol described above, except that we injected saline instead of Carbachol.
2. In two rabbits (M and N) we replicated the experimental protocol of Tan and Collewijn (1991). To that means we performed the normal protocol, but kept the rabbit in a stable environment for 2 h, instead of presenting prolonged optokinetic stimulation. Every 20 min the animal was briefly stimulated (11 periods), in order to measure its OKR gain.

Data recording and analysis

Eye position was measured with the induction coil technique (Collewijn et al, 1975), in a three-field system (Remmellabs, Ashland, USA).

Position signals of the optokinetic and vestibular apparatus, as well as both eyes were sampled at a frequency of 250 Hz (CED1401, Cambridge, UK) and stored on hard disc for offline analysis. During the prolonged optokinetic stimulation 11 periods were recorded every 20 min.

From the 11 periods that were obtained in each condition the gain of the response was determined. To that means the first period was discarded, in order to avoid responses that were due to the onset of the stimulus. From the remaining ten periods, the velocity signal of the stimulus and the eye was calculated by differentiation of the position signals. Through the stimulus signal a sinus was fitted. Fitted parameters were amplitude, frequency and phase. The amplitudes and frequencies that were thus obtained differed less than 1% from the parameters that were specified in the stimulation program.

Fast phases in the eye velocity signal were removed on the basis of a velocity/duration criterion. Through the remaining slow phase signal a sine was fitted with a frequency that was identical to the one obtained from the stimulus, leaving two free parameters: amplitude and phase. Analysis was always performed on the signals from the right eye. Only incidentally, when the induction coil in this eye was broken, we used the signal from the left eye.

The gain of the response was now defined as the ratio of eye velocity amplitude and stimulus velocity amplitude. Note that this is mathematically identical to the – more commonly used – ratio between eye and stimulus *position* amplitude, because differentiation is a linear operation. The advantage of defining gain in the velocity domain is that no signal needs to be substituted for the removed fast phases. Confidence intervals of the fitted gain-values were obtained by calculating the gain of each individual period separately and determining the variation (expressed as standard deviation) in these values.

Histological verification

After all experiments were performed, the rabbits were perfused. The skull covering the flocculus at both sides of the head was carefully removed and an injection canula was positioned into the guide tube. In all animals the tip of the canulas was located in the floccular cortex of the cerebellum, indicating that all observed effects were related to neurochemical influences inside the flocculus.

Results

Effects of Carbachol injection

In all rabbits ($N=6$) except one (rabbit G) the injections of Carbachol caused a significant increase of the gain of the OKR at the frequency that was used for prolonged optokinetic stimulation (0.1 Hz). The absolute values of this increase varied considerably between rabbits (Fig. 1). This was probably due to the fact that Carbachol injections never increased the gain to values above unity (Tan and Collewijn, 1991). Therefore rabbits with a relative high gain can only increase their performance slightly, whereas the converse is true for animals with a low gain. Fig. 1 shows that there is a negative correlation between the gain before injection, and the size of the Carbachol induced gain shift. To overcome this problem we will express gain change in this paper as a fraction of the pre-adaptation undershoot. Thus,

$$rGC = \frac{G_i - G_p}{1 - G_p}$$

where rGC is the relative gain change, and G_p and G_i are the gains of the OKR before and after injection, respectively. Note that if the Carbachol injection has no effect, rGC equals 0, whereas rGC is 1 when the injection causes the gain to become equal to unity.

The closed bars in the top panel of Fig. 2 summarize the effects of Carbachol for the different frequencies. In the means of this figure the data of rabbit G have been excluded. Since a Carbachol induced gain change is a prerequisite for this study, its data were not relevant for the rest of the paper.

As is shown in this figure, Carbachol caused an increase of the gain in all conditions tested. The increase of the gain in response to optokinetic stimulation exceeded the gain increase of the VOR. This is consistent with the literature (Tan and Collewijn, 1991). As expected, injections of saline had no effect on the oculomotor response of the animals (Fig. 2, bottom panel).

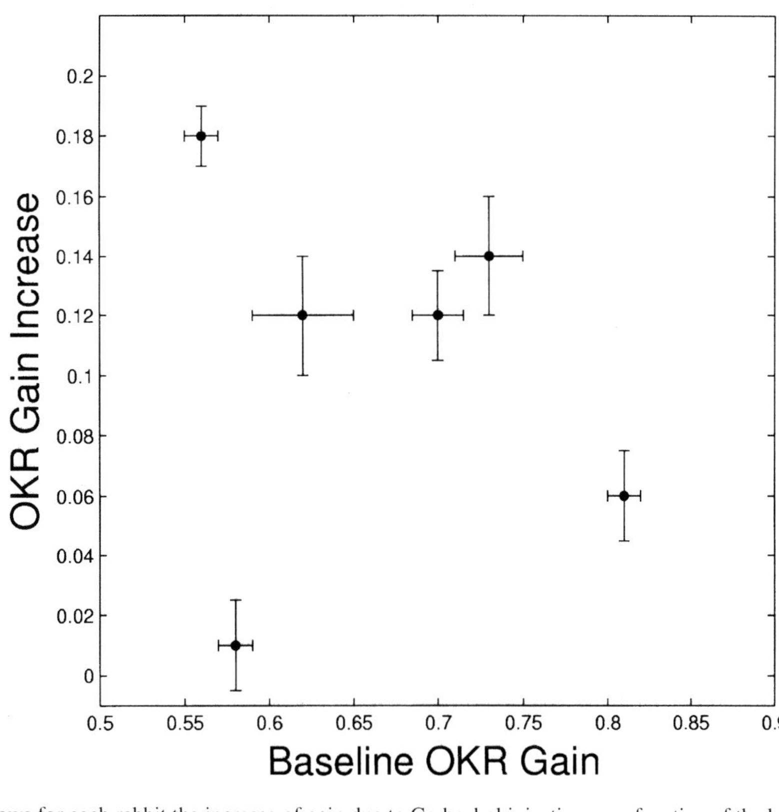

Fig. 1. This figure shows for each rabbit the increase of gain due to Carbachol injection, as a function of the baseline optokinetic gain at 0.1 Hz. Note that one rabbit (G) hardly responded to the application of Carbachol. For the other rabbits a negative correlation can be observed between baseline gain and gain increase. Errorbars represent standard deviations (see Methods).

Effects of optokinetic stimulation

When the animals were subjected to optokinetic stimulation after the floccular application of Carbachol, the gain of the OKR decreased systematically over time (Fig. 3, solid line). This was not due to a limited time range in which Carbachol was effective. When Carbachol was given, but no subsequent optokinetic stimulation was applied, the gain increase was more or less constant over the whole recording period of 120 min (Fig. 3, dashed line). When saline was injected into the flocculus, the gain of the optokinetic response of the rabbit shown in Fig. 3 did not change as a result of prolonged stimulation (dotted line). Averaged over all rabbits however, a slight increase of gain was observed (Fig. 2), consistent with earlier findings (Collewijn and Kleinschmidt, 1975).

Frequency specificity and transfer to VOR

The mean values in the post-stimulation test are shown in Fig. 2 (open bars). The gain increase of the OKR due to the injection of Carbachol is substantially reduced at 0.1 Hz as a result of prolonged optokinetic stimulation at this frequency. However smaller reductions can be observed at the other optokinetic frequencies, as well as at in the vestibular response. Therefore the adaptation is not strictly frequency or modality specific, but it shows a limited transfer to other stimulus conditions.

Discussion

Effects of Carbachol injection

In line with earlier findings by Tan and Collewijn (1991) we found that Carbachol has a stimulating

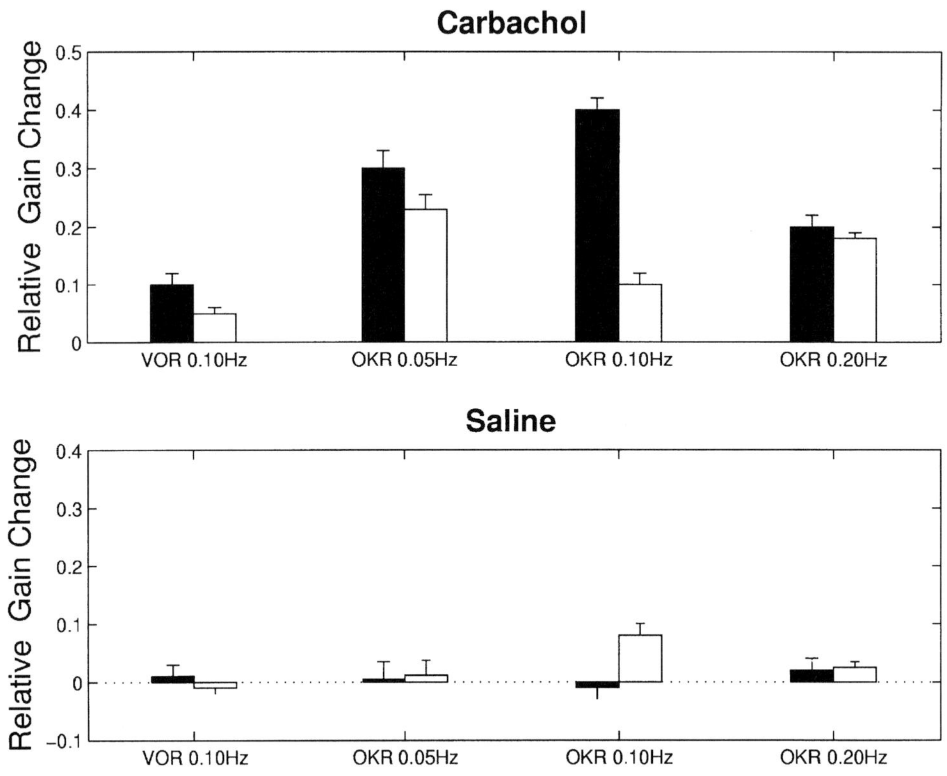

Fig. 2. Top panel: Mean relative gain changes (±SD) for each stimulus condition, directly after floccular application of Carbachol (closed bars), as well as after subsequent prolonged optokinetic stimulation (open bars). Note that a reduction is present in all gains. The largest reduction is observed at the frequency applied during the optokinetic stimulation (0.1 Hz). Bottom panel: Same graph, but with saline injections. Note that the gain at 0.1 Hz is slightly increased due to the prolonged optokinetic stimulation.

effect on the gain of the OKR. However, the absolute effect that we found is considerably smaller than what is reported in their paper (a mean absolute gain increase of 0.12 in our data, compared to a gain increase of 0.4 in their paper). Several factors may contribute to this. Firstly our rabbits on average had a somewhat larger gain of their optokinetic response before injection (means 0.69 vs. 0.63), and therefore this gain could increase less. Another factor may be that the used frequencies and amplitudes in both studies were slightly different.

Effects of optokinetic stimulation

When an effect of Carbachol was present in the eye position data, this effect remained more or less constant over a period of at least two hours (Fig. 3, dashed line), provided that the rabbit was kept in a visually stable environment. This condition is in fact a replication of the experiment of Tan and Collewijn (1991), and the results are comparable. However, when the optokinetic system of the rabbit was allowed to evaluate its gain by giving optokinetic stimulation, the gain increase that was created by Carbachol quickly reduced (Figs. 2 and 3).

The OKR can be considered to be a closed loop system. The output of the system is directly compared with its input through sensory feedback. As was discussed in the introduction, the only way to induce lasting changes in a closed loop system is to change its setpoint, which is in the case of the OKR the 'desired' amount of retinal slip. Apparently, floccular application of Carbachol does not change this setpoint value, since the reduction of retinal slip that is induced by such injections, is compensated for by a plastic response that changes the gain to its original value.

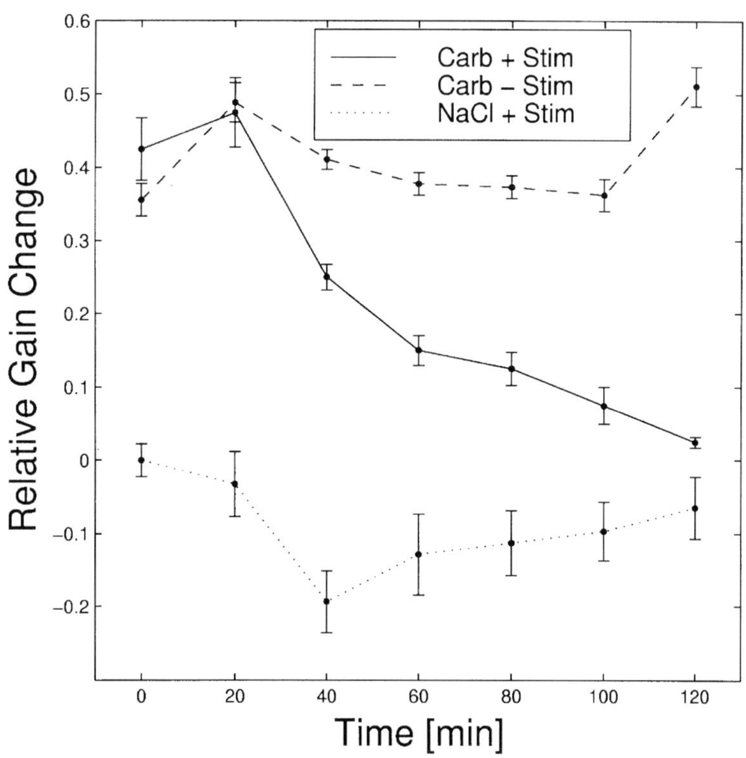

Fig. 3. Time course of adaptation (data from rabbit N). This graph shows the data of three different experiments with rabbit N. The solid line represents optokinetic stimulation after Carbachol injection. The dashed line shows the gain values after Carbachol injection, but without subsequent stimulation. The dotted line shows the effect of optokinetic stimulation after injection of saline. Time is given relative to the onset of optokinetic stimulation.

Therefore the gain change of the OKR that is neurochemically induced in this study is fundamentally different from the effects that have been found for gain changes that are due to prolonged optokinetic stimulation (Collewijn and Kleinschmidt, 1975). In the latter study a change in the response was found, despite a normal amount of slip at the onset of stimulation. This is indicative for a change of setpoint. Conversely, at the onset of stimulation in our study an unusual amount of slip was created by a gain increase, and this was not accepted by the optokinetic system, suggesting that the setpoint had not altered. This finding is somewhat surprising, since the retinal slip at the onset of stimulation was *less* than the normal slip and the gain of the OKR was closer to unity.

We therefore conclude that the undershoot that is normal for the OKR of the rabbit (expressed as a response gain < 1), is not due to the incapability of the rabbit to perform better, since plastic increase of the OKR gain can be induced (Collewijn and Kleinschmidt, 1975; Tan and Collewijn, 1991; this study). The gain is also not determined by 'the maximum error that is acceptable', which could cause plastic mechanisms to cease their work. If this were the case, the better gain that was induced by Carbachol in our study would not have been actively reduced. Therefore the normal gain undershoot must represent a true optimum for the rabbit. The value of this optimum gain may change under different conditions, for example when a certain optokinetic stimulation is given for a long period (i.e. the setpoint change in the study of Collewijn and Kleinschmidt, 1975). We can only speculate why a gain that is less than unity is optimal. It seems reasonable to assume that this is due to the fact that under physiological conditions the OKR often works in conjunction with the VOR, and their combined action should be optimal.

Is plasticity of the OKR due to the same mechanism as plasticity of the VOR?

The gain reduction that we observe for the OKR in this study may very well be due to the same mechanism that causes gain changes in the VOR (see Introduction). Both the time course of adaptation and the frequency specificity are comparable (e.g. Collewijn and Grootendorst, 1979). Also the finding that the gain of the VOR is somewhat altered by the adaptation of the OKR is in line with this hypothesis.

However, it must be stressed that the driving force for the adaptation that we observe is 'restoration of the original gain', which, in our case, *increases* the amount of retinal slip. Therefore a mechanism that is activated purely on the basis of the amount of slip, such as LTD is thought to work, cannot account for these findings. One possibility is that other mechanisms, in or outside the flocculus, must play a role as well. A likely candidate may be presynaptic long-term potentiation (LTP) that has been found at the parallel fiber/Purkinje cell synapse (Salin et al., 1996; Storm et al., 1998). Alternatively, LTD may not strive for minimum slip, but rather for 'optimal slip', similar to what we find in this study. The fact that during normal OKR there is a considerable amount of climbing fiber activity present in the floccular Purkinje cells (Graf et al., 1988), without apparent induction of plastic changes is in agreement with this notion.

Acknowledgements

We thank H. Kock, B.L.F. Weijer and E. Dalm for technical assistance and Prof. H. Collewijn for useful discussion. ALM is supported by NWO-ALW grant 805–33–174.

References

Barmack, N.H., Baughman, R.W., Eckenstein, F.P. and Wescott-Hodson, M. (1986) Cholinergic projections to the cerebellum. *Soc. Neurosci. Abst.*, 12: 577.

Barmack, N.H., Baughman, R.W. and Eckenstein, F.P. (1992a) Cholinergic innervation of the cerebellum of rat, rabbit, cat, and monkey as revealed by choline acetyltransferase activity and immunohistochemistry. *J. Comp. Neurol.*, 317: 233–249.

Barmack, N.H., Baughman, R.W., Eckenstein, F.P. and Shojaku, H. (1992b) Secondary vestibular cholinergic projection to the cerebellum of rabbit and rat as revealed by choline acetyltransferase immunohistochemistry, retrograde and orthograde tracers. *J. Comp. Neurol.*, 317: 250–270.

Collewijn, H. (1969) Optokinetic eye movements in the rabbit: input-output relations. *Vision Res.*, 9: 117–132.

Collewijn, H. and Kleinschmidt, H.J. (1975) Vestibulo–ocular and optokinetic reactions in the rabbit: changes during 24 hours of normal and abnormal interaction. In: G. Lennerstrand, Bach-Y-Rita (Eds), *Basic Mechanisms of Ocular Motility and their Clinical Implications*, pp. 477–483. Pergamon, Oxford, New York.

Collewijn, H. and Grootendorst, A.F. (1979) Adaptation of optokinetic and vestibulo–ocular reflexes to modified visual input in the rabbit. *Prog. Brain Res.*, 50: 771–781.

Collewijn, H., van der Mark, F. and Jansen, T.C. (1975) Precise recording of human eye movements. *Vision Res.*, 15: 447–450.

Demer, J.L., Porter, F.I., Goldberg, J., Jenkins, H.A. and Schmidt, K. (1989) Adaptation to telescopic spectacles: vestibulo–ocular reflex plasticity. *Invest. Ophthalmol. Vis. Sci.*, 30: 159–170.

De Zeeuw, C.I., Hansel, C., Bian, F., Koekkoek, S.K., van Alphen, A.M., Linden, D.J. and Oberdick, J. (1998) Expression of a protein kinase C inhibitor in Purkinje cells blocks cerebellar LTD and adaptation of the vestibulo–ocular reflex. *Neuron*, 20: 495–508.

Graf, W., Simpson, J.I. and Leonard, C.S. (1988) Spatial organization of visual messages of the rabbit's cerebellar flocculus. II. Complex and simple spike responses of Purkinje cells. *J. Neurophysiol.*, 60: 2091–2121.

Illing, R.B. (1990) A subtype of cerebellar Golgi cells may be cholinergic. *Brain Res.*, 522: 267–274.

Ito, M. (1970) Neurophysiological aspects of the cerebellar motor control system. *Intern. J. Neurol.*, 7: 162–176.

Ito, M., Sakurai, M. and Tongroach, P. (1982) Climbing fibre induced depression of both mossy fibre responsiveness and glutamate sensitivity of cerebellar Purkinje cells. *J. Physiol. (Lond.)*, 324: 113–134.

Jaarsma, D., Ruigrok, T.J., Caffé, R., Cozzari, C., Levey, A.I., Mugnaini, E. and Voogd, J. (1997) Cholinergic innervation and receptors in the cerebellum. *Prog. Brain Res.*, 114: 67–96.

Linden, D.J. (1994) Long-term synaptic depression in the mammalian brain. *Neuron*, 12: 457–472.

Mathoera, A.L., Frens, M.A. and Van der Steen, J. (1999) Visual-vestibular interaction during transparent optokinetic stimulation in the rabbit. *Exp. Brain Res.*. 125: 87–96.

Ojima, H., Kawajiri, S. and Yamasaki, T. (1989) Cholinergic innervation of the rat cerebellum: qualitative and quantitative analyses of elements immunoreactive to a monoclonal antibody against choline acetyltransferase. *J. Comp. Neurol.*, 290: 41–52.

Salin, P.A., Malenka, R.C. and Nicoll, R.A. (1996) Cyclic AMP mediates a presynaptic form of LTP at cerebellar parallel fiber synapses. *Neuron*, 16: 797–803.

Simpson, J.I. and Alley, K.E. (1974) Visual climbing fiber input to rabbit vestibulo–cerebellum: a source of direction-specific information. *Brain Res.*, 82: 302–308.

Storm, D.R., Hansel, C., Hacker, B., Parent, A. and, Linden, D.J. (1998) Impaired cerebellar long-term potentiation in type I adenylyl cyclase mutant mice. *Neuron*, 20: 1199–1210.

Tan, H.S., Collewijn, H. (1991) Cholinergic modulation of optokinetic and vestibulo–ocular responses: a study with microinjections in the flocculus of the rabbit. *Exp. Brain Res.*, 85: 475–481.

Van der Steen, J. and Tan, H.S. (1997) Cholinergic control in the floccular cerebellum of the rabbit. *Prog. Brain Res.*, 114: 335–345.

N.M. Gerrits, T.J.H. Ruigrok and C.I. De Zeeuw (Eds.)
Progress in Brain Research, Vol 124

CHAPTER 19

Use-dependent changes in synaptic strength at the Purkinje cell to deep nuclear synapse

Carlos D. Aizenman[1], Eric J. Huang[1], Paul B. Manis[1,2], and David J. Linden[1]*

[1]*Department of Neuroscience,* [2]*Department of Otolaryngology, Head and Neck Surgery, Johns Hopkins University School of Medicine, 725 N. Wolfe Street, Baltimore, MD 21205, USA*

Introduction

A central hypothesis of contemporary neurobiology is that memory is stored in the brain through use-dependent modifications in the properties of synaptic transmission. The vast majority of work in support of this hypothesis has focused upon two cellular processes, homosynaptic long-term potentiation and depression (LTP and LTD) of excitatory, glutamatergic synaptic transmission. LTP and LTD are typically evoked by brief, high-frequency and sustained, low-frequency stimulation respectively (Linden and Connor, 1995). These phenomena have been shown to coexist in the same set of synapses and, in some cases, to reverse each other, thus endowing these synapses with use-dependent bidirectional modification, a computationally important feature. A cellular model has emerged to explain the induction phase of LTP and LTD (Bear and Malenka, 1994; Lisman, 1994). In their most frequently studied forms, induction of LTP and LTD are dependent upon the influx of post-synaptic Ca^{2+} through NMDA receptor-associated ion channels and/or voltage-gated Ca^{2+} channels. The direction of change in synaptic strength (LTP vs. LTD) is believed to be determined by the amount of postsynaptic activity (as indexed by Ca^{2+} influx) which occurs during induction: a small amount of post-synaptic Ca^{2+} influx results in LTD, while a larger amount results in LTP (Artola et al., 1990; Mulkey and Malenka, 1992; Hansel et al., 1997). Moreover, the threshold for eliciting LTP vs. LTD may be adjusted by the previous activity of the synapse (Bear et al., 1987).

Inhibitory, GABAergic synapses are also capable of expressing LTP and LTD (Kano, 1995; Marty and Llano, 1995) and in some respects these phenomena are similar to those described for excitatory synapses. For example, all require post-synaptic Ca^{2+} influx (Kano et al., 1992; Komatsu and Iwakiri, 1993; Komatsu, 1996; McLean et al., 1996; Morishita and Sastry, 1996). However, many aspects of these phenomena are unclear. It is not known how GABAergic IPSPs drive the post-synaptic Ca^{2+} transients that might trigger LTP or LTD. Nor is it known whether a relationship exists between the degree of postsynaptic activation (as indexed by the Ca^{2+} transient) and the direction of long-term synaptic change.

We have addressed these questions by examining a brain region which has been strongly implicated in motor learning. The deep cerebellar nuclei (DCN) are a crossroads where streams of excitatory and inhibitory inputs, representing a variety of sensory-motor information, converge. They are now thought to play an increasingly important role in some forms of cerebellar-mediated motor learning (see du Lac et al., 1995; Kim and Thompson, 1997 for review). Several converging lines of

*Corresponding author. Tel.: (410) 614–1529; Fax: (410) 614–7334; e-mail: dlinden@jhmi.edu

evidence, based on lesion, inactivation and extracellular recordings, have pointed to the DCN as a possible locus for information storage during associative eyeblink conditioning, and to their analogous nuclei, the vestibular nuclei, in adaptation of the vestibulo-ocular reflex (Raymond et al., 1996; Mauk and Donegan, 1997). We have performed intracellular recordings from DCN neurons in slices of rat cerebellum in order to address the questions of (a) whether bidirectional synaptic plasticity exists in the inhibitory synapses to the DCN, supporting the role for this structure in learning, and (b) whether these changes follow general computational principles similar to those that underlie long-term plasticity in excitatory synapses.

Circuitry and anatomy of the deep cerebellar nuclei neurons

The DCN lie at the heart of the cerebellar circuitry, and receive excitatory inputs from two sources, mossy fibers, which are the axons of a large number of pre-cerebellar nuclei, and climbing fibers, which are the axons of cells of the inferior olive. The DCN are also the principal target of Purkinje neurons, which are inhibitory and are the sole output of the cerebellar cortex. Purkinje cells converge in large numbers onto DCN cells. In rats, it has been estimated that about 30 Purkinje cell axons converge onto a single DCN neuron, and that one Purkinje cell can innervete at least 20–50 DCN cells. Purkinje cell axons form synapses mainly in the proximal dendrites as well as the cell soma, axon hillock and initial segment, whereas mossy fiber collaterals synapse more distally in the dendritic tree (Chan-Palay, 1977).

In general, there are two different kinds of projection neurons in the DCN. These have been classified both by size and by neurotransmitter phenotype. The 'small' neurons, which are GABAergic, project to the inferior olive, and the 'large' neurons, which contain glutamate, send axonal fibers to several pre-motor areas, including the red nucleus and thalamus. The large projection neurons have been described as having fusiform or multipolar shaped somata, with diameters ranging from 10–35 μm, and typically 2–5, thick, primary dendrites. These neurons may be either be spiny or smooth (see Voogd et al., 1996 and Chan-Palay, 1977 for review) and are immunoreactive for glutamate (Batini et al., 1992). The small projection neurons are also fusiform or multipolar, but their cell soma size ranges from 5–20 μm in diameter, are immunoreactive for GABA (Batini et al., 1992). In addition to projection neurons there are also local GABAergic interneurons which tend to be smaller in size. To date, there has been no attempt to characterize the various cell types in the DCN by their electrophysiological phenotype. Because our recordings were performed 'blind' using microelectrodes, an important issue is the identity of the cell type(s) in our recordings. To address this point, we are currently filling DCN neurons with neurobiotin and reconstructing their morphology after recording their intrinsic electrophysiological characteristics. Preliminary results suggest that our recording technique tends to be biased towards recording from the large, glutamatergic neurons, judging from the large cell body diameters and extensive dendrites present in our reconstructed cells. An example of such a neuron is shown in Fig. 1A. Note the relatively large soma and extensive primary dendrites, as well as the axon which projects to the edge of the slice. Figure 1B shows a section of DCN immunolabeled with a monoclonal antibody directed against GAD, which, in our hands, tends to favor labeling of GABAergic synaptic terminals. The punctuate staining seen in this figure most likely corresponds to Purkinje cell terminals which clearly outline the somata of the vast majority of DCN neurons, including both large and small cells.

The role of the deep cerebellar nuclei in classical eye-blink conditioning

There are only two locations in the mammalian brain where learning and memory can be clearly understood at the level of circuits, the cerebellum and the amygdala. In the cerebellum, it is possible to assign the conditioned (CS) and unconditioned stimuli (US) in an associative eyeblink conditioning protocol to specific pathways (mossy/parallel and climbing fibers, respectively). In contrast, the hippocampus, for all of its experimental utility, receives information that is so highly processed that

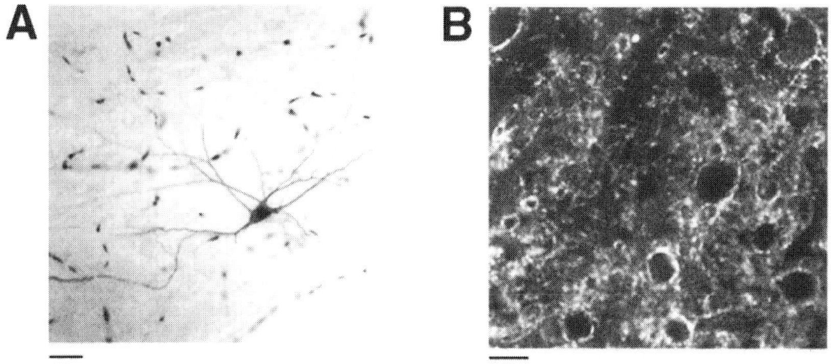

Fig. 1. Morphology of DCN neurons. (A) Example of a DCN neuron filled with neurobiotin and processed with a peroxidase/DAB reaction. Notice the large soma and the thick pyramidal dendrites as well as the axon projecting out of the field of view to the lower left. (B) DCN slice labeled with an anti-GAD antibody and a fluorescent secondary antibody. Notice the perisomatic, punctate labeling, presumably from Purkinje cell terminals. Outlines of both large and small DCN neurons can be seen. Scale bar in A is 40 μm, and in B is 20 μm.

its content cannot be easily characterized (what is the nature of the information conveyed by the perforant path?). Over the last 15–20 years, a series of experiments which have used behavioral tasks (eyeblink conditioning, VOR adaptation) together with extracellular recording, lesion and inactivation have produced a mixed set of results, with some favoring a locus of information storage at the parallel fiber-Purkinje cell synapse of the cerebellar cortex while others have implicated the synapses received by the DCN in the case of eyeblink conditioning, and their analogs, the cells of the medial vestibular nucleus, in the case of VOR adaptation (see du Lac et al., 1995; Kim and Thompson, 1997 for review).

There is extensive evidence (Thompson, 1986; Thompson et al., 1993; Thompson and Krupa, 1994) to support the involvement of cerebellar circuits in associative eyeblink conditioning. The majority of the evidence comes from several lesion studies performed primarily by Thompson and colleagues. Initially, it was observed that lesioning either the whole cerebellum (ipsilateral to the trained eye), the anterior interpositus nucleus, or the superior cerebellar peduncle, completely and selectively abolished the eye-blink conditioned response (CR) but *not* the unconditioned response (UR; but see Welsh and Harvey, 1989). Additional electrophysiological evidence showed that populations of cells in the nucleus interpositus discharge during the conditioning stimuli and, after training, fire whenever the CS alone is presented. This firing is correlated with the performance of the CR, suggesting that the CR behavior is somehow expressed in the firing rate and pattern of DCN neurons. Moreover, electrical stimulation of the pontine nuclei, or of the middle cerebellar peduncle, can serve as a CS, and this conditioning shows transfer of the CR to a tone CS (Thompson and Krupa, 1994). Reversible inactivation of the DCN with muscimol (a $GABA_A$ receptor agonist) prevented the acquisition of the eye-blink CR, whereas inactivation of the red nucleus prevented the expression of the CR during training, but not its acquisition, as evidenced by the fact that the eye-blink CR was present after inactivation (Krupa et al., 1993). Reversible inactivation of these structures, via a cooling probe, revealed similar results (Clark et al., 1992; Clark and Lavond, 1993). These studies suggest that the cerebellum and its associated projections are essential for acquisition and expression of the conditioned eye-blink response. More specifically the memory trace seems to be localized upstream of the red nucleus, perhaps in the cerebellar cortex or in the DCN.

Cerebellar LTD has been proposed as a cellular mechanism which could underlie associative eyeblink conditioning. This phenomenon, which was first described by Ito and colleagues (Ito, 1984), involves a long-term, activity dependent decrease in synaptic strength of the parallel fiber to Purkinje neuron synapse. This synaptic depression results when the climbing fiber (corresponding to the US) and parallel fiber (corresponding to the CS) inputs are activated together at low frequencies (1–4 Hz). The depression is restricted to those parallel fiber synapses activated during the pairing, and is not typically induced by parallel fiber or climbing fiber activation alone. It is long lasting, and has been observed in acute in vivo preparations, cerebellar slices, and cerebellar cultures (see Daniel et al., 1998 for review). The biochemical mechanisms involved in cerebellar LTD have been extensively studied and point to a postsynaptic site of expression. LTD in the parallel fiber to Purkinje neuron synapse would result in decreased firing of the Purkinje neuron and therefore, in disinhibition of the DCN, facilitating the excitation of the DCN by the mossy fibers. This would then result in increased firing of DCN neurons and enhanced expression of the CR.

There has been considerable debate over the specific role of the cerebellar cortex in eyeblink conditioning. Studies in which part of the cerebellar cortex was removed or reversibly inactivated after training, did not abolish the CR entirely, but affected the timing of the conditioned response (Perrett et al., 1993) as well as preventing its extinction (Perrett and Mauk, 1995; Garcia and Mauk, 1998). These experiments are consistent with the idea that basic eyeblink conditioning may be stored in at least two possible sites, one being the parallel fiber to Purkinje neuron synapse, and the other in the DCN. More recently, models have been proposed where the conditioned eye-blink response itself is sequentially stored first in the cerebellar cortex and then in the mossy fiber to DCN synapse, while the timing of the conditioned response is retained in the cerebellar cortex. The latter would occur via conventional cerebellar LTD and the former via Hebbian mechanisms occurring in the DCN (Raymond et al., 1996; Mauk and Donegan, 1997). An additional possibility, which we have explored, is that the Purkinje cell–DCN synapses can also be a locus for information storage. Alteration of a single Purkinje cell synapse could well regulate the output of thousands of parallel fibers converging onto that one Purkinje cell, while maintaining the relative strengths of an array of parallel fiber synapses.

Electrophysiological properties of DCN neurons

Spontaneous spiking properties

Initially, we characterized the responses of DCN neurons to current injection and synaptic activation using microelectrode current clamp recording in slices of 11–15 day-old rat cerebellum. The intrinsic electrophysiological properties of DCN neurons have been studied with intracellular recordings in a variety of different preparations, including isolated brainstem (Llinás and Mühlethaler, 1988), cerebellar slices (Gardette et al., 1985; Jahnsen, 1986a, b) and organotypic cultures (Muri and Knöpfel, 1994; Mouginot and Gahwiler, 1995; see Sastry et al., 1997 for review). In our hands, the time-averaged membrane potential of the cells in this study ranged between –75 and –50 mV, with an average of -58 ± 1 mV ($n = 33$). DCN neurons tonically fire fast action potentials at frequencies which are dependent upon the membrane potential of the cell. An example of a typical spontaneous firing pattern in a DCN neuron is shown in Fig. 2A. At rest, this neuron continuously fired action potentials at a rate of approximately 35 spikes/sec (top trace). The firing rate slowed as hyperpolarizing current was injected via the recording electrode to lower the membrane potential (lower traces), and finally subsided at around –70 mV. Although this was the firing pattern observed in most DCN neurons, in a subpopulation of cells (9 out of 136 cells) a different pattern was observed. Figure 2B illustrates a DCN neuron which also fires regularly at its resting potential (top trace). However, when the cell is hyperpolarized, rather than simply slowing its firing rate, it shifts firing mode to reveal brief, spontaneously occurring bursts of action potentials (lower traces). It is presently unresolved whether these modes constitute different cell types or a single neuronal population under different modulatory conditions. Recent evidence from our laboratory suggests that all regularly spiking cells can be induced to burst spontaneously by using various pharmacological manipulations (Aizenman and Linden, 1999).

Intrinsic responses to current pulses

DCN neurons can, in some cases, sustain firing frequencies ≥ 100 Hz when depolarized with pos-

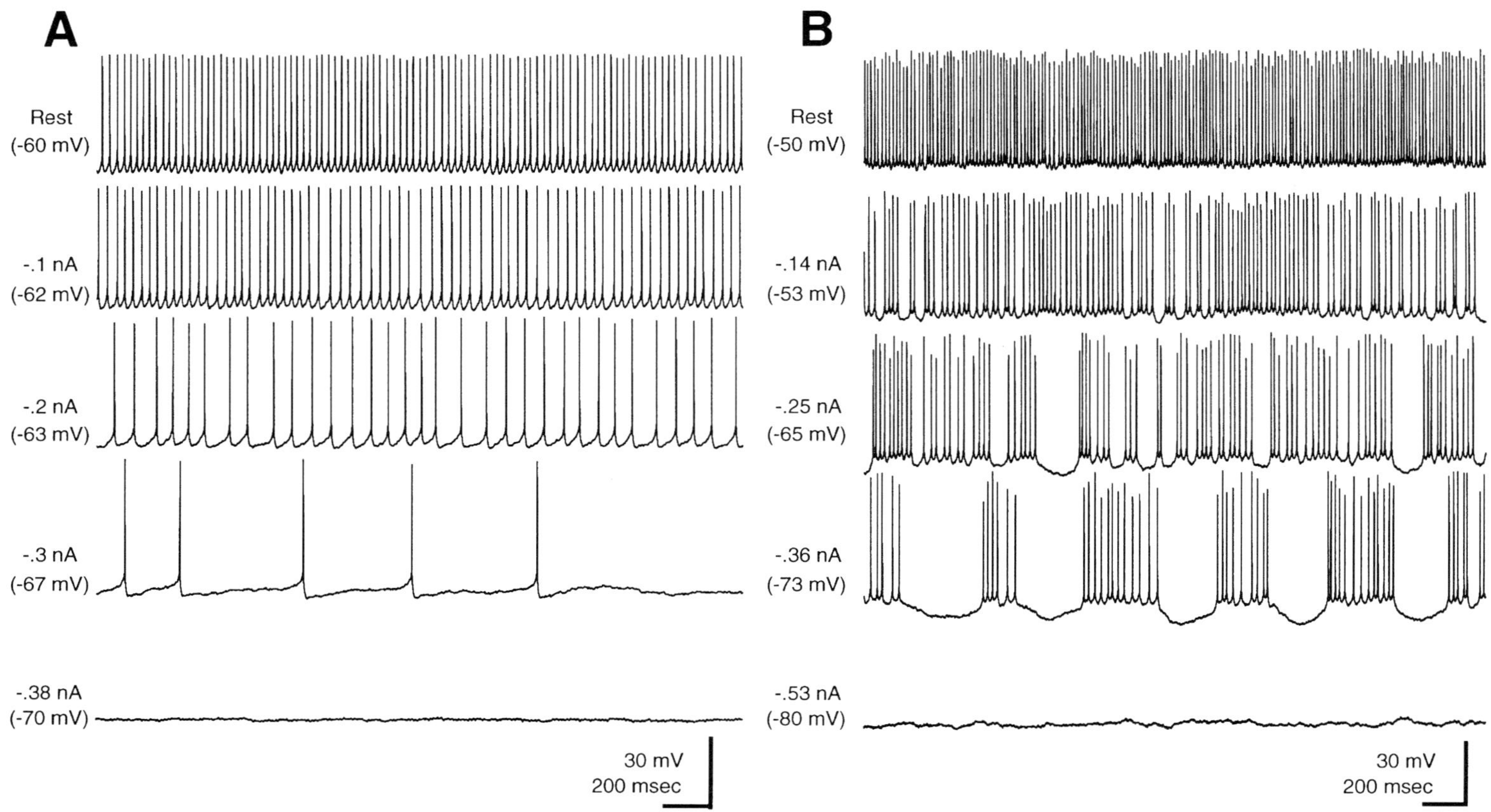

Fig. 2. Spontaneous firing patterns of DCN neurons. (A) Microelectrode recordings of spontaneous activity from a regularly-spiking DCN neuron at resting membrane potential (top trace) in the presence of the ionotropic glutamate receptor antagonist kynurenate. Lower traces represent recordings at more hyperpolarized membrane potentials. (B) Activity of a spontaneously-bursting DCN neuron at resting (top trace) and more hyperpolarized (lower traces) membrane potentials, also recorded in the presence of kynurenate.

itive current pulses (Fig. 3A, left trace). A prominent rebound depolarization (RD) can be elicited upon release from a hyperpolarizing current pulse (Jahnsen, 1986; Llinás and Mühlethaler, 1988; Muri and Knöpfel, 1994; Mouginot and Gahwiler, 1995). Typically, this RD will evoke a burst of fast action potentials (Fig. 3A, right trace). The RD depends on the membrane potential of the cell during the current pulse, and is more reliably evoked when the cell is held in current clamp at potentials positive to –60 mV, and when the hyperpolarizing pulse is sufficiently large. The underlying RD remains when Na^+ action potentials are blocked with TTX (Fig. 3B). Taken together these results suggest that the RD is evoked when the hyperpolarizing step deinactivates low-threshold, voltage-activated Ca^{2+} channels (T-type), which are then available to open upon return to the holding potential. The time-course of the RD may also be modulated by the hyperpolarization-activated cation current, I_h (McCormick and Pape, 1990; Aizenman and Linden, 1999) as well as by Ca^{2+}-dependent K^+ channels (Aizenman and Linden, 1999). Unfortunately, potent and specific antagonists of T-type Ca^{2+} channels are not presently available and the compounds that are vary considerably in their effects on different neuronal types. In our hands, a wide range of putative T-type Ca^{2+} channel antagonists (ethosuximide, 1 mM; mibefradil, 1 μM; Ni, 100 μM) failed to significantly alter the RD.

Response to synaptic stimulation

To evoke synaptic responses in DCN neurons, a stimulating electrode was placed in the white matter immediately adjacent to the DCN (Fig. 4A). Stimulation at this site evoked a monosynaptic biphasic sequence of hyperpolarization followed by depolarization. This response is likely to result from simultaneous activation of GABAergic axons of the Purkinje cells (the only inhibitory axons running in the white matter which project to the DCN) and glutamatergic axons of the mossy and/or

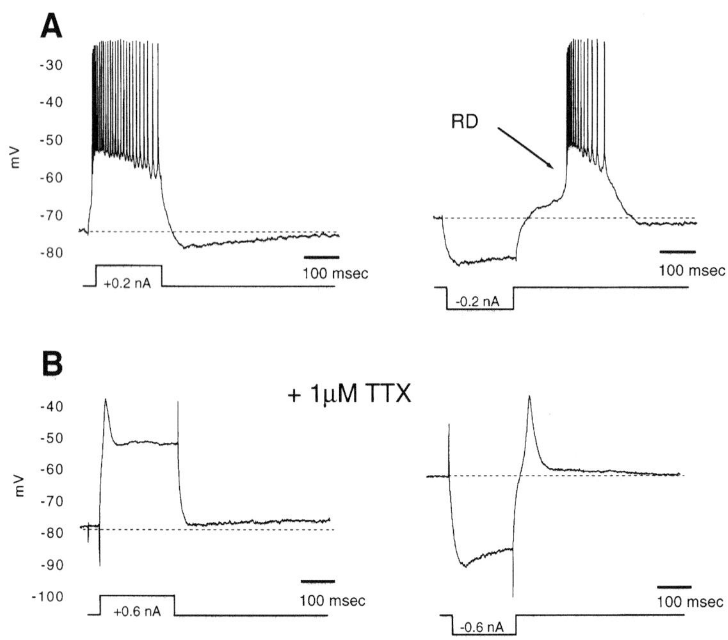

Fig. 3. Response of DCN neurons to intracellular current pulses. (A) Sample current-clamp traces from microelectrode recordings in response to a 200 ms depolarizing (left) or hyperpolarizing (right) current pulse in a cell which was held slightly below spike threshold with tonic hyperpolarizing current. Notice the robust rebound depolarization (RD) and associated spike burst that follow the hyperpolarizing pulse. (B) Sample traces from a different cell in the presence of 1 μM TTX. Although the spike burst is eliminated, the underlying RD remains.

climbing fiber collaterals, as bath application of an antagonist of ionotropic glutamate receptors (kynurenate, 2 mM) abolished the late depolarizing component. The remaining IPSP was blocked by a $GABA_A$ receptor antagonist (bicuculline, 20 μM; Fig. 4A). Although a single IPSP sometimes evokes a small RD (Fig. 4B, top trace), a brief, high-frequency train of IPSPs (100 Hz, 10 pulses) reliably evokes a large RD (middle trace). This IPSP burst-evoked RD has similar properties to an RD evoked by a hyperpolarizing pulse, in that its size, and associated spike burst, are determined by the membrane potential of the cell (bottom trace). IPSP-burst evoked RDs provide a frequency-dependent mechanism by which inhibitory inputs can drive postsynaptic excitation.

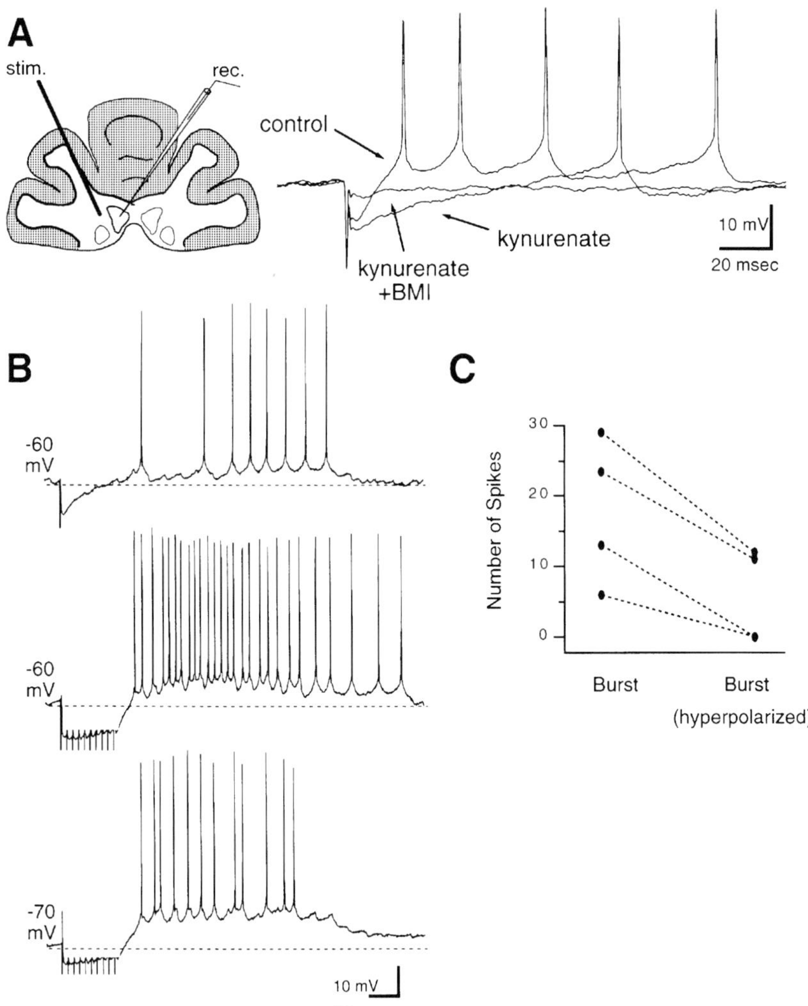

Fig. 4. Synaptically-driven RDs. (A, left) Stimulation-recording configuration in a coronal cerebellar slice. (A, right) Synaptic stimulation of white matter evoked a mixed IPSP/EPSP. Its pharmacological components could be revealed with subsequent applications of kynurenate (2 mM) and bicuculline methiodide (BMI, 20 μM). (B) Sample traces showing an example of spike firing following a single IPSP (top) and following a brief, high-frequency train of IPSPs (100 Hz × 10 pulses, middle). The bottom trace illustrates the relative reduction in spiking when the train is applied while the cell is tonically hyperpolarized to –70 mV. (C) Results from four cells measuring the reduction caused by hyperpolarization in the average number of spikes evoked after a burst of IPSPs. Each point is an average of 2 RDs.

LTP and LTD at the Purkinje cell-DCN synapse

Because postsynaptic activation is central to the induction of LTP and LTD in several model systems including excitatory (Linden and Connor, 1995) and inhibitory synapses (Kano, 1995; Marty and Llano, 1995), we tested the hypothesis that the IPSP-driven RD may provide a mechanism by which inhibitory activity may undergo use-dependent plasticity. In order to avoid contamination of the inhibitory synaptic responses by EPSPs, all recordings were performed using 2 mM kynurenate in the external saline. To induce long-term changes in the efficacy of the isolated IPSP, a burst stimulation protocol was used which consisted of 10 high-frequency trains (100 Hz, 10 pulses) applied at 2 Hz while the DCN neuron was at its resting membrane potential. This form of stimulation reliably evoked RDs (Fig. 5C, top trace) and produced LTP which persisted for at least 20 min. (Fig. 5A, open circles), the amplitude of which was 118 ± 5% of baseline (mean ± SEM, measured at t = 15–20 min, n = 9). The cumulative probability distribution of the amount of change for individual experiments is shown in Fig. 5B.

In order to test whether LTP of IPSPs was consistent with the computational rule that large amounts of postsynaptic activation induce LTP, while small amounts induce LTD, we performed two different manipulations which would reduce the RD and therefore limit the postsynaptic activity during the induction protocol. Tonically hyperpolarizing the neuron during burst stimulation reduces the strength of the RD, as assessed by the number of spikes evoked (Fig. 5C, bottom). If the same burst protocol that is used to evoke LTP is applied while the cell is tonically hyperpolarized (to −67 ± 2 mV with −400 ± 60 pA) then LTD tends to be elicited (82 ± 9% of baseline, n = 12; Fig. 5A, closed circles). The effect is significantly different from the change caused by the same protocol applied to DCN cells at the resting potential (p < 0.02, Mann-Whitney U test), as indicated by the leftward shift in the cumulative probability distribution (Fig. 5B). The second manipulation to limit postsynaptic excitation was intracellular perfusion of the voltage-gated Na^+ channel blocker QX–314. By eliminating Na^+ spikes, QX–314 attenuated the net activation caused by the burst-evoked RD without eliminating it altogether (Fig. 6C) and thereby converted the LTP induction protocol to one which elicits LTD (82 ± 10% of baseline; n = 7; Fig. 6A). The effect of the QX–314 on the synaptic gain change induced after burst stimulation is also significant when compared to that induced without QX–314 (p < 0.01). QX–314 had no effect on baseline responses over a 25 min

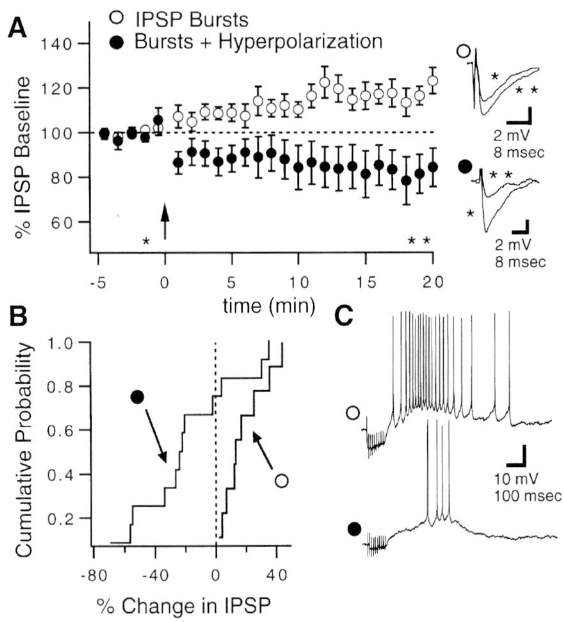

Fig. 5. Bidirectional synaptic plasticity of IPSPs in the DCN. (A) Normalized and averaged time course reflecting changes in IPSP amplitude induced by a burst-protocol (applied at arrow, see text). Open circles represent synaptic changes when the burst-protocol is applied while the cell is at resting membrane potential, effectively driving the RD and associated spike burst (see C, top trace); closed circles correspond to average changes occurring when the same protocol is applied while the cell is tonically hyperpolarized, thus reducing the amount of post-synaptic spike firing (see C, bottom trace). The inset shows sample traces taken immediately before (*) and 20 min after (**) conditioning stimulation. (B) Cumulative probability distribution of amount of change averaged 15–20 min. after conditioning stimulation, applied either while the cell was at rest, or while it was tonically hyperpolarized. While group data for LTP and LTD conditions is presented here for 20 minutes after the application of an induction protocol, these changes in synaptic strength were seen to persist for the useful life of the recording, which was 30–35 min in some cases (data not shown).

period (102 ± 2%, $n = 4$; not shown). While these results suggest that the IPSP-driven RD is necessary for induction of LTP–LTD, they do not address the issue of whether activation of the RD alone without synaptic activation is sufficient. To test this, we replaced the synaptic burst protocol with a train of ten 200 msec long steps of hyperpolarizing current (–0.5 nA) at 2 Hz, a manipulation which maximally evoked RDs (Fig. 7C, top trace). This resulted in LTP which closely matched that which was elicited by the synaptic burst stimulation (120 ± 10%, $n = 7$; Fig. 7A, open circles). Hyperpolarizing pulses delivered to QX–314-loaded cells (see Fig. 7C, bottom trace) failed to produce any significant change in synaptic strength (96 ± 5%; $n = 6$; Fig. 7A, closed circles), indicating that Na^+ spiking is necessary for this process.

NMDA receptor-dependent LTP and LTD of excitatory synapses (Lynch et al., 1983; Mulkey and Malenka, 1992), as well as LTP and LTD of GABAergic inhibitory synapses (Kano et al., 1992; Komatsu and Iwakiri, 1993; McLean et al., 1996; Morishita and Sastry, 1996) are dependent upon postsynaptic Ca^{2+} flux. To determine whether this is also true for the present form of LTP and LTD, DCN cells were loaded with the Ca^{2+} chelator BAPTA. The time-averaged resting membrane potential in BAPTA-loaded cells was lower than normal (–67 ± 2 mV; $n = 7$), however no noticeable difference in the IPSPs, nor their ability to drive the RD was observed. No significant change in synaptic efficacy (105 ± 3% of baseline; $n = 7$) was induced by the synaptic burst protocol in this condition, indicating that an elevation in postsynaptic Ca^{2+} is necessary for plasticity to occur.

Relationship between postsynaptic spiking, calcium levels and polarity of synaptic plasticity

The RD can vary considerably among different DCN neurons. However, for any given cell, tonic hyperpolarization always reduces the strength of the RD relative to that evoked while the cell is at

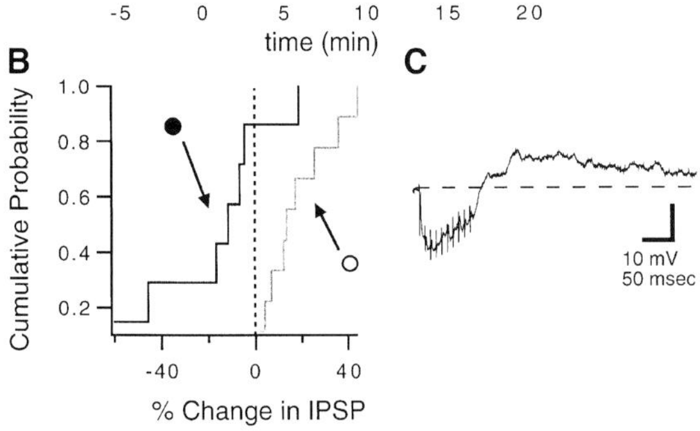

Fig. 6. LTD induction synaptic burst stimulation in QX–314-filled neurons. (A) Averaged time course and (B) cumulative probability distribution (solid line) of changes induced by burst-stimulation in cells intracellularly perfused with voltage-gated Na^+ channel blocker, QX–314. (C) Intracellular perfusion with QX–314 results in a reduction of postsynaptic excitation during the burst-induced RD.

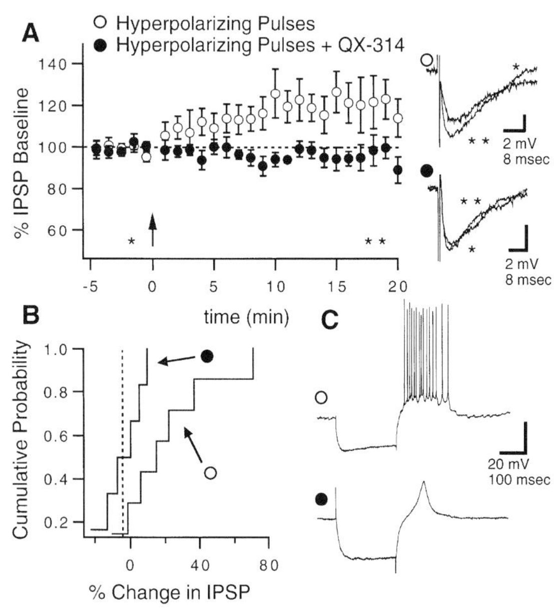

Fig. 7. LTP induction by hyperpolarizing pulse-evoked RDs. (A) Averaged time course and (B) cumulative probability distribution of changes induced by postsynaptic application of 200 ms long hyperpolarizing pulses which reliably drive the RD and associated spike burst (open circles, see top trace in C), without synaptic stimulation. Filled circles represent the same condition but given while the cell was perfused with QX–314 (see bottom trace in C).

rest. Cell-to-cell variability in the RD may account for some of the variability in the amount of LTP or LTD elicited by different stimulation conditions. Tetanic bursts of IPSPs strongly activate the postsynaptic cell via the RD, eliciting LTP. However, in the conditions which limit the postsynaptic activity (tonic hyperpolarization, QX–314), the same protocol tends to elicit LTD instead. In these conditions LTD induction was less consistent, sometimes yielding LTP (3/19 cases, for both conditions combined). This suggests that there is either variability in the ability for IPSPs to evoke an RD, or perhaps different modification thresholds across the population of neurons. Nevertheless, if all the experiments across the different conditions are taken together, a clear pattern emerges. Figure 8A illustrates the change in IPSP amplitude produced by burst or hyperpolarizing step stimulation as a function of the average number of RD-driven spikes during the initial part of the induction protocol. This reveals a relationship between the number of spikes evoked, and the polarity of the synaptic change. Whereas LTP tends to be induced under conditions in which robust RD-driven spiking is present ($113 \pm 6\%$, $n = 18$), LTD tends to be induced when the RD is elicited ($87 \pm 6\%$, $n = 23$), but drives little or no spiking (≤ 5 spikes). The same data is replotted as a cumulative probability histogram (Fig. 8B), where each group is plotted as a separate curve, highlighting the statistically significant ($p = 0.0015$) difference between the two groups. However, for those cells which showed robust RD-driven spiking, there was only a weak correlation between the number of spikes evoked and the amount of change in synaptic strength ($r = +0.452$, $p = 0.06$, $n = 18$, Pearson r test).

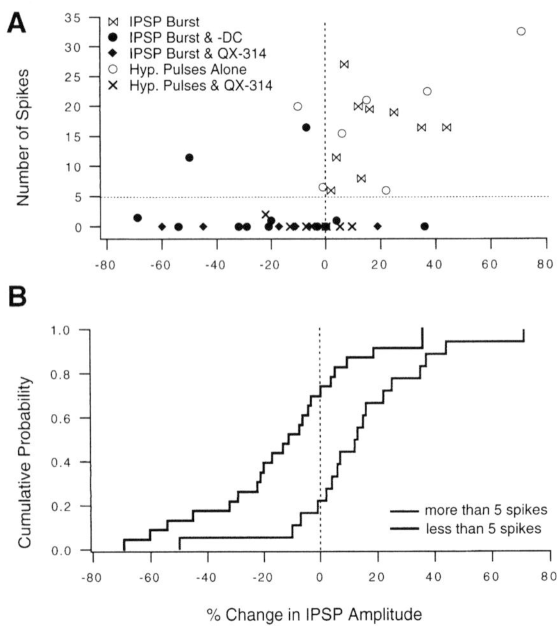

Fig. 8. RD-driven spike firing correlates with the polarity of long-term synaptic change. (A) The amount of change caused by conditioning stimulation (averaged 15–20 min post conditioning) is plotted as a function of the average number of RD-driven spikes evoked during the conditioning, across all of the conditions described in Figs 5–7. (B) The same data as in (A), replotted as a cumulative probability histogram. The right line represents experiments that showed an average of more than 5 RD-driven spikes, the left line represents experiments with 5 or less RD-driven spikes.

The relationship between RD-evoked spike firing and LTP–LTD, combined with the observation that postsynaptic Ca^{2+} is required for synaptic plasticity to occur, prompted us to measure intracellular Ca^{2+} transients while systematically varying RD-driven spiking. Figure 9A shows the response of a representative Fluo-3-loaded cell to 200 msec hyperpolarizing current pulses, followed by a return to either the holding level or slightly depolarized levels for an additional 300 msec. Upon return to the holding level, a subthreshold depolarization is produced (Fig. 9A1). When the membrane potential following the hyperpolarizing step is systematically increased, an increasingly strong train of action potentials is evoked (Fig. 9A2–A4). The corresponding fluorescence changes observed at the soma and in a proximal dendrite are also shown (Fig. 9B1, B2). With a subthreshold RD, a small, weak Ca^{2+} influx was observed (in 2/5 Fluo–3 loaded cells); however, when the cell discharged action potentials, a much larger intracellular Ca^{2+} transient was detected whose amplitude was proportional to the degree of spiking. In every case (5/5 cells loaded with Fluo–3 and 2/2 cells loaded with calcium green–1, the latter of which are not shown) the Ca^{2+} levels increased with increasing discharge rates, and only small or undetectable signals were associated with the subthreshold RD. These results confirm the notion that the RD alone is not providing a large Ca^{2+} influx, whereas higher intracellular Ca^{2+} concentrations are provided by the opening of Ca^{2+} channels during action potentials. Furthermore, they suggest that the amplitude of RD-evoked Ca^{2+} transients is a critical factor in determining whether LTP or LTD of IPSPs is induced.

Summary and discussion

LTP and LTD of Purkinje to DCN Synapses

The main finding of this study is that Purkinje cell-DCN inhibitory synapses can express long-term, bidirectional synaptic plasticity. LTP is induced by high-frequency trains of IPSPs which reliably evoke a RD, an associated spike burst and a large Ca^{2+} transient. Conditions which limit RD-evoked spiking and the associated Ca^{2+} transients will result in LTD. RD-driven spiking appears to be both necessary and sufficient to induce LTP. Thus, a major computational feature of excitatory synapses, that moderate postsynaptic activation evokes LTD while strong postsynaptic activation evokes LTP, is also present in this inhibitory synapse.

The amplitudes of LTP and LTD induced herein are smaller than those typically observed in excitatory synapses. However, because inhibitory synapses have a much more powerful net effect on cell firing than excitatory synapses, relatively small changes in the strength of inhibitory synapses will have large effects on the integrated output of the postsynaptic cell. The power of inhibitory synapses derives from their location, interposed between excitatory synapses and the spike initiation zone, thereby allowing for the negation of a large number of excitatory synapses by shunting. This has been shown by many investigators, using recording in different brain regions, as well as in modeling

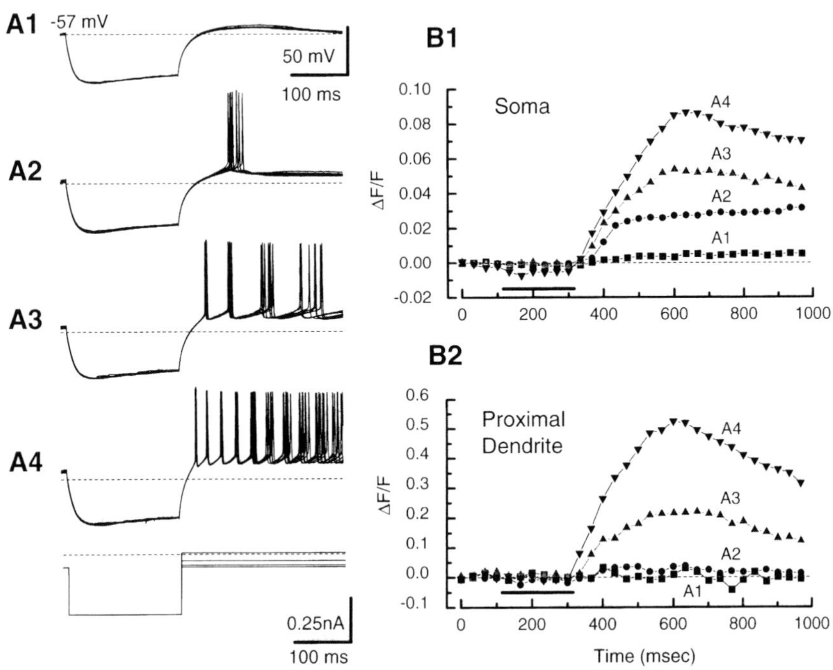

Fig. 9. Intracellular Ca^{2+} transients are evoked by RD-driven spike firing. (A) Superimposed recordings of membrane potential during 5–8 trials, as a function of an increasing post-hyperpolarizing current injection, as shown in the current traces below A4. (B).B1: Fluorescence transients recorded from the soma of the cell (each traces is the average of 5–8 trials). Bar below traces shows the duration and timing of the hyperpolarizing current pulses in A. Small transients are seen with subthreshold RDs (■, data taken simultaneously with traces in panel A1). Larger transients are seen with more action potentials (●, ▲, ▼ correspond to data taken simultaneously with traces in A2, A3, and A4 respectively). B2 shows data taken from the proximal dendrite, approximately 30 microns from the soma.

studies (Koch et al., 1983). For example, in paired recordings, single IPSPs initiated by perisomatic inhibitory cells could suppress the repetitive discharge of Na spikes in a hippocampal pyramidal cell (Miles et al., 1996). Thus, it is likely that the LTP and LTD seen herein will have a much larger effect on the output of the postsynaptic cell than would typical LTP (150% of baseline) of an excitatory hippocampal synapse.

A model for synaptic plasticity in DCN neurons

It is interesting to consider the degree to which the cellular mechanisms for use-dependent bidirectional modification differ at excitatory versus inhibitory synapses. One common element is that both excitatory (Lynch et al., 1983; Mulkey and Malenka, 1992) and inhibitory (Kano et al., 1992; Komatsu and Iwakiri, 1993; McLean et al., 1996; Morishita and Sastry, 1996) synapses require postsynaptic Ca^{2+} entry in order to induce synaptic plasticity. In our experiments, suppression of plasticity by postsynaptic BAPTA indicates that Ca^{2+} is also required. In excitatory synapses, postsynaptic Ca^{2+} enters via the activation of NMDA receptors (Collingridge et al., 1983) and/or voltage-gated Ca^{2+} channels (Grover and Teyler, 1990). In the present report, ionotropic glutamate receptors were routinely blocked, and the remaining IPSPs drove postsynaptic excitation by evoking an RD. The RD in DCN cells is thought to be triggered by a low-threshold Ca^{2+} spike which results from the opening of T-type Ca^{2+} channels (Jahnsen, 1986a, b; Llinás and Mühlethaler, 1988; Mouginot and Gahwiler, 1995) and which has been shown to transiently increase intracellular Ca^{2+} concentrations (Muri and Knöpfel, 1994). Indeed, it has been seen in hippocampal pyramidal cells and neocortical neurons that a significant fraction of Ca^{2+} entry during subthreshold excitatory synaptic stimulation is mediated by low-threshold Ca^{2+} channels (Miyakawa et al., 1992; Markram and Sakmann, 1994; Magee et al., 1995). However, this is not likely to be the only source of Ca^{2+} influx during the RD as the RD can evoke Na^+ spiking. While Na^+ spiking does not produce a direct Ca^{2+} influx, Na^+ action potentials, which are triggered in the initial segment of the axon, can backpropagate via active conductances into the dendritic arbor causing large transient increases in intracellular Ca^{2+} mediated by opening of high-threshold voltage-gated Ca^{2+} channels (Stuart et al., 1997). Our experiments, as well as previous work (Muri and Knöpfel, 1994) suggest that in the cells of the DCN, a majority of depolarization-evoked Ca^{2+} entry is associated with Na^+ spikes. Furthermore, it has been shown that LTP of EPSPs in pyramidal neurons can be elicited by pairing postsynaptic spikes with low-frequency synaptic stimulation (Magee and Johnston, 1997; Markram et al., 1997) Backpropagating spikes can therefore act as a 'coincidence' signal that can sum pre- and postsynaptic activation through a contribution to the dendritic Ca^{2+} transient (see Linden, 1999 for review).

Figure 10 illustrates our current working hypothesis for use-dependent plasticity in the Purkinje cell-DCN synapse: (a) High-frequency trains of IPSPs will cause low-threshold (LVA) Ca^{2+} channels to open upon release from hyperpolarization, leading to a small Ca^{2+} influx in the dendritic arbor. This would be sufficient to induce LTD. (b) If this depolarization is sufficiently large, it will depolarize the cell soma past the threshold for Na^+ spiking. (c) The Na^+ spikes then backpropagate into the

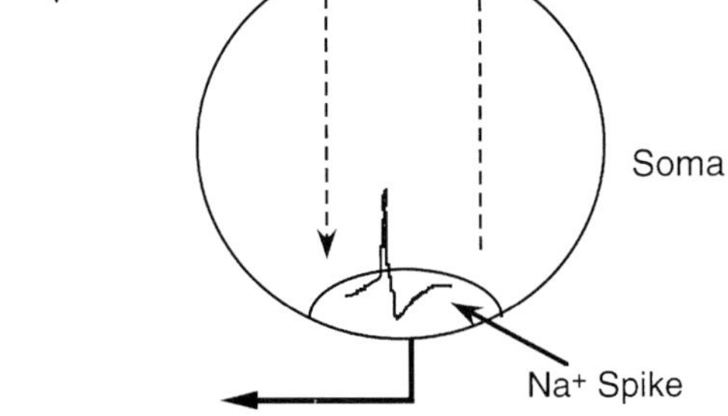

Fig. 10. A working model for the induction of synaptic plasticity in Purkinje cell-DCN synapses. Please see the text for a narrative explanation. Abbreviations: LVA, low threshold Ca^{2+} channels; HVA, high threshold Ca^{2+} channels.

dendritic arbor, where they induce the activation of high-threshold (HVA) Ca^{2+} channels, further increasing the intracellular Ca^{2+} transient, and reaching the threshold for LTP induction. Our data support this model, because conditioning which strongly evokes the RD and Na^{+} spiking (IPSP trains at resting membrane potential, hyperpolarizing pulses alone) induces LTP as well as large intracellular Ca^{2+} transients, while conditioning given under limiting conditions (tonic hyperpolarization, QX–314) reduces Na^{+} spiking but does not eliminate the underlying RD, and thus tends to induce LTD. Due to a lack of a specific and potent antagonist, it is not presently possible to test directly the role of low-threshold Ca^{2+} channels. It is worthwhile to note that the key computational feature of this model, that RD-evoked spiking correlates with both the amplitude of the postsynaptic Ca^{2+} transient and the induction of LTP–LTD, would remain intact even if additional sources of Ca^{2+} were found to contribute to the transient.

In our hands, hyperpolarizing pulses given in the presence of QX–314 (which completely abolished spiking), resulted in no alteration of synaptic strength (Fig. 7). We show in Fig. 9 that a hyperpolarizing pulse-evoked rebound depolarization that is subthreshold for spike generation produces little or no Ca^{2+} transient. Thus, it is consistent with our model that no change in synaptic strength is evoked. This finding contrasts with the observation that synaptic stimulation in the presence of QX–314 resulted in LTD (Fig. 6) suggesting that the RD generated by synaptic inputs (which are all over the cell) is probably qualitatively different than that induced by an electrode (a point source) at the soma. Perhaps a protocol could be developed in which hyperpolarizing pulses would produce minimal spiking, and a small dendritic Ca^{2+} transient, and thereby induce LTD. However, it should be emphasized that this could be difficult due to the variability in dendritic Ca^{2+} transients produced by somatic hyperpolarization (Fig. 9).

Our data suggest a novel solution for a long-standing problem in cellular synaptic plasticity: how do inhibitory synapses produce postsynaptic Ca^{2+} signals that are required for use-dependent plasticity when they do not directly gate a Ca^{2+} conductance? A number of studies in various brain regions have found that LTP or LTD of inhibitory synapses can be blocked by postsynaptic application of a Ca^{2+} chelator. However, the proposed source(s) of this Ca^{2+} transient have varied widely across preparations. One explanation has been that the Ca^{2+} transient is not directly triggered by the IPSP. For example, in 'rebound potentiation' in the cerebellar interneuron-Purkinje cell synapse, the Ca^{2+} transient is driven by coincident activation of excitatory climbing fibers, resulting in the subsequent activation of voltage-gated Ca^{2+} channels (Kano et al., 1992). Likewise, it has been shown that LTD of IPSPs in neonatal hippocampus (McLean et al., 1996) and LTD of the Purkinje cell-DCN IPSPs in 5–9 days old rat cerebellum (Morishita and Sastry, 1996) requires coincident activation of NMDA receptors. A second explanation, which comes from studies of developing visual cortex, is that activation of postsynaptic $GABA_B$ receptors potentiates PI turnover and subsequent Ca^{2+} mobilization driven by adrenergic innervation, thus inducing LTP of inhibition (Komatsu, 1996). Finally, it has been shown that in early development of the hippocampus ($\leq$P5) depolarizing IPSPs can directly activate voltage-sensitive Ca^{2+} channels thereby contributing to LTP of these synapses (McLean et al., 1996). Because the experiments described herein were conducted with excitatory synapses blocked, it became possible to study the capacity for inhibitory synapses to undergo use-dependent modification in isolation. Our finding that the IPSP-driven rebound depolarization and subsequent spiking is the key determinant of LTP and LTD induction in this synapse is computationally significant in that it describes a mechanism for inhibitory synapses to undergo LTP and LTD in a manner which does not require interaction with non-inhibitory synapses, and which can function throughout the lifespan of the animal (in mature synapses with hyperpolarizing IPSPs). In addition, the IPSP burst-driven RD can also be a mechanism by which DCN cells can differentiate between two different forms of inputs. For example, the DCN cell might be able to differentiate between a continuous, tonic, mossy fiber-driven Purkinje cell input, and an IPSP burst followed by pause, which would be driven by a climbing fiber input to the Purkinje cell (Granit and Phillips, 1956 see Ito, 1984 for discussion) and which would presumably evoke an RD.

Future Directions

There are a number of avenues for future experimentation that are opened by the present observations. At a cellular level, it will be interesting to determine the biochemical mechanisms that couple RD-driven Ca^{2+} influx to LTP and LTD of IPSPs. In glutamatergic synapses, low and high postsynaptic Ca^{2+} signals have been suggested to differentially activate protein phosphatase 1 and Ca^{2+}/calmodulin-dependent protein kinase II respectively (Lisman, 1989; Mulkey et al., 1993; Mulkey et al., 1994) resulting in dephosphorylation and phosphorylation of postsynaptic AMPA receptors. Similar mechanisms may be operative in inhibitory synapses. It has been shown that blockade of protein phosphatase activity can attenuate 10 Hz-induced LTD in Purkinje cell-DCN synapses (Morishita and Sastry, 1996). Inhibition of Ca^{2+}/calmodulin-dependent protein kinase II has also been shown to block LTP of GABAergic interneuron-Purkinje cell synapses in the cerebellar cortex (Kano et al., 1992; Kano et al., 1996). However, this may not be a common mechanism for all GABAergic synapses, as the alpha isoform of this enzyme is not found at GABAergic synapses in the neocortex and thalamus and hippocampus (Liu and Jones, 1996, 1997)

At a phenomenological level, it will be interesting to determine if the present forms of LTP and LTD are saturable, reversible and input-specific. One prediction of our model would be that LTD would tend to have a greater degree of input specificity than LTP due to the broad invasion of the dendritic arbor by backpropagating spikes in the latter case. It remains to be determined whether subthreshold Ca^{2+} signals evoked locally by IPSP burst-induced RDs can be detected, and whether these might be sufficient to induce LTD or even LTP under appropriate conditions. Another prediction is that crosstalk may occur between glutamatergic and GABAergic inputs to the cells of the DCN. This could provide a mechanism by which mossy fiber driven Ca^{2+} signals could produce LTP–LTD of GABAergic synapses or vice-versa and is consistent with previous observations that NMDA receptor activation can trigger LTD of IPSPs in the DCN (Morishita and Sastry, 1996). The details of these phenomena will be of great importance in building models of cerebellar motor learning that incorporate synaptic information storage at both cortical and DCN loci (Raymond et al., 1996; Mauk, 1997).

An interesting interpretation of these data is that LTP and LTD of IPSPs may serve as a synaptic 'homeostatic' mechanism to regulate the overall excitability and dynamic range of the DCN cell. Thus, large amounts of postsynaptic spiking will potentiate IPSPs, making the cell less likely to fire more action potentials, whereas LTD tends to be induced when the postsynaptic activity is limited, increasing the probability of postsynaptic spiking. Moreover, by regulating the membrane potential, changes in the overall level of inhibition might serve the function of switching the firing mode of the DCN cell from regular spiking to spontaneously bursting (Fig. 2). This last interpretation is intriguing, due to the fact that both olivary and Purkinje cells, both of which are interconnected with the DCN, have the intrinsic capabilities of firing RDs and spontaneously bursting (Llinás and Sugimori, 1980; Llinás and Sugimori, 1980; Llinás and Yarom, 1981), forming perhaps the basis of a reverberating circuit, such as has been proposed for certain types of coordinated movements and tremor (Llinás, 1984; Llinás, 1985), and has been extensively described in the thalamus (Bal et al., 1995).

Acknowledgements

Portions of this chapter, including Figs 3–9 have been adapted from Aizenman et al. (1998), reprinted by permission of the copyright holder, Cell Press. This work was supported by the Develbiss Fund and NIMH grant K02 MH01590 (DJL), a HHMI Predoctoral Fellowship (CDA), and NIDCD grant R01 DC00425 (PBM).

List of Abbreviations

BAPTA	bis-(*O*-aminophenoxy)-*N*,*N*,*N*′,*N*′-tetraacetic acid
CF	climbing fiber
CR	conditioned response

CS	conditioned stimulus
DCN	deep cerebellar nuclei
EPSP	excitatory postsynaptic potential
GABA	γ-Aminobutyric acid
GAD	glutamic acid decarboxylase
HVA	high-voltage activated Ca^{2+} channels
IPSP	inhibitory postsynaptic potential
LTD	long-term depression
LTP	long-term potentiation
LVA	low-voltage activated Ca^{2+} channels
MF	mossy fiber
NMDA	*N*-Methyl-D-Aspartate
PF	parallel fiber
RD	rebound depolarization
UR	unconditioned response
US	unconditioned stimulus
VOR	vestibulo–ocular reflex

References

Aizenman, C. and Linden, D. Regulation of the rebound depolarization and spontaneous firing patterns of deep nuclear neurones in slices of rat cerebellum. *J. Neurophysiol.*, 82: 1697–1709.

Artola, A., Brocher, S. and Singer, W. (1990) Different voltage-dependent thresholds for inducing long-term depression and long-term potentiation in slices of rat visual cortex. *Nature*, 347: 69–72.

Bal, T., von Krosigk, M. and McCormick, D.A. (1995) Synaptic and membrane mechanisms underlying synchronized oscillations in the ferret lateral geniculate nucleus in vitro. *J. Physiol. (Lond.)*, 483: 641–663.

Batini, C., Compoint, C., Buissert-Delmas, C., Daniel, H. and Guegan, M. (1992) Cerebellar nuclei and the nucleocortical projections in the rat: retrograde tracing coupled to GABA and glutamate immunohistochemistry. *J. Comp. Neurol.*, 315: 74–84.

Bear, M.F., Cooper, L.N. and Ebner, F.F. (1987) A physiological basis for a theory of synapse modification. *Science*, 237: 42–48.

Bear, M.F. and Malenka, R.C. (1994) Synaptic plasticity: LTP and LTD. *Curr. Opin. Neurobiol.*, 4: 389–399.

Chan-Palay, V. (1977) *Cerebellar Dentate Nucleus*, Springer-Verlag, Berlin.

Clark, R.E. and Lavond, D.G. (1993) Reversible lesions of the red nucleus during acquisition and retention of a classically conditioned behavior in rabbits. *Behav. Neurosci.*, 107: 264–270.

Clark, R.E., Zhang, A.A. and Lavond, D.G. (1992) Reversible lesions of the cerebellar interpositus nucleus during acquisition and retention of a classically conditioned behavior. *Behav. Neurosci.*, 106: 879–888.

Collingridge, G.L., Kehl, S.J. and McLennan, H. (1983) Excitatory amino acids in synaptic transmission in the Schaffer collateral-commissural pathway of the rat hippocampus. *J. Physiol. (Lond.)*, 334: 33–46.

Daniel, H., Levenes, C. and Crepel, F. (1998) Cellular mechanisms of cerebellar LTD. *Trends Neurosci.* 21: 401–407.

du Lac, S., Raymond, J.L., Sejnowski, T.J. and Lisberger, S.G. (1995) Learning and memory in the vestibulo–ocular reflex. *Ann. Rev. Neurosci.*, 18: 409–441.

Garcia, K.S. and Mauk, M.D. (1998) Pharmacological analysis of cerebellar contributions to the timing and expression of conditioned eyelid responses. *Neuropharmacology*, 37: 471–480.

Gardette, R., Debono, M., Dupont, J.L. and Crepel, F. (1985) Electrophysiological studies on the postnatal development of intracerebellar nuclei neurons in rat cerebellar slices maintained in vitro. II. Membrane conductances. *Brain Res.*, 352: 97–106.

Granit, R. and Phillips, C. (1956) Excitatory and inhibitory processes acting upon individual Purkinje cells of the cerebellum in cats. *J. Physiol. (Lond.)*, 133: 520–547.

Grover, L.M. and Teyler, T.J. (1990) Two components of long-term potentiation induced by different patterns of afferent activation. *Nature*, 347: 477–479.

Hansel, C., Artola, A. and Singer, W. (1997) Relation between dendritic Ca^{2+} levels and the polarity of synaptic long-term modifications in rat visual cortex neurons. *Eur. J. Neurosci.*, 9: 2309–2322.

Ito, M. (1984). *The Cerebellum and Neural Control*, Raven Press, New York.

Jahnsen, H. (1986a) Electrophysiological characteristics of neurones in the guinea-pig deep cerebellar nuclei in vitro. *J. Physiol. (Lond.)*, 372: 129–147.

Jahnsen, H. (1986b) Extracellular activation and membrane conductances of neurones in the guinea-pig deep cerebellar nuclei in vitro. *J. Physiol. (Lond.)*, 372: 149–168.

Kano, M. (1995) Plasticity of inhibitory synapses in the brain: a possible memory mechanism that has been overlooked. *Neurosci. Res.*, 21: 177–182.

Kano, M., Kano, M., Fukunaga, K. and Konnerth, A. (1996) Ca^{2+}-induced rebound potentiation of gamma-aminobutyric acid-mediated currents requires activation of Ca^{2+}/calmodulin-dependent kinase II. *Proc. Natl. Acad. Sci. USA*, 93: 13351–13356.

Kano, M., Rexhausen, U., Dreessen, J. and Konnerth, A. (1992) Synaptic excitation produces a long-lasting rebound potentiation of inhibitory synaptic signals in cerebellar Purkinje cells. *Nature*, 356: 601–604.

Kim, J.J. and Thompson, R.F. (1997) Cerebellar circuits and synaptic mechanisms involved in classical eyeblink conditioning. *Trends Neurosci.*, 20: 177–181.

Koch, C., Poggio, T. and Torre, V. (1983) Nonlinear interactions in a dendritic tree: localization, timing and role in information processing. *Proc. Natl. Acad. Sci. USA*, 80: 2799–2802.

Komatsu, Y. (1996) $GABA_B$ receptors, monoamine receptors, and postsynaptic inositol trisphosphate-induced Ca^{2+} release are involved in the induction of long-term potentiation at visual cortical inhibitory synapses. *J. Neurosci.*, 16: 6342–6352.

Komatsu, Y. and Iwakiri, M. (1993) Long-term modification of inhibitory synaptic transmission in developing visual cortex. *NeuroReport*, 4: 907–910.

Krupa, D.J., Thompson, J.K. and Thompson, R.F. (1993) Localization of a memory trace in the mammalian brain. *Science*, 260: 989–991.

Linden, D. (1999) The return of the spike: postsynaptic action potentials and the induction of LTP and LTD. *Neuron*, 22: 661–666.

Linden, D.J. and Connor, J.A. (1995) Long-term synaptic depression. *Annu. Rev. Neurosci.*, 18: 319–357.

Lisman, J. (1994) The CaM kinase II hypothesis for the storage of synaptic memory. *Trends Neurosci.*, 17: 406–412.

Lisman, J. (1989) A mechanism for the Hebb and the anti-Hebb processes underlying learning and memory. *Proc. Natl. Acad. Sci. USA*, 86: 9574–9578.

Liu, X.-B. and Jones, E.G. (1997) Localization of alpha-type II calcium/calmodulin-dependent protein kinase specifically at glutamatergic synapses in rat hippocampus. *NeuroReport*, 8: 1475–1479.

Liu, X.B. and Jones, E.G. (1996) Localization of alpha type II calcium calmodulin-dependent protein kinase at glutamatergic but not gamma-aminobutyric acid (GABAergic) synapses in thalamus and cerebral cortex. *Proc. Natl. Acad. Sci. USA*, 93: 7332–7336.

Llinás, R. (1984) Functional significance of the basic cerebellar circuit in motor coordination. In: J.R. Bloedel, J. Dichgans and W. Precht, (Eds), *Cerebellar Functions*, Springer-Verlag, Berlin, Heidelberg.

Llinás, R. (1985) Rebound excitation as the physiological basis for tremor: a biophysical study of the oscillatory properties of mammalian central neurones in vitro. In: L.J. Findley and R. Capildeo (Eds), *Movement Disorders: Tremor*, Macmillan, London, pp. 165–182.

Llinás, R. and Mühlethaler, M. (1988) Electrophysiology of guinea-pig cerebellar nuclear cells in the in vitro brain stem-cerebellar preparation. *J. Physiol. (Lond.)*, 404: 241–258.

Llinás, R. and Sugimori, M. (1980) Electrophysiological Properties of in vitro Purkinje cell dendrites in mammalian cerebellar slices. *J. Physiol. (Lond.)*, 305: 197–213.

Llinás, R. and Sugimori, M. (1980) Electrophysiological Properties of in vitro Purkinje cell somata in mammalian cerebellar slices. *J. Physiol. (Lond.)*, 305: 171–195.

Llinás, R. and Yarom, Y. (1981) Properties and distribution of ionic conductances generating electroresponsiveness of mammalian inferior olivary neurones in vitro. *J. Physiol. (Lond.)*, 315: 569–584.

Lynch, G., Larson, J., Kelso, S., Barrionuevo, G. and Schottler, F. (1983) Intracellular injections of EGTA block induction of hippocampal long-term potentiation. *Nature*, 305: 719–721.

Magee, J.C., Christofi, G., Miyakawa, H., Christie, B., Lasser-Ross, N. and Johnston, D. (1995) Subthreshold synaptic activation of voltage-gated Ca^{2+} channels mediates a localized Ca^{2+} influx into the dendrites of hippocampal pyramidal neurons. *J. Neurophysiol.*, 74: 1335–1342.

Magee, J.C. and Johnston, D. (1997) A synaptically controlled, associative signal for Hebbian plasticity in hippocampal neurons. *Science*, 275: 209–213.

Markram, H., Lubke, J., Frotscher, M. and Sakmann, B. (1997) Regulation of synaptic efficacy by coincidence of postsynaptic APs and EPSPs. *Science*, 275: 213–215.

Markram, H. and Sakmann, B. (1994) Calcium transients in dendrites of neocortical neurons evoked by single subthreshold excitatory postsynaptic potentials via low-voltage-activated calcium channels. *Proc. Natl. Acad. Sci. USA*, 91: 5207–5211.

Marty, A. and Llano, I. (1995) Modulation of inhibitory synapses in the mammalian brain. *Curr. Opin. Neurobiol.*, 5: 335–341.

Mauk, M.D. (1997) Roles of cerebellar cortex and nuclei in motor learning: contradictions or clues? *Neuron*, 18: 343–346.

Mauk, M.D. and Donegan, N.H. (1997). A model of Pavlovian eyelid conditioning based on the synaptic organization of the cerebellum. *Learn. and Mem.*, 3: 130–158.

McCormick, D.A. and Pape, H.-C. (1990) Properties of a hyperpolarization-activated cation current and its role in rhythmic oscillation in thalamic relay neurones. *J. Physiol. (Lond.)*, 431: 291–318.

McLean, H.A., Caillard, O., Ben-Ari, Y. and Gaiarsa, J.L. (1996) Bidirectional plasticity expressed by GABAergic synapses in the neonatal rat hippocampus. *J. Physiol. (Lond.)*, 496: 471–477.

Miles, R., Tóth, K., Gulyás, A., Hájos, N. and Freund, T. (1996) Differences between somatic and dendritic inhibition in the hippocampus. *Neuron*, 16: 815–823.

Miyakawa, H., Ross, W.N., Jaffe, D., Callaway, J.C., Lasser-Ross, N., Lisman, J.E. and Johnston, D. (1992) Synaptically activated increases in Ca^{2+} concentration in hippocampal CA1 pyramidal cells are primarily due to voltage-gated Ca^{2+} channels. *Neuron*, 9: 1163–1173.

Morishita, W. and Sastry, B.R. (1996) Postsynaptic mechanisms underlying long-term depression of GABAergic transmission in neurons of the deep cerebellar nuclei. *J. Neurophysiol.*, 76: 59–68.

Mouginot, D. and Gahwiler, B.H. (1995) Characterization of synaptic connections between cortex and deep nuclei of the rat cerebellum in vitro. *Neuroscience*, 64: 699–712.

Mulkey, R.M., Endo, S., Shenolikar, S. and Malenka, R.C. (1994) Involvement of a calcineurin/inhibitor–1 phosphatase cascade in hippocampal long-term depression. *Nature*, 369: 486–488.

Mulkey, R.M., Herron, C.E. and Malenka, R.C. (1993) An essential role for protein phosphatases in hippocampal long-term depression. *Science*, 261: 1051–1055.

Mulkey, R.M. and Malenka, R.C. (1992) Mechanisms underlying induction of homosynaptic long-term depression in area CA1 of the hippocampus. *Neuron*, 9: 967–975.

Muri, R. and Knöpfel, T. (1994) Activity induced elevations of intracellular calcium concentration in neurons of the deep cerebellar nuclei. *J. Neurophysiol*, 71: 420–428.

Perrett, S. and Mauk, M. (1995) Extinction of conditioned eyelid response requires the anterior lobe of the cerebellar cortex. *J. Neurosci.*, 15: 2074–2080.

Perrett, S., Ruiz, B. and Mauk, M. (1993) Cerebellar cortex lesions disrupt learning-dependent timing of conditioned eyelid responses. *J. Neurosci.*, 13: 1708–1718.

Raymond, J.L., Lisberger, S.G. and Mauk, M.D. (1996) The cerebellum: a neuronal learning machine? *Science*, 272: 1126–1131.

Sastry, B.R., Morishita, W., Yip, S. and Shew, T. (1997) GABA-ergic transmission in deep cerebellar nuclei. *Prog. Neurobiol.*, 53: 259–271.

Stuart, G., Spruston, N., Sakmann, B. and Hausser, M. (1997). Action potential initiation and backpropagation in neurons of the mammalian CNS. *Trends Neurosci.*, 20: 125–131.

Thompson, R. (1986) The neurobiology of learning and memory. *Science*, 233: 941–947.

Thompson, R. and Krupa, D. (1994) Organization of memory traces in the mammalian brain. *Ann. Rev. Neurosci.*, 17: 519–549.

Thompson, R., Thompson, J., Kim, J., Krupa, D., Nordholm, A. and Chen, C. (1993) Synaptic plasticity of memory storage. In: M. Baudry, R. Thompson and J. Davis (Eds), *Synaptic Plasticity*, MIT Press, Cambridge, MA; London, England, pp. 231–259.

Voogd, J., Jaarsma, D. and Marani, E. (1996) The cerebellum: chemoarchitecture and anatomy. In: L.W. Swanson, A. Björklund and T. Hökfelt, (Eds), *Integrated Systems of the CNS, Part III: Cerebellum, Basal Ganglia, Olfactory System*, Elsevier, Amsterdam..

Welsh, J.P. and Harvey, J.A. (1989) Cerebellar lesions and the nictitating membrane reflex: performance deficits of the conditioned and unconditioned response. *J. Neurosci.*, 9: 299–311.

N.M. Gerrits, T.J.H. Ruigrok and C.I. De Zeeuw (Eds.)
Progress in Brain Research, Vol 124

CHAPTER 20

Time window control: a model for cerebellar function based on synchronization, reverberation, and time slicing

Werner M. Kistler*[1], J. Leo van Hemmen[2], and Chris I. De Zeeuw[3]

[1] *Center for Neuromimetic Systems, Swiss Federal Institute of Technology, 1015 Lausanne EPFL, Switzerland*
[2] *Department of Physics, Technical University of Munich, 85747 Garching Munich, Germany*
[3] *Department of Anatomy, Erasmus University Rotterdam, PO Box 1738, 3000 DR Rotterdam, The Netherlands*

Introduction

Almost 100 years after the internal structure of the cerebellar cortex (CBCX) has been elucidated by Ramón y Cajal (Ramón y Cajal, 1911) the function of the cerebellum is still a matter of fervent debate. Inspired by the beauty and simplicity of its neuronal wiring, several attempts have been made to explain cerebellar function by means of reductionist models at the level of individual neurons (Braitenberg, 1967; Marr, 1969; Albus, 1971; Ito, 1984). Here we propose a hypothesis of cerebellar function that relies on the new concept of time window control in conjunction with reverberating loops connecting the output of the cerebellum with its major inputs, viz., the mossy fiber and climbing fiber pathway. Periodic inhibitory input to a neuron results in a restriction of firing to those intervals of time where inhibition is weakest, and hence, to well-defined time windows. Here we pose, and answer, the question of whether the notion of time window can be relevant to the cerebellar system and present a model that comprises, apart from the cerebellar cortex, both deep cerebellar nuclei (DCN) and precerebellar brain stem nuclei. It is consistently based on a time coding paradigm that entrusts the control of granule cell firing to cerebellar inhibitory interneurons.

There are three key ideas to the present model, the morphological substrates of which are depicted in Fig. 1. First, we show by means of a detailed single-neuron model that the granule cell/Golgi cell system can operate as a potent gating device admitting only those spikes to the parallel fibers that arrive within short and well-defined time windows. This mechanism can be exploited so as to organize neuronal activity in discrete 'time slices' and to discern meaningful information from background noise. Second, we argue that neurons from the DCN are involved in a delayed-reverberating loop that relays cerebellar output-activity via precerebellar nuclei back to the mossy fibers. We show by means of a detailed neuron model that post-inhibitory rebound firing in DCN neurons can function as a robust timing mechanism that reliably produces delays of about 100 ms. The third key element of the model is provided by the dynamics of the network made up of inferior olive (IO), deep cerebellar nuclei, and mesodiencephalic junction (MDJ). We propose that intrinsic oscillatory properties of IO neurons in conjunction with delayed-reverberating loops (see Fig. 1) result in long transients of neuronal activity that are triggered by external input to the IO. This transient activity comes in the form of a sequence of spike

*Corresponding author. Tel.: +41(21) 693 3907; Fax: +41(21) 693 5263; e-mail: Werner.Kistler@epfl.ch

patterns that may act as a 'neuronal clock' (Llinás, 1991; Welsh et al., 1995).

A synthesis of the cerebellar time windows and the delayed reverberations in both the mossy fiber and the climbing fiber system result in rich oscillatory dynamics. For instance, synchronous Purkinje cell activity in one cycle is reverberated to the mossy fibers during the next cycle 100 ms later. Provided that the reverberated activity arrives with the correct timing relative to the opening of the granule cell time window, parallel fiber activity in one cycle is, at least partially, determined by the Purkinje cell activity during the previous cycle. Plasticity in the parallel fiber-Purkinje cell system allows the system to learn to react to a certain instance of these activity patterns; that is, to react

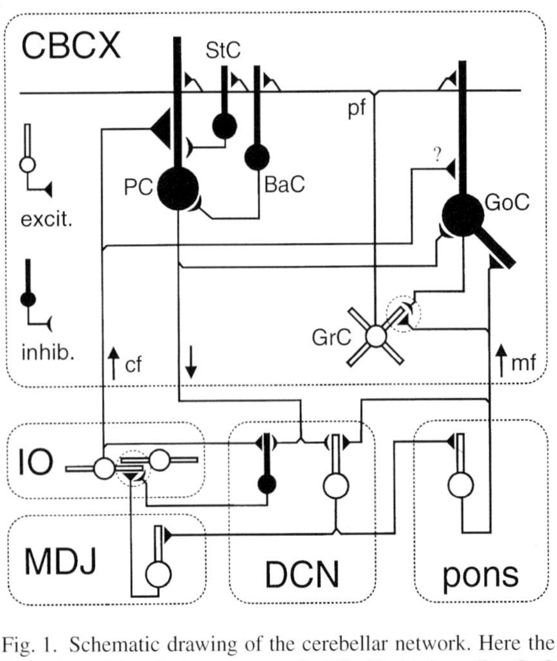

Fig. 1. Schematic drawing of the cerebellar network. Here the following abbreviations are used: PC (Purkinje cell), GoC (Golgi cell), GrC (granule cell), BaC (basket cell), StC (stellate cell), pf (parallel fiber), mf (mossy fiber), cf (climbing fiber), CBCX (cerebellar cortex), IO (inferior olive), MDJ (meso-diencephalic junction), and DCN (deep cerebellar nucleus). There are three loops related to the key ideas of the present model. The first loop is formed by granule cells and Golgi cells, and is responsible for the time-window mechanism. The second loop comprises PC, DCN, pons, and its mossy fiber system. The third loop is made up of PC, DCN, MDJ, IO, and its climbing fiber system. The latter two loops are responsible for the delayed reverberation of activities.

with a certain temporal relation to external events. This provides a novel explanation of how the cerebellum solves timing tasks on a time scale of several hundreds of milliseconds, which is much longer than the natural millisecond-time scale of ordinary neuronal dynamics.

Granule cell time windows

Cerebellar granule cells receive excitatory input from about four mossy fibers and a similar number of inhibitory projections from adjacent Golgi cells (Palay and Chan-Palay, 1974) (Fig. 1). These excitatory and inhibitory synapses meet in a structure called mossy fiber glomerulus so that the mossy fiber synapses can hardly evoke their excitatory effects without the inhibitory control of an accompanying Golgi cell synapse. Golgi cells have recently been shown to fire synchronously, and, less consistently, periodically (Vos et al., 1999; Vos, personal communication). The granule cell may thus receive a periodically modulated inhibition, and may be able to fire an action potential only in a small time interval shortly before the next set of Golgi cell spikes arrive, i.e. when inhibition is weakest. If the Golgi cells are firing asynchronously, inhibitory postsynaptic potentials (IPSPs) overlap partially and the granule cell is probably subject to a more or less constant level of inhibition and therefore unable to fire whenever the mossy fiber spikes arrive. A similar time window mechanism is probably also present in the basket-stellate cell/Purkinje cell system, because excitatory parallel-fiber input to the Purkinje cells is immediately followed by strong basket-cell inhibition delivered through sophisticated synapses that allow Purkinje cells to fire only during a short interval of time (Korn and Axelrad, 1980; Axelrad and Korn, 1982).

We have investigated the excitability of granule cells for various timings of excitatory and inhibitory input activities using a modified version of a detailed multi-compartment model developed for turtle granule cells (Gabbiani et al., 1994). We have extended the original model so as to include $GABA_A$-ergic synapses with a maximum conductance of 1.5 pS per synapse, a reversal potential of -75 mV, and a bi-exponential decay with time constants of 5 ms (peak amplitude $g_{fast} = 1.0$ pS) and 50 ms (peak amplitude $g_{slow} = 0.5$ pS) as reported for rat granule cells (Tia et al., 1996). Full details are given in Appendix (A).

The simulations shown in Fig. 2 confirm the predicted time window behavior. The granule cell receives inhibition from four Golgi cells firing at 10 Hz each and a volley of four synchronous mossy fiber spikes. If the Golgi cells are firing synchronously, then there is a short time window in which the mossy fiber volley can trigger an action potential (Fig. 2A). If the volley arrives too early, inhibition is still too strong and the firing threshold cannot be reached (Fig. 2B). On the other hand, if the volley arrives too late, then there is not enough time to reach the threshold and the action potential is clipped by the next set of Golgi cell spikes (Fig. 2C). If the Golgi cells are firing asynchronously, then no action potential can be triggered whenever the mossy fiber spikes arrive (Fig. 2D).

In addition to the time window for triggering spikes in granule cells we observe a second effect related to the timing of the resulting granule cell spike. As can be seen in Fig. 3, the granule cell spikes are 'focused' to a very narrow interval shortly before the arrival of the Golgi cell spikes. The timing of the granule cell spikes is thus relatively independent of the arrival time of the mossy fibers within the granule cell time window. That is to say, the timing of granule cell spikes can be more precise than the arrival time of mossy fiber spikes.

Figure 3 also demonstrates the dependence of the granule cell time window upon synaptic properties, in particular, on the relative amplitudes of the slow and the fast component of the inhibitory post-synaptic current (IPSC). The slow component in the IPSC is the result of the α6 subunit of the $GABA_A$ receptor, which is a peculiarity of the cerebellar cortex (Farrant and Cull-Candy 1993). It has been shown that the concentration of the α6 subunit and thereby the time course of the IPSC is subject to a systematic modification during development (Tia et al., 1996). Since the expression of the α6 subunit is regulated by the activation of the GABA receptor itself (Ueno et al., 1996; Carlson et al., 1997) we propose that these observations are related to a self-organized process which controls

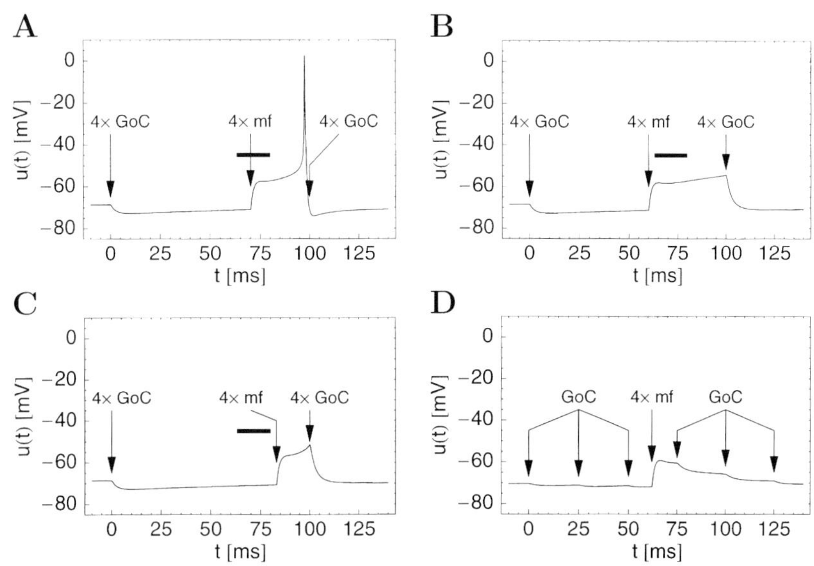

Fig. 2. Granule-cell time window. **A**, **B**, **C**, membrane potential (mV) as a function of time (ms) in a simulation of a realistic model of turtle granule cells. The vertical arrows indicate the arrival time of four synchronous Golgi cell spikes at $t = 0$ ms and $t = 100$ ms, and four mossy fiber spikes at $t = 70$ ms (A), $t = 60$ ms (B), and $t = 83$ ms (C). The horizontal bar marks the interval during which mossy fiber spikes can trigger an action potential. **D**, if the same number of Golgi cell spikes arrives asynchronously, here one spike every 25 ms, then mossy fiber spikes are unable to trigger an action potential, whatever their arrival time.

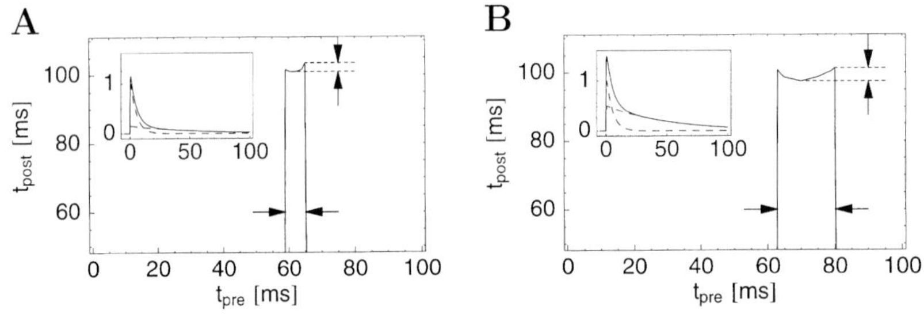

Fig. 3. Focusing of granule cell spikes. As in Figs 2A-C, the granule cell receives four synchronous Golgi cell spikes every 100 ms. The graphs give the firing time of the granule cell t_{post} as a function of the arrival time t_{pre} of mossy fiber spikes. Insets show the time course of the conductivity (pS) of the GABAergic synapse as a function of time (ms); slow and fast component are indicated by dashed lines. **A**, postsynaptic firing time induced by three synchronous mossy fiber spikes; $g_{slow} = 0.15$ pS, $g_{fast} = 1$ pS. **B**, postsynaptic firing time induced by four synchronous mossy fiber spikes; $g_{slow} = 0.5$ pS and $g_{fast} = 1$ pS.

the parameters that are critical to the time-window mechanism.

Delayed reverberation

Both the excitatory and inhibitory neurons in the DCN receive inhibitory input from Purkinje cells (Fig. 4) (De Zeeuw and Berrebi, 1995; Teune et al., 1998) and presumably both exhibit a strong post-inhibitory rebound firing (Jahnsen, 1986a; Llinás and Mühlethaler, 1988a; Tempia/Ruigrok personal communication). While the inhibitory DCN neurons project exclusively to the inferior olive (De Zeeuw et al., 1989), the excitatory DCN neurons project to various brain stem nuclei, such as the nucleus reticularis tegmenti pontis in the pons, which is a major source of cerebellar mossy fibers (Murakami et al., 1981; Tsukahara et al., 1983; Verveer et al., 1997).

In order to study the rebound firing mechanism in detail we have devised a single-compartment active-membrane model for DCN neurons; see Appendix (B). Thalamic relay neurons have electrophysiological properties similar to those of DCN neurons (Llinás, 1988). We have modified an existing model for thalamic relay neurons (McCormick and Huguenard, 1992) so as to reproduce in vitro measurements of current-step responses and gain function of DCN neurons (Llinás and Mühlethaler, 1988a). Furthermore, the model exhibits low-threshold Ca^{2+} spikes and non-inactivating Na^{+} plateaus, which are characteristic features of DCN neurons (Jahnsen, 1986b; Llinás and Mühlethaler, 1988b).

We have investigated the timing of rebound spikes as a function of the level and the temporal structure of the preceeding inhibition. Figure 5 shows that once there is a rebound spike, it occurs 100 ms after the onset of inhibition, irrespectively of the number of involved IPSPs and their degree of synchrony. The rebound spike can be even more precisely timed than an individual Purkinje cell spike, because several Purkinje cell spikes are required to trigger a rebound spike. Noise is efficiently reduced because the timing of a rebound spike depends predominantly on the average of the arrival times of the Purkinje cell spikes at a DCN neuron. We thus observe an effect similar to the focusing of granule cell spikes, namely that the precision of a postsynaptic neuron's firing time can be higher than the presynaptic spike arrival times.

Plasticity at parallel fiber/Purkinje cell synapses

Combination of time windows and delayed reverberation yields interesting dynamics, if we allow the reverberated activity to return at the CBCX within the granule cell time window. This way the CBCX can lock in on its own output and learn, store, and recall spatiotemporal patterns of neuronal spike activity by adjusting the weights of parallel fiber/Purkinje cell synapses.

Activities of different climbing fibers can arrive synchronously in sagittal zones of the CBCX (Sugihara et al., 1993). The climbing fibers elicit so-called all-or-none complex spikes in the Purkinje cells (Eccles et al., 1966), and in addition they might affect the activity of Golgi cells via climbing fiber collaterals (Palay and Chan-Palay, 1974; see the question mark in Fig. 1). Recurrent axon

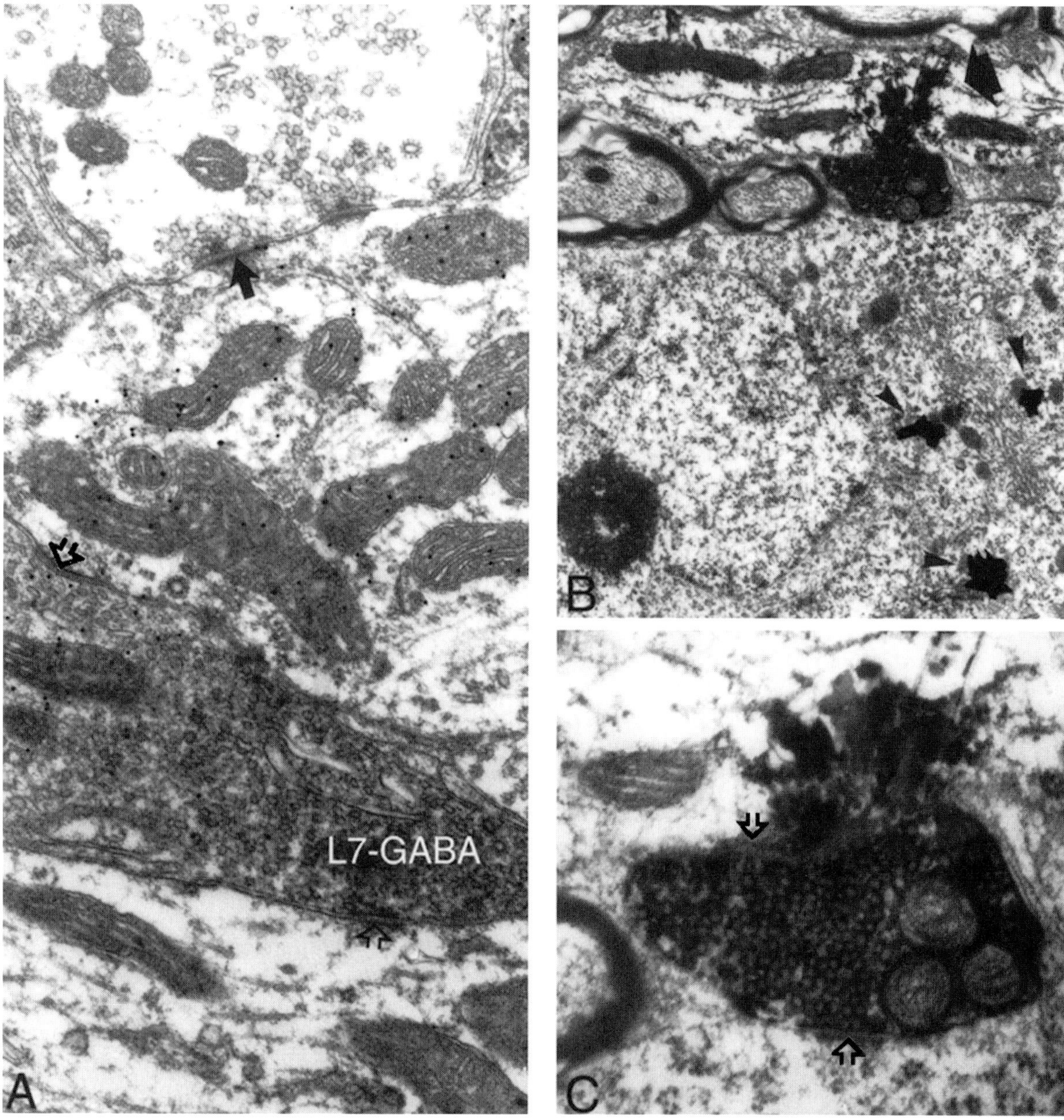

Fig. 4. Characteristic synaptology in deep cerebellar nuclei (DCN). **A**, a double L7-GABA-labeled Purkinje cell terminal contacts both a GABAergic (inhibitory) and a non-GABAergic (excitatory) dendrite in a cerebellar nucleus. The GABA-labeling is visualized by means of 15 nm gold particles (black dots), while the Purkinje cell specific L7-protein is visualized with the use of dark diamino benzidine reaction products (L7-GABA; for details, see De Zeeuw and Berrebi, 1995). **B** and **C**, a biotinylated dextran amine labeled Purkinje cell terminal contacts both a dendrite retrogradely labeled with WGA-HRP (thick arrow in B) from the mesodiencephalic junction and a cellbody retrogradely labeled with gold-lectin (arrowheads) from the inferior olive (for details, see Teune et al., 1998). Open arrows in A and C indicate symmetric synapses; closed arrow in A indicates asymmetric synapse.

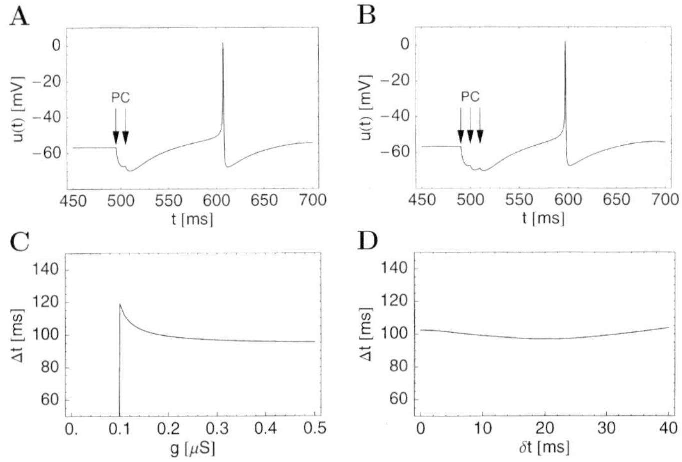

Fig. 5. Post-inhibitory rebound in DCN neurons. **A** and **B**, membrane potential of a model for DCN neurons in response to two (A) and three (B) inhibitory Purkinje cell spikes (arrows). The resulting rebound spike occurs after a delay of approximately 100 ms, irrespectively of the number of Purkinje cell spikes. **C**, delay Δt of the rebound spike after a volley of Purkinje cell spikes as a function of the total synaptic conductivity, i.e. the conductivity per synapse times the number of activated synapses. **D**, delay of the rebound spike after a volley of three Purkinje cell spikes versus the degree of their synchronicity δt (width of the interval containing all three spikes) measured from the center of the Purkinje cell spike volley. Note the plateaus in C and D.

collaterals of the Purkinje cells in turn will inhibit the Golgi cells and thereby allow the granule cells to recover from inhibition (Schulman and Bloom, 1981). Hence the granule cell window will be open after a delay of several tens of milliseconds. In the mean time, those DCN neurons that were inhibited by Purkinje cells fire their rebound spikes, which reverberate to the CBCX via brainstem nuclei and the mossy fiber system. We assume that the delayed opening of the granule cell time window corresponds to the delay imposed by the rebound firing and the reverberating loop. That is to say, the rebound activity caused by the complex spikes returns to the CBCX just in time so as to pass the granule cell/Golgi cell system after the inhibition of granule cells has decayed. The rebound activity is thus relayed to the parallel fibers and can in turn trigger certain subsets of Purkinje cells and Golgi cells. The Golgi cell spikes inhibit the granule cells and close the granule cell time window again. Each pattern of Purkinje cell activity may trigger a new set of Purkinje cells in the next cycle depending on the wiring of the reverberating loop and on the weight matrix of the parallel fiber/Purkinje cell synapses. Thus, the oscillation can continue until external events alter the firing pattern of the inhibitory interneurons.

We show by means of large-network simulations that the biologically plausible mechanisms of plasticity at the parallel fiber/Purkinje cell synapses, i.e. long term potentiation (LTP) and long term depression (LTD), allows the system to learn *arbitrary* sequences of spike patterns. The network is made up of spike response neurons – a kind of a generalized integrate-and-fire neuron (Gerstner and van Hemmen, 1992, 1994; Kistler et al., 1997). This set-up allows for a qualitative description of different types of neuronal behavior such as adaptation and post-inhibitory rebound. The connectivity is within morphological constraints random so that each pair of neurons in a simulation is assigned an arborization function that gives the probability of their being 'connected' as a function of their distance; see Appendix (C).

The nature of the plasticity at the parallel fiber/ Purkinje cell synapses may depend on the relative timing of the activities in the climbing fibers and parallel fibers. Recent experimental evidence (Sakurai, 1987; Hirano, 1990; Schreurs et al., 1996; Lev-Ram et al., 1997; see however, Bell et al., 1997) suggests that these synapses are potentiated, if a complex spike occurs within about 10 ms *after* the corresponding 'parallel fiber spike', and that they are depressed if a complex spike arrives within about 100 ms *before* a 'parallel fiber spike'. In this way only those synapses are potentiated that have contributed to the simple-spike firing of the Purkinje cell shortly before or after the arrival of a climbing fiber spike. Other synapses that would make the Purkinje cell fire at an other moment of time are depressed (cf. Hebb's rule; Hebb, 1949)

In the present model climbing fiber spikes are attributed a two-fold function. First, climbing fiber activities are interpreted as 'teacher signals' that indirectly tell Purkinje cells when precisely they have to fire simple spikes. This notion, which is consistent with their immediate triggering of complex spikes, is different from that of climbing fiber spikes serving as error signals (Albus, 1971). Second, climbing fiber spikes might have a synchronizing effect on cerebellar interneurons and initialize the granule cell time window, as we have seen above.

Figure 6A shows the spike trains of neurons involved in a network simulation. We start with an uncorrelated rest activity in the mossy fibers. The pattern the Purkinje cells are about to learn is presented to them by climbing fiber spikes. In the example shown in Fig. 6, climbing fibers have been divided in four groups that subsequently deliver synchronous action potentials every 120 ms. Experimental evidence has been provided by Welsh et al. (1995) who showed that spatio-temporal patterns of climbing fiber activity can be phase-locked to ongoing movements.

As shown in Fig. 6A, the first set of climbing fiber spikes arriving at $t = 400$ ms forces the corresponding Purkinje cells to fire complex spikes that inhibit DCN neurons and cause a rebound volley in the mossy fiber pathway at $t = 518$ ms. This rebound volley is paired with the second set of climbing fiber spikes that arrive at $t = 520$ and again will trigger complex spikes at certain Purkinje cells; and so on. After a few repetitions parallel fiber-Purkinje cell synapses are modified in such a way that a characteristic activity pattern of parallel fibers suffices to trigger the corresponding Purkinje cell, even without a climbing fiber acitivation. The teacher signal is no longer needed.

Figure 6B shows that after 10 repetitions the pattern has been learned successfully and can be recalled by the first climbing fiber event. We note

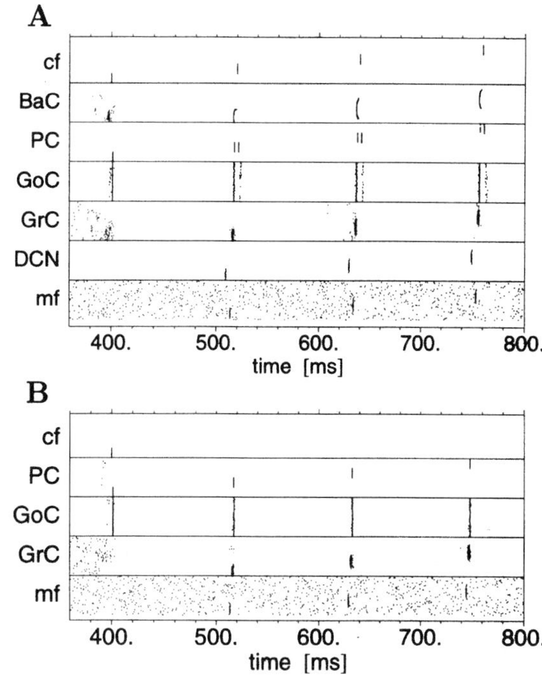

Fig. 6. Cerebellum learning spatiotemporal patterns. The diagrams show the spike trains of 200 deep cerebellar nuclei neurons (DCN), 1000 granule cells (GrC), 80 Golgi cells (GoC), 200 Purkinje cells (PC), 400 basket or stellate cells (BaC), and the spike trains delivered by 200 mossy fibers (mf) and 200 climbing fibers (cf) as a raster diagram over time. Each short vertical line corresponds to an action potential, but because of limited printing resolution, synchronous spikes of neighboring neurons may merge into one line. **A**. The climbing fibers have been divided into four groups that subsequently deliver synchronous action potentials every 120 ms (uppermost panel). The following climbing fiber spikes are paired with mossy fiber events that are produced by a rebound mechanism depending on Purkinje cell activity in the previous cycle (see text). The role of the granule cell/Golgi cell system is clearly visible. After synchronization of the Golgi cells at $t = 400$ ms, mossy fiber activity is admitted to the parallel fibers (GrC-panel) only during narrow time windows shortly before the Golgi cells fire their action potentials. This results in a concentration of the activity of the various types of neurons in discrete 'time slices' centered around the climbing fiber volleys. **B**. Similar diagram as in A but after the pattern has been learned successfully. A single climbing fiber volley at $t = 400$ ms suffices to recall the learned pattern without any further climbing fiber spikes being involved. Note that the complex spike synchrony levels presented in A and B are at the highest end of those that can occur in vivo (for details see De Zeeuw et al., 1996). In addition it should be noted that the average firing frequency of the simple spike activity presented in this diagram is relatively low due to the high level of inhibition. The BaC and DCN cells are not shown in B for clarity of presentation.

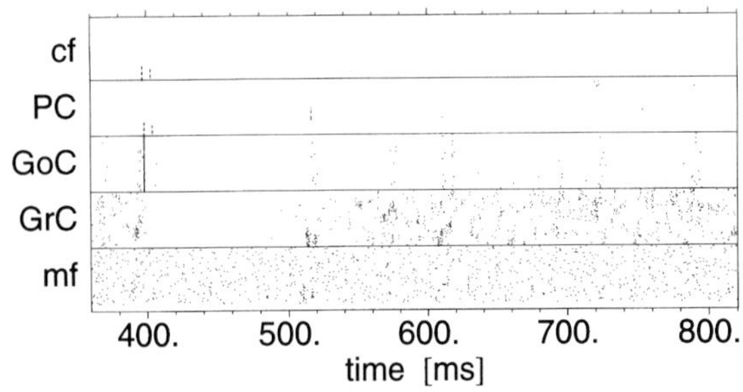

Fig. 7. Simulation of the same network as in Fig. 6B demonstrating the importance of climbing fiber delay tuning. If the climbing fiber spikes arrive at the cerebellar cortex with a jitter of as small as 5 ms, the pattern fails to be recalled.

that the recall process is stable despite the high level of background noise in the mossy fiber input and the presence of inherent noise due to the stochastic spike triggering process. The present network model also illustrates the importance of climbing fiber synchrony, as has been found experimentally (Bell and Kawasaki, 1972; Llinás et al., 1974; Llinás and Sasaki, 1989; De Zeeuw et al., 1996). Figure 7 shows that if the climbing fiber spikes arrive asynchronously in the cerebellar cortex with a jitter of only 5 ms, then the pattern fails to be recalled. Thus, synchrony of the climbing fiber activity is essential to the recall function.

Self-organization of climbing fiber activity

In the previous section we have seen that the cerebellar network can learn spatiotemporal patterns of spike activity. This set-up required climbing fibers to deliver a teacher signal every 100 ms. The signal is generated by an until now unspecified external mechanism. In this section we will show that electrophysiological properties of inferior olivary (IO) neurons in conjunction with topographic projections between IO, DCN, and mesodiencephalic junction (MDJ) can produce precisely timed sequences of climbing fiber spike patterns. These sequences are hard-wired in the olivo-cerebellar system and may serve as a 'neuronal clock' that basically extends the system's short-term memory beyond several hundreds of milliseconds (see also Ruigrok and Voogd, 1995).

Inferior olivary neurons have oscillatory properties with an intrinsic frequency of about 10 Hz (Llinás and Yarom, 1986; Llinás 1991). The electrophysiological correlates of this behavior are a calcium-dependent potassium current and two different voltage-dependent calcium currents. One of them, the low-threshold somatic calcium current, is also responsible for rebound activation similar to that seen in DCN neurons (Llinás and Yarom, 1981a; Llinás and Yarom, 1981b). Presumably, the subthreshold oscillations and the strong coupling of neighboring neurons via gap junctions (Fig. 8) can result in synchronous climbing fiber activity (Llinás et al., 1974; De Zeeuw et al., 1989) and in a precise timing of the complex spikes (Lampl and Yarom, 1993).

Albeit oscillatory properties exist, there is no sustained super-threshold oscillatory activity in the IO, at least in the absence of pharmacological stimulation (Llinás and Yarom, 1986). Thus, any activity in the IO triggered by some external events finally will die out. We hypothesize that the decay of IO activity is a relatively slow process comprising several periods of 100 ms each (Ruigrok and Voogd, 1995). The sequence of spike patterns that is generated in the wake of this decay can serve as a neuronal clock or as a counter of the periods that have elapsed since the oscillation has been started by some external event.

Delayed reverberating loops between the IO and the DCN also contribute to the slowly decaying oscillations (Fig. 9). In fact, they determine the sequence of firing patterns. One of the loops includes IO neurons that send excitatory collaterals to the inhibitory neurons in the DCN, which in turn project back to the IO (De Zeeuw et al., 1997). Inhibition of IO neurons that is produced by small neurons in the complex of cerebellar and vestibular nuclei (see also Fig. 8), may result in post-inhibitory rebound with a proper delay of about 100 ms (Llinás, 1991). Another reverberating loop comprises the inhibitory input of the Purkinje cells to the DCN neurons, which in turn produce rebound spikes with a certain delay after the complex spike activity of the Purkinje cells. The rebound activity of many of the excitatory neurons in the DCN is relayed through a disynaptic excitatory pathway via the MDJ back to the IO (Figs. 1 and 10). Both loops are topographic in that particular subsets of neurons are directly or indirectly, reciprocally connected to one another (De Zeeuw et al., 1998b). The result is that both pathways form positive feedback loops that deliver excitation with a delay of about 100 ms to the neurons that have been active in the previous cycle. However, in both loops the inputs to the IO, i.e. the inhibitory input from the DCN and the excitatory input from the MDJ, directly synapse onto the olivary dendrites that are electrotonically coupled to each other (Fig. 8), and presumably both inputs can dynamically regulate the domains of the coupled olivary neurons (Llinás and Sasaki, 1989; De Zeeuw et al., 1998b). Thus depending on the precise wiring scheme and status of coupling in the inferior olive a pattern of spike activity in one cycle may induce a new pattern of spike activity in the next cycle 100 ms later; and so forth.

Theoretical considerations (Kistler, 1999) show that a slowly decaying oscillatory activity requires a tuning of the firing threshold of the neurons involved and of the amount of convergence and divergence present in the reverberating loop. Simulations indicate that this tuning can be achieved by means of a self-organized process called anti-Hebbian elimination. In short, initially there are more synaptic connections than required for a slowly decaying activity. During a training phase oscillations are triggered by external inputs to IO neurons and those synapses that are contributing *repetitively* to the firing of their postsynaptic targets are gradually eliminated with a certain (small) probability (Fig. 9). The elimination process stops when a critical connectivity is left so that oscillations are no longer self-sustained but decay (slowly).

The sequence of spike patterns produced by the resulting connectivity can be used to solve timing tasks as shown in Fig. 11. An external event

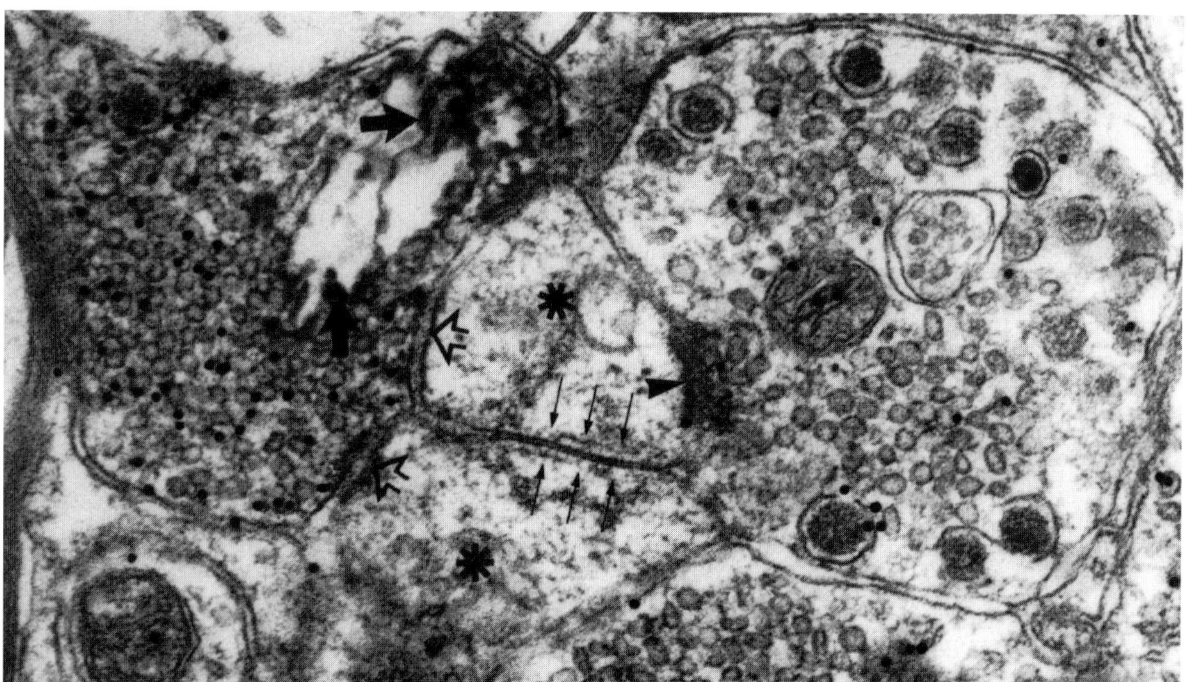

Fig. 8. Characteristic synaptology in inferior olive (IO). The electron micrograph shows a small olivary glomerulus containing a core of two dendritic spines (asterisks) coupled by a gap junction (small arrows). The spines are innervated by a double WGA–HRP/GABA labeled terminal derived from the complex of cerebellar and vestibular nuclei in the hindbrain following a combination of anterograde tracing of WGA–HRP and postembedding immunocytochemistry (for details see De Zeeuw et al., 1989). Open arrows and arrow head indicate symmetric and asymmetric synapses, respectively. Closed thick arrow indicates WGA–HRP reaction product. Small black dots are 15 nm gold particles linked to a GABA-antibody.

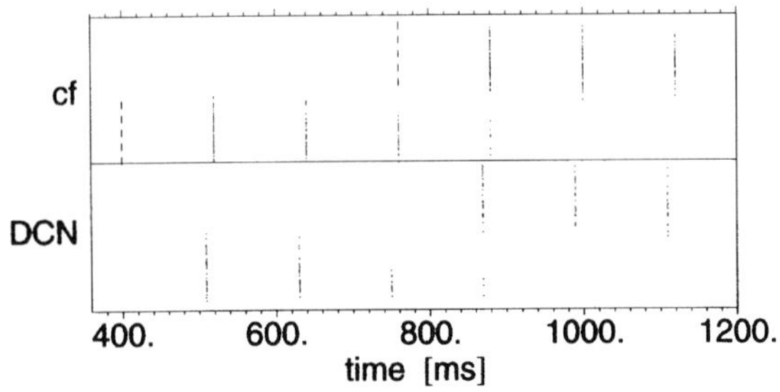

Fig. 9. Transient activity in reverberating loops connecting DCN, MDJ, and IO. Two external events trigger IO neurons at $t = 400$ and $t = 560$, respectively. The connectivity between the involved nuclei is the result of anti-Hebbian elimination (see text) and is responsible for the slow decay of activity in the climbing fiber system (upper panel) and DCN neurons (lower panel) shown here in a simulation similar to Fig. 6.

triggers a subpopulation of IO neurons that gives rise to a climbing fiber event at time $t = 400$ ms. The subsequent series of mossy fiber-spike patterns that is produced by the ongoing activity in the reverberating loop of DCN, MDJ, and IO is paired with a second climbing fiber event at time $t = 760$ ms (Fig. 11A). Purkinje cells thus learn to respond to the first event with a proper delay of 360 ms, even if the second climbing fiber event is omitted (Fig. 11B). Such mechanisms might for example play a role in particular timing sensitive tasks during classical conditioning (for details see Yeo and Hesslow, 1998).

DISCUSSION

We have presented a functional model of the cerebellar network comprising CBCX, DCN, precerebellar nuclei, and IO. The distinguishing feature of the present model is that cerebellar interneurons can be attributed a function that goes beyond a mere regulation of the mean firing rate of the principal neurons (Marr, 1969; Albus, 1971) in that they generate narrow and well-defined time windows for triggering spikes. We have shown that the granule cell/Golgi cell system can discern incoming mossy fiber spikes by their arrival times. This has several interesting consequences. First, time windows allow for an intricate parallel processing, if information for different tasks is organized in different time slices. Second, the overall signal-to-noise ratio is significantly improved if, on the one hand, meaningful spike activity is concentrated in a certain time slice, and background noise, on the other hand, is spread evenly in time.

The second key ingredient of the present model is a delayed reverberating loop that conveys post-inhibitory rebound activity of DCN neurons via precerebellar nuclei back to the mossy fiber system. We have shown by means of a detailed model for DCN cells that post-inhibitory rebound is a robust mechanism to produce precisely timed delays of about 100 ms. The aforementioned granule cell time window thus allows the cerebellar network to lock in on its own output provided that the rebound delay matches the delayed opening of the granule cell time window.

Combination of time windows and delayed reverberation results in a closed system that, in conjunction with long-term plasticity at the parallel fiber/Purkinje cell synapses, is able to learn, store, and recall spatiotemporal patterns of spike activity. Large network simulations (Kistler, 1999; Kistler and van Hemmen, 1999) reflecting the morphological structure of the CBCX predict that the LTD and LTP mechanisms underlying long-term plasticity attribute to climbing fibers the function of a 'teacher signal' that tells Purkinje cells when to fire and evoke rebound spikes in the DCN.

Whether plasticity at the parallel fiber/Purkinje cell synapses depends on the timing of climbing fiber and parallel fiber activities is still a controversial issue (see for example Llinás et al., 1997; Bell et al., 1997; Lev-Ram et al., 1997). In the present model a synapse is depressed, if a parallel fiber potential arrives a few tens of milliseconds *after* a climbing fiber spike. If the timing is the other way around, i.e. if a parallel fiber action potential arrives a few milliseconds *before* the climbing fiber activity, then the corresponding synapse is potentiated. The present version of plasticity is a generalization of Hebb's learning rule (Hebb, 1949), because synapses are strengthened only if they are contributing to the firing of the postsynaptic neuron (Gerstner et al., 1996; Markram et al., 1997; Kistler and van Hemmen, 2000).

Apart from learning spatiotemporal patterns the network can solve timing tasks, i.e. the network can learn to respond to an external stimulus with a variable delay of up to several hundreds of milliseconds. The underlying idea is that reverberating loops comprising DCN, MDJ, and IO can produce long sequences of varying spike patterns that are hard-wired in the reverberating circuit. Synaptic plasticity in the reverberating loop made up of Purkinje cells, DCN, pons, and mossy fiber system allows Purkinje cells to learn when to fire in response to an external event that has started the sequence. This provides – to our knowledge – a novel explanation of how the cerebellum solves timing tasks on a time scale of hundreds of milliseconds (McCormick et al., 1982; Perret et al., 1993; Welsh et al., 1995; Yeo and Hesslow, 1998, De Zeeuw et al., 1998a).

Relation to other models

Every system that deals with temporal patterns inevitably requires a kind of internal clock or

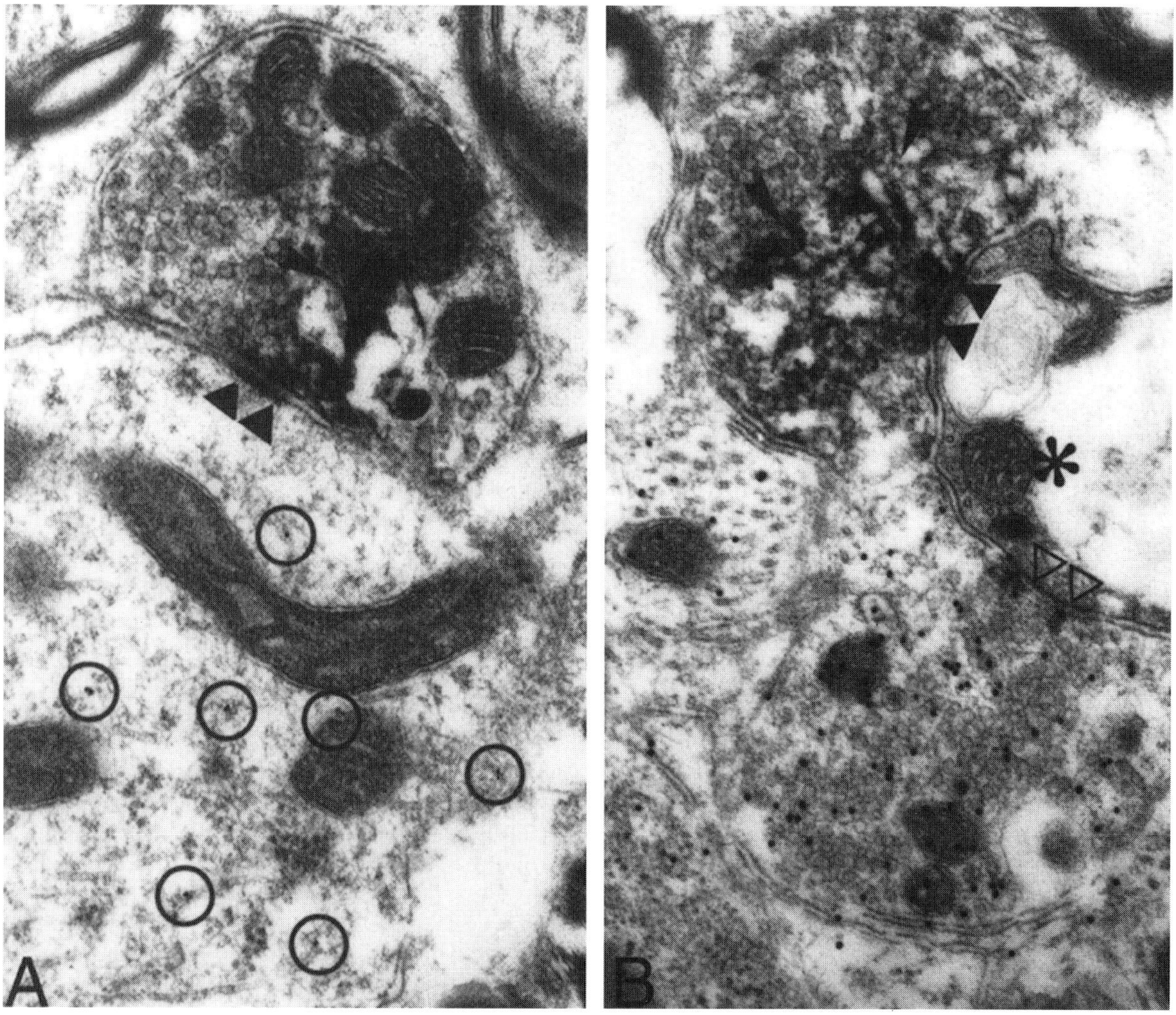

Fig. 10. Characteristic synaptology in the mesodiencephalic junction (MDJ). The electron micrographs show two examples of terminals in the MDJ anterogradely labeled with WGA–HRP (arrowheads) from the cerebellar nuclei. One terminal (A) contacts a proximal dendrite with gold particles (5–10 nm in diameter; indicated by circles) that are transported retrogradely from the inferior olive, while the other terminal (B) contacts a peripheral dendrite (asterisk) of a non-GABAergic cell that is also contacted by a GABAergic terminal from another source. Closed and open triangles indicate asymmetric and symmetric synapses, respectively.

timing device. Previous models of cerebellar function that explicitely include time – the classical models of Marr (Marr, 1969) and Albus (Albus, 1971) are static – can be classified according to their specific timing mechanisms. An early theory of cerebellar function is that of Braitenberg (1967). Here parallel fibers are used as delay lines and different spatial distances are employed to compensate for different arrival times of mossy fiber spikes so as to discern spatiotemporal spike patterns in the mossy fiber system (Braitenberg, 1993; Braitenberg et al., 1997). A similar idea underlies the work of Meek (1992), who exploits differences in the propagation time of parallel fiber spikes to translate short time differences into a spatial map. To overcome the limited range of attainable delays of only a few tens of milliseconds, different timing mechanisms have been proposed. Moore et al. (1989) postulate the existence of external delay lines in the brain stem and Fiala et al. (1996) employ slow metabotropic glutamate receptors in order to arrive at delays long enough to explain the role of the cerebellum in solving timing tasks. All these models have in common that they require an array of different delays that cover the whole range of delays that are to be learned. Different in this respect is a model of Buonomano (1994) that relies on the granule cell population vector, which is subject to a temporal evolution even for a stationary mossy fiber input.

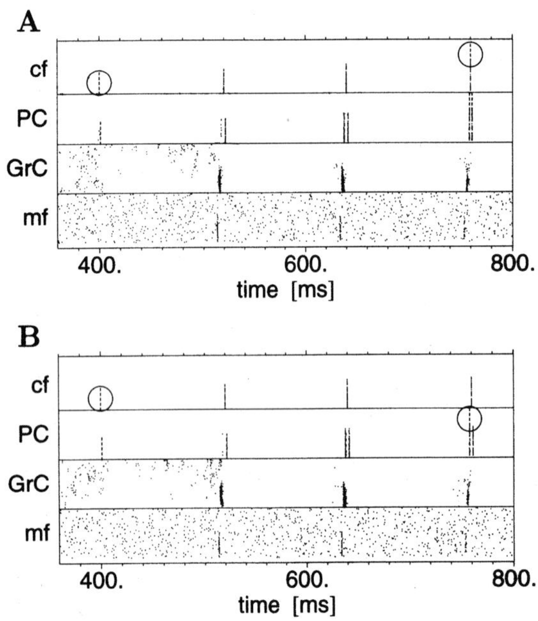

Fig. 11. Cerebellar network performing a timing task. **A**, two external events trigger certain subpopulations of IO neurons giving rise to two climbing fiber events at $t=400$ and $t=560$, respectively (circles). Transient oscillatory activity in the reverberating loops of DCN, MDJ, and IO produce characteristic spike patterns in the mossy fiber system every 120 ms. **B**, after a few repetitions of the external stimuli, Purkinje cells are able to recognize the mossy fiber pattern that has been paired with the second climbing fiber event. The network is thus able to learn the time difference between the two stimuli and to respond with the correct timing (circle at $t=560$), even if the first stimulus only (circle at $t=400$) is given.

One of the major delay-generating mechanisms in the present model is post-inhibitory rebound in DCN neurons. Rebound firing seems to be a concept that is ubiquitously used in various parts of the nervous system in order to reliably produce well-defined delays of several tens of milliseconds. Prominent examples are the auditory system of bats (Olsen and Suga, 1991, Carr, 1993) and the visual system of vertebrates, where delays of such a long duration are required for the processing of moving stimuli (Mastronarde, 1987).

In a similar way to Buonomano (1994), the present model does not require an array of different delays but uses the time-dependent granule cell activity as a 'neuronal clock'. The vital difference between the two models is the manner in which the temporal evolution is imposed upon the granule cell population. Buonomano (1994) assumes a random connectivity between mossy fibers, granule cells, and Golgi cells and relies on the resulting dynamics that is generated by the inner-cortical connections and the mossy fiber input. Severe problems arise, however, from the lacking stability of the dynamics with respect to noise (Buonomano, 1994). We believe that the chaotic nature of the granule cell-Golgi cell dynamics disqualifies it for the proposed task. The present model, in contrast, is based on stable oscillations of neuronal activity between the CBCX, DCN, and pre-cerebellar nuclei.

Stability properties

The stability of the present model to noise has been demonstrated by simulations that include both internal (stochastic spike triggering) and external (noisy, mossy fiber input) noise. Insensitivity to external noise is due to the restriction of firing times to narrow time windows. External noise is effectively blocked by the system's admitting only those mossy fiber spikes that arrive with the correct timing and rejecting all spikes that arrive outside the granule cell time window. The proviso is that there is a fixed temporal relation between the input signal and the internal state of the cerebellum. At least for the reverberated signals this condition can be fulfilled by tuning the delay after which Purkinje cell spikes reappear in the cerebellar cortex.

The robustness with respect to intrinsic noise results from the quasi-digital mode of information processing and the auto-associative way in which spatiotemporal patterns are stored. Time is divided into well-separated discrete 'slices' defined by the time windows during which granule cells and Purkinje cells can fire. Purkinje cells have learned to recognize certain parallel fiber events in an associative and error-tolerant manner. Since the Purkinje cell activity in one cycle is reverberated to the parallel fibers in the next cycle, the whole succession of spike patterns is stable (Gerstner et al., 1993).

Two effects that have been revealed by our simulations of detailed neuron models of granule cells and of DCN neurons further enhance the overall stability to noise. In the time-window mode of operation of granule cells, the firing time of granule cells is insensitive to the arrival time of the mossy fiber volley provided the latter arrives within the time window. Granule cell spikes are thus 'focused' into an interval that is even shorter than the granule cell time window. As presented above, a similar effect has been found in the model of DCN neurons that fire a rebound spike in response to a couple of inhibitory Purkinje cell spikes. The rebound spike occurs with a fixed delay of about 100 ms after the Purkinje cell spike volley and is irrespective of the number of spikes in the volley. The timing of a rebound spike is even independent of the temporal spreading of the volley and thus determined by an average over all spikes of the volley. This effectively reduces noise in the rebound firing time. Both effects are crucial to a safe operation of the present model (and of the cerebellum). Otherwise small deviations in spike timing would grow until synchronization is so poor that all neuronal activity in a reverberating loop would cease to exist.

Summary

We present a new hypothesis of cerebellar function that is based on synchronization, delayed reverberation, and time windows for triggering spikes. Our model suggests that granule cells admit mossy fiber activity to the parallel fibers only if the Golgi cells are firing synchronously and if the mossy-fiber spikes arrive within short and well-defined time windows. The concept of time window control organizes neuronal activity in discrete 'time slices' that can be used to discern meaningful information from background noise. In particular, Purkinje cell activity can trigger rebound spikes in deep cerebellar nuclei cells, which project via brain stem nuclei and mossy fibers back to the cerebellar cortex. Using a detailed model of deep cerebellar nuclei cells, we demonstrate that the delayed firing of rebound spikes is a robust mechanism so as to ensure that the reverberated activity re-arrives in the mossy fibers just during the granule-cell time window. Large network simulations reveal that synaptic plasticity (LTD and LTP) at the parallel fiber/Purkinje cell synapses that relies on the timing of the parallel fiber and climbing fiber activities allows the system to learn, store, and recall spatiotemporal patterns of spike activity. Climbing fiber spikes function both as teacher and as synchronization signals. The temporal characteristics of the climbing fiber activity are due to intrinsic oscillatory properties of inferior olivary neurons and to reverberating projections between deep cerebellar nuclei, the mesodiencephalic junction, and the inferior olive. Thus, the reverberating loops of the mossy fiber system and climbing fiber system may interact directly with the time windows provided by the circuitry of the cerebellar cortex so as to generate the appropriate spatio-temporal firing patterns in the deep cerebellar nuclei neurons that control premotor systems. In future studies the model will be extended in that high frequency simple spike activities will be included and that their relevance for motor control will be addressed.

Acknowledgements

It is a great pleasure to the authors to thank Herbert Axelrad and Rodolfo Llinás for stimulating discussions and helpful advice, and Fabrizio Gabbiani and Ulrich Hillenbrand for help with the detailed neuron models. W.M.K. gratefully acknowledges financial support from the Boehringer Ingelheim Fonds. C.I.D.Z. gratefully acknowledges support from the HFSP, the Life Sciences Foundation (SLW; no. 805–33.310-p; C.I.D.Z.), which is subsidized by the Netherlands Organization for Scientific Research (NWO), and the support by a NWO project grant (no. 903–68–361). Furthermore, C.I.D.Z. thanks the Graduiertenkolleg 'Multisensorische Interaktion' for a pleasant and fruitful stay in Munich, where this manuscript was finished.

References

Albus, J.S. (1971) A theory of cerebellar function. *Math. Biosci.*, 10: 25–61.

Axelrad, H. and Korn, H. (1982) Field effect and chemical transmission: dual inhibition action of basket cells in the rat cerebellar cortex. *Exp. Brain Res. Suppl.*, 6: 412–439.

Bell, C.C. and Kawasaki, T. (1972) Relations among climbing fiber responses of nearby Purkinje cells. *J. Neurophysiol.*, 35: 155-169.

Bell, C.C., Han, V.Z., Sugawara, Y. and Grant, K. (1997) Synaptic plasticity in a cerebellum-like structure depends on temporal order. *Nature*, 387: 278–281.

Braitenberg, V. (1967) Is the cerebellar cortex a biological clock in the millisecond range? In: C.A. Fox and R.S. Snider (Eds.), *Progress in Brain Research* 25: 334–346.

Braitenberg, V. (1993) The cerebellar network: attempt at a formalization of its structure. *Network*, 4: 11–17.

Braitenberg, V., Heck, D. and Sultan, F. (1997) The detection and generation of sequences as a key to cerebellar function, experiments and theory. *Behav. Brain Sci.*, 20: 229–277.

Buonomano, D.V. (1994) Neural network model of the cerebellum: temporal discrimination and the timing of motor responses. *Neural Comput.*, 6: 38–55.

Carlson, B.X., Belhage, B., Hansen, G.H., Elster, L., Olsen, R.W. and Schousboe, A. (1997) Expression of the $GABA_A$ receptor α6 subunit in cultured cerebellar granule cells is developmentally regulated by activation of $GABA_A$ receptors. *J. Neurosci. Res.*, 50: 1053–1062.

Carr, C.E. (1993) Processing of temporal information in the brain. *Annu. Rev. Neurosci.*, 16: 223–243.

De Zeeuw, C.I. and Berrebi, A.S. (1995) Postsynaptic targets of Purkinje cell terminals in the cerebellar and vestibular nuclei of the rat. *Eur. J. Neurosci.*, 7(11): 2322–2333.

De Zeeuw, C.I., Holstege, J.C., Ruigrok, T.J. and Voogd, J. (1989). Ultrastructural study of the GABAergic, cerebellar, and mesodiencephalic innervation of the cat medial accessory olive: anterograde tracing combined with immunocytochemistry. *J. Comp. Neurol.*, 284(1): 12–35.

De Zeeuw, C.I., Lang, E.J., Sugihara, I., Ruigrok, T.J., Eisenman, L. M., Mugnaini, E. and Llinás, R. (1996). Morphological correlates of bilateral synchrony in the rat cerebellar cortex. *J. Neurosci.*, 16(10): 3412–3426.

De Zeeuw, C.I., Alphen van, A.M., Hawkins, R.K. and Ruigrok, T.J.H. (1997) Climbing fiber collaterals contact neurons in the cerebellar nuclei that provide a GABAergic feedback to the inferior olive. *Neuroscience*, 80: 981–986.

De Zeeuw, C.I., Hansel, C., Bian, F., Koekkoek, S.K., van Alphen, A.M., Linden, D.J. and Oberdick, J. (1998a) Expression of a protein kinase C inhibitor in Purkinje cells blocks cerebellar LTD and adaptation of the vestibulo-ocular reflex. *Neuron*, 20(3): 495–508.

De Zeeuw, C.I., Simpson, J.I., Hoogenraad, C.C., Galjart, N., Koekkoek, S.K.E. and Ruigrok, T.J.H. (1998b) Microcircuitry and function of the inferior olive. *Trends Neurosci.*, 21: 391–400.

Eccles, J.C., Llinás, R., Sasaki, K. and Voorhoeve, P.E. (1966) Interaction experiments on the responses evoked in Purkinje cells by climbing fibres. *J. Physiol. (Lond.)*, 182(2): 297–315.

Farrant M. and Cull-Candy S. (1993) GABA receptors, granule cell and genes. *Nature*, 361: 302–303.

Fiala, J.C., Grossberg, S. and Bullock, D. (1996) Metabotropic glutamate receptor activation in cerebellar Purkinje cells as substrate for adaptive timing of the classically conditioned eye-blink response. *J. Neurosci.*, 16: 3760–3774.

Gabbiani, F., Midtgaard, J. and Knoepfl, T. (1994) Synaptic integration in a model of cerebellar granule cells. *J. Neurophys.*, 72: 999–1009.

Gerstner, W. and van Hemmen, J.L. (1992) Associative memory in a network of 'spiking' neurons. *Network*, 3: 139–164.

Gerstner, W. and van Hemmen, J.L. (1994) Coding and information processing in neural networks. In : E. Domany, J.L. van Hemmen and K. Schulten (Eds.), *Models of Neural Networks II*. Springer, New York.

Gerstner, W., Kempter, R., van Hemmen, J.L. and Wagner H. (1996) A neuronal learning rule for sub-millisecond temporal coding. *Nature*, 384: 76–78.

Gerstner, W., Ritz, R. and van Hemmen, J.L., (1993) Why spikes? Hebbian learning and retrieval of time-resolved excitation patterns. *Biol. Cybern.*, 69:503–515.

Hebb, D.O. (1949) *The Organization of Behavior*. Wiley, New York.

Hirano, T. (1990) Depression and potentiation of the synaptic transmission between a granule cell and a Purkinje cell in rat cerebellar culture. *Neurosci. Lett.*, 119: 141–144.

Ito, M. (1984) *The Cerebellum and Neural Control*, Raven Press, New York.

Jahnsen, H. (1986a) Electrophysiological characterisitcs of neurones in the guinea-pig deep cerebellar nuclei in vitro. *J. Physiol. (Lond.)*, 372: 129–147.

Jahnsen, H. (1986b) Extracellular activation and membrane conductances of neurones in the guinea-pig deep cerebellar nuclei in vitro. *J. Physiol. (Lond.).*, 372: 149–168.

Kistler, W.M. (1999) Ein neuronales Modell des Kleinhirns basierend auf Zeitkodierung und synaptischer Plastizität (Munich, Dept. of Physics, TU), Thesis.

Kistler, W.M., Gerstner, W. and van Hemmen, J.L. (1997) Reduction of the Hodgkin Huxley equations to a single-variable threshold model. *Neural. Comput.*, 9: 1015–1045.

Kistler, W.M. and van Hemmen, J.L. (1999) Delayed reverberation through time windows as a key to cerebellar function. *Biol. Cybern.*, 81: 373–380.

Kistler, W.M. and van Hemmen, J.L. (2000) Modeling synaptic plasticity in conjunction with the timing of pre- and postsynaptic action potentials. *Neural Comput.*, 12: 385–405.

Korn, H. and Axelrad, H. (1980) Electrical inhibition of Purkinje cells in the cerebellum of the rat. *Proc. Natl. Acad. Sci.*, 77(10): 6244–6247.

Lampl, I. and Yarom, Y. (1993) Subthreshold oscillations of the membrane potential: a functional synchronizing and timing device. *J. Neurophysiol.*, 70: 2181–2186.

Lev-Ram, V., Jiang, T., Wood, J., Lawrence, D.S. and Tsien, R.Y. (1997) Synergies and coincidence requirements between NO, cGMP, and Ca^{++} in the induction of cerebellar long-term depression. *Neuron*, 18: 1025–1038.

Llinás, R.R., Baker, R. and Sotelo, C. (1974) Electrotonic coupling between neurons in cat inferior olive. *J. Neurophysiol.*, 37: 560–571.

Llinás, R. (1988) The intrinsic electrophysiological properties of mammalian neurons: insights into central nervous system function. *Science*, 242: 1654–1664.

Llinás, R. (1991) The non-continuous nature of movement execution. In: D.R. Humphrey and H.-J. Freund (Eds.), *Motor Control: Concepts and Issues*, J. Wiley & Sons Ltd, pp. 223–242.

Llinás, R. and Mühlethaler, M. (1988a) An electrophysiological study of the in vitro, perfused brain stem-cerebellum of adult guinea-pig. *J. Physiol. (Lond.)*, 404: 215–240.

Llinás, R. and Mühlethaler, M. (1988b) Electrophysiology of guinea-pig cerebellar nuclear cells in the in vitro brain stem-cerebellar preparation. *J. Physiol. (Lond.)*, 404: 241–258.

Llinás, R. and Sasaki, K. (1989) The functional organization of the olivo–cerebellar system as examined by multiple Purkinje cell recordings. *Eur. J. Neurosci.*, 1: 587–602.

Llinás, R. and Yarom, Y. (1981a) Electrophysiology of mammalian inferior olivary neurones in vitro. different types of voltage-dependent ionic conductances. *J. Physiol. (Lond.)*, 315: 549–567.

Llinás, R. and Yarom, Y. (1981b) Properties and distribution of ionic conductances generating electroresponsiveness of mammalian inferior olivary neurones in vitro. *J. Physiol. (Lond.)*, 315: 569–584.

Llinás, R. and Yarom Y. (1986) Oscillatory properties of guinea-pig inferior olivary neurones and their pharmacological modulation: an in vitro study. *J. Physiol. (Lond.)*, 376: 163–182.

Llinás, R., Lang, E., and Welsh, J.P. (1997) The cerebellum, LTD, and memory: alternative views. *Learn. Mem.* 3: 445–455.

Markram, H., Lübke, J., Frotscher, M. and Sakmann, B. (1997) Regulation of synaptic efficacy by coincidence of post-synaptic APs and EPSPs. *Science*, 275: 213–215.

Marr, D. (1969) A theory of cerebellar cortex. *J. Physiol. (Lond.)*, 202: 437–470.

Mastronarde, D.N. (1987) Two classes of single-input X-cells in cat lateral geniculate nucleus. II. Retinal inputs and the generation of receptive-field properties. *J. Neurophysiol.*, 57: 381–413.

McCormick, D.A., Clark, G.A., Lavond, D.G. and Thompson, R.F. (1982) Initial localization of the memory trace for a basic form of learning. *Proc. Natl. Acad. Sci.*, 79: 2731–2735.

McCormick, D.A. and Huguenard, J.R. (1992) A model of the electrophysiological properties of thalamocortical relay neurons. *J. Neurophysiol.*, 68: 1384–1400.

Meek J. (1992) Why run parallel fibers parallel? Teleostean Purkinje cells as possible coincidence detectors in a timing device subserving spatial coding of temporal differences. *Neuroscience*, 48: 249–283.

Moore, J.W., Desmond, J.E. and Berthier, N.E. (1989) Adaptively timed conditioned response and the cerebellum: a neural network approach. *Biol. Cybern.*, 62: 17–28.

Murakamil, F., Ozawa, N.N., Katsumaru, H. and Tsukahara, H. (1981) Reciprocal connections between the nucleus interpositus of the cerebellum and precerebellar nuclei. *Neurosci. Lett.*, 25: 209–213.

Olsen, J.F. and Suga, N. (1991) Combination-sensitive neurons in the medial geniculate body of the mustached bat: encoding of target range information. *J. Neurophysiol.*, 65: 1275–1296.

Palay, S.L. and V. Chan-Palay, V. (1974) *Cerebellar Cortex*, Springer-Verlag, Berlin. Heidelberg. New York.

Perret, S.P., Ruiz, B.P. and Mauk, M.D. (1993) Cerebellar cortex lesions disrupt learning-dependent timing of conditioned eyelid response. *J. Neurosci.*, 13: 1708–1718.

Ramón y Cajal, S. (1911) *Histologie du Système Nerveux de l'homme et des Vertébres*, Maloine, Paris.

Ruigrok, T.J. and Voogd, J. (1995) Cerebellar influence on olivary excitability in the cat. *Eur. J. Neurosci.*, 7(4): 679–693.

Sakurai, M. (1987) Synaptic modification of parallel fiber-Purkinje cell transmission in in vitro guinea-pig cerebellar slices. *J. Physiol. (Lond.)*, 394: 463–480.

Schreurs, B.G., Oh, M.M. and Alkon, D.L. (1996) Pairing-specific long-term depression of Purkinje cell excitatory postsynaptic potentials results from a classical conditioning procedure in the rabbit cerebellar slice. *J. Neurophysiol.*, 75(3): 1051–1060.

Schulman, J.A. and Bloom, F.E. (1981) Golgi cells of the cerebellum are inhibited by inferior olive activity. *Brain Res.*, 210: 350–355.

Sugihara, I., Lang, E.J. and Llinas, R. (1993). Uniform olivo–cerebellar conduction time underlies Purkinje cell complex

spike synchronicity in the rat cerebellum. *J. Physiol. (Lond.).*, 470: 243–271.
Teune, T.M., van der Burg, J., De Zeeuw, C.I. Voogd, J. and Ruigrok, T.J.H. (1998) Single Purkinje cell can innervate multiple classes of projection neurons in the cerebellar nuclei of the rat: a light microscopic and ultrastructural triple-tracer study in the rat. *J. Comp. Neurol.*, 392(2): 164–178.
Tia, S., Wang, J.F., Kotchabhakdi, N. and Vicini, S. (1996) Developmental changes of inhibitory synaptic currents in cerebellar granule cells: role of $GABA_A$ receptor α6 subunit. *J. Neurosci.*, 16: 3630–3640.
Tsukahara, N., Bando, T., Murakami, F. and Oda, Y. (1983) Properties of cerebello-precerebellar reverberating circuits. *Brain Res.*, 274: 249–259.
Ueno, S., Zempel, J.M. and Steinbach, J.H. (1996) Differences in the expression of $GABA_A$ receptors between functionally innervated and non-innervated granule neurons in neonatal cerebellar cultures. *Brain Res.*, 714: 49–56.
Verveer, C., Hawkins, R.K., Ruigrok, T.J.H. and Zeeuw, C.I. De (1997) Ultrastructural study of the GABAergic and cerebellar input to the nucleus reticularis tegmenti pontis. *Brain Res.*, 766: 289–296.
Vos, B.P., Maex, R., Volny-Luraghi, A. and De Schutter, E.D. (1999) Parallel fibers synchronize spontaneous activity in cerebellar Golgi cells. *J. Neurosci.*, 19(11): RC6.
Welsh, J.P., Lang, E.J., Suglhara, I. and Llinas, R. (1995). Dynamic organization of motor control within the olivo–cerebellar system. *Nature*, 374(6521): 453–457.
Yeo, C.H. and Hesslow, G. (1998) Cerebellum and conditioned reflexes. *Trends Cognit. Sci.*, 2: 322–330.

APPENDIX

(A) Granule cells

We have used a realistic model of cerebellar granule cells that consists of several passive dendritic compartments and one active somatic compartment and contains various Na^+, K^+, and Ca^{2+} currents. Our granule cell model is identical to that of Gabbiani et al. (1994) except that (a) we have simplified the Ca^{2+} dynamics as proposed by Traub et al. (1991) – see below – and (b) we have included $GABA_A$-controlled ion channels with a maximum conductance of 1.5 pS per synapse, a reversal potential of −75 mV, and a bi-exponential decay with time constants 5 ms (peak amplitude $\bar{g}_{\rm fast} = 1.0$ pS) and 50 ms (peak amplitude $\bar{g}_{\rm slow} = 0.5$ pS), as reported for rat granule cells (Tia et al., 1996). Excitatory synaptic currents are made up of a combination of NMDA and AMPA components as described by Gabbiani et al. (1994).

The innercellular Ca^{2+} concentration $[Ca^{2+}]_i$ is given by an ordinary differential equation,

$$\frac{d}{dt}[\mathrm{Ca}^{2+}]_i = \phi i_{\mathrm{Ca}} - [\mathrm{Ca}^{2+}]_i/\tau_{\mathrm{Ca}}, \qquad (1)$$

with $i_{\rm Ca}$ being the Ca^{2+} influx as described in (Gabbiani et al. 1994) and $\phi = 16\ \mathrm{mM\mu A^{-1}ms^{-1}}$, a constant that scales the amplitude of Ca^{2+} transients produced by an action potential to about 800 nM. The time constant $\tau_{\rm Ca} = 130$ ms describes the decay of a Ca^{2+} transient as it is caused by diffusion.

For the sake of completeness and because of a couple of typos in the original work we have summarized the parameters for the activation and inactivation functions for the various types of ion currents in Table 1.

(B) Deep cerebellar nuclei neurons

Deep cerebellar nuclei cells are described by a single compartment model that is actually a modification of a model developed for thalamic relay neurons (McCormick and Huguenard 1992). The membrane potential υ is given by

$$C\frac{d}{dt}\upsilon(t) = I_{\rm syn}(t) + \sum_x I_x(t), \qquad (2)$$

with $C = 0.15$ nF being the membrane capacity, $I_{\rm syn}$ and I_x being the synaptic current and the current through channel x, respectively.

Sodium and potassium currents are described by Ohmic equations (Hodgkin and Huxley 1952) of the form

$$I_x = \bar{g}_x m^p h(v - E_x), \qquad (3)$$

or

$$I_x = \bar{g}_x m^p (v - E_x) \qquad (4)$$

in the case of a non-inactivating ion current. The calcium currents $I_{\rm L}$ and $I_{\rm T}$ (Huguenard and McCormick 1992) are given by

$$I_x = \bar{g}_x \frac{[\mathrm{Ca}^{2+}]_e e^{-\upsilon/13.3} - [\mathrm{Ca}^{2+}]_i}{1 - e^{-\upsilon/13.3}} m^p h\upsilon \qquad (5)$$

Activation and inactivation variables m and h evolve according to the usual relaxation equations

TABLE 1

Parameters of a detailed model for turtle granule cells (Gabbiani et al. 1994) as used in this paper. In case of the currents I_{Na} and $I_{K(A)}$, the time constants $\tau_x(\upsilon) = 1/[\alpha_x(\upsilon) + \beta_x(\upsilon)]$ are subject to a lower bound as it is given in the last column of the table.

Channel, state	variable	$\bar{g}_x$ $[\frac{mS}{cm^2}]$	E_x [mV]	p	α_x $[ms^{-1}]$	β_x $[ms^{-1}]$	τ_{min} [ms]
I_{Na}	m	70	55	3	$1.5e^{0.081(\upsilon+39)}$	$1.5e^{-0.066(\upsilon+39)}$	0.05
	h			1	$0.12e^{-0.089(\upsilon+50)}$	$0.12e^{0.089(\upsilon+50)}$	0.225
I_{KDR}	m	19	− 90	4	$0.17e^{0.073(\upsilon+38)}$	$0.17e^{0.018(\upsilon+38)}$	
	h			1	$[70+6.5e^{-0.08\,\min(0,\upsilon+46)}]\cdot 10^{-5}$	$1.1\cdot 10^{-3}/[1+e^{0.0807(\upsilon-44)}]$	
$I_{K(A)}$	m	3.67	− 90	3	$0.35e^{0.039(\upsilon+70)}$	$0.35e^{-0.091(\upsilon+70)}$	0.16
	h			1	$0.175e^{-0.02(\upsilon+80)}$	$0.175e^{0.18(\upsilon+80)}$	6
$I_{Ca(HVA)}$	m	2.91	80	2	$1.6/[1+e^{-0.072(\upsilon-5)}$	$-0.02(\upsilon+8.9)/[1-e^{0.2(\upsilon+8.9)}]$	
	h			1	$0.005e^{-0.05\,\max(0,\upsilon+60)}$	$0.005[1-e^{-0.05\max(0,\upsilon+60)}]$	
$I_{K(Ca)}$	m	80	− 90	1	$2.5/\{1+1.5e^{-0.085\upsilon}/[Ca^{2+}]_i\}$	$1.5/\{1+[Ca^{2+}]_i/(0.15e^{-0.077\upsilon})\}$	
I_h	m	0.09	− 42	1	$0.0008\,e^{-0.0909(\upsilon+75)}$	$0.0008\,e^{0.0909(\upsilon+75)}$	

TABLE 2

Detailed model for DCN neurons, part 1.

Channel, state	variable	$\bar{g}_x$ [μS]	E_x [mV]	p	m_x^∞, (h_x^∞)	τ_x^m, (τ_x^h) [ms]
I_{Na}	m	6.25	41	3	$[1+e^{-0.2(v+39.9)}]^{-1}$	$\left[\frac{0.379(v+38)}{1-e^{-0.2(v+38)}} - \frac{0.258(v+38)}{1-e^{0.2(v+38)}}\right]^{-1}$
	h			1	$\left[1+\frac{129.4e^{(v+55)/15}}{1+e^{(17-v)/21}}\right]^{-1}$	$\left[0.0667e^{-(v+55)/15}+\frac{8.62}{1+e^{(17-v)/21}}\right]^{-1}$
I_{NaP}	m	$3.64 \cdot 10^{-3}$	41	1	$[1+e^{-0.2(v+49)}]^{-1}$	$\left[\frac{0.379(v+38)}{1-e^{-0.2(v+38)}} - \frac{0.258(v+38)}{1-e^{0.2(v+38)}}\right]^{-1}$
I_{A1}	m	0.314	− 85	4	$[1+e^{-(v+60)/8.5}]^{-1}$	$0.0888+\frac{0.240}{e^{-(v+79.7)/12.7}+e^{(v+35.8)/19.7}}$
	h			1	$[1+e^{(v+78)/6}]^{-1}$	$\begin{cases} \frac{0.240}{e^{(v+46)/5}+e^{-(v+238)/37.5}} & v<-63 \\ 4.56 & v>-63 \end{cases}$
I_{A2}	m	0.210	− 85	4	$[1+e^{-(v+36)/20}]^{-1}$	$0.0888+\frac{0.240}{e^{-(v+79.7)/12.7}+e^{(v+35.8)/19.7}}$
	h			1	$[1+e^{(v+78)/6}]^{-1}$	$\begin{cases} \frac{0.240}{e^{(v+46)/5}+e^{-(v+238)/37.5}} & v<-73 \\ 14.4 & v>-73 \end{cases}$
I_{K2a}	m	0.314	− 85	4	$[1+e^{-(v+43)/17}]^{-1}$	$2.38+\frac{0.240}{e^{(v-81)/25.6}+e^{-(v+132)/18}}$
	h			1	$[1+e^{(v+58)/10.6}]^{-1}$	$28.8+\frac{0.240}{e^{(v-1329)/200}+e^{-(v+130)/7.1}}$
I_{K2b}	m	0.210	− 85	4	$[1+e^{-(v+43)/17}]^{-1}$	$2.38+\frac{0.240}{e^{(v-81)/25.6}+e^{-(v+132)/18}}$
	h			1	$[1+e^{(v+58)/10.6}]^{-1}$	$\begin{cases} 28.8+\frac{0.240}{e^{(v-1329)/200}+e^{-(v+130)/7.1}} & v<-70 \\ 2136 & v>-70 \end{cases}$

TABLE 3

Detailed model of DCN neurons, part 2. The table gives maximum conductivity $\bar{g}$, reversal potential E, equilibrium values (m_x^∞, h_x^∞), and time constants $\tau_x^{m,h}$ for activation (m) and inactivation variables (h). The parameters are compensated for 36.5 centigrade. The capacity of the neuron is $C = 150$ pF.

Channel, state	variable	$\bar{g}_x$ [μS]	E_x [mV]	p	m_x^∞, (h_x^∞)	τ_x^m, (τ_x^h) [ms]
$I_{K(Ca)}$	m	0.520	− 85	1	$\left[1 + \frac{e^{-\upsilon/12}}{2.5\cdot 10^6[Ca^{2+}]_L}\right]^{-1}$	$\frac{2.40}{e^{-\upsilon/24} + 2.5\cdot 10^6[Ca^{2+}]_L e^{\upsilon/24}}$
I_h	m	0.010	− 43	1	$[1 + e^{(\upsilon+92)/7}]^{-1}$	35.9
I_L	m	0.645 μS/mM		2	$[1 + e^{-(\upsilon+10)/10}]^{-1}$	$\left[\frac{6.67}{1 + e^{-0.072(\upsilon-5)}} + \frac{0.0833\,(\upsilon - 1.31)}{e^{(\upsilon-1.31)/5.36} - 1}\right]^{-1}$
I_T	m	0.181 μS/mM		2	$[1 + e^{(-\upsilon+49)/6.2}]^{-1}$	$0.118 + \frac{0.193}{e^{-(\upsilon+127)/16.7} + e^{(\upsilon+11.8)/18.2}}$
	h			1	$[1 + e^{(\upsilon+78)/4}]^{-1}$	$\begin{cases} 0.263 e^{(\upsilon+462)/66.6} & \upsilon < -75 \\ 0.263\,[28 + e^{-(\upsilon+17)/10.5}] & \upsilon > -75 \end{cases}$
$I_{Na,leak}$		0.0012	41			
$I_{K,leak}$		0.0034	− 85			

with voltage-dependent equilibrium values and time constants (Hodgkin and Huxley 1952, Traub et al. 1991) which are summarized in Tables 2 and 3. The calcium dynamics is identical to that of McCormick and Huguenard.

(C) Network model

In the present network simulations we have used a neuron model – the spike response model – that is in many aspect similar to but on the other hand much more general than the standard integrate-and-fire model (Gerstner and van Hemmen 1992, Gerstner 1995, Kistler et al. 1997). The membrane potential is given by a combination of pre- and postsynaptic contributions described by a response kernel ϵ that gives the form of an elementary postsynaptic potential and a kernel η that accounts for refractoriness and has the function of an afterpotential, so that

$$h_i(t) = \sum_{j,f} J_{ij}\, \epsilon_{ij}(t - t_j^f - \Delta_{ij}) - \sum_f \eta_i(t - t_i^f). \tag{6}$$

Here, h_i is the membrane potential of neuron i, J_{ij} is the strength of the synapse connecting neuron j to neuron i, Δ_{ij} is the corresponding axonal delay from j to i, and $\{t_j^f, f = 1, 2, \ldots\}$ are the firing times of neuron j. Causality is respected, if both $\epsilon(s)$ and $\eta(s)$ vanish identically for $s < 0$. Depending on the choice for ϵ and η, the spike response model can either reproduce the standard integrate-and-fire model or even mimic complex Hodgkin-Huxley type neuron models (Kistler et al., 1997). A typical choice for ϵ_{ij} is

$$\epsilon_{ij}(t) = \varepsilon(t, \tau_1, \tau_2) \equiv \frac{1}{c}\,(e^{-t/\tau_2} - e^{-t/\tau_1})\Theta(t), \tag{7}$$

and

$$c = \left(\frac{\tau_1}{\tau_2}\right)^{\frac{\tau_1}{\tau_2 - \tau_1}} - \left(\frac{\tau_1}{\tau_2}\right)^{\frac{\tau_2}{\tau_2 - \tau_1}} \tag{8}$$

where $\tau_{1,2}$ are time constants for the rising and decaying phase of the postsynaptic potential (cf. Table 4) and Θ is the Heaviside step function with $\Theta(t) = 1$ for $t > 0$ and $\Theta(t) = 0$ elsewhere. Post-inhibitory rebound in DCN and IO neurons is described by using an IPSP that includes a depolarizing component (low-threshold Ca^{2+} spike),

$$-\epsilon_{PIR}(t) = -\varepsilon(t, \tau_1, \tau_2) + \kappa\varepsilon(t - \Delta_{PIR}, \tau_3, \tau_4). \tag{9}$$

The function ε has been defined in the previous equation; Δ_{PIR} is the delay after which the low-threshold Ca^{2+} spike occurs ($\Delta_{PIR} = 105$ ms and $\Delta_{PIR} = 112$ ms for DCN and IO neurons, respectively), and κ is its amplitude ($\kappa = 0.2$ and $\kappa = 0.22$ for DCN and IO neurons, respectively). The time constants $\tau_3 = 2$ and $\tau_4 = 5$ give the shape of the Ca^{2+} spike. The fact that the reversal potential of chloride ions may be close to the resting potential is taken care of by assuming that IPSPs do not superpose additively. Instead, only the IPSP produced by the very last inhibitory action potential is included; cf. Table 4.

As for the refractory function η_i, a simple exponential decay can be used,

$$\eta_i(t) = \begin{cases} 0 & t < 0 \\ -\infty & 0 < t < t_0 \\ -\eta_0 e^{-(t - t_0)/\tau} & t > t_0 \end{cases} \tag{10}$$

Here, t_0 determines the duration of the absolute refractory period where no further spikes can be triggered at all, and η_0 and τ describe the form of the afterpotential; cf. Table 4.

Firing times are generated by means of a stochastic process with the *probability density* of having a spike at time t being a nonlinear function of the membrane potential $h(t)$,

$$\text{Prob}\,\{\text{spike in } [t, t + dt\,[\,\} = \tau_0^{-1} \exp[\beta(h - \vartheta)]\, dt, \tag{11}$$

with $\tau_0 = 2.5$ ms, noise parameter $\beta^{-1} = 0.01$, and firing threshold ϑ. This particular ansatz reflects the fact that firing an action potential is more or less a threshold process. Noise, however, can trigger an action potential even before the threshold is actually reached.

The firing threshold of the various types of neuron is also given in Table 4. Instead of including

TABLE 4

Parameters used to describe the different types of neuron involved in the network simulations. Inhibitory interneurons as there are stellate cells and basket cells have been lumped together and are – for the sake of simplicity – referred to as BaC (* see text.)

		pons	DCN	GrC	GoC	PC	BaC	IO
Threshold:	ϑ	0.03*	1	1.5	10	10	5	2.5*
EPSP:	τ_1/ms	2	5	2	2	5	2	5
	τ_2/ms	5	10	5	5	10	5	10
IPSP:	τ_1/ms		2	2	2	1	2	2
	τ_2/ms		50*	40	20	15	20	50*
IPSPs additive			no	no	yes	yes	yes	no
AHP:	τ/ms	5	5	5	10	5	5	5
	t_0/ms	2	2	2	2	2	2	2
	η_0	20	100	100	100	20	100	100

external excitatory input to neurons of the pons, a high mossy fiber background activity is obtained by a low firing threshold ($\vartheta = 0.03$). Subthreshold oscillations in IO neurons are mimicked by periodically reducing the firing threshold of IO neurons to $\vartheta = 1.5$ every 120 ms.

The connectivity of the network is defined by means of arborization functions that give the probability of two neurons being 'connected' as a function of of the type of the neurons and of their distance. Before a simulation is started a particular connectivity is generated according to these arborization functions (cf. Table 5). The synaptic strength J_{ij} and the axonal and synaptic delay Δ_{ij} that are assigned to these connections are summarized in Table 6.

Plasticity at the parallel fiber/Purkinje cell synapses is described by an differential equation of the form

$$\frac{\mathrm{d}J}{\mathrm{d}t} = [1 - J(t)]S_{\mathrm{cf}}(t) \int \mathrm{d}t' \, \kappa_{\mathrm{LTD}}(t')S_{\mathrm{pf}}(t - t').$$

$$- J(t)S_{\mathrm{pf}}(t) \int \mathrm{d}t' \, \kappa_{\mathrm{LTD}}(t')S_{\mathrm{cf}}(t - t'). \quad (12)$$

Here, the kernel $\kappa_{\mathrm{LTP}}(t')$ describes the amount of weight change that is induced by a climbing fiber spike that arrives t' ms after a parallel fiber spike Kistler and van Hemmen 1999). As for $\kappa_{\mathrm{LTD}}(t')$ the situation is similar except that the role of climbing fiber and parallel fiber spikes are exchanged. We use exponentially decaying kernels,

$$\kappa_{\mathrm{LTP/D}}(t) = c_{\mathrm{LTP/D}} \exp(-t/\tau_{\mathrm{LTP/D}})\Theta(t), \quad (13)$$

with time constants $\tau_{\mathrm{LTD}} = 200$ ms and $\tau_{\mathrm{LTP}} = 20$ ms, and amplitude $c_{\mathrm{LTP/D}} = 0.2$. With this ansatz a parallel fiber/Purkinje cell synapse is potentiated if

TABLE 5

Arborization functions used to generate the connectivity that is used in the network simulations. The probability of two neurons i and j being connected depends on the type of the neurons and their distance r_{ij}. The probability is given by Prob $\{J_{ij} \neq 0\} = p[1 + \mathrm{e}^{(r_{ij} - \varrho)/\delta}]^{-1}$, where p, ϱ and δ describe the overall probability, the range and the shape of the arborization, respectively.

			presynaptic neurons						
			pons	IP	pf	GoC	PC	BaC	IO
postsynaptic neurons	**pons**	ϱ		20					
		δ		1					
		p		0.05					
	IP	ϱ					20		
		δ					1		
		p					0.07		
	GrC	ϱ	20			100			
		δ	1			10			
		p	*			*			
	GoC	ϱ	200		200		20		200
		δ	10		10		10		10
		p	0.1		0.1		0.1		0.1
	PC	ϱ			200			500	0*
		δ			10			10	
		p			1			0.2	1
	BaC	ϱ			20			50	
		δ			10			10	
		p			0.5			0.05	
	IO	ϱ		20					20
		δ		1					1
		p		0.08					0.08

TABLE 6

Synaptic weights J and axonal and synaptic delays Δ as they are used in the network simulations.

			presynaptic neurons						
			pons	IP	pf	GoC	PC	BaC	IO
postsynaptic neurons	**pons**	J		1					
		Δ		3					
	IP	J					-10		
		Δ					1		
	GrC	J	1			-3.5			
		Δ	2			2			
	GoC	J	5		1		-10		10
		Δ	2		1		1		1
	PC	J			$0 \ldots 1$			-5	1000
		Δ			1			1	1
	BaC	J			0.75			-1	
		Δ			1			1	
	IO	J		1					-10
		Δ		5					5

the parallel fiber spike arrives within a about 20 ms milliseconds before the climbing fiber spike and the synapse is depressed, if the timing is the other way round. If the parallel fiber spike arrives long before the climbing fiber spike, the synapse is not modified.

In the simulation shown in Fig. 11 equation (12) is supplemented by a third term, $-J(t)S_{\mathrm{cf}}(t)\int \mathrm{d}t'\,\tilde{\kappa}_{\mathrm{LTD}}(t')S_{\mathrm{pf}}(t-t')$, with $\tilde{\kappa}_{\mathrm{LTD}}(t)=0.2\exp(-t/200\ \mathrm{ms})\,\Theta(t)$, and κ_{LTP} replaced by $\kappa_{\mathrm{LTP}}(t)=0.35\exp(-t/20\ \mathrm{ms})\,\Theta(t)$. In this case, a parallel fiber/Purkinje cell synapse is potentiated only if the parallel fiber spike arrives within a few milliseconds before the climbing fiber spike. Otherwise, i.e. if the parallel fiber spike arrives after, or long before the climbing fiber spike, the synapse is depressed.

Appendix references

Gabbiani, F., Midtgaard, J. and Knoepfl, T. (1994) Synaptic integration in a model of cerebellar granule cells. *J. Neurophysiol.*, 72: 999–1009. Corrigenda have been published in *J. Neurophysiol.* (1996) 75(6), without, however, covering all typing errors.

Gerstner, W. (1995) Time structure of the activity in neural network models. *Phys. Rev. E.*, 51: 738–758.

Gerstner, W. and van Hemmen, J.L. (1992) Associative memory in a network of 'spiking' neurons. *Network*, 3: 139–164.

Hodgkin, A.L. and Huxley, A.F. (1952) A quantitative description of ion currents and its applications to conduction and excitation in nerve membranes. *J. Physiol. (Lond.)*, 117: 500–544.

Huguenard, J.R. and McCormick, D.A. (1992) Simulation of the currents involved in rythmic asocillations in thalamic relay neurons. *J. Neurophysiol.*, 68: 1373–1383.

Kistler, W.M., Gerstner, W. and van Hemmen J.L. (1997) Reduction of the Hodgkin–Huxley equations to a single-variable threshold model. *Neural Comput.*, 9: 1015–1045.

Kistler, W.M. and van Hemmen, J.L., (2000) Modeling synaptic plasticity in conjunction with the timing of pre- and postsynaptic action potentials. *Neural Comput.*, 12: 385–405.

McCormick, D.A. and Huguenard, J.R. (1992) A model of the electrophysiological properties of thalamocortical relay neurons. *J. Neurophysiol.*, 68: 1384–1400.

Tia, S., Wang, J.F., Kotchabhakdi, N. and Vicini, S. (1996) Developmental changes of inhibitory synaptic currents in cerebellar granule cells: role of $GABA_A$ receptor $\alpha 6$ subunit. *J. Neurosci.*, 16: 3630–3640.

Traub, R.D., Wong, R.K.S., Miles, R. and Michelson H. (1991) A model of a CA3 hippocampal pyramidal neuron incorporating voltate-clamp data on intrinsic conductances. *J. Neurophysiol.*, 66: 635–650.

N.M. Gerrits, T.J.H. Ruigrok and C.I. De Zeeuw (Eds.)
Progress in Brain Research, Vol 124

CHAPTER 21

Electrophysiological behavior of Purkinje cells and motor coordination in calretinin knock-out mice

Guy Cheron[1,4,*], Stéphane Schurmans[3], Ann Lohof[5], Pablo d'Alcantara[2,3], Michael Meyer[6], Jean-Philippe Draye[1], Marc Parmentier[3] and Serge N. Schiffmann[2]

[1] *Laboratory of Movement Biomechanics*, [2] *Brain Research Unit CP601, and* [3] *IRIBHN, Université Libre de Bruxelles, 1000 Brussels, Mons, Belgium*
[4] *Laboratory of Neurophysiology, Université de Mons, Mons, Belgium*
[5] *Laboratoire de Neurobiologie, CNRS URA 1857, Ecole Normale Supérieure, Paris, France*
[6] *Department of Neurochemistry, Max Planck Institute for Psychiatry, Am Klopferspitz 18a, D–82152 Planegg, Germany*

Introduction

In recent years, there has been considerable interest in the different forms of calcium signalling which may involve the entire neuron (Llano et al., 1994) or be spatially limited to a dendritic microdomain (Llinás et al., 1992). Variations in intracellular calcium concentration ($[Ca^{2+}]_i$) (Sugimori and Llinás, 1990; Usowicz et al., 1992) play a crucial role in information processing in Purkinje cells (PC), since they modulate the efficacy of both excitatory and inhibitory inputs (Sakurai, 1990; Kano et al., 1992) and are involved in the induction of synaptic plasticity (Ito, 1989; Lisberger, 1998; Kano et al., 1992; Tank et al., 1995). The control of $[Ca^{2+}]_i$ is crucial for the neuron and is performed by pumps, exchangers, organelle sequestration and cytoplasmic Ca^{2+} binding proteins (Clapham, 1995). Among these, calbindin-D28k and calretinin, which are highly conserved across mammalian species (Parmentier et al., 1987; Parmentier and Lefort, 1991), are both particularly enriched in the cerebellum: calbindin-D28k in PC and calretinin in granule, unipolar brush and Lugaro cells (Resibois and Rogers, 1992). Calbindin-D28k controls Ca^{2+} homeostasis by buffering Ca^{2+} and affects the time course and spatial spreading of intracellular Ca^{2+} transients (Lledo et al., 1992; Chard et al., 1993). Accordingly, marked changes in synaptically-evoked Ca^{2+} transients have been observed in PC of calbindin-D28k-deficient mice (Airaksinen et al., 1997). Alterations in Ca^{2+} homeostasis in cerebellar neurons would be expected to dramatically affect information processing in the cerebellum and to give new insights in the understanding of its physiology. Since calretinin is highly expressed in granule cells and the parallel fiber-PC synapse is central to cerebellar physiology, calretinin gene inactivation would likely perturb cerebellar function. In the present paper we review the outcome of four sets of investigation (functional morphology, slice and in vivo electrophysiology and motor coordination) recently performed on calretinin knock-out mice (Schurmans et al., 1997; Schiffmann et al., 1999).

Results

The generation of calretinin-deficient ($Cr^{-/-}$) mice using a classical gene knock-out strategy has been previously reported (Schurmans et al., 1997). These

*Corresponding author. Tel.: 32/2/6502187;
Fax: 32/2/6503745; e-mail: gcheron@ulb.ac.be

mice grow and reproduce normally and have no increased mortality rate.

Paradoxical calretinin-immunoreactivity in cerebellum of $Cr^{-/-}$ mice

Immunocytochemistry revealed that no calretinin-immunoreactivity was observed in the brain of $Cr^{-/-}$ mice, as expected for the creation of a null mutation, with the exception of the cerebellum, where PC bodies and dendritic trees (Fig. 1D), as well as axons crossing the granule cell layer and terminals in deep cerebellar and vestibular nuclei (data not shown) were strongly calretinin-immunoreactive. In granule cells of $Cr^{-/-}$ mice (Fig. 1D), no calretinin-immunoreactivity was seen, in contrast to $Cr^{+/+}$ mice in which these cells and parallel fibers in the molecular layer are calretinin-immunoreactive (Fig. 1A). The cerebellar

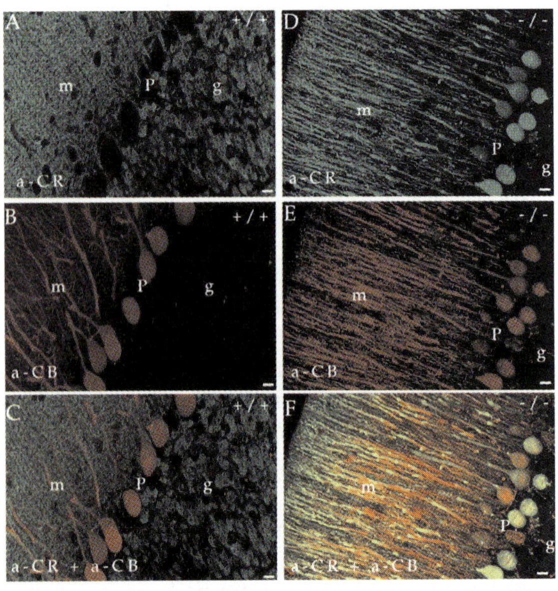

Fig. 1. Confocal microscopy showing the colocalization of calretinin- and calbindin-D28k-immunoreactivity in Purkinje cells of $Cr^{-/-}$ mice. In double-labeling experiments, calretinin (green)- and calbindin-D28k (red)-immunoreactivity were detected respectively in $Cr^{+/+}$ granule cells (A,C) and Purkinje cells (B,C), and were strictly colocalized (yellow) in $Cr^{-/-}$ Purkinje cells (D-F). Scale bars, 10 μm. Abbreviations: a-CB, anti-calbindin-D28k antiserum; a-CR, anti-calretinin antiserum; g, granule cell layer; m, molecular layer; P, Purkinje cell layer.

localization of calbindin-D28k and other related proteins such as calmodulin, S–100 protein and parvalbumin (data not shown) were similar in $Cr^{+/+}$ and $Cr^{-/-}$ mice. The calretinin-immunoreactivity was still detected in PC of 18–24 month-old $Cr^{-/-}$ mice and no apparent change in its level was observed whereas the number of PC had similarly decreased in both genotypes. The nature of the calretinin-immunoreactivity in PC was further investigated by different biochemicals techniques (see Schiffmann et al., 1999) which have confirmed that in $Cr^{-/-}$ cerebellum, calbindin-D28k is recognized by the anti-calretinin antiserum. This was explained by the fact that several anti-Ca^{2+} binding proteins antibodies preferentially recognize peculiar calcium-induced protein conformations in a calcium-concentration dependent manner (Winsky and Kuznicki, 1996). This anti-calretinin antiserum therefore recognizes calbindin-D28k only in a calcium-induced conformation and leads to the hypothesis that in PC of $Cr^{-/-}$ mice, calbindin-D28k binds more Ca^{2+} ions as compared to $Cr^{+/+}$ mice.

Histological analysis showed that PC bodies and granule cells have normal morphologies, densities and locations within the different layers of the $Cr^{-/-}$ cerebellum. Electron microscopy revealed that synaptic contacts on dendritic spines in the molecular layer appeared morphologically normal (Fig. 2) and in a normal density with 11.03 ± 3.35 and 8.99 ± 2.38 synapses per 100 μm^2 for $Cr^{+/+}$ and $Cr^{-/-}$ mice, respectively ($p > 0.05$).

The Purkinje cell firing behavior is altered in alert $Cr^{-/-}$ mice

The firing behavior of identified PC were analyzed in 15 alert mice, eight young mice (4 $Cr^{-/-}$ and 4 $Cr^{+/+}$) aged from 2 to 4 months, and seven older mice (4 $Cr^{-/-}$ and 3 $Cr^{+/+}$) aged from 2 to 2.5 years. For all these PC the micropipette was optimally positioned in order to record simultaneously simple spikes (SS) and complex spikes (CS) of opposite polarities (see Cheron et al., 1997, for more details). The spontaneous SS firing rate was significantly enhanced in $Cr^{-/-}$ mice when compared with age-matched $Cr^{+/+}$ mice (Figs 3A–D). For the 81 PC recorded in the young $Cr^{+/+}$ mice,

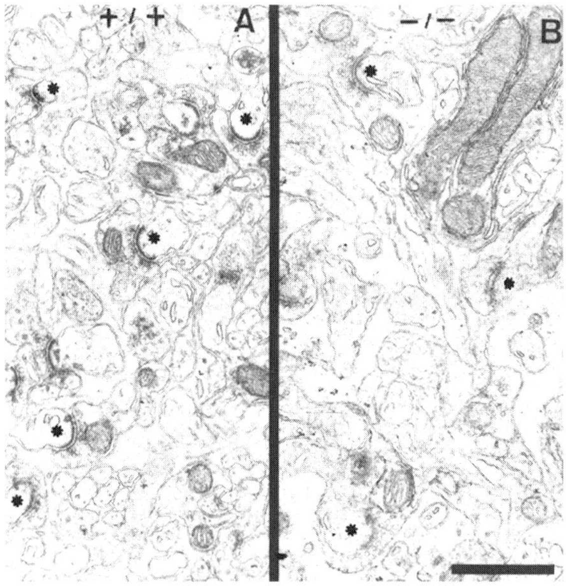

Fig. 2. Ultrastructure of parallel fiber synapses onto purkinje cells is indistinguishable in $Cr^{+/+}$ and $Cr^{-/-}$ mice. Electron micrographs of the molecular layer of wild-type (A) and calretinin-deficient (B) cerebella. Asterisks indicate postsynaptic dendritic spines. Scale bar, 1 μm.

the spontaneous SS firing rate was 66.2 ± 24.5 (mean ± SD) Hz. This frequency was significantly higher in young $Cr^{-/-}$ mice and reached a value of 124.0 ± 45.5 Hz for the 58 PC analyzed in this group. Similar results were obtained for old animals, with a spontaneous SS firing rate of 42.9 ± 19.6 Hz in $Cr^{+/+}$ mice and 102.4 ± 38.8 Hz in $Cr^{-/-}$ mice. In contrast, the mean spontaneous firing rate of CS was not statistically different in $Cr^{-/-}$ and $Cr^{+/+}$ mice (Figs 3A′–D′): the mean spontaneous CS firing rates were 0.61 ± 0.3 Hz and 0.52 ± 0.23 Hz for young and old $Cr^{+/+}$ mice, respectively and 0.54 ± 0.24 Hz and 0.59 ± 0.23 Hz for young and old $Cr^{-/-}$ mice, respectively. The duration of the CS was significantly reduced in calretinin-deficient mice (Fig. 4E–H). In young and old $Cr^{+/+}$ mice, the CS durations were 9.4 ± 2.3 ms and 12.5 ± 3.2 ms, respectively. They were reduced in young and old $Cr^{-/-}$ mice to 6.4 ± 1.8 ms and 6.3 ± 2.1 ms, respectively.

For all the recorded PC a transient pause (complete suppression) in SS firing followed a spontaneous CS but the duration of the pause was greatly different in $Cr^{-/-}$ mice as compared to $Cr^{+/+}$ mice (Fig. 4A–D). To record this pause, we averaged the SS firing during a 100 ms epoch (with a pre-trigger time of 10 ms) triggered by the spontaneous occurrence of a CS. In $Cr^{+/+}$ mice the PC was well characterized by a complete suppression of the SS discharge following the CS. For the young $Cr^{+/+}$ mice the mean duration of this pause was 24.9 ± 6.6 ms, and this effect was well preserved in old $Cr^{+/+}$ mice with a mean pause duration of 31.4 ± 14.2 ms the pause was significantly decreased in $Cr^{-/-}$ mice to 7.9 ± 5.6 and 5.4 ± 6.1 ms for young and old mice, respectively.

Excitatory synaptic transmission and short-term plasticity are normal in $Cr^{-/-}$ cerebellar slices

To determine whether the electrophysiological differences seen in vivo resulted from direct modifications of the PC or their afferent synapses, extracellular voltage or whole-cell patch-clamp recordings were made from PC in slices. Standard procedures were used for cerebellar slice electrophysiology (Llano et al., 1991). Parallel fibers or climbing fibers were focally stimulated by applying square pulses of 0.1 ms and 0.01–10 mA.

SS frequency, resting potential and input resistance were not statistically different in $Cr^{+/+}$ and $Cr^{-/-}$ PC (Table 1). Although calretinin is normally expressed in presynaptic parallel fiber terminals, parallel fiber excitatory postsynaptic currents (EPSC) appeared unchanged in $Cr^{-/-}$ mice. Magnitude or interval-dependence of paired-pulse facilitation (Fig. 5) of the parallel fiber current were not different in $Cr^{+/+}$ and $Cr^{-/-}$ mice (Fig. 5). The slope of parallel fiber EPSC in response to increasing stimulation intensity was similar in $Cr^{+/+}$ and $Cr^{-/-}$ mice (Table 1), indicating no detectable difference in parallel fiber recruitment. The duration of the CS elicited by climbing fiber stimulation and the number of peaks in this spike were not significantly different (Table 1). This stimulation evoked all-or-none EPSC in both genotypes. Finally, the magnitude or interval-dependence of the climbing fiber current paired-pulse depression (Fig. 6) were not different in $Cr^{+/+}$ and $Cr^{-/-}$ mice. These results strongly suggest that the alterations in PC activity observed in alert $Cr^{-/-}$ mice cannot be attributed to a difference in PC excitability or abnormal function of its afferent synapses.

Impaired motor coordination in $Cr^{-/-}$ mice

The general behavior of ~6 week-old calretinin-deficient ($Cr^{-/-}$) and wild-type ($Cr^{+/+}$) mice previously analyzed in a series of tests was not different (Schurmans et al., 1997). Motor coordination was now tested on the runway, the horizontal stationary rod and the wheel running test. In the latter test, mice were put inside a rotating drum (14 cm in diameter, 7 cm wide). The test was performed both in unperturbed condition and in presence of a sensory perturbation produced by a

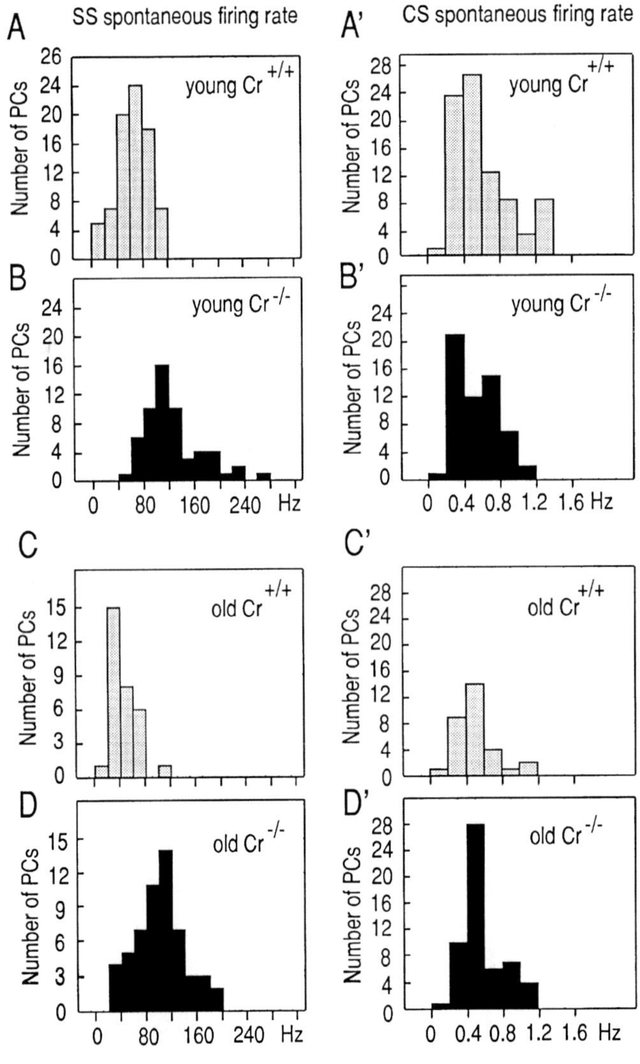

Fig. 3. Alteration of the firing behavior of purkinje cells in alert $Cr^{-/-}$ mice. The extracellular activity in the uvula and the nodulus was explored and the spontaneous simple spike (A-D) and complex spike (A′-D′) firing rate of identified Purkinje cells were analyzed in 4 $Cr^{+/+}$ (A, A′) and 4 $Cr^{-/-}$ (B, B′) 2–4 month-old and 3 $Cr^{+/+}$ (C, C′) and 4 $Cr^{-/-}$ (D, D′) 2–2.5 year-old alert mice. The spontaneous simple spike firing rate was enhanced in both young and old $Cr^{-/-}$ mice as compared with the age-matched $Cr^{+/+}$ mice whereas mean spontaneous firing rates of the complex spike were similar in both genotypes. Abbreviations: CS, complex spike; PCs, Purkinje cells; SS, simple spike.

plastic tape around the shank of the right hindlimb. Locomotion was analyzed during two successive periods of 60 sec using the optoelectronic ELITE System analyzing in the sagittal plane the trajectory of a retroreflective marker placed on the right hindlimb foot. Neither of the first two classical tests revealed significant differences between 4–6 week-old $Cr^{+/+}$ and $Cr^{-/-}$ mice. In contrast, in the wheel running test, $Cr^{+/+}$ mice performed a well-reproduced cyclic movement (Fig. 7A), whereas $Cr^{-/-}$ mice made only few small displacements (Fig. 7B). The number of steps/min and the total horizontal

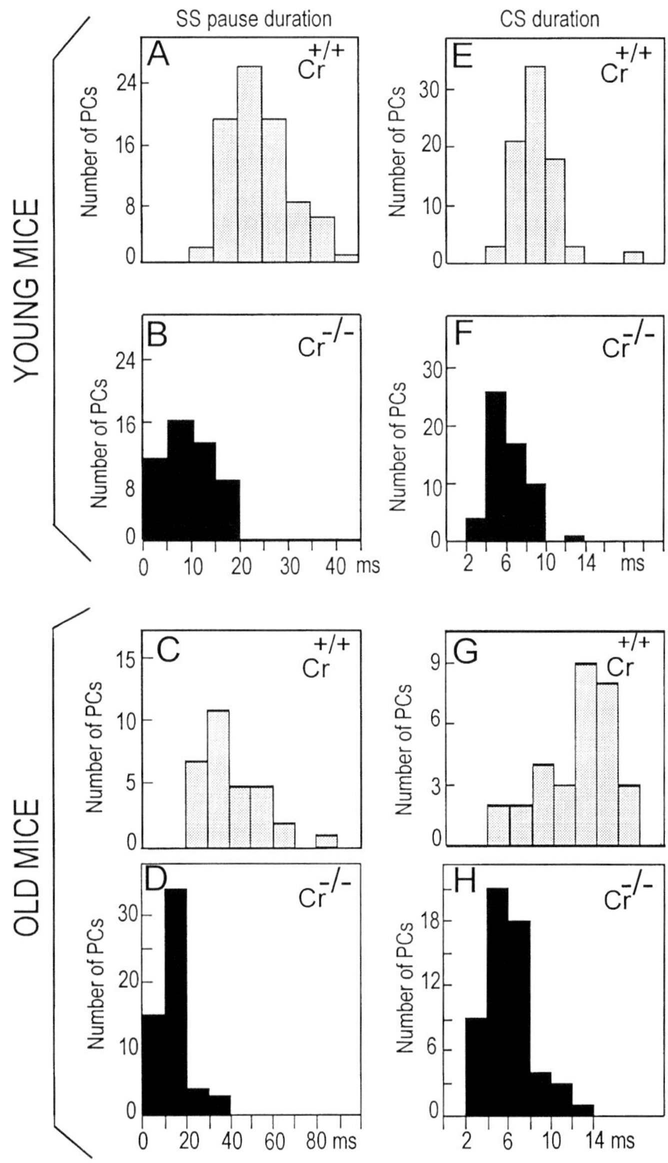

Fig. 4. The duration of complex spikes and the duration of the transient pause in simple spike discharge are reduced. The histograms summarize the results for the simple spike pause duration and complex spike duration, respectively. Complex spike duration and simple spike pause duration were determined in 2–4 month-old (A, B, E, F) and 2–2.5 year-old (C, D, G, H) mice and were reduced in $Cr^{-/-}$ mice. (B, F) and (D, H).

TABLE 1

Purkinje cell firing behavior in $Cr^{+/+}$ and $Cr^{-/-}$ cerebellar slices

Parameter	$Cr^{+/+}$	$Cr^{-/-}$ mice
Resting potential (mV)	-50.87 ± 8.7 ($n = 15$ cells)	-49.24 ± 12.23^{NS} ($n = 21$ cells)
Input resistance (MΩ)	217.0 ± 179.6 ($n = 15$ cells)	181.7 ± 127.8^{NS} ($n = 21$ cells)
SS firing rate (spikes/min)	71.76 ± 4.76 ($n = 95$ cells)	80.06 ± 5.03^{NS} ($n = 73$ cells)
Slope of PF recruitment (fC/mA)	34.37 ± 4.27 ($n = 33$ cells)	37.62 ± 5.39^{NS} ($n = 29$ cells)
CS duration (ms)	39.06 ± 2.89	44.88 ± 3.44^{NS}
Number of peaks in CS	3.55 ± 0.17	3.48 ± 0.15^{NS}
Multiple climbing fiber inervations	No	No

^{NS} not significant as compared to age-matched wild-type animals, one way analysis of variance (ANOVA). Abbreviations: CS, complex spike; PF, parallel fiber; SS, simple spike.

displacement of the hindlimb foot were significantly different (122.2 ± 58.5 and 4.50 ± 2.80 m, respectively for $Cr^{+/+}$ mice and 32.4 ± 23.8 and 0.38 ± 0.24 m for $Cr^{-/-}$ mice). In the presence of a sensory perturbation locomotion of $Cr^{+/+}$ mice was not significantly modified, with 104.4 ± 76.1 steps/min and 3.78 ± 3.2 m of horizontal displacement (Fig. 7C) and locomotion of $Cr^{-/-}$ mice remain very poor, with 31.4 ± 14.8 steps/min and 0.35 ± 0.22 m. However, in $Cr^{-/-}$ mice elevations of the hindfoot was induced (Fig. 7D) and their number/min was significantly different (0.6 ± 0.96 and 10.45 ± 6.1 for $Cr^{+/+}$ and $Cr^{-/-}$ mice, respectively).

The motor coordination of aged (18–24 month-old) $Cr^{+/+}$ and $Cr^{-/-}$ mice was also examined, since the motor discoordination of calbindin-D28k-deficient mice has been shown to dramatically worsen with age (Airaksinen et al., 1997). The feet of aged mutants slipped significantly more often than those of the wild-type in the runway test and the time spent on the rod was significantly lower for aged $Cr^{-/-}$ mice in the horizontal stationary rod test (see Schiffmann et al., 1999, for more details). In the wheel running test, aged $Cr^{-/-}$ mice presented the same deficit as young animals in the unperturbed condition and frequently fell following hindfoot elevations in the presence of the sensory perturbation.

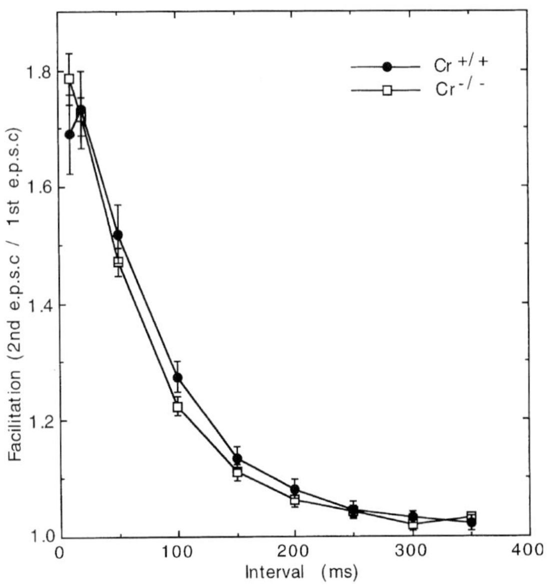

Fig. 5. No alteration of paired-pulse facilitation in $Cr^{-/-}$ mice. Paired-pulse facilitation of the parallel fiber current was recorded in cerebellar slices at different inter-pulse intervals. No differences in the magnitude or interval-dependence of this facilitation were observed between $Cr^{+/+}$ and $Cr^{-/-}$ mice.

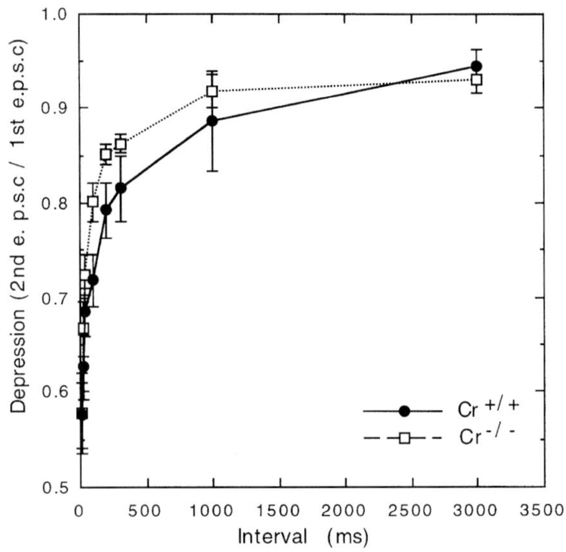

Fig. 6. No alteration of paired-pulse depression in $Cr^{-/-}$ mice. Paired-pulse depression of the climbing fiber current was recorded in cerebellar slices at different inter-pulse intervals. No differences in the magnitude or interval-dependence of this depression were observed.

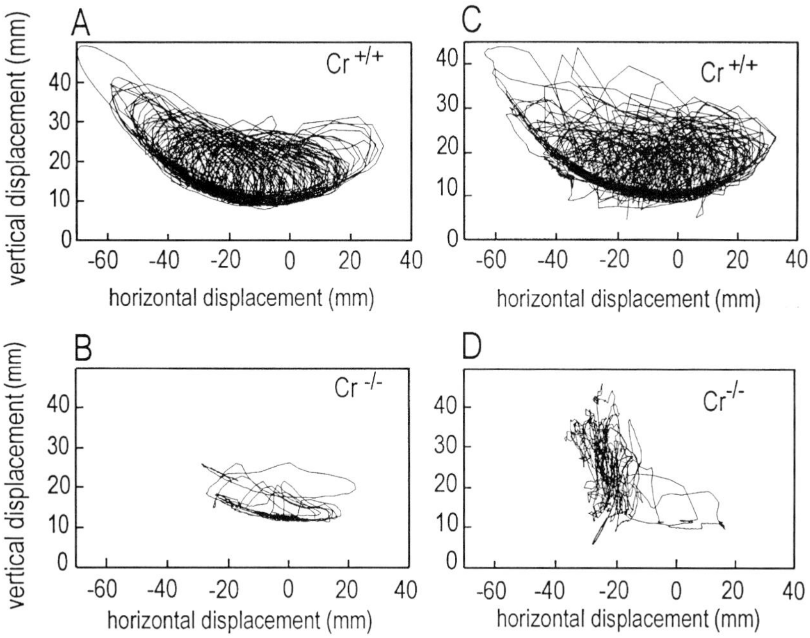

Fig. 7. Locomotion of young $Cr^{+/+}$ (A,C) and $Cr^{-/-}$ mice (B, D) was analyzed using the optoelectronic ELITE System in the wheel running test. Locomotion of $Cr^{-/-}$ mice (B,D) was reduced both in unperturbed (B) and in the presence of a sensory perturbation (D) which induced an oscillatory elevation of the hindlimb foot (D).

Discussion

In this chapter we have shown that the cerebellar cytoarchitecture of $Cr^{-/-}$ mice was normal. PC morphology, the formation of PC synapses and the localization or level of expression of Ca^{2+} binding proteins examined so far were unaltered. However, the absence of calretinin expression results in the detection of an unexpected calretinin-like immunoreactivity in PC, in marked abnormalities in CP firing with alterations of both simple and complex spikes, and in impaired motor coordination.

Calbindin-D28k in Purkinje cells of $Cr^{-/-}$ mice exhibits a higher saturation in Ca^{2+}

Using biochemical techniques such as Western blot, two-dimensional gel electrophoresis and immunoprecipitation, Schiffmann et al., 1999 have demonstrated that the immunoreactivity detected in $Cr^{-/-}$ PC by the anti-calretinin antiserum corresponded to the closely related protein calbindin-D28k. This hypothesis was definitely demonstrated by the generation and analysis of $Cr^{-/-}Cb^{-/-}$ double knock-out mice, as no immunoreactivity was detected in PC of these mice which have an unaltered cerebellar architecture. This paradoxical immunoreactivity does not reflect an increased expression of calbindin-D28k either at the protein or mRNA level, but results from the appearance of a specific Ca^{2+}-induced conformation of this protein. In immunoblotting experiments using the same anti-calretinin antiserum, recombinant calbindin-D28k was detected only when highly saturated with Ca^{2+}. A higher resting and/or evoked $[Ca^{2+}]_i$ in $Cr^{-/-}$ Purkinje cells represents the most reasonable hypothesis to explain this higher saturation of calbindin-D28k. This increase of $[Ca^{2+}]_i$ in $Cr^{-/-}$ Purkinje cells could result from the alterations in synaptic activity and a consequent overflow of the calcium buffering capacity of the cell. Although, as far as we know, an increased affinity of some or all sites of calbindin-D28k for Ca^{2+} has never been described in vivo, this might also result in a higher calbindin-D28k saturation in Ca^{2+} and cannot be definitely ruled out.

Excitatory synaptic transmission in cerebellar slices is normal while in vivo firing properties of Purkinje cells are impaired in $Cr^{-/-}$ mice

The impairment of Ca^{2+} homeostasis in PC as well as the absence of calretinin-mediated Ca^{2+} buffering in parallel fibers would predict modifications of the excitatory synaptic transmission in cerebellar slices. Unexpectedly, neither the SS frequency, the parallel fibers recruitment nor the CS duration were altered in slices taken from $Cr^{-/-}$ mice up to two months of age, strongly suggesting that intrinsic properties of $Cr^{-/-}$ PC are not notably affected. In contrast, in alert $Cr^{-/-}$ mice where signalling through cerebellar afferents as well as inhibitory networks are preserved, severe impairment of PC firing properties was observed with a doubling of the spontaneous SS firing rate, a halving of the duration of CS and a 70–80% reduction in the duration of the SS pause following the CS. These results demonstrate that marked modifications of the neuronal firing behavior undetectable in brain slices can be unmasked by in vivo recording. In the most likely scenario the primary change occurs in granule cells and the increase in SS firing rate could either result from an increased responsiveness to mossy fibers or from an increased release activity at the parallel fiber-PC synapse due to the absence of calretinin either in granule cells bodies or terminals, respectively. Since calretinin is also expressed in subsets of mossy fibers, their principal afference, and in less frequent cerebellar cell types such as unipolar brush and Lugaro cells (Resibois and Rogers, 1992), we cannot claim that the observed changes are exclusively due to changes in granule cells.

The hypothesis of an increased release activity at the parallel fiber-PC synapse is less likely since in slices its release capacity appeared normal. Paired-pulse facilitation, a presynaptic phenomenon (Zucker, 1989) believed to be strongly dependent on residual $[Ca^{2+}]_i$ in the presynaptic terminal, was also unaltered. It is surprising that the absence of calretinin did not affect paired-pulse facilitation whilst overexpression of calbindin-D28k in hippocampal neurons suppressed post-tetanic potentiation, a similar short-term plasticity (Chard et al., 1995). However, at the parallel fiber-PC synapse, it seems that the residual $[Ca^{2+}]_i$ dynamic is not the only mechanism involved and that it mainly depends on calcium extrusion and not on binding to fast endogenous Ca^{2+} buffers (Atluri and Regehr, 1996). In addition, Ca^{2+} binding protein(s) other than calretinin might be sufficient to assure the fast endogenous Ca^{2+} buffering in parallel fibers.

How can we explain the modifications in the durations of CS and SS pause? The response of PC dendrites to parallel fiber stimulation consists of a monosynaptic EPSP followed by a disynaptic inhibitory postsynaptic potential (IPSP) resulting from the activation of inhibitory interneurons such as stellate and basket cells. This GABA-mediated IPSP decreases the amplitude and duration of the climbing fiber-evoked response (Campbell et al., 1983; Midtgaard, 1992; Callaway et al., 1995) and strongly reduces the dendritic climbing fiber-induced $[Ca^{2+}]_i$ increase (Callaway et al., 1995). Parallel fiber stimulation may therefore importantly regulate the CS duration (Campbell et al., 1983). In $Cr^{-/-}$ mice, the large increase in SS firing rate could therefore induce the observed reduction in CS duration. This would result in a large decrease in Ca^{2+} influx and, as a consequence, an attenuated Ca^{2+}-dependent potassium channel activation, leading to the reduction in the transient pause in SS firing which normally follows the CS. Similar changes in CS and transient pause durations have been described following direct stimulation of stellate cells in vitro (Midtgaard, 1992). Following the interpretations of this author, the presumed increased activation of inhibitory interneurons in $Cr^{-/-}$ mice would result in an increased excitability of PC, supporting the view that the reduction in CS duration may in turn have an effect on the SS firing rate in $Cr^{-/-}$ PC. Consistent with this hypothesis, decreased Ca^{2+} influx (Callewaert et al., 1996) or abolition of CS generation (Colin et al., 1980) have been shown to increase SS firing by 80–100%. A similar 90–135% increase was observed in $Cr^{-/-}$ mice, whereas the CS duration was only reduced by 30–50%. This suggests that the shortening of the CS could not account for the entire increase in SS firing.

Alternative or additional mechanisms leading to an increased SS firing rate in $Cr^{-/-}$ PC could be related to a probable decrease in Ca^{2+}-induced long-lasting rebound potentiation of IPSP which normally follows the CS (Kano et al., 1992) or to a chronic Ca^{2+}-induced decrease in GABA-activated currents (Marchenko, 1991).

$Cr^{-/-}$ mice displayed an age-related motor discoordination

The motor deficit of $Cr^{-/-}$ mice in classical tests (runway and stationary rod tests) was only detected in aged animals, an observation reminiscent of the worsened motor discoordination of aged $Cb^{-/-}$ mice (Airaksinen et al., 1997). In aged animals, a similar loss of Purkinje cells occurred in both genotypes and may render aged $Cr^{-/-}$ mice more sensitive to cerebellar dysfunction resulting in the increased motor deficit. A significant alteration was observed in young $Cr^{-/-}$ mice during the wheel running test which subjects mice to a more complex situation for which they must produce self-generated locomotor activity against an unrestrained ground. In addition, a sensory perturbation induced an abnormal motor pattern in $Cr^{-/-}$ mice while it did not alter the locomotor performance of wild-type animals. This pattern likely results from an alteration of sensory data processing in the cerebellum (Gao et al., 1996) and a failure of motor behavior adaptation (Raymond et al., 1996; Mauk, 1997).

The cerebellum is important for the control (Ito, 1984) and fine tuning of movements (Welsh et al., 1995; Raymond et al., 1996; Mauk, 1997). Unlike mice deficient for protein kinase Cγ (Chen et al., 1995)) or metabotropic glutamate receptor mGluR1 (Conquet et al., 1994), $Cb^{-/-}$ (Airaksinen et al., 1997) and $Cr^{-/-}$ mice are not ataxic in standard environments (Schurmans et al., 1997; Schiffmann et al., 1999) but display impaired coordination when challenged. The cerebellum is required for adaptation of automated compound movements to novel environments, suggesting therefore that this function is particularly disturbed in $Cr^{-/-}$ and $Cb^{-/-}$ mice. Climbing fibers are believed to play a crucial role in the refinement of the motor program for a compound movement (Welsh et al., 1995) since they organize movement in time, with a CS duration being critical for optimal motor coordination. Thus the impairment of the climbing fiber-evoked CS detected in alert $Cr^{-/-}$ mice may represent one of the major cellular correlates to the behavioral data.

Acknowledgements

The continuous support and interest of Profs. G. Vassart, J.E. Dumont and J.-J. Vanderhaeghen is gratefully acknowledged. We thank A. Konnerth and T. Plant for their participation in initial experiments on slices, S. Dieudonné for helpful discussion and B. Robaye for help with two-dimensional gel electrophoresis. The expert assistance of R. Menu, M. Authelet, E. Marion, M.-P. Dufief for immunocytochemistry, in situ hybridization, electron microscopy, southern blot and in vivo electrophysiology, respectively and that of J.-L. Conreur and H. N'Guyen for photography and confocal microscopy is also acknowledged. This work was supported by the Belgian Program on Interuniversity Poles of Attraction, Prime Minister's Office, Science Policy Programming. It was also supported by the Queen Elizabeth Medical Foundation (FMRE-Neurobiology 96–98), the Fund for Medical Scientific Research (FRSM-Belgium) (3.4582.94, 9.4524.96F) and Boerhinger Ingelheim. S. Schurmans is Research Associate of the National Fund for Scientific Research (Belgium).

References

Airaksinen, M.S., Eilers, J., Garaschuk, O., Thoenen, H., Konnerth, A. and Meyer, M. (1997) Ataxia and altered dendritic calcium signaling in mice carrying a targeted null mutation of the calbindin D28k gene. *Proc. Natl. Acad. Sci. USA*, 94: 1488–1493.

Atluri, P.P. and Regehr, W.G. (1996) Determinants of the time course of facilitation at the granule cell to Purkinje cell synapse. *J. Neurosci.*, 16: 5661–5671.

Callaway, J.C., Lasser Ross, N. and Ross, W.N. (1995) IPSPs strongly inhibit climbing fiber-activated $[Ca^{2+}]_i$ increases in the dendrites of cerebellar Purkinje neurons. *J. Neurosci.*, 15: 2777–2787.

Callewaert, G., Eilers, J. and Konnerth, A. (1996) Axonal calcium entry during fast 'sodium' action potentials in rat cerebellar Purkinje neurones. *J. Physiol. (Lond.)*, 495: 641–647.

Campbell, N.C., Ekerot, C.F. and Hesslow, G. (1983) Interaction between responses in Purkinje cells evoked by climbing fibre impulses and parallel fibre volleys in the cat. *J. Physiol. (Lond.)*, 340: 225–238.

Chard, P.S., Bleakman, D., Christakos, S., Fullmer, C.S. and Miller, R.J. (1993) Calcium buffering properties of calbindin D28k and parvalbumin in rat sensory neurones. *J. Physiol. (Lond.)*, 472: 341–357.

Chard, P.S., Jordan, J., Marcuccilli, C.J., Miller, R.J., Leiden, J.M., Roos, R.P. and Ghadge, G.D. (1995) Regulation of excitatory transmission at hippocampal synapses by calbindin D28k. *Proc. Natl. Acad. Sci. USA*, 92: 5144–5148.

Chen, C., Kano, M., Abeliovich, A., Chen, L., Bao, S., Kim, J.J., Hashimoto, K., Thompson, R.F. and Tomegawa, S. (1995). Impaired motor coordination correlates with persistent multiple climbing fiber innovation in PKC and mutant mice. *Cell*, 83: 1233–1242.

Cheron, G; Dufief, M.P., Gerrits, N.M. Draye, J.P. and Godaux, E. (1997) Behavioural analysis of purkinje cell outpout from the horizontal zone of the cat flocculus. In: C.I. de Zeeuw, P. strata and J. Voogd (Eds), *Progress in Brain Research, Vol 114*, Elsevier Science, pp. 347–356.

Clapham, D.E. (1995) Calcium signaling. *Cell*, 80: 259–268.

Colin, F., Manil, J. and Desclin, J.C. (1980) The olivocerebellar system. I. Delayed and slow inhibitory effects: an overlooked salient feature of cerebellar climbing fibers. *Brain Res.*, 187: 3–27.

Conquet, F., Bashir, Z.I. and Davies, C.H. et al. (1994) Motor deficit and impairment of synaptic plasticity in mice lacking mGluR1. *Nature*, 372: 237–243.

Gao, J.H., Parsons, L.M., Bower, J.M., Xiong, J., Li, J. and Fox, P.T. (1996) Cerebellum implicated in sensory acquisition and discrimination rather than motor control. *Science*, 272: 545–547.

Ito, M. (1984) *The Cerebellum and Neural Control*. Raven Press, New York.

Ito, M. (1989) Long-term depression. *Annu. Rev. Neurosci.*, 12: 85–102.

Kano, M., Rexhausen, U., Dreessen, J. and Konnerth, A. (1992) Synaptic excitation produces a long-lasting rebound potentiation of inhibitory synaptic signals in cerebellar Purkinje cells. *Nature*, 356, 601–604.

Lisberger, S.G. (1998) Cerebellar LTD: a molecular mechanism of behavioral learning? *Cell*, 92: 701–704.

Llano, I., Marty, A., Armstrong, C.M. and Konnerth, A. (1991) Synaptic- and agonist-induced excitatory currents of Purkinje cells in rat cerebellar slices. *J. Physiol. (Lond.)*, 434: 183–213.

Llano, I., Dipolo, R. and Marty, A. (1994) Calcium-induced calcium release in cerebellar Purkinje cells. *Neuron*, 12: 663–673.

Lledo, P.M., Somasundaram, B., Morton, A.J., Emson, P.C. and Mason, W.T. (1992) Stable transfection of calbindin-D28k into the GH3 cell line alters calcium currents and intracellular calcium homeostasis. *Neuron*, 9: 943–954.

Llinás, R., Sugimori, M. and Silver, R.B. (1992) Microdomains of high calcium concentration in a presynaptic terminal. *Science*, 256: 677–679.

Marchenko, S.M. (1991) Mechanism of modulation of GABA-activated current by internal calcium in rat central neurons. *Brain Res.*, 546: 355–357.

Mauk, M.D. (1997) Roles of cerebellar cortex and nuclei in motor learning: contradictions or clues? *Neuron*, 18: 343–346.

Midtgaard, J. (1992) Stellate cell inhibition of Purkinje cells in the turtle cerebellum in vitro. *J. Physiol. (Lond.)*, 457: 355–367.

Parmentier, M., Lawson, D.E. and Vassart, G. (1987) Human 27-kDa calbindin complementary DNA sequence. Evolutionary and functional implications. *Eur. J. Biochem.*, 170: 207–215.

Parmentier, M. and Lefort, A. (1991) Structure of the human brain calcium-binding protein calretinin and its expression in bacteria. *Eur. J. Biochem.*, 196: 79–85.

Raymond, J.L., Lisberger, S.G. and Mauk, M.D. (1996) The cerebellum: a neuronal learning machine? *Science*, 272: 1126–1131.

Resibois, A. and Rogers, J.H. (1992) Calretinin in rat brain: an immunohistochemical study. *Neuroscience*, 46: 101–134.

Sakurai, M. (1990) Calcium is an intracellular mediator of the climbing fiber in induction of cerebellar long-term depression. *Proc. Natl. Acad. Sci. USA*, 87: 3383–3385.

Schiffmann, S.N., Cheron, G.,Lohof, A., D'Alcantrara, P., Meyer, M., Parmentier, M. and Schurmans, S. (1999) Impaired motor coordination and Purkinje cell excitability in mice lacking calretinin. *Proc. Natl. Acad. Sci. USA*, 96: 5257–5262.

Schurmans, S., Schiffmann, S.N., Gurden, H., Lemaire, M., Lipp, H.P., Schwam, V., Pochet, R., Imperato, A., Bohme, G.A. and Parmentier, M. (1997) Impaired long-term potentiation induction in dentate gyrus of calretinin-deficient mice. *Proc. Natl. Acad. Sci. USA*, 94: 10415–10420.

Sugimori, M. and Llinas, R.R. (1990) Real-time imaging of calcium influx in mammalian cerebellar Purkinje cells in vitro. *Proc. Natl. Acad. Sci. USA*, 87: 5084–5088.

Tank, D.W., Regehr, W.G. and Delaney, K.R. (1995) A quantitative analysis of presynaptic calcium dynamics that contribute to short-term enhancement. *J. Neurosci.*, 15: 7940–7952.

Usowicz, M.M., Sugimori, M., Cherksey, B. and Llinas, R. (1992) P-type calcium channels in the somata and dendrites of adult cerebellar Purkinje cells. *Neuron*, 9: 1185–1199.

Welsh, J.P., Lang, E.J., Suglhara, I. and Llinas, R. (1995) Dynamic organization of motor control within the olivocerebellar system. *Nature*, 374: 453–457.

Winsky, L. and Kuznicki, J. (1996) Antibody recognition of calcium-binding proteins depends on their calcium-binding status. *J. Neurochem.*, 66: 764–771.

Zucker, R.S. (1989) Short-term synaptic plasticity. *Annu. Rev. Neurosci.*, 12: 13–31.

N.M. Gerrits, T.J.H. Ruigrok and C.I. De Zeeuw (Eds.)
Progress in Brain Research, Vol 124

CHAPTER 22

Contralateral cerebellar involvement in conditioned eyeblink responses

M. Ivarsson*, P. Svensson and G. Hesslow

Department of Physiological Sciences, Section for Neuroscience; Lund University,S–223 62 Lund, Sweden.

Introduction

Classical conditioning is a simple form of associative learning where a neutral *conditioned stimulus* (CS) is paired with an *unconditioned stimulus*, (US) that elicits a specific reflex response, the *unconditioned response* (UR). By pairing the CS and US, the previously neutral stimulus will acquire the ability to elicit a new response, termed the *conditioned response* (CR).

The study of CRs has been shown to be an objective tool to study associative learning and has been applied to a wide variety of different responses, ranging from autonomic responses to precise motor responses such as the eyeblink (Gormezano and Moore, 1976). In the 1960s Gormezano and associates developed an animal preparation where the movement of the nictitating membrane (NM) of the rabbit was used as the behavioral measure (see Gormezano, et al., 1983). A tone (CS) was paired with an airpuff (US), which reflexively activated the retractor bulbi muscle. This retracts the eyebulb causing the NM to sweep passively over the cornea. After repeated pairings of the tone and the airpuff, the tone acquires the ability to elicit a NM movement and eyeblink. Studies of classical eyeblink conditioning have used both the movement of the NM or the eyelid and recorded EMG directly from the orbicularis oculi muscle (m.o.o.) to measure the CR (Mccormick et al., 1981, 1982; Yeo et al., 1985a, b; Gruart et al., 1995; Ivarsson and Hesslow, 1993; Hesslow and Ivarsson, 1996; Ivarsson et al., 1997).

Thompson and associates made a major breakthrough in the search for the CR learning site in the beginning of the 1980s (see Thompson and Krupa, 1994). Large unilateral lesions of the cerebellum abolished the CRs on the lesioned side without any effect on the eyeblink UR (McCormick et al., 1981). The lesions had different effects on the CRs and URs, which suggested that the lesions did not abolish the CRs because of a general motor performance deficit, but rather by affecting a neural circuit specifically involved in CR production. Thus it was proposed that the cerebellum has a critical role in the expression of a CR and might even be the site for the memory formation in classical conditioning. The importance of the cerebellum in conditioning was further strengthened by studies showing that selective lesions of the pathway from the cerebellum to the motor nuclei (that performed the CR) had the same effect as lesioning the ipsilateral cerebellum (McCormick et al., 1982; Rosenfield and Moore, 1983). Since then a substantial number of papers have confirmed the importance of the cerebellum in classical conditioning (see Yeo and Hesslow, 1998).

*Corresponding author. M. Ivarsson,University College London, Dept. of Anatomy and Developmental Biology, Gower Street, Foster Court, London WC1E 6BT, UK. Tel.: INT +44 (0) 171-5042891; Fax: INT +44 (0) 171-4193390; e-mail: m.ivarsson@ucl.ac.uk

Early lesion studies all confirmed that restricted lesions of the interpositus nucleus were sufficient to abolish a previously learned CR and prevent acquisition and reacquisition (Lincoln et al., 1982; Clark et al., 1984; McCormick and Thompson, 1984a, b; Yeo et al., 1985a). On the other hand restricted lesions of the cerebellar cortex have yielded conflicting results. Yeo and associates reported a complete loss of CRs after removal of the entire lobule HVI, whereas Thompson and colleagues found that cortical lesions only caused changes in CR topography (McCormick and Thompson 1984a, b; Yeo et al., 1985b; Hardiman and Yeo, 1992). Additional studies have shown that large unilateral cortical lesions, including lobule HVI, either substantially impaired or temporarily abolished CRs, but with extended post-lesion training the CRs recovered to varying degrees depending on the extension of the lesion (Lavond et al., 1987; Yeo 1991; Yeo and Hardiman, 1992; Harvey et al., 1993; Thompson and Krupa, 1994).

Do these findings suggest that the interpositus nucleus is the critical site for learning, since cerebellar cortical lesions have been less efficient in abolishing CRs?

There are several alternatives that could explain the expression of CRs after cortical lesions. (a) several eyeblink related microzones in different lobes have been shown to exist in the cerebellar cortex of cat and ferret, which individually could be involved in the expression of CRs (Hesslow 1994a, b; Ivarsson et al., 1996); (b) the lesions discussed above have all been unilateral and the variable results after cortical lesions could be caused by contralateral cerebellar influences as suggested by Yeo and Hardiman (1992). This is indirectly suggested by the finding of bilateral CRs during initial unilateral acquisition (Disterhoft et al., 1977) and the more pronounced effect of bilateral lesions on CRs by Skelton (1988). (c) Mauk and colleagues have suggested that plasticity related to CR could be located in both the cerebellar cortex and nuclei and therefore lesions of cerebellar cortex, after the learning has been established, do not affect the performance of the CR (Mauk and Donegan 1997).

Since there exist some data that indirectly suggest a bilateral influence of unilaterally expressed CRs, we decided to study this possible bilateral interaction further. Two main questions were addressed: (a) is there a pathway via which one cerebellar hemisphere can generate muscle activity in the contralateral m.o.o. (Ivarsson and Hesslow 1993); (b) if a bilateral pathway exists is it involved in expressing contralaterally elicited CRs (Ivarsson et al., 1997)?

Is there a pathway via which one cerebellar hemisphere can activate the contralateral orbicularis oculi muscle?

By recording short latency climbing fiber input and receptive fields, the eyeblink related area in the c3 zone of the cerebellar cortex in decerebrate ferrets was located. It has previously been shown that stimulation of the cortical areas receiving short latency climbing fiber input can elicit muscular activity in the m.o.o. (Hesslow 1994a, b). When a train of stimuli was applied to the c3 eyeblink area of the cortex, activity in both the ipsilateral and contralateral m.o.o. could be recorded. Usually the contralateral response had a longer onset latency, but occasionally the opposite occurred. Bilateral activation of the m.o.o. was also recorded when stimulating the brachium conjunctivum (BC) unilaterally. By tracking through the BC, the site with the lowest threshold for eliciting muscle activity in the ipsilateral m.o.o. was located (<10 μA). Stimulation of this site elicited activity in both m.o.o.s' but the contralateral response was always smaller in size and had a longer onset latency, approximately 4 ms. These findings show that there is a pathway from one cerebellar hemisphere to the contralateral m.o.o. which potentially could be involved expressing the contralateral CRs.

A bilateral pathway exists but is it involved in contralaterally elicited CRs?

Bilateral independent CRs. Decerebrate ferrets were trained to express eyeblink CRs bilaterally to a forelimb CS (300 ms, 50 Hz train of electrical stimuli, 1 mA) and a bilateral periorbital US (50 Hz train of three electrical stimuli, 3–4 mA). The first question we addressed was whether it was possible for these animals to generate CRs unilaterally. In four well-trained ferrets the US was withheld on one side and the size of the bilateral CRs was studied during unilateral extinction training. When the CRs on the non-reinforced side began to decrease in size and frequency, it was noted that the CRs on the reinforced side were depressed. With continued unilateral extinction training the CRs on the non-reinforced side were extinguished while the CRs on the reinforced side recovered. This clearly shows that CRs can be expressed independently on the two sides even though the animals are trained to bilaterally applied US. However the decrease in frequency of CRs on the reinforced side implies that there might be a bilateral interaction in the output from the cerebellar hemispheres. This was further studied by blocking cerebellar outflow either reversibly or irreversibly.

Unilateral reversible and irreversible blocking of the BC. Sites in the BC that elicited eyeblinks on weak stimulation were located in eight well-trained ferrets and small amounts of lignocaine were injected (0.5–1.0 μl of 4% lignocaine). The CR on the ipsilateral side was abolished, which is consistent with previous studies (for review see Yeo, 1991; Thompson and Krupa, 1994), but in addition the unilateral injection consistently suppressed the CR on the contralateral side too. The CRs recovered gradually after the injection and reached the same size as before the blockade, but the lignocaine effect on the contralateral side was of shorter duration. The bilateral effect on the CRs with unilateral reversible blockade was seen in all animals tested. In four ferrets the unilateral injection of lignocaine was repeated at least once with the similar effects (Ivarsson et al., 1997). The reversible lesions show that output from one cerebellar hemisphere is involved in the expression of a contralateral CR, but it does not say anything about whether the contralateral contribution is necessary. This was tested by mechanically lesioning the BC on one side in three animals. Both the ipsi- and the contralateral CRs were abolished immediately, but the CRs on the contralateral side reappeared within 1–2.5 hours. The completeness of the BC lesion was confirmed histologically and was also implied by the lack of CRs on the ipsilateral side for the rest of the experiment.

Effect on UR by blockade of BC. One possibility is that the effect on the CRs we record after reversible and irreversible lesions of the BC is caused by a general decrease in facilitation of motor structures downstream from the BC. The motoneurones in the facial nucleus are the final common pathway for both the blink reflex and the eyeblink CR and if the blockade caused a general decrease in excitability this would most likely be recorded in the URs as well. To test the effect on URs without interaction by the CS and CR, US alone trials after unilateral blockade were used. In the four animals tested the effects on the URs were variable, but the findings suggest that a general decrease in excitability of facial motoneurones are unlikely to be the cause behind the suppression of CRs on the contralateral side.

We have shown that a pathway from one cerebellar hemisphere to the contralateral m.o.o. is present in the ferret (Ivarsson and Hesslow, 1993) and that it is involved in the expression of CRs bilaterally (Ivarsson et al., 1997). There are several possible locations for the crossed pathway (a) a bilateral trigeminal projection to the motoneurones or pre-motoneurones in the facial nucleus; (b) there could be reciprocal connections between the facial nuclei; (c) the contralateral blink reflex and the bilateral cerebellar output could be separate and there could be a contralaterally projecting pathway somewhere between the BC and the facial nucleus to pre-motoneurones on the contralateral side. At this point it is unclear which pathway mediates the output to the contralateral side. The findings discussed above suggest that the bilateral pathway facilitates the contralateral CR pathway, mainly because a permanent lesion only cause a temporary suppression of the contralateral CR and that this is located before the facial nucleus due to the lack of effects on the contralateral URs. The crossed facilitatory pathway projecting to a pre-facial site has been inserted in Fig. 1.

In the remaining part of the paper, data will be presented that suggests a possible involvement of the crossed facilitatory pathway in extinction of bilateral CRs, which further supports a role for the bilateral pathway in expressing CRs.

Contralateral cerebellar involvement in the extinction of eyeblink CR

It has been shown that the cerebellum is critical for the extinction of CRs (Ramnani and Yeo, 1996) and also that output from one cerebellar hemisphere facilitates the expression of a contralateral CR (Ivarsson et al., 1997). The aim of the following experiments was to investigate a possible bilateral interaction between the cerebellar hemispheres in the extinction of the eyeblink CR.

Ferrets, decerebrated rostral to the superior colliculi, were trained in a classical conditioning paradigm. The CS was a train of electrical stimuli (15 pulses, 50 Hz, 1 mA) applied to the left forelimb. The US was a train of electrical stimuli (3 pulses, 50 Hz, 3–4 mA) applied either unilaterally or bilaterally to the periorbital region. The CRs were studied by recording EMG from the m.o.o. on both sides. EMG responses were rectified and integrated during predetermined time periods in every trial. The EMG activity during a 100 ms interval before CS-onset was used to measure the spontaneous activity of the m.o.o.. The size of the CR was calculated from EMG activity during the interval 80–298 ms after CS-onset. In the plots the

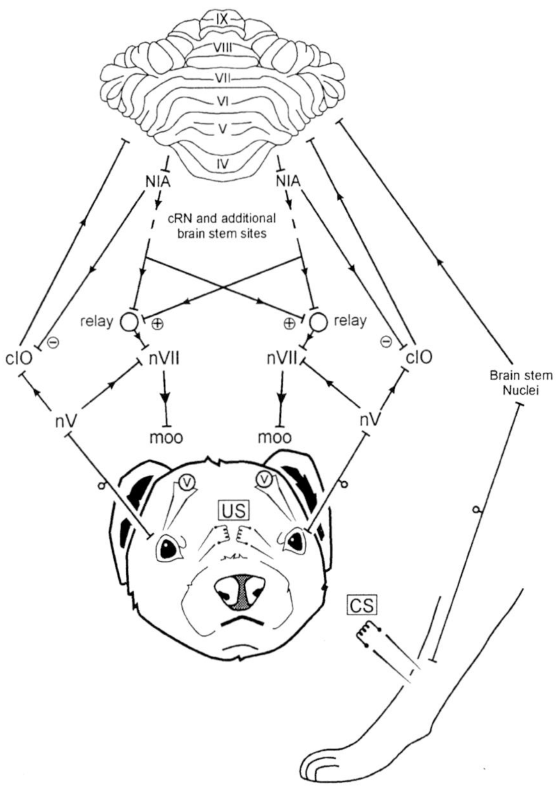

Fig. 1. Simplified wiring diagram of the circuit important for expression of CRs. NIA: anterior interpositus nucleus; cIO: contralateral inferior olive; nV: trigeminal nucleus; nVII: facial nucleus; cRN: contralateral red nucleus; CS: conditioned stimulus; US: unconditioned stimulus.

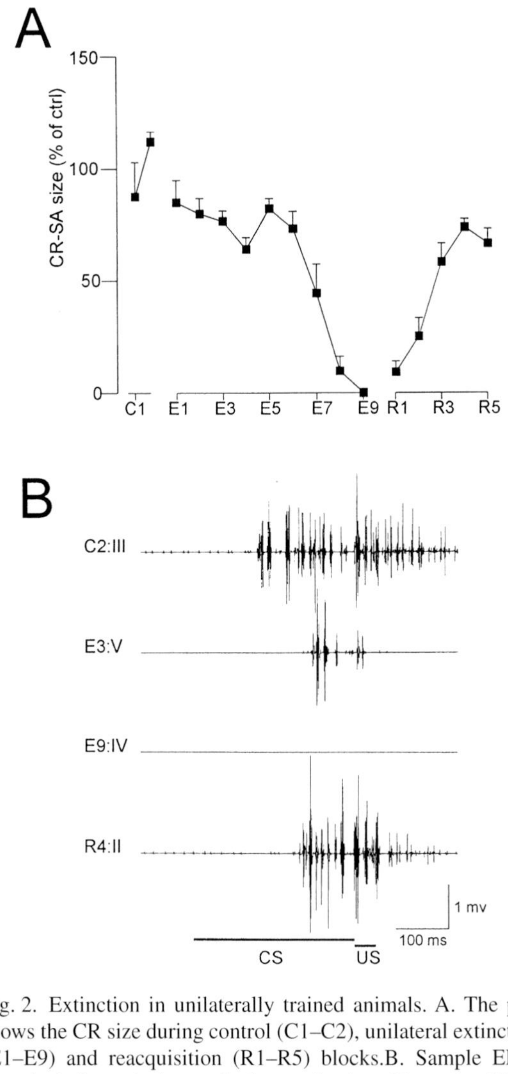

Fig. 2. Extinction in unilaterally trained animals. A. The plot shows the CR size during control (C1–C2), unilateral extinction (E1–E9) and reacquisition (R1–R5) blocks.B. Sample EMG records before, during and after extinction. Example: C2:III corresponds to the third trial (III) in the second control block (C2).

spontaneous activity in each trial was subtracted from the CR size in that trial. When the animals had elicited CRs in more than 9/10 of the trials for at least one hour the US was either withheld on one side (unilateral extinction) or on both sides (bilateral extinction). The number of presented trials necessary to reduce the frequency of CRs to 1/9 was counted.

In order to study the interaction of the two hemispheres in classical eyeblink conditioning we compared unilateral and bilateral extinction training in bilaterally trained animals and unilaterally trained animals. If there is a difference in the number of trials necessary for extinction between unilateral and bilateral extinction procedures it could be suggested that the extinction process also involves both hemispheres.

The plot in Fig. 2 shows a representative example of the extinction of CRs in a unilaterally trained animal. Every point in the plot represents the average size of 10 consecutive trials (±SEM). EMG records from the left m.o.o. during different periods of the extinction process are shown below the plot.

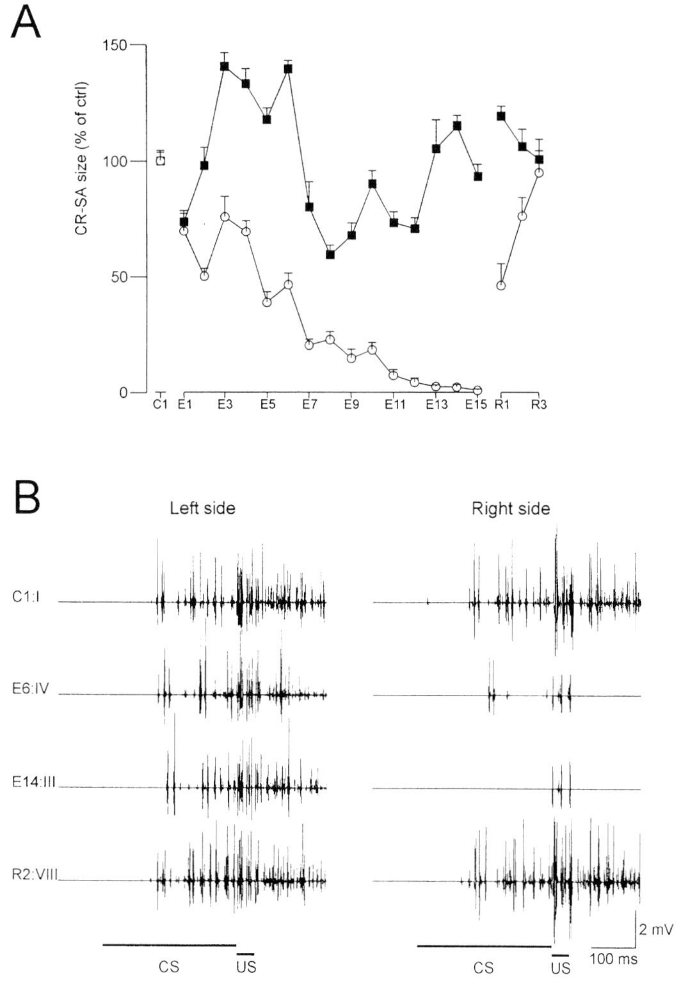

Fig. 3. Unilateral extinction in bilaterally trained animals.A. The plot shows the CR size during control (C1), unilateral extinction (E1–E15) and reacquisition (R1–R3) blocks. The CRs on the left reinforced side are represented by filled squares and the right non-reinforced side by open circles. Error bars show SEM.B. Sample EMG records from the left and right side before, during and after extinction.

By comparing unilateral extinction and bilateral extinction in bilaterally trained animals, the interaction between the two sides can be studied. Figure 3 shows unilateral extinction in an animal that has acquired CRs bilaterally. The CRs are extinguished on the non-reinforced side and on the contralateral reinforced side the CRs are of the same size as during the control sessions. Every point in the plot represents the average size of 10 consecutive trials (±SEM). EMG records from left and right m.o.o. during different stages of the unilateral extinction training are also shown.

The plot in Fig. 4 shows bilateral extinction of CRs in a bilaterally trained animal. In this animal the responses are extinguished after only eight trials. Note that every point in this plot is the average size of five recordings (±SEM). Sample EMG records from both m.o.o.s during extinction are shown.

Figure 5 shows the average number of trials necessary for extinction in unilaterally and bilaterally trained animals in the different extinction

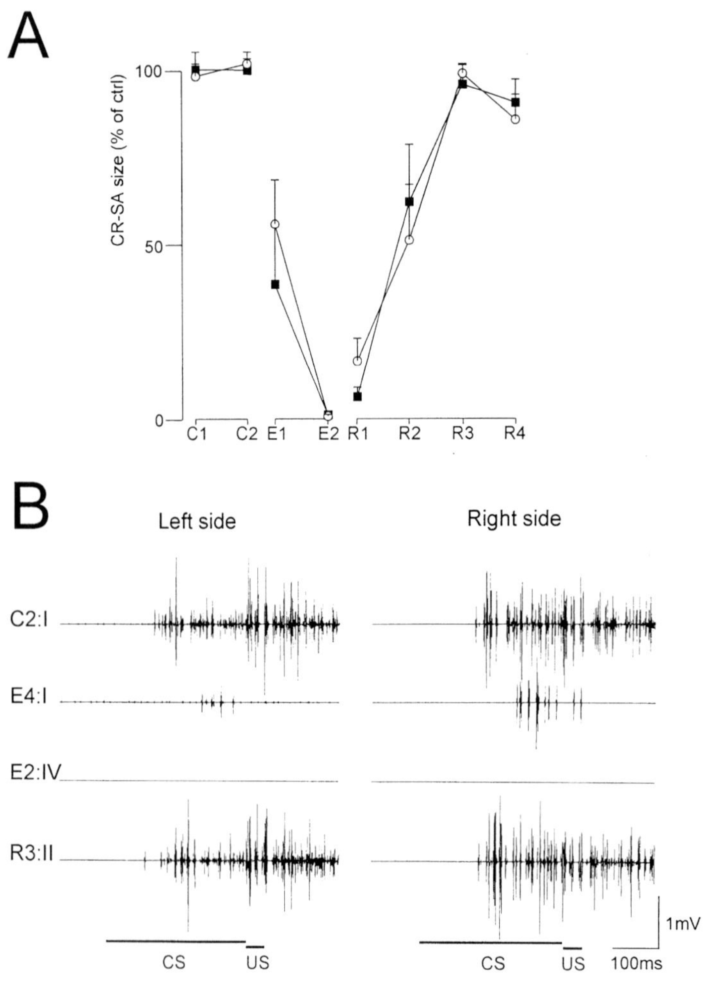

Fig. 4. Bilateral extinction in bilaterally trained animals. The plot shows the CR size during control (C1–C2), bilateral extinction (E1–E2) and reacquisition (R1–R5) blocks. Note that each block only consists of 5 trials. The CRs on the left side are represented by filled squares and the right side by open circles. Error bars show SEM.B. Sample EMG records from the left and right side before, during and after extinction.

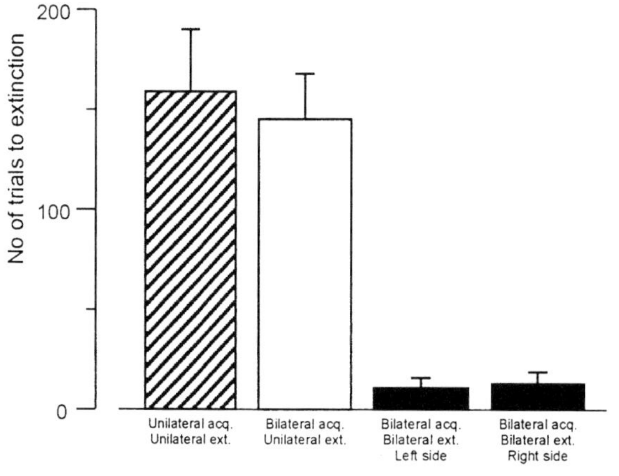

Fig. 5. Pooled data of the number of trials to extinction in all animals.

groups. The number of trials needed for bilateral extinction is (average left side: 11 ± 5; $n = 4$; right side: 13 ± 6; $n = 4$) and for unilateral extinction (average 145 ± 25; $n = 6$) in the bilaterally trained animals and (average 159 ± 31; $n = 3$) in unilaterally trained animals.

Together these findings show that there is a marked difference in the number of trials necessary to extinguish the CR between bilateral extinction training and unilateral extinction training. These findings could be explained by an interaction between the two cerebellar sides during the extinction process, which is in agreement with the findings of a crossed facilitatory pathway. Animals with unilaterally acquired CRs show a similar number of extinction trials as the bilaterally trained animals during unilateral extinction. The difference in the number of trials to extinction between unilateral and bilateral extinction in bilaterally trained animals could be due to both cerebellar hemispheres being involved in the expression of a CR. If the cerebellar hemispheres facilitate the contralateral CR pathways it is conceivable that during bilateral extinction training the removal of bilateral facilitation will speed up the extinction process. This is suggested by our data, which is based on the bilateral interaction between the cerebellar hemispheres during extinction. This implies that extinction is an active process and that both hemispheres are involved in its control.

General conclusions

We have shown that the CRs recorded after unilateral cerebellar cortical lesions, could be generated by the contralateral cerebellar hemisphere and thus the previous unilateral cerebellar lesion studies are inconclusive. Specifically, we showed that a bilateral pathway exists via which the contralateral eyeblink CRs could be mediated (Ivarsson and Hesslow, 1993) and that this pathway is actively involved in the control of the contralateral eyeblink CRs (Ivarsson et al., 1997). This is in agreement with a study showing that unilateral cortical lesions do not cause a permanent loss of CRs, while bilateral lesions do (Gruart and Yeo, 1995). Yeo and colleagues (1997) have also shown that blocking the ipsilateral cerebellum do not completely prevent acquisition of a new CR; they suggest that this new CR might have a contralateral origin. The extinction experiments presented here further imply a bilateral interaction of the cerebellar hemispheres in the expression of CRs.

Acknowledgements

This study was supported by grants from the Swedish Medical Research Council (project no. 09899), The Segerfalk Foundation, The Swedish Institute and The Foundation Blanceflor Boncompagni-Ludovisi, neé Bildt.

References

Clark, G.A., McCormick, D.A., Lavond, D.G. and Thompson, R.F. (1984) Effects of lesions of cerebellar nuclei on conditioned behavioral and hippocampal neuronal responses. *Brain Res.*, 291: 125–136.

Disterhoft, J.F., Kwan, H.H. and Lo, W.D. (1977) Nictitating membrane conditioning to tone in the immobilized albino rabbit. *Brain Res.*, 137: 127–143.

Gormezano, I. and Moore, J.W. (1976). Classical conditioning. In: M.H. Marx (Ed.), *Learning: Processes*, Macmillan, New York.

Gormezano, I., Kehoe, E.J. and Marshall, B.S. (1983) Twenty years of classical conditioning research with the rabbit. *Progress in Psychobiology and Physiological Psychology. Vol. 10*; New York; Academic Press; pp. 197–275.

Gruart A. and Yeo C.H. (1995) Cerebellar cortex and eyeblink conditioning – bilateral regulation of conditioned responses. *Exp. Brain Res.*, 104: 431–448.

Gruart, A. Blázquez, P. and Delgado-Garca, J.M. (1995) Kinematics of spontaneous, reflex, and conditioned eyelid movements in the alert cat. *J. Neurosci.*, 74: 226–248.

Harvey, J.A., Welsh, J.P., Yeo C.H. and Romano, A.G. (1993) Recoverable and non-recoverable deficits in conditioned responses after cerebellar cortical lesions. *J. Neurosci.*, 13: 1624–1635.

Hesslow, G. (1994a) Correspondence between climbing fibre input and motor output in eyeblink related areas in cat cerebellar cortex. *J. Physiol.*, 476: 229–244.

Hesslow, G. (1994b) Inhibition of classically conditioned eyeblink responses by stimulation of the cerebellar cortex in the cat. *J. Physiol.*, 476: 245–256.

Hesslow, G. and Ivarsson, M. (1996) Inhibition of the inferior olive during conditioned responses in the decerebrate ferret. *Exp. Brain Res.*, 110: 36–46.

Ivarsson, M. and Hesslow, G. (1993) Bilateral control of the orbicularis oculi muscle from one cerebellar hemisphere in the ferret. *NeuroReport*, 4: 1127–1130.

Ivarsson, M., Svensson, P. and Hesslow, G. (1997) Bilateral disruption of conditioned responses after unilateral blockade of cerebellar output in the decerebrate ferret. *J. Physiol.*, 502: 189–201.

Ivarsson, M., Svensson, P. and Hesslow, G. (1996) Relative contributions of the C1 and C3 zones of the cerebellar cortex to eyeblink conditioning in the ferret. *Soc. Neurosci. Abstr.*, 638.1

Lavond, D.G., Steinmetz, J.E., Yokaitis, M.H. and Thompson, R.F. (1987) Reacquisition of classical conditioning after removal of cerebellar cortex. *Exp. Brain Res.*, 67: 569–593.

Lincoln, J.S., McCormick, D.A. and Thompson, R.F. (1982) Ipsilateral cerebellar lesions prevent learning of the classically conditioned nictitating membrane/eyelid response. *Brain Res.*, 242:190–193.

Mauk, M.D. and Donegan, N.H. (1997) A model of Pavlovian eyelid conditioning based on the synaptic organization of the cerebellum. *Learn. Mem.*, 3: 130–158.

McCormick, D.A., Lavond, D.G., Clark, G.A. Kettner, R.E., Rising, C.E. and Thompson, R.F (1981) The engram found? Role of the cerebellum in classical conditioning of nictitating membrane and eyelid responses. *Bull. Psych. Soc.*, 18: 103–105.

McCormick, D.A., Clark, G.A., Lavond, D.G. and Thompson, R.F. (1982) Initial localization of the memory trace for a basic form of learning. *Proc. Nat. Acad. Sci.*, 79: 2731–2735.

McCormick, D.A. and Thompson, R.F. (1984a) Neuronal responses of the rabbit cerebellum during acquisition and performance of a classically conditioned nictitating membrane-eyelid response. *J. Neurosci.*, 4: 2811–2822.

McCormick, D.A. and Thompson, R.F. (1984b) Cerebellum: essential involvement in the classically conditioned eyelid response. *Science*, 223: 296–299.

Ramnani, N. and Yeo, C.H. (1996) Reversible inactivations of the cerebellum prevent the extinction of conditioned nictitating membrane responses in rabbits. *J. Physiol.*, 495: 159–168.

Rosenfield, M.E. and Moore, J.W. (1983) Red nucleus lesions disrupt the classically conditioned nictitating membrane response in rabbits. *Behav. Brain Res.*, 10: 393–398.

Skelton, R.W. (1988) Bilateral cerebellar lesions disrupt conditioned eyelid responses in unrestrained rats. *Behav. Neurosci.*, 102: 586–590.

Thompson, R.F. and Krupa, D.J. (1994) Organization of memory traces in the mammalian brain. *Ann. Rev. Neurosci.*, 17: 519–549.

Yeo, C.H., Hardiman, M.J. and Glickstein, M. (1985a) Classical conditioning of the nictitating membrane response of the rabbit. I. Lesions of the cerebellar nuclei. *Exp. Brain Res.*, 60: 87–98.

Yeo, C.H., Hardiman, M.J. and Glickstein, M. (1985b) Classical conditioning of the nictitating membrane response of the rabbit. II. Lesions of the cerebellar cortex. *Exp. Brain Res.*, 60: 99–113.

Yeo, C.H. (1991). Cerebellum and classical conditioning of motor responses. In: J.R. Wolpaw, J.T. Schmidt and T.M. Vaughan (Eds), *Activity Driven CNS Changes in Learning and Development*, Annals of New York Academy of Science, 627: pp. 292–304.

Yeo, C.H. and Hardiman, M.J. (1992) Cerebellar cortex and eyeblink conditioning: a re-examination. *Exp. Brain Res.*, 88: 623–638.

Yeo, C.H., Lobo, D.H and Baum, A. (1997) Acquisition of a new-latency conditioned nictitating membrane response – major but not complete, dependence on the ipsilateral cerebellum. *Learn. Mem.*, 3: 557–577.

Yeo, C.H. and Hesslow, G. (1998) Cerebellum and conditioned reflexes. *TICS*, 2: 322–330.

N.M. Gerrits, T.J.H. Ruigrok and C.I. De Zeeuw (Eds.)
Progress in Brain Research, Vol 124

CHAPTER 23

Involvement of the cerebellum in a new temporal property of the conditioned eyeblink response

Pär Svensson*, Magnus Ivarsson and Germund Hesslow

Department of Physiological Sciences, Section for Neurophysiology, Lund University, Sölvegatan 19, S-223 62 Lund, Sweden

Introduction

A critical aspect of classical conditioning is the adaptive timing of the conditioned response (CR). In the eyeblink conditioning paradigm the conditioned eyeblink response is elicited just in time to peak at the onset of the unconditioned stimulus (US). If the interval between the onset of the conditioned stimulus (CS) and the US is increased, the latency of the eyeblink CR will increase. This is functional since the eyeblink CR then tends to protect against the aversive US. Here, we describe a new temporal property of the eyeblink CR. A sudden increase in the CS intensity decreased the eyeblink CR latency. When the high CS intensity was maintained over several trials the short CR latency gradually increased until the CR was again elicited at the US onset. This temporal property was first observed in decerebrate ferret, but is also present in the intact rabbit, suggesting that it is a general feature of the conditioned eyeblink response. Experimental work revealed that pre-rubral and pre-cerebellar structures are not involved in this timing property, indicating that cerebellum and or post-cerebellar structures are be responsible.

*Corresponding author. Tel.: +46 46 2220327;
Fax: +46 46 2224546;
e-mail: Par.Svensson@mphy.lu.se

Classical conditioning

In the classical conditioning paradigm a conditioned stimulus (CS), say a tone, is presented just prior to a reflex-eliciting unconditioned stimulus (US), such as a brief puff of air to the eye. The response elicited by the US, in this case an eyeblink, is called the unconditioned response (UR). Before training, the tone will not elicit any eyeblink. However, after several paired presentations with the airpuff, the tone will have acquired the ability to elicit an eyeblink. This learned response is called the conditioned response (CR; for references see Pavlov, 1927; Marx, 1969; Macintosh, 1974).

Timing of the conditioned response

One important feature of the eyeblink CR is the timing. The eyeblink CR is normally timed so that it starts just prior to, and peaks at, the US onset. For instance, if the animal is trained with an interval between the CS and the US (Interstimulus interval, ISI) of 300 ms, the CR onset latency will typically be around 200 ms. If the ISI is 500 ms, the latency will normally be a couple of hundred milliseconds longer. In this way, the CR gives maximal protection against the aversive US. Furthermore, the eyeblink CR normally terminates just after the peak. In this way the CR will not be elicited longer than necessary, but only during the shortest time required.

More complex temporal tasks can be acquired. For example, when two different ISIs of 200 ms and 700 ms were presented in a pseudorandom order, the rabbits learned to elicit CRs with two peaks. A CS of 700 ms elicited a double-peak CR with the first peak around 200 ms and the second around 700 ms while the CS of 200 ms evoked a CR with only one peak around 200 ms (Millenson et al., 1977). Furthermore, individual ISIs can be associated to different CSs as shown by Mauk and Ruiz (1992). One and the same rabbit was conditioned to two different CS tone frequencies. A 400 Hz tone was only presented on training trials with an ISI of 150 ms and an 8 kHz tone only with a 750 ms ISI. When presenting the tones on test trials the 400 Hz tone elicited eyeblink CRs with a peak around 150 ms and the 8 kHz elicited eyeblink CRs peaking at 750 ms (Mauk and Ruiz, 1992).

Consequently, the ISI determines when the CR will be elicited. However, the ISI specific CR latency is not fixed once learned, but can change according to new temporal requirements. For example, if the ISI is suddenly increased the eyeblink CR will not be elicited at the time of the US any longer. After further training to the new ISI, however, the eyeblink CR will finally be elicited just prior to and peak at the new US onset (Boneau 1958; Ebel and Prokasy 1963; Prokasy and Papsdorf 1965; Leonard and Theios, 1967; Coleman and Gormezano 1971; Norman et al., 1974). This is functional since the protective value of the eyeblink CR will be regained.

A new timing property

We have relatively recently described a new timing property of the eyeblink CR (Svensson et al., 1997). Decerebrate ferrets were conditioned to an electrical stimulation of the forelimb as CS (300 ms duration, 1 mA, 50 Hz, 0.5 ms square wave pulses) and of the periorbital area as US (3 pulse, 3 mA, 50 Hz, 1 ms square wave pulses). When the CS intensity was suddenly increased from 1 mA to 2 mA, the eyeblink CR was elicited roughly 100 ms earlier (Fig. 1A) and as the CS intensity was reset to the original 1 mA, the CR latency returned to training levels. If the higher intensity was maintained at 2 mA for several trials (range: 25–110 trials), the CR latency gradually increased back to initial levels (Fig. 1D). We also analyzed the effect of an increase in the CS train frequency from 50 Hz to 100 Hz while maintaining the current amplitude at 1 mA. Similar changes in CR latency were observed. Both the 100 ms decrease and the long-term adaptation of the CR latency were observed when the CS train frequency was increased to 100 Hz and was kept constant for several trials (range: 10–105 trials). These results show, firstly, that an increase in CS intensity decreases the eyeblink CR latency and, secondly, that a shortened CR latency is not maintained if the ISI is kept constant but will instead gradually increase so that the CR eventually is elicited at the US onset.

To exclude the possibility that the timing effect was an experimental artifact due to the reduced preparation, similar experiment was performed in intact rabbits with a tone as CS. It was not improbable that such experiments could show a similar effect as in the ferret experiments since changes in eyeblink CR latency have been observed in other studies. For example, rabbits conditioned to two different CS tone intensities, elicited CRs with shorter latencies to the high intensity CS compared to the low intensity CS (Leonard and Monteau, 1971). Furthermore, rabbits conditioned to one CS intensity elicited CRs with longer latency when the CS intensity was decreased (Frey, 1969). However, the effects of increased tone intensity or maintained high CS intensity over several trials have not been studied previously. Replication in intact rabbits of the CR latency changes should reveal whether that the timing phenomenon is limited to the decerebrate ferret or the electrical CS but that the timing is a general phenomenon. Preliminary results show that rabbits, conditioned to a 1 kHz tone of 65 dB (background noise level was 42 dB) as CS and electrical periorbital stimulation as US, elicited eyeblink CRs with shorter latency when the CS intensity was suddenly increased from 65 dB to 75 dB (Fig. 1B). When the intensity was returned to the original training level of 65 dB, the CR latency immediately increased back to training levels. As illustrated in Fig. 1E, the gradual increase in CR latency when the CS was kept at 75 dB for many trials was also demonstrated (Svensson et al., unpublished data).

Are the eyeblink CR timing properties relevant for the behaving animal or are the observed changes in CR latency an epiphenomenon? It may seem dysfunctional to blink before the US arrives. However, another possibility is that the intensity of a sound which may be a warning signal for an aversive stimulus, indicates the distance to the stimulus. Although speculative, this may be an important feature of classical conditioning. One example under natural conditions may be when a fly approaches the eye. Not only the fly but also the sound of the fly can elicit an eyeblink as protection. A louder buzz would predict that the fly is closer to the eye and will probably hit it sooner. In this case an earlier eyeblink CR would protect the eye in contrast to an eyeblink CR with an unaltered latency.

Localization of the mechanism behind the timing property

What structures or neurons mediate the shorter CR latency and the gradual latency increase? One plausible hypothesis is that the neurons responsible

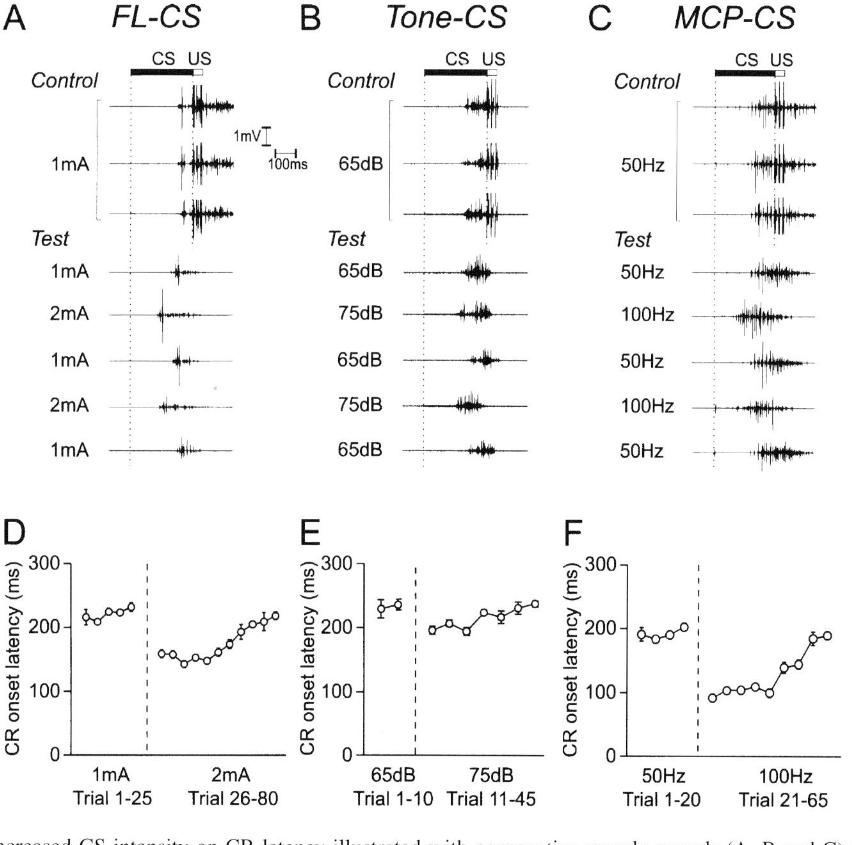

Fig. 1. Effect of increased CS intensity on CR latency illustrated with consecutive sample records (A, B and C) and average onset latency plotted against trial number (E, F and G; each circle represents the average of latency of CRs in five consecutive trials and the SEM). The high-pass filtered EMG records from the orbicular oculi muscle show that the CR onset latency was decreased in decerebrate ferrets when the forelimb CS intensity suddenly was increased from 1 mA to 2 mA (A), in rabbit when the tone CS was increased from 65 dB to 75 dB (B) and in decerebrate ferrets when the frequency of the middle cerebellar peduncle CS (MCP–CS) suddenly was increased from 50 Hz to 100 Hz (C). The graphs (average ± SEM of the CR latency in five consecutive trials) show that the CR onset latency gradually was increased to initial values when the high forelimb CS intensity was maintained at 2 mA (D), the high tone intensity was maintained at 75 dB (E) and the high MCP–CS train frequency was maintained at 100 Hz (F).

for the acquisition and generation of the CR also are responsible for the changes in CR latency. However, alternatives exist. Neurons transmitting the CS from the periphery to the CR generation site may be involved as well as neurons projecting from the CR generation site to the effector organ. Also, structures non-essential for the CS–CR pathway may influence temporal properties of the eyeblink CR.

Although this view is still opposed by some authors, the site responsible for the learning of the association between the CS and the US is most likely the cerebellum (see Thompson and Krupa, 1994; Yeo and Hesslow, 1998; Ivarsson, 1998). For example, lesions of the nucleus interpositus abolish and prevent acquisition (McCormick et al., 1982; McCormick and Thompson 1984; Yeo et al., 1985a) while inactivation of the superior cerebellar peduncle blocks expression but not acquisition of the eyeblink CR (Krupa and Thompson, 1995). The CS has, in analogy with the cerebellar learning models proposed by Marr (1969) and Albus (1971), been suggested to be transmitted through the cerebellar mossy fibers to the cerebellum (Yeo et al., 1985b; McCormick et al., 1985). Consistent with this, direct stimulation of the mossy fibers in the middle cerebellar peduncle (MCP) or to their origins, such as the pontine nuclei and the lateral reticular nuclei, has been shown to acquire the ability to elicit eyeblink CRs (Steinmetz et al., 1986, 1987; Lavond et al., 1987; Steinmetz, 1990; Tracy et al., 1998; Perrett, 1998). We have recently presented almost conclusive evidence that the CS is transmitted through mossy fibers in the MCP (Hesslow et al., 1999). Firstly, in decerebrate ferrets, trained to blink to a forelimb CS, direct stimulation of the MCP could elicit eyeblink CRs without any further training (Fig. $2A_1$, $2A_2$ and 2B). Secondly, the CRs elicited by MCP were 'true' CRs since they were extinguished with unpaired presentation of the MCP–CS and the US (Fig. $2A_2$ and 2C). The forelimb CS was not able to elicit any CRs after such extinction. The opposite relation was also tested showing that the after extinction of the forelimb elicited CR, the MCP–CS could not elicit any CRs. Thirdly, inactivation of the MCP did not abolish the CRs elicited by the MCP–CS, thus excluding the possibility of antidromic activation of a CR generation site in the brainstem (Fig. $2A_3$, 2D and 2E). During the inactivation it was not possible to elicit CRs from the forelimb suggesting that the correct fibers were inactivated (Fig. 2C and 2D). Fourthly, it is possible that the CS just passed through the cerebellum and triggered the eyeblink CR from a post-cerebellar brainstem site. However, this was not very likely since eyeblink CRs were never elicited with stimulation of the output pathway from the interpositus nucleus, the superior cerebellar peduncle (Fig. $2A_3$ and 2F).

If structures in the pre-cerebellar CS pathway, such as the pontine nuclei, were responsible for the changes in eyeblink CR timing, one would not expect similar CR latency changes when increasing the frequency of a CS train applied to the MCP. However, as illustrated in Fig. 1C, an increase in frequency of the MCP–CS from 50 Hz to 100 Hz (0.2 ms square pulses; 18–100 μA) caused a decrease in the eyeblink CR latency which closely paralleled the effects of forelimb CS changes. Also, when the frequency was kept at 100 Hz for several trials (range: 20–45 trials) the eyeblink CR latency gradually increased back to training levels (Fig. 1E; Svensson et al., 1997).

Neurons in the post-cerebellar eyeblink CR pathway may also be involved in the timing behavior. Even though such neurons probably are not required for the acquisition of an eyeblink CR (Krupa and Thompson, 1995), it is still possible that they adjust the CRs temporally. For example, it has been suggested that rubro-cortical projections may be involved in temporal adaptation of the eyeblink CR (Rosenfield and Moore, 1995). Whether post-cerebellar sites are involved in the eyeblink CR timing has not yet been tested.

The effect of lesions and cellular recordings have suggested that structures not directly involved in the CS–CR pathway may still be involved in the temporal aspects of classical conditioning. One such structure is the hippocampus. Hippocampectomy induced deficits in the timing of the eyeblink CR in the trace-conditioning paradigm (Port et al., 1986) and hippocampal activity is highly correlated with the eyeblink CR topography (Berger and Thompson, 1978). These studies, together with the finding that nucleus interpositus lesions make hippocampal activity less eyeblink-CR related,

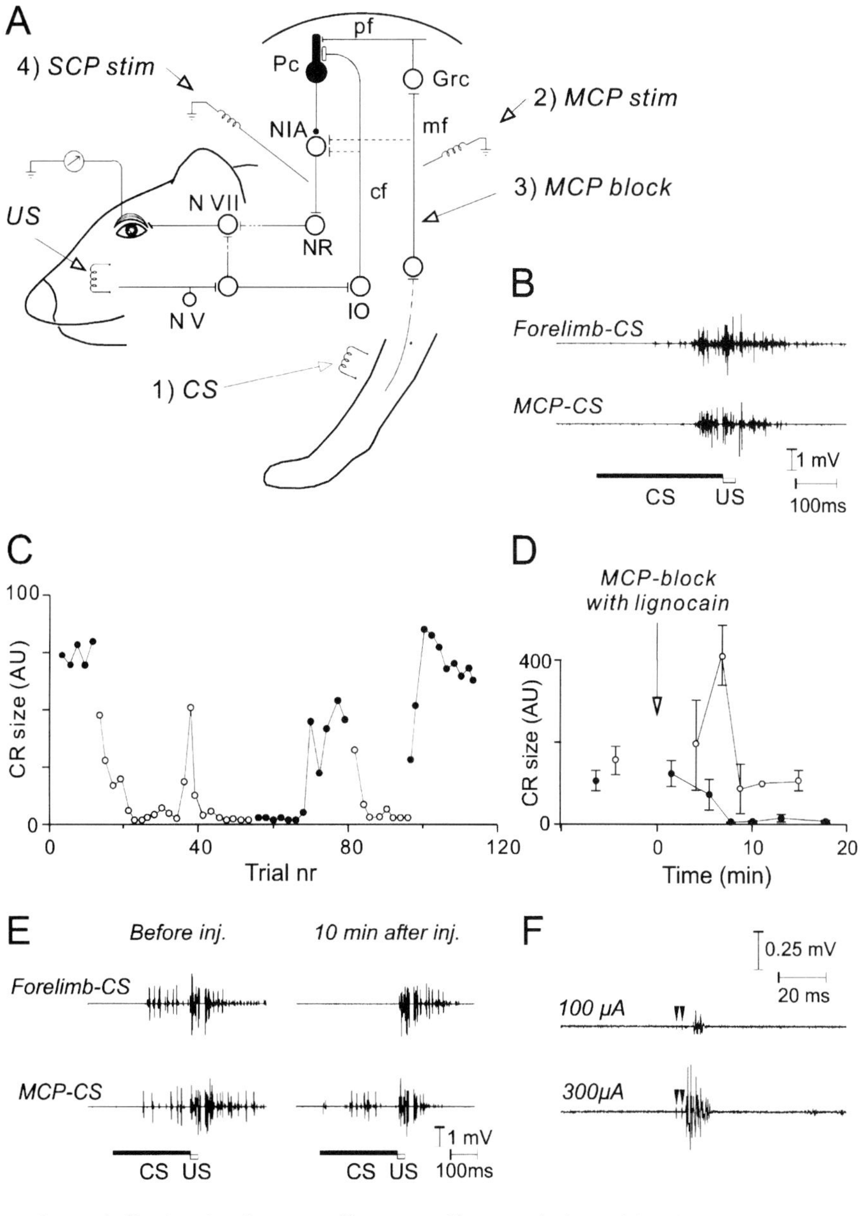

Fig. 2. A set of experiments indicating that the mossy fibers normally transmit the peripheral CS to the cerebellum, from where the CR is generated. (A) Experimental setup and hypothetical wiring diagram. The CS is hypothesized to reach the blink related cerebellar Purkinje cells (Pc) via the mossy fibers (mf), the Granulae cells (GrC) and their axons, the parallel fibers (pf). The US is hypothesized to reach the Pc via the trigeminal nucleus (NV), the neurons in inferior olive (IO) and their axons, the climbing fibers (cf). The Pc send the information from the cortex via the nucleus interpositus anterior (NIA), the nucleus ruber (NR), and the facial nucleus (NVII) to the eyelid. The ferrets were conditioned to an electrical stimulation as CS (A_1). (B) Single high pass filtered EMG records from the orbicular oculi muscle elicited by forelimb CS and by the first reinforced middle cerebellar peduncle CS (MCP–CS; A_2). (C) The CRs were extinguished when MCP–CS was applied unpaired with the US (unfilled circles) and reacquired when the MCP–CS was paired with the US (filled circles). Each circle represents the size of one CR. (D) Injection of lignocain into the MCP (A_3) abolished the CRs elicited by the forelimb CS (filled circles) but not the CRs elicited by the MCP–CS (unfilled circles). Average and SEM of five trials. (E) Sample records of EMG activity evoked by forelimb CS and MCP–CS before lignocain injection and 10 min after. (F) Sample EMG records from the orbicular oculi muscle evoked by stimulation of the superior cerebellar peduncle (A_4).

have inspired the proposal that 'at least one function of the cerebellar-hippocampal interaction is to produce the most adaptively timed and executed classically conditioned response possible' (Sears and Steinmetz, 1990). Another structure is the neostriatum. On the basis of the presence of conditioning related activity in the striatum it was suggested that this activity 'may be projected to cerebellar or other brain stem areas where it could modulate timing or amplitude functions in those circuits essential for initiating and performing the learned response' (White et al., 1994). However, since the new temporal CR behavior presented here was observed in decerebrate ferrets as well as intact rabbits, as described above, it seems that rostral structures such as hippocampus and the neostriatum are not required for the temporal CR property in question.

Conclusions

A new functional timing property of the conditioned response is described. The CR onset latency was decreased when the CS intensity was suddenly increased and that the short latency was gradually increased back to initial levels if the high CS intensity is maintained over several trials. This temporal property seems to be a general phenomenon since it is not limited to the decerebrate preparation, the ferret or the forelimb-CS. The structure and the neurons responsible for the property are under current delineation. So far, it has been shown that neither the pre-cerebellar CS pathways nor rostral structures, such as the hippocampus or the neostriatum, are required. Whether the cerebellum or the post-cerebellar pathway is involved has yet to be tested.

Acknowledgements

These studies were supported by grants from the Medical Faculty, Lund University, the Swedish Medical Research Council (project no. 09899) and the Knut and Alice Wallenberg foundation in cooperation with Astra AB. We thank Henrik Jörntell for writing the computer software used for acquisition and analysis of data.

References

Albus, J.S. (1971) A theory of cerebellar function. *Math. Biosci.*, 10: 25–61.

Berger, T.W. and Thompson, R.F. (1978) Identification of pyramidal cells as the critical elements in hippocampal neuronal plasticity during learning. *Proc. Natl. Acad. Sci. USA*, 75: 1572–1576.

Boneau, C.A. (1958) The interstimulus interval and the latency of the conditioned eyelid response. *J. Exp. Psychol.*, 56: 464–472.

Coleman, S.R. and Gormezano, I. (1971) Classical conditioning of the rabbit's (Oryctolagus cuniculus) nictitating membrane response under symmetrical CS–US interval shifts. *Comp. Physiol. Psychol.*, 77: 447–455.

Ebel, H.C. and Prokasy, W.F. (1963) Classical conditioning as a function of sustained and shifted interstimulus intervals. *J. Exp. Psychol.*, 65: 52–58.

Frey, P.W. (1969) Differential rabbit eyelid conditioning as a function of age, interstimulus interval, and cue similarity. *J. Exp. Psychol.*, 81: 326–333.

Hesslow, G., Svensson P. and Ivarsson, M. (1999) Learned movements elicited by direct stimulation of cerebellar mossy fiber afferents. *Neuron.*, 24: 179–185.

Ivarsson, M. (1998) *Cerebellar Control of Classical Conditioning*. Thesis, Lund University, Sweden.

Krupa, D.J. and Thompson, R.F. (1995) Inactivation of the superior cerebellar peduncle blocks expression but not acquisition of the rabbit's classically conditioned eye-blink response. *Proc. Natl. Acad. Sci. USA*, 92: 5097–5101.

Lavond, D.G., Knowlton, B.J., Steinmetz, J.E. and Thompson, R.F. (1987) Classical conditioning of the rabbit eyelid response with a mossy-fiber stimulation CS: II. Lateral reticular nucleus stimulation. *Behav. Neurosci.*, 101: 676–682.

Leonard, D.W. and Monteau, J.E. (1971) Does CS intensity determine CR amplitude? *Psychon. Sci.*, 23: 369–371.

Leonard, D.W. and Theios, J. (1967) Effect of CS–US interval shift on classical conditioning of the nictitating membrane in the rabbit. *J. Comp. Physiol. Psychol.*, 63: 355–358.

Mackintosh, N.J. (1974) *The Psychology of Animal Learning*. Academic Press. London.

Mauk, M.D. and Ruiz, B.P. (1992) Learning-dependent timing of Pavlovian eyelid responses: differential conditioning using multiple interstimulus intervals. *Behav. Neurosci.*, 106: 666–681.

Marr, D. (1969). A theory of cerebellar cortex. *J. Physiol.*, 202: 437–470.

Marx, M.H. (1969) *Learning: Processes*, The Macmillan Company, Collier-Macmillan Ltd, London.

McCormick, D.A., Guyer, P.E. and Thompson, R.F. (1982) Superior cerebellar peduncle lesions selectively abolish the ipsilateral classically conditioned nictitating membrane/eyelid response of the rabbit. *Brain Res.*, 244: 347–350.

McCormick, D.A. and Thompson, R.F. (1984) Neuronal responses of the rabbit cerebellum during acquisition and performance of a classically conditioned nictiating membrane-eyelid response. *J. Neurosci.*, 4: 2811–2822.

McCormick, D.A., Steinmetz, J.E. and Thompson, R.F. (1985) Lesion of the inferior olivary complex cause extinction of the classically conditioned eyeblink response. *Brain. Res.*, 359: 120–130.

Millenson, J.R., Kehoe, E.J. and Gormezano, I. (1977) Classical conditioning of the rabbit's nictitating membrane response under mixed CS–US intervals. *Learn. Motivat.*, 8: 351–366.

Norman, R.J., Villablanca, J.R., Brown, K.A., Schwafel, J.A. and Buchwald, J.S. (1974) Classical eyeblink conditioning in the bilaterally hemispherectomized cat. *Exp. Neurol.*, 44: 363–380.

Pavlov, I.P. (1927) *Conditioned Reflexes.* (GV Anrep, Trans.) London: Oxford University Press.

Perrett, S.P. (1998) Temporal discrimination in the cerebllar cortx during conditioned eyelid responses. *Exp. Brain Res.*, 121: 115–124.

Port, R.L., Romano, A.G., Steinmetz, J.E., Mikhail, A.A. and Patterson, M.M. (1986) Retention and acquisition of classical trace conditioned responses by rabbits with hippocampal lesions. *Behav. Neurosci.*, 100: 745–752.

Prokasy, W.F. and Papsdorf, J.D. (1965) Effects of increasing the interstimulus interval during classical conditioning of the albino rabbit. *J. Comp. Physiol. Psychol.*, 60(2): 249–252.

Rosenfield, M.E. and Moore, J.W. (1995) Connections to cerebellar cortex (Larsell's HVI) in the rabbit: a WGA–HRP study with implications for classical eyeblink conditioning. *Behav. Neurosci.*, 109: 1106–1118.

Sears, L.L., and Steinmetz, J.E. (1990) Acquisition of classically conditioned-related activity in the hippocampus is affected by lesions of the cerebellar interpositus nucleus. *Behav. Neurosci.*, 104: 681–692.

Steinmetz, J.E., Rosen, D.J., Chapman, P.F., Lavond, D.G. and Thompson, R.F. (1986) Classical conditioning of the rabbit eyelid response with a mossy-fiber stimulation CS: I. Pontine nuclei and middle cerebellar peduncle stimulation. *Behav. Neurosci.*, 100: 878–887.

Steinmetz, J.E., Logan, C.G., Rosen, D.J., Thompson, J.K., Lavond, D.G. and Thompson, R.F. (1987) Initial localization of the acustic conditioned stimulus projection system to the cerebellum essential for classical eyelid conditioning. *Proc. Natl. Acad. Sci. USA*, 84: 3531–3535.

Steinmetz, J.E. (1990) Neuronal activity in the rabbit interpositus nucleus during classical NM-conditioning with a pontine-nucleus-stimulation CS. *Psychol. Sci.*, 1: 378–382.

Svensson, P., Ivarsson, M. and Hesslow, G. (1997) Effect of varying the intensity and train frequency of forelimb and cerebellar mossy fibre conditioned stimuli on the latency of conditioned eye-blink responses in decerebrate ferrets. *Learn. Mem.*, 3: 105–115.

Thompson, R.F. and Krupa, D.J. (1994) Organization of memory traces in the mammalian brain. *Ann. Rev. Neurosci.*, 17: 519–549.

Tracy, J.A., Thompson, J.K., Krupa, D.J. and Thompson, R.F. (1998) Evidence of plasticity in the pontocerebellar conditioned stimulus pathway during classical conditioning of the eyeblink response in the rabbit. *Behav. Neurosci.*, 112: 267–285.

White, I.M., Miller, D.P., White, W., Dike, G.L., Rebec, G.V. and Steinmetz, J.E. (1994) Neuronal activity in rabbit neostriatum during classical eyelid conditioning. *Exp. Brain. Res.*, 99: 179–190.

Yeo, C.H. and Hesslow, G. (1998) Cerebellum and conditioned responses. *TICS*, 2: 322–330.

Yeo, C.H., Hardiman M.J. and Glickstein, M. (1985a) Classical conditioning of the nictitating membrane response of the rabbit. I. Lesions of the cerebellar nuclei. *Exp. Brain Res.*, 60: 87–98.

Yeo, C.H., Hardiman, M.J. and Glickstein, M. (1985b) Classical conditioning of the nictitating membrane response of the rabbit. III. Connections of cerebellar lobule HVI. *Exp. Brain Res.*, 60: 114–126.

Subject Index